Electrodynamics for Physicists

An Introduction, Emphasizing Special Relativity

Electrodynamics for Physicists

An Introduction, Emphasizing Special Relativity

Torsten Fliessbach

University of Siegen, Germany

World Scientific

NEW JERSEY • LONDON • SINGAPORE • BEIJING • SHANGHAI • HONG KONG • TAIPEI • CHENNAI • TOKYO

Published by

World Scientific Publishing Co. Pte. Ltd.
5 Toh Tuck Link, Singapore 596224
USA office: 27 Warren Street, Suite 401-402, Hackensack, NJ 07601
UK office: 57 Shelton Street, Covent Garden, London WC2H 9HE

Library of Congress Control Number: 2024950798

British Library Cataloguing-in-Publication Data
A catalogue record for this book is available from the British Library.

This book is the English translation to the 7th Edition of the original German language version, published by Springer-Verlag GmbH Germany, a part of Springer Nature (Copyright © 2022), under the title "Elektrodynamik: Lehrbuch zur Theoretischen Physik II".

ELECTRODYNAMICS FOR PHYSICISTS
An Introduction, Emphasizing Special Relativity

ISBN 978-981-98-0014-8 (hardcover)
ISBN 978-981-98-0015-5 (ebook for institutions)
ISBN 978-981-98-0016-2 (ebook for individuals)

For any available supplementary material, please visit
https://www.worldscientific.com/worldscibooks/10.1142/14024#t=suppl

Desk Editor: Muhammad Ihsan Putra

Typeset by Stallion Press
Email: enquiries@stallionpress.com

Preface

This book entitled *Electrodynamics for Physicists: An Introduction, emphasizing Special Relativity* is the English version of my German textbook *Elektrodynamik* (7th edition) [2]. This book is the second one of a series of textbooks [1–4] which all developed from repeatedly held lectures. These textbooks are complemented by *Workbook for Theoretical Physics: Repetitorium and Exercises* [5] where the solutions of the numerous exercises are presented in detail.

The presentation is on an average university level in theoretical physics. The approach is rather intuitive instead of deductive. Formal derivations and proofs are carried out without emphasis on mathematical rigor. The material is divided into chapters which roughly correspond to a one to two hour lecture. Each chapter starts with a short abstract and is devoted to a specific subject. As far as possible, each chapter is kept self-contained.

This book first discusses some mathematical prerequisites, in particular the vector (or tensor) calculus (Part I). For understanding Maxwell's equations, the vector operations gradient, divergence and curl should be understood not only mathematically but also with their meaning in a physics context. Part I discusses also the formal tensor (vector) definition including the Lorentz tensors (needed for the special relativistic formulation). Furthermore, distributions (δ-function) are introduced in an intuitive way.

Parts II and III present the electrostatics and the magnetostatics, the corresponding field equation and their standard applications. For defining the charge as a measuring quantity, one must choose a system of physical units. That most common systems are the Gaussian unit system and the SI

(Système International d'Unités). We prefer the Gaussian system because then the relativistic structure of electrodynamics becomes especially transparent (relativistic effects are accompanied by a factor v/c). This choice is motivated in more detail in Chapter 5, and an Appendix presents the most important formulae in the SI. The field equations for electrostatics and magnetostatics are special cases of Maxwell's equations. We investigate various solutions of these field equations. In this context, we introduce a set of complete orthogonal functions. The applications include the electric and magnetic dipoles.

The title of this book says "emphasizing special relativity". Electrodynamic is *the* relativistic theory par excellence. In Part IV, we establish Maxwell's equations and their relativistic form. In this covariant form, the equations are form invariant under Lorentz transformations (LT). Thereby, we learn how the fields transform when changing the inertial system (IS). For example, a charge at rest in an IS has a simple Coulomb field. In a different IS$'$, the charge is moving and exhibits a magnetic field which may be determined by a LT of the Coulomb field. This is a prime example for the application of the relativistic structure. Various other applications are presented in Part V. A rather prominent application is the dipole radiation of an oscillating charge source.

For slowly varying fields, one may use a quasi-static approximation of Maxwell's equations. Often, the quasi-static approximation is set equal to the neglect of the displacement current. Using the simple example of a oscillating circuit, Chapter 26 shows that the neglect of the displacement current is complementary to that of the (Faraday) induction term.

Our electrodynamics in matter (Part VI) follows the modern concept that an external perturbation (index ext) leads to a response (index ind, for induced field) of the system (the undisturbed fields are marked by an index 0). The dielectric function (ratio between the induced and the external field) is thus a response function. The dielectric functions of various materials (in particular of water) are discussed in some detail.

Finally, Part VII discusses some basic element of optics (like, for example, the refraction and reflection laws).

All over the world, one finds different conventions for the typesetting of formulas or mathematical expressions. This book follows the European style that numbers are written as upright (roman) letters, for example,

Euler's number $\mathrm{e} = 2.71...$, the imaginary unit $\mathrm{i} = \sqrt{-1}$ or the number $\mathrm{\pi} = 3.14....$ In contrast to this, physical quantities (like the elementary charge e or the time t) are written in italic (slanted fonts). Physical quantities consist of a number times a (physical) unit (like $t = 3$ s, units like s for second are written in roman). If the physical quantity is a vector, then it is written in bold italic. In the US, one normally finds $\mathrm{e} = 2.71...$ (roman) but $\pi = 3.14...$ (italic), and physical quantities are written in italic except for capital Greek letters (roman) or vectors (bold roman). The derivative of a function $f(t)$ may be found as df/dt or $\mathrm{d}f/\mathrm{d}t$. The d in $\mathrm{d}t$ is written in roman in order to point out to the mathematical limit (d stands for differential). Alternatively (from a more physical point of view), one may consider df/dt just as a fraction with a sufficiently small interval dt; then dt is a physical quantity and is written in italic. This book uses the form df/dt. For my figures, I use a Latex package named pictex. The modest graphical capacities of this package are sufficient for my purposes. Its full integration into Latex has the advantage that the symbols in the figures are identical to those in the text and displayed equations.

I want to thank Dr. Andrea Wolf (former acquisitions editor of World Scientific) who initiated the translations of my German textbooks. Last but not least, many readers of various German editions helped improve my book.

Torsten Fliessbach
June 2024

Preface of the 7th German Edition

This book is of a lecture elaboration [1–4] of the cycle Theoretical Physics I–IV. It gives the material of my lecture on Electrodynamics [2]. This lecture is often presented to physics students in their 4rd semester.

The presentation is at the average level of a standard lecture in Theoretical Physics. The approach is rather intuitive instead of deductive; formal derivations and proofs are presented without mathematical meticulousness.

In close reference to the text, but also to its continuation and supplementation of the text, over 100 exercises are presented. These exercises only fulfill their purpose if they are worked out by the student as independently as possible. This work should be done before reading the sample solutions, which can be found in the *Workbook for Theoretical Physics: Repetitorium and Exercises* [5]. In addition to the solutions, the workbook contains a compact revision of the material covered in the textbooks [1–4].

The scope of this book goes slightly beyond the matter that can be usually covered during a semester at German universities. The material is divided into chapters, which on average correspond to a double lecture hour. Of course, different chapters build on each other. However, we have tried to organize the individual chapters in such a way that they are as self-contained as possible. On one hand, this makes it possible to select chapters for a specific course (e.g., in a bachelor program) in which the material is more limited. On the other hand, the student can more easily read the chapters that are of specific interest to them.

In theoretical physics, it is of fundamental importance that Maxwell's equations are relativistic. The cover of this book emphasizes this by reproducing the inhomogeneous Maxwell equations in covariant form (i.e., expressed by Lorentz tensors and vectors). Previous knowledge of special relativity is helpful for understanding this (for example, in Part IX of my *Mechanics* [1]). However, the most important points are presented also in this book (the Lorentz tensors in Chapter 4, and the relativity principle in Chapter 18).

In mechanics, the Lagrange formalism plays a central role; different Lagrange functions are established for different systems. In electrodynamics, on the other hand, there is exactly one *Lagrangian density* for the electromagnetic fields; the associated Lagrangian equations are then the Maxwell equations (Chapter 19). The covariance of Maxwell's equations is not only a central theoretical property but also the starting point for numerous applications. In particular, it results in the transformation formulas of the fields for a transition from an inertial frame to another. In practice, this can be used to calculate the fields of moving charges.

There are many good presentations of electrodynamics that are suitable for in-depth study. I only list a few books here, which I myself have preferred to consult and which are occasionally cited in the text. As a standard work, I would like to cite *Classical Electrodynamics* by Jackson [6]. Recommended is also the *Classical field theory* of the textbook series by Landau–Lifschitz [7]. In addition, every physics student should once read the *Feynman Lectures* [8].

I would like to thank the numerous readers of the previous editions for valuable hints. Error messages, comments and hints are always welcome, for example, via the contact link on my homepage www2.uni-siegen.de/~flieba/. On this homepage, you will also find correction lists.

Torsten Fließbach
November 2021

References

[1] T. Fliessbach, *Mechanics for Physicists: An Introduction, Including Special Relativity*, 1st edition, World Scientific Publishing Company, Singapore, 2024; T. Fließbach, *Mechanik*, 8th edition, Springer Spektrum, Heidelberg, 2020 (German original).

[2] T. Fliessbach, *Electrodynamics for Physicists: An Introduction, Emphasizing Special Relativity*, 1st edition, World Scientific, 2025 (this book); T. Fließbach, *Elektrodynamik*, 7th edition, Springer Spektrum, Heidelberg, 2022 (German original).

[3] T. Fliessbach, *Quantum Mechanics: An Introduction for Physicists*, 1st edition, World Scientific, 2025; T. Fließbach, *Quantenmechanik*, 6th edition, Springer Spektrum, Heidelberg, 2018 (German original).

[4] T. Fliessbach, *Statistical Physics* (in planning); T. Fließbach, *Statistische Physik*, 6th edition, Springer Spektrum, Heidelberg, 2018 (German original).

[5] T. Fliessbach and H. Walliser, *Workbook for Theoretical Physics: Repetitorium and Exercises* (in planning); T. Fließbach and H. Walliser, *Arbeitsbuch zur Theoretischen Physik — Repetitorium und Übungsbuch*, 4th edition, Springer Spektrum, Heidelberg, 2020 (German original).

[6] J. D. Jackson, *Classical Electrodynamics*, 3rd edition, Wiley, New York, 1998, 2nd edition may be found in https://archive.org/details/ClassicalElectrodynamics2nd/page/n3/mode/2up.

[7] L. D. Landau, *The Classical Theory of Fields*, various editions, 2nd Volume of the Course on Theoretical Physics, may also be found in the internet.

[8] *The Feynman Lectures on Physics*, various editions or https://www.feynmanlectures.caltech.edu/.

Contents

Introduction

In point mechanics, the focus lies on the trajectories of particles. The fundamental laws (e.g., Newton's laws) are ordinary differential equations for these trajectories. In contrast to this, *fields* are the fundamental quantities in electrodynamics. The concept of fields is known from simple applications of continuum mechanics (such as the vibration of a string, Part VIII in [1]). The electromagnetic fields $\boldsymbol{E}(\boldsymbol{r}, t)$ and $\boldsymbol{B}(\boldsymbol{r}, t)$ are defined by the force $\boldsymbol{F}$ that they exert on a charge q:

$$\boldsymbol{F} = q\,\boldsymbol{E}(\boldsymbol{r}, t) + q\,\frac{\boldsymbol{v}}{c} \times \boldsymbol{B}(\boldsymbol{r}, t) \quad \text{(definition of the fields)}$$

Here $\boldsymbol{r}$ is the position and $\boldsymbol{v}$ the velocity of the of the considered charged particle. The Gaussian unit system is used; c is the velocity of light.

The equations of motion for fields are called *field equations*. They are partial differential equations that determine the spatial-temporal behavior of the fields. The field equations of the electromagnetic fields are the *Maxwell equations*:

$$\operatorname{div}\boldsymbol{E} = 4\pi\varrho, \quad \operatorname{curl}\boldsymbol{E} + \frac{1}{c}\frac{\partial \boldsymbol{B}}{\partial t} = 0, \quad \operatorname{div}\boldsymbol{B} = 0, \quad \operatorname{curl}\boldsymbol{B} - \frac{1}{c}\frac{\partial \boldsymbol{E}}{\partial t} = \frac{4\pi}{c}\boldsymbol{j}$$

The charge density $\varrho(\boldsymbol{r}, t)$ and the current density $\boldsymbol{j}(\boldsymbol{r}, t)$ are sources of the field. These Maxwell equations are the basic equations of electrodynamics. Their justification, their properties, implications and solutions are investigated in this book.

The vector operation (here div, curl) may all be expressed by the nabla operator. We mostly prefer verbal notations by grad, div and curl because they may be connected with the intuitive understanding of their meaning (Chapter 1).

Just like other laws of nature or fundamental equations of physics, Maxwell's equations cannot be derived or proven. They can be either stated as a postulate or as a generalization of some key experiments (such as the measurement of the Coulomb force). In the following, we take the second approach, which follows an idealized historical development. Numerous conclusions are derived from Maxwell's equations. These conclusions can be tested experimentally. Thereby, the Maxwell equations can be verified or falsified experimentally.

For static phenomena, the time derivatives in the Maxwell equations fall away. The Maxwell equations then decompose into two independent pairs of equations, namely, on one hand, $\operatorname{div} \boldsymbol{E} = 4\pi\,\varrho$ and $\operatorname{curl} \boldsymbol{E} = 0$, and, on the other hand, $\operatorname{curl} \boldsymbol{B} = (4\pi/c)\boldsymbol{j}$ and $\operatorname{div} \boldsymbol{B} = 0$. These are the field equations of the separate areas of *electrostatics* (Part II) and *magnetostatics* (Part III).

For time-dependent processes, electric and magnetic fields are coupled with each other. This coupling is expressed by Faraday's law of induction and by Maxwell's displacement current. Furthermore, the fields $\boldsymbol{E}$ and $\boldsymbol{B}$ transform into each other when one goes from one inertial frame to another. The splitting into electric and magnetic phenomena depends therefore partly on the observer. We show that the *Lorentz transformation* (LT) known from mechanics (Part IX in [1]) determines the transformation of the electromagnetic fields. The Maxwell's equations themselves do not change their form under LT. These basic properties of Maxwell's equations are examined in Part IV. The general solution for finite charge and current distributions is presented.

Part V deals with the most important applications of Maxwell's equations. These include, in particular, electromagnetic waves, the radiation fields of accelerated or oscillating charges, the scattering of light by atoms and the oscillating circuit.

Maxwell's equations also apply to matter (Part VI). The fields and the sources are divided according to the external perturbation and the reaction of the matter. We start with the microscopic Maxwell equations and the corresponding response functions. Then, the transition is to the macroscopic Maxwell equations and response functions are made. The response functions are calculated in some simple models. The properties of electromagnetic waves in matter and the associated phenomena (like dispersion and absorption) are studied.

Part VII introduces basic principles of optics. The derivation of Huygens' principle is discussed. Interference and diffraction as well as reflection and refraction are examined. Finally, we briefly discuss the geometrical optics.

PART I

Tensor Analysis

Chapter 1

Gradient, Divergence and Curl

In Part I, which begins here, the necessary mathematical requisites are compiled. The main focus lies on the tensor analysis (or tensor calculus) which deals with the differentiation and integration of tensor fields.

Chapter 1 introduces the coordinate independent and descriptive definitions of the vector operations gradient, divergence and curl. In Chapter 2, tensor fields are defined by their behavior under orthogonal transformations; in addition, the practical calculation with the vector operations is demonstrated. In Chapter 3, the δ-function is introduced, the relationship $\delta(1/r) = -4\pi\,\delta(\boldsymbol{r})$ is derived and a vector field is expressed by its sources and vortices. Chapter 4 deals with Lorentz tensor fields.

We consider an arbitrary scalar field $\Phi(\boldsymbol{r})$ and an arbitrary vector field $\boldsymbol{V}(\boldsymbol{r})$ in the three-dimensional space. The formal definition of the properties *scalar* and *vector* is given in Chapter 2. We assume the differentiability of the occurring functions. For the partial derivatives examined in the following, a possible time dependence in $\Phi(\boldsymbol{r}, t)$ and $\boldsymbol{V}(\boldsymbol{r}, t)$ does not matter; it is therefore suppressed in the notation.

We define the following differential operations:

1. The *gradient* of a scalar field $\Phi(\boldsymbol{r})$ is denoted by $\operatorname{grad}\Phi$. The component of the vector field $\operatorname{grad}\Phi$ in the direction of an arbitrary unit vector $\boldsymbol{n}$ is given by

$$\boxed{\boldsymbol{n}\cdot\operatorname{grad}\Phi(\boldsymbol{r}) = \lim_{\Delta r\to 0}\frac{\Phi(\boldsymbol{r}+\boldsymbol{n}\,\Delta r)-\Phi(\boldsymbol{r})}{\Delta r}} \tag{1.1}$$

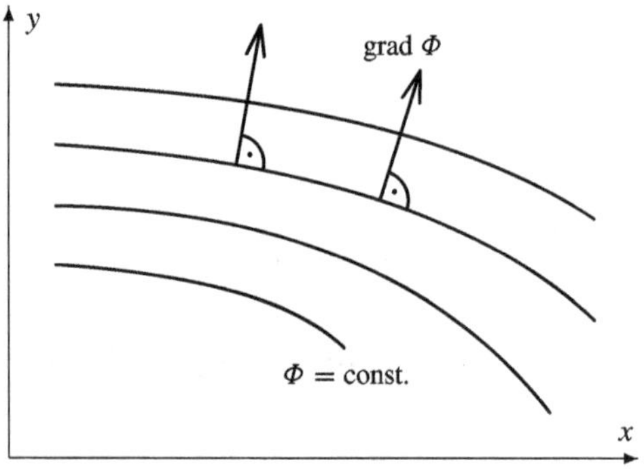

Figure 1.1 The illustration shows some equipotential contour lines $\Phi(x, y) = \text{const.}$ In three dimensions, these contour lines become the surfaces $\Phi(\boldsymbol{r}) = \Phi(x, y, z) = \text{const.}$ The gradient of Φ is perpendicular to this surfaces. It points in the direction of the steepest increase of Φ; its magnitude is proportional to this increase.

The quantity $\boldsymbol{n} \cdot \text{grad}\,\Phi$ is the derivative of Φ in this direction. The geometric meaning of the gradient is illustrated in Figure 1.1.

2. The *divergence* of a vector field $\boldsymbol{V}(\boldsymbol{r})$ is denoted by div $\boldsymbol{V}$. The scalar field div $\boldsymbol{V}$ is defined by

$$\text{div}\,\boldsymbol{V}(\boldsymbol{r}) = \lim_{\Delta\mathcal{V}\to 0} \frac{1}{\Delta\mathcal{V}} \oint_{\Delta A} d\boldsymbol{A} \cdot \boldsymbol{V} \tag{1.2}$$

For this purpose, a volume element[1] $\Delta\mathcal{V}$ is considered at $\boldsymbol{r}$; over its surface ΔA, the scalar product $\boldsymbol{V} \cdot d\boldsymbol{A}$ is summed up. The geometric meaning of the divergence is illustrated in Figure 1.2.

3. The *curl* (also called rotor) of a vector field $\boldsymbol{V}(\boldsymbol{r})$ is denoted by curl $\boldsymbol{V}$. The component of the vector field curl $\boldsymbol{V}$ in the direction of any unit vector $\boldsymbol{n}$ is defined by

$$\boldsymbol{n} \cdot \text{curl}\,\boldsymbol{V}(\boldsymbol{r}) = \lim_{\Delta A\to 0} \frac{1}{\Delta A} \oint_{\Delta C} d\boldsymbol{r} \cdot \boldsymbol{V}, \quad \boldsymbol{n} = \frac{\Delta\boldsymbol{A}}{\Delta A} \tag{1.3}$$

[1]Usually, the volume is denoted by the letter V. In formulas in which a vector field $\boldsymbol{V}$ or its components V_i occur, however, we use the slightly different letter $\mathcal{V}$ for the volume.

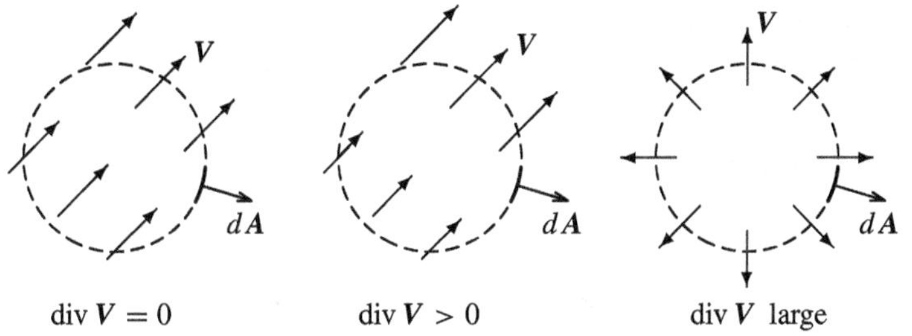

Figure 1.2 To calculate the divergence (1.2), a small volume $\Delta\mathcal{V}$ is considered (here specifically a sphere). If the vector field $\boldsymbol{V}$ is constant in the region of the volume (left), the divergence disappears. If, on the other hand, $\boldsymbol{V}$ increases in the field direction (center), then div $\boldsymbol{V}$ is positive. The divergence is maximum when the vector field is consistently parallel to the normal vector $d\boldsymbol{A}$ of the surface of $\Delta\mathcal{V}$ (right). The vector field has a source here; in general, div $\boldsymbol{V}$ is a measure for the source strength.

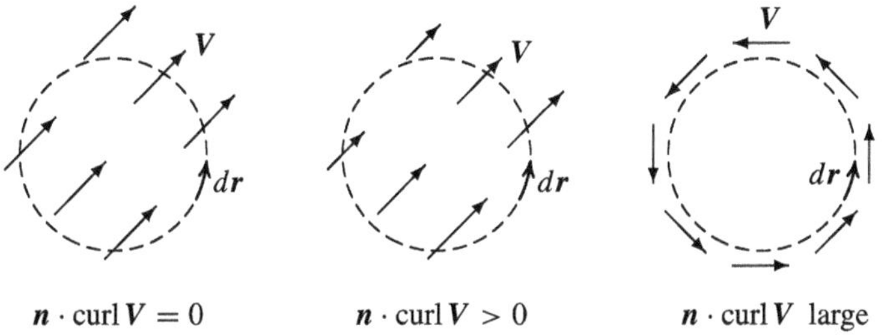

Figure 1.3 To calculate the curl (1.3), a small surface ΔA (here specifically a circular area) is considered. Its the normal vector $\boldsymbol{n}$ that is perpendicular to the image plane. If the vector field $\boldsymbol{V}$ is constant in the area of the surface (left), the curl disappears. If, on the other hand, $\boldsymbol{V}$ increases transversely to the field direction (center), then $\boldsymbol{n} \cdot$ curl $\boldsymbol{V}$ is not equal to zero. The curl is maximum when the vector field is consistently parallel to the line element $d\boldsymbol{r}$ of the boundary of ΔA (right). The vector field has a vortex here; in general, $|\text{curl}\,\boldsymbol{V}|$ is a measure for the vorticity of the field.

For this purpose, a surface element $\Delta A \parallel \boldsymbol{n}$ is considered at $\boldsymbol{r}$; over its boundary ΔC, the scalar product $\boldsymbol{V} \cdot d\boldsymbol{r}$ is summed up. The geometric meaning of the curl is illustrated in Figure 1.3.

The definitions (1.1)–(1.3) have the following advantages:

- They illustrate the meaning of the differential operations for physical fields.

- They are independent of the choice of coordinates.
- Important integrals theorems follow immediately from them.

Cartesian Coordinates

For Cartesian coordinates, the infinitesimal displacement vector is

$$d\boldsymbol{r} = dx\,\boldsymbol{e}_x + dy\,\boldsymbol{e}_y + dz\,\boldsymbol{e}_z \tag{1.4}$$

This quantity is also referred to as *line element* or length element.

We evaluate the definitions (1.1) to (1.3) for Cartesian coordinates. In (1.1), we choose $\boldsymbol{n} = \boldsymbol{e}_x$; then, $d\boldsymbol{r} = dx\,\boldsymbol{e}_x$. Furthermore, we set $\Phi(\boldsymbol{r}) = \Phi(x, y, z)$:

$$\boldsymbol{e}_x \cdot \text{grad}\,\Phi = \frac{\Phi(x + dx, y, z) - \Phi(x, y, z)}{dx} = \frac{\partial \Phi}{\partial x} \tag{1.5}$$

This is the x-component of the vector grad Φ. Overall, we obtain

$$\text{grad}\,\Phi = \frac{\partial \Phi}{\partial x}\,\boldsymbol{e}_x + \frac{\partial \Phi}{\partial y}\,\boldsymbol{e}_y + \frac{\partial \Phi}{\partial z}\,\boldsymbol{e}_z \tag{1.6}$$

In the following, we also use the abbreviations

$$\partial_x \Phi = \frac{\partial \Phi}{\partial x} \quad \text{or} \quad \partial_x = \frac{\partial}{\partial x} \tag{1.7}$$

By

$$\boldsymbol{\nabla}\,\Phi(\boldsymbol{r}) \equiv \text{grad}\,\Phi(\boldsymbol{r}) \tag{1.8}$$

we define the *nabla operator* $\boldsymbol{\nabla}$. The comparison with (1.6) results in

$$\boldsymbol{\nabla} = \boldsymbol{e}_x\,\partial_x + \boldsymbol{e}_y\,\partial_y + \boldsymbol{e}_z\,\partial_z \tag{1.9}$$

To evaluate div $\boldsymbol{V}$, we consider the volume element $\Delta x\ \Delta y\,\Delta z$ at $\boldsymbol{r} = (x, y, z)$, Figure 1.4. We evaluate $\oint d\boldsymbol{A} \cdot \boldsymbol{V}$ for the areas marked in the figure:

$$\oint d\boldsymbol{A} \cdot \boldsymbol{V} = \Delta y\ \Delta z\,\Big(V_x(x + \Delta x, \overline{y}, \overline{z}) - V_x(x, \overline{y}, \overline{z})\Big) + \cdots \tag{1.10}$$

The integrand $V_x(x + \Delta x, y, z) - V_x(x, y, z)$ was drawn in front of the integral. According to the mean value theorem of integral calculus, it must then be taken at a suitable (unknown) point $\overline{y}$, $\overline{z}$ in the considered integration range. We realize the limiting case $\Delta \mathcal{V} \to 0$ by $\Delta x \to 0$,

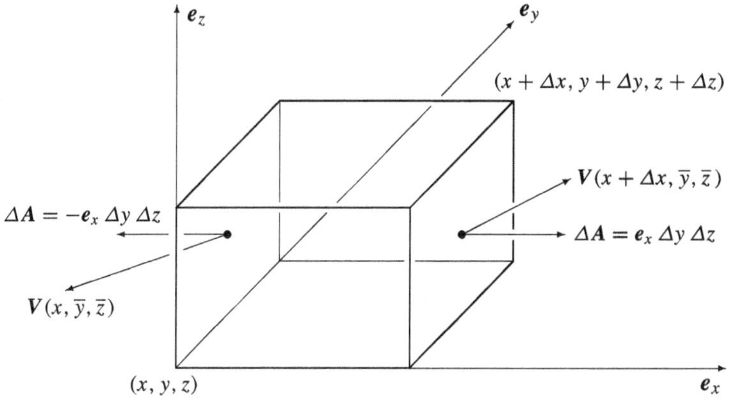

Figure 1.4 For the calculation of div $\boldsymbol{V}$ in Cartesian coordinates, a cuboidal volume $\Delta\mathcal{V} = \Delta x\ \Delta y\ \Delta z$ is considered. The contributions of all six sides to the surface integral $\oint d\boldsymbol{A} \cdot \boldsymbol{V}$ are added up and divided by $\Delta\mathcal{V}$.

$\Delta y \to 0$ and $\Delta z \to 0$. Then, the unknown arguments $\overline{y}$ and $\overline{z}$ go against y and z and we get

$$\frac{1}{\Delta x\ \Delta y\ \Delta z}\ \Delta y\ \Delta z \Big(V_x(x + \Delta x, \ldots) - V_x(x, \ldots) \Big) \stackrel{\Delta\mathcal{V}\to 0}{\longrightarrow} = \frac{\partial V_x}{\partial x} \tag{1.11}$$

The summation over the whole surface results in

$$\operatorname{div} \boldsymbol{V} = \frac{\partial V_x}{\partial x} + \frac{\partial V_y}{\partial y} + \frac{\partial V_z}{\partial z} = \boldsymbol{\nabla} \cdot \boldsymbol{V} \tag{1.12}$$

The divergence can therefore be expressed by the scalar product with the nabla operator. The differential operators in $\boldsymbol{\nabla}$ act on all quantities on the right.

To evaluate curl $\boldsymbol{V}$, we consider the surface element $\Delta \boldsymbol{A} = \Delta x\ \Delta y\ \boldsymbol{e}_z$ at $\boldsymbol{r} = (x, y, z)$, Figure 1.5. We evaluate $\oint d\boldsymbol{r} \cdot \boldsymbol{V}$ for the boundaries marked in the figure:

$$\oint d\boldsymbol{r} \cdot \boldsymbol{V} = \Delta x \Big(V_x(\overline{x}, y, z) - V_x(\overline{x}, y + \Delta y, z) \Big) + \cdots \tag{1.13}$$

The mean value theorem of integral calculus is used again. Dividing by $\Delta A = \Delta x\ \Delta y$ and taking the limit $\Delta x \to 0$ and $\Delta y \to 0$ result in the term

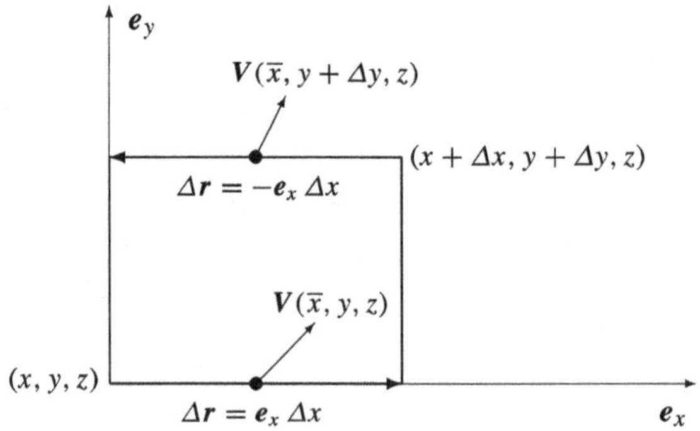

Figure 1.5 For the calculation of $e_z \cdot \text{curl}\, V$ in Cartesian coordinates, a rectangular area $\Delta A = \Delta x\, \Delta y$ is considered. The contributions of all four sides to the line integral $\oint dr \cdot V$ are added up and divided by ΔA.

$-\partial V_x/\partial y$. Overall, we obtain

$$(\text{curl}\, V)_z = \frac{\partial V_y}{\partial x} - \frac{\partial V_x}{\partial y} \quad \text{and} \quad \text{curl}\, V = \nabla \times V \tag{1.14}$$

The curl can be expressed by the vector product with the nabla operator (1.9).

Orthogonal Coordinates

Definitions (1.1)–(1.3) are coordinate independent; they can therefore be evaluated for any coordinates. Practically important cases are orthogonal coordinates; these include spherical, cylindrical and elliptical coordinates. For Cartesian, cylindrical and spherical coordinates, the explicit forms of the differential operators are given in Appendix C.

We denote the coordinates by q_1, q_2 and q_3. For locally orthogonal coordinates, the three vector elements dr that show from the point (q_1, q_2, q_3) to $(q_1 + dq_1, q_2, q_3)$ and $(q_1, q_2 + dq_2, q_3)$ and $(q_1, q_2, q_3 + dq_3)$ form an orthogonal tripod. We define e_1, e_2, e_3 as the orthonormalized base vectors of this tripod; for spherical coordinates, these are the vectors e_r, e_θ and e_ϕ. The essential difference to Cartesian coordinates is that these basis vectors depend on the considered point, i.e., on the coordinates q_1, q_2, q_3. Therefore, one speaks of *locally* orthogonal coordinates. Cartesian coordinates, on the other hand, are globally orthogonal.

According to its definition, $\boldsymbol{e}_1$ points from (q_1, q_2, q_3) to $(q_1 + dq_1, q_2, q_3)$. The associated line element is proportional to $dq_1\,\boldsymbol{e}_1$; we denote the proportionality factor by h_1. The vector $d\boldsymbol{r}$ from (q_1, q_2, q_3) to $(q_1 + dq_1, q_2 + dq_2, q_3 + dq_3)$ is then of the form

$$d\boldsymbol{r} = \sum_{i=1}^{3} h_i\, dq_i\, \boldsymbol{e}_i \tag{1.15}$$

Especially for Cartesian $(x,\ y,\ z)$, cylindrical $(\rho,\ \varphi,\ z)$ and spherical coordinates (r, θ, ϕ), the line element is

$$d\boldsymbol{r} = \begin{cases} dx\, \boldsymbol{e}_x + dy\, \boldsymbol{e}_y + dz\, \boldsymbol{e}_z \\ d\rho\, \boldsymbol{e}_\rho + \rho\, d\varphi\, \boldsymbol{e}_\varphi + dz\, \boldsymbol{e}_z \\ dr\, \boldsymbol{e}_r + r\, d\theta\, \boldsymbol{e}_\theta + r\, \sin\theta\, d\phi\, \boldsymbol{e}_\phi \end{cases} \tag{1.16}$$

From this, we read off the h_i:

$$(h_1,\ h_2,\ h_3) = \begin{cases} (1,\ 1,\ 1) & \text{(Cartesian coordinates)} \\ (1,\ \rho,\ 1) & \text{(cylindrical coordinates)} \\ (1,\ r,\ r\sin\theta) & \text{(spherical coordinates)} \end{cases} \tag{1.17}$$

Fields can be expressed as functions of these coordinates and their base vectors:

$$\Phi(\boldsymbol{r}) = \Phi(q_1, q_2, q_3), \quad \boldsymbol{V}(\boldsymbol{r}) = \sum_{i=1}^{3} V_i(q_1, q_2, q_3)\, \boldsymbol{e}_i \tag{1.18}$$

The derivation of the vector operations from (1.1)–(1.3) is carried out analogously to that in Cartesian coordinates. Due to the local orthogonality, $\Delta\mathcal{V}$ in (1.2) can be represented as a cuboid and ΔA in (1.3) as a rectangle. The side lengths are given by $h_i\, dq_i$. This yields

$$\text{grad}\,\Phi = \boldsymbol{\nabla}\Phi = \sum_{i=1}^{3} \frac{1}{h_i} \frac{\partial \Phi}{\partial q_i}\, \boldsymbol{e}_i \tag{1.19}$$

$$\text{div}\,\boldsymbol{V} = \frac{1}{h_1 h_2 h_3} \left[\frac{\partial(h_2 h_3 V_1)}{\partial q_1} + \frac{\partial(h_1 h_3 V_2)}{\partial q_2} + \frac{\partial(h_1 h_2 V_3)}{\partial q_3} \right] \tag{1.20}$$

$$\text{curl}\,\boldsymbol{V} = \frac{1}{h_2 h_3} \left[\frac{\partial(h_3 V_3)}{\partial q_2} - \frac{\partial(h_2 V_2)}{\partial q_3} \right] \boldsymbol{e}_1 + \text{cyclic} \tag{1.21}$$

The nabla operator follows from (1.19)

$$\boldsymbol{\nabla} = \sum_{i=1}^{3} \boldsymbol{e}_i \, \frac{1}{h_i} \, \frac{\partial}{\partial q_i} \tag{1.22}$$

All vector operations can be expressed by the nabla operator:

$$\boxed{\operatorname{grad} \Phi = \boldsymbol{\nabla} \Phi, \quad \operatorname{div} \boldsymbol{V} = \boldsymbol{\nabla} \cdot \boldsymbol{V}, \quad \operatorname{curl} \boldsymbol{V} = \boldsymbol{\nabla} \times \boldsymbol{V}} \tag{1.23}$$

Since the operations on the right-hand sides (such as the vector product) are defined independent of coordinates, it is sufficient to show the validity of any statement for Cartesian coordinates. For arbitrary coordinates, the first relation in (1.23) defines the nabla operator. Then, the divergence and the curl can then be defined as in (1.23) (instead of (1.2) and (1.3)). On must take into account, however, that the partial derivatives $\partial/\partial q_i$ also act on the basis vectors in $\boldsymbol{V}$, too; in general, $\partial \boldsymbol{e}_i/\partial q_j \neq 0$.

According to Equation (1.23), we may use the notation with the nabla operator instead of grad, div and curl. The notation with the nabla operator is preferably used in practical calculation. This notation is, moreover, independent from language specific uses; it is widely used in the English speaking community. In our discussions, the verbal notations (grad, div, curl) are often preferred because they are connected with the intuitive understanding of their meaning: The divergence might be imagined as a divergent vector field (right in Figure 1.2); the curl or rotor might be imagined as the rotational vector field (right in Figure 1.3); the gradient is connected to the ascent of a scalar field.

Another important differential operator is the *Laplace operator*

$$\Delta = \operatorname{div} \operatorname{grad} = \frac{1}{h_1 h_2 h_3} \left[\frac{\partial}{\partial q_1} \left(\frac{h_2 h_3}{h_1} \, \frac{\partial}{\partial q_1} \right) + \text{cyclic} \right] \tag{1.24}$$

By $\Delta = \operatorname{div} \operatorname{grad}$, the Laplace operator is defined independently of specific coordinates.

Vector operations (such as the scalar and vector product, the gradient, the divergence and curl) can be defined independently of the choice of coordinates. Therefore, vector equations (for example, $\operatorname{curl} \operatorname{grad} \Phi = 0$ for any Φ) are invariant under coordinate transformations. To prove such an equation, it is therefore sufficient to show its validity in specific coordinates. To do this, we choose the coordinates for which the proof is easiest to perform; these are usually Cartesian coordinates. In Chapter 2, the calculation with the nabla operator in Cartesian coordinates is described in more detail.

Integral Theorems

Using the definition of divergence (1.2), an integration over a finite volume $\mathcal{V}$ yields

$$\boxed{\int_{\mathcal{V}} d\mathcal{V}\, \operatorname{div} \boldsymbol{V} = \oint_A d\boldsymbol{A} \cdot \boldsymbol{V} \quad \text{Gauss theorem}} \tag{1.25}$$

Here $A = A(\mathcal{V})$ is the (smooth) boundary surface of the volume $\mathcal{V}$.

From the definition of curl (1.3) follows by integration over a finite area A

$$\boxed{\int_A d\boldsymbol{A} \cdot \operatorname{curl} \boldsymbol{V} = \oint_C d\boldsymbol{r} \cdot \boldsymbol{V} \quad \text{Stokes theorem}} \tag{1.26}$$

Here $C = C(A)$ is the (smooth) boundary contour of the area A.

We prove these two integral theorems. For the derivation of (1.25), we think of the volume $\mathcal{V}$ as being decomposed into small subvolumes:

$$\int_{\mathcal{V}} d\mathcal{V}\, \operatorname{div} \boldsymbol{V} = \sum_i \Delta\mathcal{V}_i\, \operatorname{div} \boldsymbol{V}(\boldsymbol{r}_i) = \sum_i \oint_{\Delta A_i} d\boldsymbol{A} \cdot \boldsymbol{V} \tag{1.27}$$

The first step follows from the meaning of an integral, and the second from (1.2). In the last expression, each surface that lies between two partial volumes occurs twice, each time with a different orientation. These surface contributions therefore cancel each other out; only the surface area A of the volume $\mathcal{V}$ survives, i.e., the right-hand side of (1.25). The derivation of (1.26) is carried out accordingly, whereby (1.3) is considered instead of (1.2).

For a vector field, vortex freedom is equivalent to the path independence of the line integral:

$$\operatorname{curl} \boldsymbol{V} = 0 \quad \longleftrightarrow \quad \int_1^2 d\boldsymbol{r} \cdot \boldsymbol{V} = \text{path independent} \tag{1.28}$$

The left-hand side shall hold in an arbitrary (simply connected) space; e.g., in the entire three-dimensional space. The paths considered in the right part lie in this domain. To prove this statement, consider two paths C_1 and C_2, which lead from 1 to 2. If the paths are different and do not intersect, they include an area A with the boundary contour C so that

$$\oint_C d\boldsymbol{r}\cdot\boldsymbol{V}=\int_{1,C_1}^{2} d\boldsymbol{r}\cdot\boldsymbol{V}-\int_{1,C_2}^{2} d\boldsymbol{r}\cdot\boldsymbol{V} \tag{1.29}$$

If curl $\boldsymbol{V}=0$, then, according to Stokes' theorem, the left-hand side is zero; therefore, the line integrals for any paths C_1 and C_2 are equal. This proves that the right side in (1.28) follows from the left side.

We now start with right-hand side of (1.28). It can be used for an infinitesimal surface yielding $\oint \boldsymbol{V}\cdot d\boldsymbol{r}=0$. According to definition (1.3), this implies curl $\boldsymbol{V}=0$. This proves the equivalence both sides in (1.28).

Green's Theorems

We consider two arbitrary scalar fields $\Phi(\boldsymbol{r})$ and $G(\boldsymbol{r})$. We insert the vector field $\boldsymbol{V}=\Phi\,(\boldsymbol{\nabla}G)$ into the Gauss theorem:

$$\int_{\mathcal{V}} d\mathcal{V}\,\big((\boldsymbol{\nabla}\Phi)\cdot(\boldsymbol{\nabla}G)+\Phi\,\Delta G\big)=\oint_A d\boldsymbol{A}\cdot\Phi\,(\boldsymbol{\nabla}G) \quad \text{(1st Green theorem)} \tag{1.30}$$

Thereby, $\boldsymbol{\nabla}\cdot(\Phi\,\boldsymbol{\nabla}G)=(\boldsymbol{\nabla}\Phi)\cdot(\boldsymbol{\nabla}G)+\Phi\,\Delta G$ was used. We write on the corresponding result for $\boldsymbol{V}=G\,(\boldsymbol{\nabla}\Phi)$ and subtract both equations from each other. This results in

$$\int_{\mathcal{V}} d\mathcal{V}\,(\Phi\,\Delta G-G\,\Delta\Phi)=\oint_A d\boldsymbol{A}\cdot(\Phi\,\boldsymbol{\nabla}G-G\,\boldsymbol{\nabla}\Phi) \quad \text{(2nd Green theorem)} \tag{1.31}$$

Here $A=A(\mathcal{V})$ is the (at least piece-wise smooth) surface of the volume $\mathcal{V}$. It is assumed that the second partial derivatives of the functions Φ and G are continuous.

Exercises

1.1 *Verification of Stokes' theorem*

Verify Stokes' theorem for the vector field

$$\boldsymbol{V}=(4x/3-2y)\,\boldsymbol{e}_x+(3y-x)\,\boldsymbol{e}_y$$

and the surface area

$$A = \{ \boldsymbol{r} \ \ (x/3)^2 + (y/2)^2 \le 1, \ z = 0 \}$$

1.2 *Verification of Gauss' theorem*

Verify Gauss' theorem for the vector field

$$\boldsymbol{V} = a\,x\,\boldsymbol{e}_x + b\,y\,\boldsymbol{e}_y + c\,z\,\boldsymbol{e}_z$$

and the sphere $x^2 + y^2 + z^2 \le R^2$.

1.3 *Elliptic cylindrical coordinates*

The transformation

$$x = q_1 q_2 \quad y = \sqrt{\left(q_1^2 - \ell^2\right)\left(1 - q_2^2\right)}, \quad z = q_3 \qquad (1.32)$$

defines the elliptic cylindrical coordinates q_i. The transformation depends on a parameter ℓ; the coordinate values are restricted by $q_1 \ge \ell > 0$, $|q_2| \le 1$, $|q_3| < \infty$.

Sketch the coordinate lines $q_1 = \text{const.}$ and $q_2 = \text{const.}$ in the x-y plane. Show that these are orthogonal coordinates. Specify h_1, h_2, h_3 and express the unit vectors $\boldsymbol{e}_1, \boldsymbol{e}_2, \boldsymbol{e}_3$ by $\boldsymbol{e}_x, \boldsymbol{e}_y, \boldsymbol{e}_z$.

1.4 *Curl for orthogonal coordinates*

Start from definition (1.3) of the curl. Show for orthogonal coordinates

$$\text{curl}\,\boldsymbol{V} = \frac{1}{h_2 h_3}\left[\frac{\partial(h_3 V_3)}{\partial q_2} - \frac{\partial(h_2 V_2)}{\partial q_3}\right]\boldsymbol{e}_1 + \text{cyclic}$$

1.5 *Divergence for orthogonal coordinates*

Start from definition (1.2) of the divergence. Show for orthogonal coordinates

$$\text{div}\,\boldsymbol{V} = \frac{1}{h_1 h_2 h_3}\left[\frac{\partial(h_2 h_3 V_1)}{\partial q_1} + \frac{\partial(h_3 h_1 V_2)}{\partial q_2} + \frac{\partial(h_1 h_2 V_3)}{\partial q_3}\right]$$

Chapter 2

Tensor Fields

Tensors can be expressed by their components in Cartesian coordinates. The tensor property is formally defined by the behavior of these components under orthogonal transformations. The differentiation of tensor fields plays an important role in electrodynamics. The use of the differential vector operations is demonstrated in a number of examples.

Orthogonal Transformations

In the common three-dimensional space, we introduce a Cartesian coordinate system S with the coordinates $(x_1, x_2, x_3) = (x, y, z)$ and the orthonormalized base vectors $(\boldsymbol{e}_1, \boldsymbol{e}_2, \boldsymbol{e}_3) = (\boldsymbol{e}_x, \boldsymbol{e}_y, \boldsymbol{e}_z)$. In the same space, we can use a another Cartesian coordinate system S′. The systems S and S′ should have the same origin, but their axes are rotated relative to each other. The coordinates of S′ are denoted by (x'_1, x'_2, x'_3) and the basis vectors by $(\boldsymbol{e}'_1, \boldsymbol{e}'_2, \boldsymbol{e}'_3)$.

A physical vector does not depend on a specific coordinate system. Physical vectors are, for example, the electric field $\boldsymbol{E}$, the position vector $\boldsymbol{r} = \boldsymbol{r}_\text{P}$ of a particle P (see Figure 2.1) or its velocity $\boldsymbol{v} = d\boldsymbol{r}_\text{P}/dt$. Such a vector can be expanded to the basis vectors $\boldsymbol{e}_i$ of S or the $\boldsymbol{e}'_i$ of S′:

$$\boldsymbol{r} = \sum_{i=1}^{3} x_i\, \boldsymbol{e}_i = \sum_{i=1}^{3} x'_i\, \boldsymbol{e}'_i \tag{2.1}$$

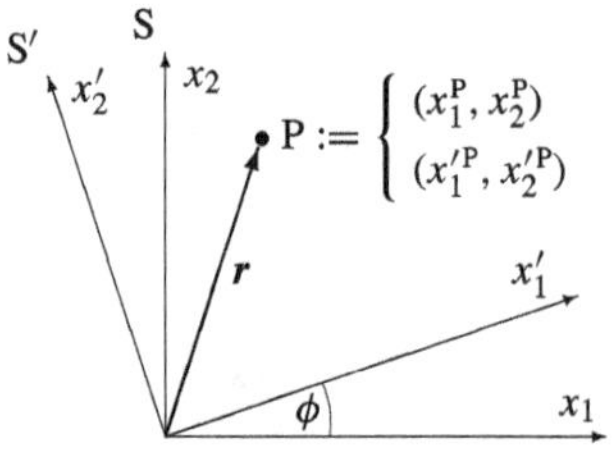

Figure 2.1 In the system S (shown here in two dimensions), a mass point P has the coordinate values $(x_1^{\mathrm{P}}, x_2^{\mathrm{P}})$, and in a rotated S′ other values $(x_1'^{\mathrm{P}}, x_2'^{\mathrm{P}})$. These coordinate values are connected by an orthogonal transformation of form (2.2).

If we multiply $\boldsymbol{r}$ by a factor, the components x_i and x_i' also receive this factor. Therefore, the relationship between the x_i' and the x_i must be linear:

$$x_j' = \sum_{i=1}^{3} \alpha_{ji}\, x_i \quad \text{(orthogonal transformation)} \tag{2.2}$$

For the two expansions in (2.1), we calculate the scalar product:

$$\boldsymbol{r}\cdot\boldsymbol{r} = \boldsymbol{r}^2 = \begin{cases} \sum_i x_i^{\,2} \\ \sum_j x_j'^{\,2} = \sum_{j,i,k} \alpha_{ji}\,\alpha_{jk}\, x_i\, x_k \end{cases} \tag{2.3}$$

Both expressions must be equal for arbitrary values of the x_i. From this follows

$$\sum_{j=1}^{3} \alpha_{ji}\,\alpha_{jk} = \delta_{ik} \quad \text{(orthogonality)} \tag{2.4}$$

Here, *Kronecker symbol* δ_{ik} is defined by

$$\delta_{ik} = \begin{cases} 1 & \text{for} \quad i = k \\ 0 & \text{for} \quad i \neq k \end{cases} \tag{2.5}$$

From the 3 × 3 numbers α_{ik}, the matrix $\alpha = (\alpha_{ik})$ can be formed, and from the three numbers x_i, the column vector $x = (x_i)$. Then, (2.2) reads in matrix notation

$$\begin{pmatrix} x_1' \\ x_2' \\ x_3' \end{pmatrix} = \begin{pmatrix} \alpha_{11} & \alpha_{12} & \alpha_{13} \\ \alpha_{21} & \alpha_{22} & \alpha_{23} \\ \alpha_{31} & \alpha_{32} & \alpha_{33} \end{pmatrix} \begin{pmatrix} x_1 \\ x_2 \\ x_3 \end{pmatrix} \quad \text{or} \quad x' = \alpha\, x \tag{2.6}$$

Relationship (2.4) becomes

$$\alpha^{\mathrm{T}}\alpha = 1 \quad \text{or} \quad \alpha^{\mathrm{T}} = \alpha^{-1} \tag{2.7}$$

Such a matrix is called *orthogonal*; the associated transformations are also called *orthogonal*. The transformation between S and S′ sketched in Figure 2.1 is a rotation around the x_3 axis by the angle ϕ. It is described by the matrix

$$\alpha = \begin{pmatrix} \cos\phi & \sin\phi & 0 \\ -\sin\phi & \cos\phi & 0 \\ 0 & 0 & 1 \end{pmatrix} \tag{2.8}$$

From $\alpha^{\mathrm{T}}\alpha = 1$, it follows that the column vectors (as well as the row vectors) of an orthogonal matrix are orthonormalized. This is easily verified for example (2.8).

If we multiply $x' = \alpha x$ from the left by α^{T} we get the reverse transformation $x = \alpha^{\mathrm{T}}x'$. In component notation, this reads

$$x_i = \sum_{j=1}^{3} \alpha_{ji}\, x_j' \tag{2.9}$$

In the following, we mainly use the component notation.

Tensor Definition

We now formally define the property "tensor". A tensor off rank N is an N-fold indexed quantity $T_{i_1 i_2 \ldots i_N}$, which transforms component-wise like the position vector, i.e.,

$$T'_{i_1\ldots i_N} = \sum_{j_1=1}^{3} \cdots \sum_{j_N=1}^{3} \alpha_{i_1 j_1} \cdot \ldots \cdot \alpha_{i_N j_N}\, T_{j_1\ldots j_N} \quad \text{(tensor definition)} \tag{2.10}$$

A zero-rank tensor is called a *scalar*, and a first-rank tensor *vector*. By *tensor*, we refer to the entirety of the indexed object; the individual elements $T_{i_1\ldots i_N}$ themselves are the components of the tensor. Thus, the x_i are the components of the vector $\boldsymbol{r}$ or of the column vector $x = (x_i)$. In a simplified way of speaking, we also call the indexed quantity itself a vector or tensor. In the equations, however, one must of course distinguish between $\boldsymbol{r}$, x and x_i.

From definition (2.10) follows various ways for constructing new tensors. If S and T are tensors, then the following applies:

1. Addition: $a\, S_{i_1 \ldots i_N} + b\, T_{i_1 \ldots i_N}$ is a tensor of rank N; here a and b are numbers.
2. Multiplication: $S_{i_1 \ldots i_N}\, T_{j_1 \ldots j_M}$ is a tensor of the rank $N + M$.
3. Contraction: $\sum_j S_{i_1 \ldots j \ldots j \ldots i_N}$ is a tensor of rank $N - 2$. In particular, $\sum_i V_i\, W_i$ and $\boldsymbol{r}^2 = \sum_i x_i^{\,2}$ are scalars.

In (2.5), the Kronecker symbol δ_{ik} was defined by a number assignment. This definition is independent of the S; for any other S$'$, we have $\delta'_{ik} = \delta_{ik}$. The so-defined quantity δ_{ik} fulfills the tensor definition (2.10) because

$$\delta'_{ik} \overset{(2.10)}{=} \sum_{n=1}^{3} \sum_{l=1}^{3} \alpha_{in}\, \alpha_{kl}\, \delta_{nl} = \sum_{n=1}^{3} \alpha_{in}\, \alpha_{kn} \overset{(2.4)}{=} \delta_{ik} \tag{2.11}$$

The Kronecker symbol δ_{ik} is therefore a tensor; it is also known as *unit tensor*. The matrix (δ_{ik}) is the unit matrix.

By the number assignment

$$\epsilon_{ikl} = \begin{cases} +1 & \text{if } (i,k,l) \text{ even permutation of } (1,2,3) \\ -1 & \text{if } (i,k,l) \text{ odd permutation of } (1,2,3) \\ 0 & \text{otherwise} \end{cases} \tag{2.12}$$

we define the so-called *Levi-Civita tensor* ϵ_{ikl}; it is also called *totally antisymmetric tensor*. Initially, like δ_{ik}, it is a quantity defined by fixed numbers, i.e., independent of the S.

If, on the right-hand side of (2.10), a factor $\det\alpha$ appears, then T is a *pseudotensor*; here $\det\alpha$ is the determinant of α. From (2.7), it follows that $\det\alpha = \pm 1$. A minus sign results from reflections, for example, the transformation $x_1 \to -x_1$ is an orthogonal transformation, too, but it implies $\det\alpha = -1$. The Levi-Civita tensor is such a pseudotensor:

$$\epsilon'_{i_1 i_2 i_3} \overset{\text{pseudotensor}}{=} \det\alpha \sum_{j_1, j_2, j_3} \alpha_{i_1 j_1}\, \alpha_{i_2 j_2}\, \alpha_{i_3 j_3}\, \epsilon_{j_1 j_2 j_3} = (\det\alpha)^2\, \epsilon_{i_1 i_2 i_3} = \epsilon_{i_1 i_2 i_3} \tag{2.13}$$

The vector product $(\boldsymbol{V} \times \boldsymbol{W})$ can be defined by

$$\big(\boldsymbol{V} \times \boldsymbol{W}\big)_i = \sum_{k=1}^{3} \sum_{l=1}^{3} \epsilon_{ikl}\, V_k\, W_l \tag{2.14}$$

If V_i and W_i are vectors, then $(\boldsymbol{V} \times \boldsymbol{W})_i$ is a pseudovector. First, $\epsilon_{ikl}\, V_m\, W_n$ is a pseudotensor of the 5th rank; then a double contraction results in (2.14).

Tensor Fields and Their Differentiation

For electrodynamics, tensor *fields* are of particular interest, i.e., tensors that depend on the position. A tensor field is defined as an indexed, coordinate-dependent quantity $T_{i_1,\ldots,i_N}$ which transforms like

$$T'_{i_1,\ldots,i_N}(x') = \sum_{j_1=1}^{3} \cdots \sum_{j_N=1}^{3} \alpha_{i_1 j_1} \ldots \alpha_{i_N j_N}\, T_{j_1,\ldots,j_N}(x) \quad \text{(tensor field)} \tag{2.15}$$

In the argument, the position coordinates are abbreviated by $x = (x_1, x_2, x_3)$. The arguments are co-transformed according to (2.2). For a scalar field and a vector field, (2.15) becomes

$$\Phi'(x') = \Phi(x), \qquad V_i'(x') = \sum_{j=1}^{3} \alpha_{ij}\, V_j(x) \tag{2.16}$$

In addition to the known ways for constructing new tensors (addition, multiplication, contraction), the derivatives of tensor fields come into play. Here, $\partial_i = \partial/\partial x_i$ behaves like a vector because

$$\partial_i' = \frac{\partial}{\partial x_i'} = \sum_{j=1}^{3} \frac{\partial x_j}{\partial x_i'} \frac{\partial}{\partial x_j} = \sum_{j=1}^{3} \alpha_{ij}\, \partial_j \tag{2.17}$$

From a scalar field Φ and a vector field V_i the following tensor fields can be formed by differentiation:

$$\big(\text{grad}\, \Phi(\boldsymbol{r})\big)_i = \partial_i\, \Phi \qquad \text{(vector field)} \tag{2.18}$$

$$\text{div}\, \boldsymbol{V}(\boldsymbol{r}) = \sum_{i=1}^{3} \partial_i\, V_i \qquad \text{(scalar field)} \tag{2.19}$$

$$\big(\text{curl}\, \boldsymbol{V}(\boldsymbol{r})\big)_i = \sum_{k,l=1}^{3} \epsilon_{ikl}\, \partial_k\, V_l \quad \text{(pseudovector field)} \tag{2.20}$$

The tensor property of the resulting quantity follows from the tensor properties of the constituent quantities (Φ, V_i, ∂_i, ϵ_{ikl}) and from the calculation rules for forming new tensors (multiplication and contraction).

Covariance

The tensors are defined by their behavior under orthogonal transformations; hereby we have referred to Cartesian coordinate systems. It follows from the definition that tensor equations are *covariant*, which means *form invariant*, under orthogonal transformations. As an example, we consider the equation

$$V_i = \sum_{j=1}^{3} S_{ij}\, W_j \tag{2.21}$$

for the tensors V_i, S_{ij} and W_j; the components refer to a specific S. We write (2.21) in the matrix form $V = S\,W$, multiply this from the left with α and insert $\alpha\,\alpha^{\mathrm{T}} = 1$. This results in $\alpha\,V = \alpha\,S\,W = \alpha\,S\,\alpha^{\mathrm{T}}\alpha\,W$, i.e., $V' = S'\,W'$ or

$$V_i' = \sum_{j=1}^{3} S_{ij}'\, W_j' \tag{2.22}$$

In a rotated S′, the equation thus has the same form as in S; the tensor equation is *form invariant*. A physical example is the relationship $L_i = \sum_j \Theta_{ij}\,\omega_j$ between the angular momentum L_i, the inertia tensor Θ_{ij} and the angular velocity ω_j.

Fundamental physical laws are formulated in inertial systems (IS). By inertial system, we mean an inertial frame of reference with coordinates like x, y, z and time t. One finds experimentally that all spatial directions in an IS are equivalent; this symmetry is called isotropy of space. Due to the isotropy of space, fundamental laws do not depend on the orientation of the IS. The laws are then either invariant (for $\boldsymbol{L} = \widehat{\Theta} \cdot \boldsymbol{\omega}$ with the dyad $\widehat{\Theta}$) or form invariant (for $L_i = \sum_j \Theta_{ij}\,\omega_j$) under rotations. These points are already known from mechanics (Chapters 5, 11 and 21 in [1]). In electrodynamics, the covariance of Maxwell's equations under the Lorentz transformations plays a central role (Chapter 18).

Calculating with the Nabla Operator

The vector operations gradient, divergence and curl have been defined independent on the choice of coordinates (Chapter 1). Therefore, vector equations such as

$$\mathrm{div}\,(\Phi \boldsymbol{V}) = \boldsymbol{V} \cdot \mathrm{grad}\,\Phi + \Phi\ \mathrm{div}\,\boldsymbol{V} \tag{2.23}$$

are invariant under coordinate transformations, e.g., a transformation to curvilinear coordinates. Therefore, it is sufficient to perform a proof of such a relation for specific coordinates. Cartesian coordinates are well suitable for this. For a number of examples, we show how such relationships can be systematically proven in Cartesian coordinates.

First of all, grad, div and rot are expressed by the ∇ operator. All vectors are represented by their components, and vector equations such as $\boldsymbol{V} = \boldsymbol{W}$ are expressed in the form $V_i = W_i$. Any scalar products can be performed by contracting the associated indices. Thus, (2.23) can be proven as follows:

$$\begin{aligned}
\nabla \cdot (\Phi \boldsymbol{V}) &= \sum_{i=1}^{3} \partial_i (\Phi V_i) = \sum_{i=1}^{3} (\partial_i \Phi) V_i + \sum_{i=1}^{3} \Phi (\partial_i V_i) \\
&= \boldsymbol{V} \cdot (\nabla \Phi) + \Phi (\nabla \cdot \boldsymbol{V})
\end{aligned} \tag{2.24}$$

For

$$\text{div}\,\text{curl}\,\boldsymbol{V} = 0 \tag{2.25}$$

we give a detailed, exemplary proof:

$$\begin{aligned}
\text{div}\,\text{curl}\,\boldsymbol{V} &\overset{1.}{=} \sum_{i,k,l} \partial_i\, \epsilon_{ikl}\, \partial_k\, V_l \overset{2.}{=} \sum_{i,k,l} \epsilon_{ikl}\, \partial_i\, \partial_k\, V_l \overset{3.}{=} \sum_{k,i,l} \epsilon_{kil}\, \partial_k\, \partial_i\, V_l \\
&\overset{4.}{=} \sum_{k,i,l} \epsilon_{kil}\, \partial_i\, \partial_k\, V_l \overset{5.}{=} -\sum_{k,i,l} \epsilon_{ikl}\, \partial_i\, \partial_k\, V_l \\
&\overset{6.}{=} -\sum_{i,k,l} \epsilon_{ikl}\, \partial_i\, \partial_k\, V_l \overset{7.}{=} 0
\end{aligned} \tag{2.26}$$

The individual steps in this demonstration are as follows:

1. Due to the invariance under coordinate transformation, it is sufficient to prove the statement in Cartesian coordinates.
2. The partial derivative ∂_i does not act on ϵ_{ikl} because this quantity consists of constant numbers.
3. The name of summation indices (so-called bound variables) is arbitrary. Therefore, we can rename i, k, l to k, i, l.
4. Due to the assumed differentiability of $V_l(x)$, the partial derivatives can be interchanged. Here, the twofold differentiability is assumed.
5. For the totally antisymmetric tensor, $\epsilon_{kil} = -\epsilon_{ikl}$ holds.

6. The order of the summations is reversed. This is always possible for finite sums.
7. The result differs from the third expression in the first line only by a minus sign. So, it is zero.

The reader may analogously prove

$$\text{curl grad}\, \Phi = 0 \tag{2.27}$$

Another differential operator is the Laplace operator:

$$\Delta = \text{div grad} \overset{(1.8,1.9,1.12)}{=} \sum_{i=1}^{3} \frac{\partial^2}{\partial x_i^2} = \sum_{i=1}^{3} \partial_i \, \partial_i \tag{2.28}$$

The right-hand side applies to Cartesian coordinates. The coordinate independent form div grad has originally be defined for a scalar field. The differential operator on the right-hand side of (2.28) can, however, also be applied to a vector field $\boldsymbol{V}$. The quantity $\Delta \boldsymbol{V}$ defined in this way can be expressed by coordinate independent vector operations:

$$\Delta \boldsymbol{V} = \text{grad}\,(\text{div}\, \boldsymbol{V}) - \text{curl}\,(\text{curl}\, \boldsymbol{V}) \tag{2.29}$$

To prove this relationship, we evaluate the right-hand side in Cartesian coordinates:

$$\begin{aligned}
\big(\text{grad div}\, \boldsymbol{V} - \text{curl curl}\, \boldsymbol{V}\big)_i &= \sum_j \partial_i \, \partial_j \, V_j - \sum_{k,l,m,n} \epsilon_{ikl} \, \partial_k \, \epsilon_{lmn} \, \partial_m \, V_n \\
&= \sum_j \partial_i \, \partial_j \, V_j - \sum_{k,m,n} \big(\delta_{in} \, \delta_{kn} - \delta_{in} \, \delta_{km}\big) \, \partial_k \, \partial_m \, V_n \\
&= \sum_j \partial_i \, \partial_j \, V_j - \sum_n \partial_n \, \partial_i \, V_n + \sum_k \partial_k \, \partial_k \, V_i = \Delta V_i = (\Delta \boldsymbol{V})_i
\end{aligned}$$

In the last step, it was used that the basis vectors are constant.

Exercises

2.1 *Calculating with gradient, divergence and curl*

Show div $(\boldsymbol{V} \times \boldsymbol{W}) = \boldsymbol{W} \cdot \text{curl}\, \boldsymbol{V} - \boldsymbol{V} \cdot \text{curl}\, \boldsymbol{W}$ by an evaluation in Cartesian components. Analogously verify the expressions curl $(\Phi \boldsymbol{V})$, curl $(\boldsymbol{V} \times \boldsymbol{W})$ and grad $(\boldsymbol{V} \cdot \boldsymbol{W})$.

2.2 *Second rank tensor*

The equation $V_i = \sum_j S_{ij} W_j$ is valid in any Cartesian coordinate system. It shall be known that V_i and W_j are vectors. Show that S_{ij} is a 2nd rank tensor.

2.3 *Levi-Civita tensor*

Show that the Levi-Civita tensor is a pseudotensor of 2nd rank, i.e.,

$$\epsilon'_{ijk} = \det \alpha \sum_{l,m,n} \alpha_{il}\, \alpha_{jm}\, \alpha_{kn}\, \epsilon_{lmn} = \epsilon_{ijk}$$

2.4 *Product rule for the nabla operator*

Show $\boldsymbol{\nabla} \cdot (\boldsymbol{r}\, \Phi) = 3\, \Phi + \boldsymbol{r} \cdot \operatorname{grad} \Phi$.

2.5 *Curl of a gradient*

Prove $\operatorname{curl} \operatorname{grad} \Phi = 0$ in Cartesian components; cross products shall be written by using the Levi-Civita tensor.

2.6 *Contraction of two Levi-Civita tensors*

Verify the relation

$$\sum_l \epsilon_{ikl}\, \epsilon_{lmn} = \sum_l \epsilon_{ikl}\, \epsilon_{mnl} = \delta_{im}\delta_{in}\, \delta_{kn} - \delta_{in}\, \delta_{km}$$

Chapter 3

Distributions

In electrodynamics, the Laplace equation $\Delta\Phi = 0$ has to be evaluated for the potential $\Phi = e/|\boldsymbol{r}|$ of a point charge. In spherical coordinates, we obtain

$$\Delta\,\frac{1}{r} = \operatorname{div}\operatorname{grad}\,\frac{1}{r} \overset{(C.4,C.5)}{=} \left(\frac{\partial^2}{\partial r^2} + \frac{2}{r}\frac{\partial}{\partial r}\right)\frac{1}{r} = \begin{cases} 0 & r \neq 0 \\ ? & r = 0 \end{cases} \tag{3.1}$$

For $r = 0$, the result appears to be undefined. We identify $\Delta(1/r)$ as the so-called δ-function. The δ-function and related objects are introduced in the following.

We consider the step function $d_\ell(x)$, which depends on a parameter ℓ:

$$d_\ell(x) = \begin{cases} 1/\ell & -\ell/2 \le x \le \ell/2 \\ 0 & \text{other} \end{cases} \tag{3.2}$$

For arbitrary continuous functions $f(x)$, we calculate the convolution integral with $d_\ell(x)$ and then let ℓ go to 0:

$$\begin{aligned} \lim_{\ell\to 0} \int_{-\infty}^{\infty} dx\; d_\ell(x - x_0)\, f(x) &= \lim_{\ell\to 0} \frac{1}{\ell} \int_{x_0-\ell/2}^{x_0+\ell/2} dx\; f(x) \\ &= \lim_{\ell\to 0} f(\overline{x}) = f(x_0) \end{aligned} \tag{3.3}$$

For the integral, the mean value theorem of integral calculus was used; then $f(\overline{x})$ is at an (unknown) point between $x_0 - \ell/2$ and $x_0 + \ell/2$. Figure 3.1 illustrates this step.

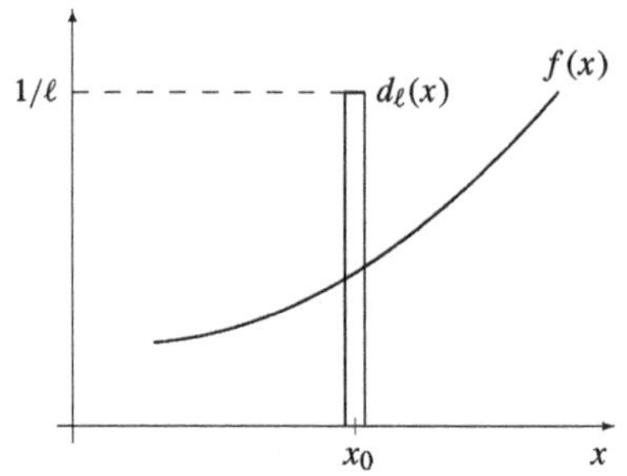

Figure 3.1 The function $d_\ell(x - x_0)$ is only non-zero in an interval of length ℓ. For $\ell \to 0$, the convolution with a test function $f(x)$ results in the value $f(x_0)$.

One can also construct other functions $d_\ell(x)$ for which (3.3) applies. It is sufficient that the function $d_\ell(x)$ is positive in an ℓ-vicinity around $x = 0$, and it must be normalized to 1. An example is

$$d_\ell(x) = \frac{1}{\sqrt{2\pi}\,\ell}\,\exp\left(-\frac{x^2}{2\ell^2}\right) \tag{3.4}$$

For convolution (3.3), we use the abbreviated notation

$$\lim_{\ell\to 0}\int_{-\infty}^{\infty} dx\; d_\ell(x - x_0)\, f(x) = \int_{-\infty}^{\infty} dx\; \delta(x - x_0)\, f(x) \tag{3.5}$$

with the so-called δ-function (or delta function)

$$\delta(x) = \lim_{\ell\to 0} d_\ell(x) \quad \text{(limit only after integration)} \tag{3.6}$$

The limit in (3.6) is to be understood as a rule in the sense of (3.5); i.e., it is to be executed after the integration. If omitted, (3.6) would be zero for $x \neq 0$ and infinity for $x = 0$. The integral over it would then be zero (or not defined, depending on the specific integral definition) and not, as required, equal to 1.

The supplement in (3.6) means that $\delta(x)$ *is not a function* (in the common sense). In mathematics, $\delta(x)$ and related objects are referred to as *distributions*. The distribution $\delta(x)$ is generally defined by requiring that

$$\int_{-\infty}^{\infty} dx\; \delta(x - x_0)\, f(x) = f(x_0) \tag{3.7}$$

holds for arbitrary (but continuous and integrable) test functions $f(x)$. For our purposes, the more specific definition is given by (3.5) suffices.

For physical applications, we might use functions like $d_\ell(x)$ with a sufficiently small ℓ. For example, we might choose $\ell = 10^{-20}$ cm; then, the value of integral $\int dx\, d_\ell(x - x_0)\, f(x)$ with a physical quantity $f(x)$ does not depend on ℓ. The independence of ℓ is a simplification, which is expressed by using the delta function. The δ-function was introduced by the physicist Dirac and also called Dirac delta function.

From (3.5) follows that the usual calculation rules of integration apply to $\delta(x)$. We examine the convolution of the derivative $\delta'(x) = d\delta(x)/dx$ with an arbitrary test function $f(x)$ by performing a partial integration:

$$\int_{-\infty}^{\infty} dx\ \delta'(x - x_0)\, f(x) = -\int_{-\infty}^{\infty} dx\ \delta(x - x_0)\, f'(x) = -f'(x_0) \quad (3.8)$$

Analogous to (3.7), this result defines the distribution $\delta'(x)$. Specifically, $\delta'(x)$ may be represented by the derivative of the function (3.4) with arbitrarily small ℓ.

The condition "localized at $x = 0$ with an integration area 1" is not sufficient to determine $\delta(x)$. This condition would, for example, it also be fulfilled by $\delta(x) + \delta'(x)$ (because $\int dx\, \delta'(x) = 0$).

The integral over the δ-function gives the *step* or Θ-function:

$$\int_{-\infty}^{x} dx'\ \delta(x' - x_0) = \Theta(x - x_0) = \begin{cases} 0 & \text{for} \quad x < x_0 \\ 1 & \text{for} \quad x > x_0 \end{cases} \quad (3.9)$$

Thus, $\delta(x)$ can be written as the derivative of the Θ-function,

$$\delta(x - x_0) = \Theta'(x - x_0) \quad (3.10)$$

The step function can be understood as a common function; for Θ', on the other hand, the reservations made following (3.6) should be noted. The integral over the Θ-function yields a continuous function. These statements are illustrated in Figure 3.2 for a slightly more general function.

In the theory of the Fourier transform, the following relations are used:

$$f(x) = \frac{1}{\sqrt{2\pi}} \int_{-\infty}^{\infty} dk\ g(k)\ \exp(ikx) \quad (3.11)$$

$$g(k) = \frac{1}{\sqrt{2\pi}} \int_{-\infty}^{\infty} dx\ f(x)\ \exp(-ikx) \quad (3.12)$$

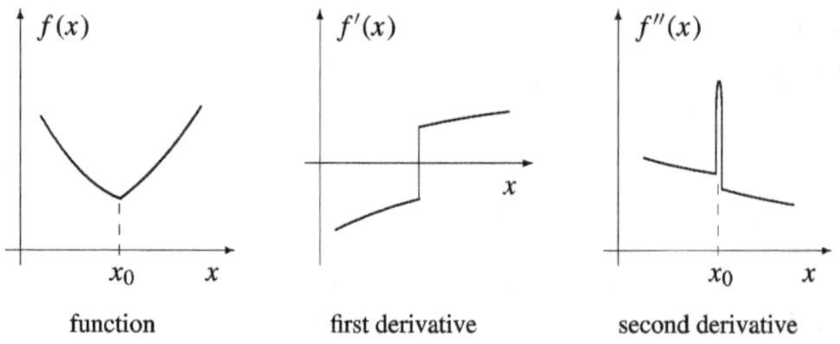

Figure 3.2 If the first derivative $f'(x)$ of a function has a jump, then $f(x)$ is continuous at this point and has a kink there. The second derivative shows a δ-function at this point. For this sketch, the $\delta(x)$ function is represented by $d_\ell(x)$ with finite ℓ.

Let $f(x)$ be continuous and square-integrable. We insert $g(k)$ into the first equation and denote the integration variable in (3.12) by x_0:

$$f(x) = \frac{1}{2\pi} \int_{-\infty}^{\infty} dx_0 \, f(x_0) \int_{-\infty}^{\infty} dk \, \exp\big(\mathrm{i}k(x - x_0)\big) \tag{3.13}$$

This gives us a representation of the δ-function:

$$\delta(x - x_0) = \frac{1}{2\pi} \int_{-\infty}^{\infty} dk \, \exp\big(\mathrm{i}k(x - x_0)\big) \tag{3.14}$$

Corresponding representations of the δ-function are obtained for each complete system of functions (see (9.8)).

The three-dimensional δ-function

$$\delta(\boldsymbol{r}) = \delta(x)\,\delta(y)\,\delta(z) \tag{3.15}$$

is defined such that for an arbitrary test function $f(\boldsymbol{r})$, the following holds:

$$\int d^3r \; f(\boldsymbol{r})\,\delta(\boldsymbol{r} - \boldsymbol{r}_0) = f(\boldsymbol{r}_0) \tag{3.16}$$

We introduce some useful properties of the δ-function, which are derived from definitions (3.5)–(3.7) and the integration calculus. The δ-function is symmetrical, i.e.,

$$\delta(x) = \delta(-x), \qquad \delta(\boldsymbol{r}) = \delta(-\boldsymbol{r}) \tag{3.17}$$

The following also applies

$$f(x)\,\delta(x) = f(0)\,\delta(x), \qquad x\,\delta(x) = 0, \qquad \boldsymbol{r}\,\delta(\boldsymbol{r}) = 0 \tag{3.18}$$

The function $h(x)$ shall have a zero at x_0. At x_0, we then obtain

$$\delta(h(x)) = \frac{\delta(x - x_0)}{|h'(x_0)|} \tag{3.19}$$

Each zero of $h(x)$ provides such a contribution. Hereby, we do not allow for coincident zeros because they would lead to the undefined product of two δ-functions at the considered point. From (3.19) follows

$$\delta(ax) = \frac{\delta(x)}{|a|}, \qquad \delta(x^2 - a^2) = \frac{\delta(x-a) + \delta(x+a)}{|2a|} \tag{3.20}$$

Density Distribution of Point Particles

We show that

$$\varrho_m(\boldsymbol{r}) = m\, \delta(\boldsymbol{r} - \boldsymbol{r}_0) \tag{3.21}$$

is the mass density of a point mass m at $\boldsymbol{r}_0$. Due to $\int d^3r\, \delta(\boldsymbol{r}) = 1$, $\delta(\boldsymbol{r})$ has the dimension of an inverse volume; thus, (3.21) has the correct dimension mass/volume. Furthermore, the following applies

$$\int_{\Delta V} d^3r\; \varrho_m(\boldsymbol{r}) = \begin{cases} m & \boldsymbol{r}_0 \in \Delta V \\ 0 & \text{sonst} \end{cases} \tag{3.22}$$

Thus, $\varrho_m(\boldsymbol{r})$ is a mass density that describes a localized mass of size m at $\boldsymbol{r}_0$. Analogous to this, the expression

$$\varrho(\boldsymbol{r}) = \sum_{i=1}^{N} q_i\; \delta(\boldsymbol{r} - \boldsymbol{r}_i) \tag{3.23}$$

stands for the *charge density* of N point charges of size q_i at $\boldsymbol{r}_i$.

Green Function of the Laplace Operator

We now return to the starting point (3.1) and prove

$$\delta\, \frac{1}{|\boldsymbol{r} - \boldsymbol{r}_0|} = -4\pi\, \delta(\boldsymbol{r} - \boldsymbol{r}_0) \tag{3.24}$$

For a somewhat more general conceptualization, instead of Δ we first consider a differential operator D_{op}, which acts on the coordinate $\boldsymbol{r}$.

The solution $G(\boldsymbol{r}, \boldsymbol{r}_0)$ of the differential equation

$$D_{\text{op}}\, G(\boldsymbol{r}, \boldsymbol{r}_0) = \delta(\boldsymbol{r} - \boldsymbol{r}_0) \tag{3.25}$$

is defined as the *Green function* G of the operator D_{op}. In this notation, $G(\boldsymbol{r}, \boldsymbol{r}_0) = -1/(4\pi\,|\boldsymbol{r} - \boldsymbol{r}_0|)$ is the Green function of the Laplace operator. If you know Green's function, you can calculate the solution of the differential equation with an arbitrary source term (instead of the δ-function).

To prove (3.24), we start from the second Green's theorem (1.31), i.e., from

$$\int_{\mathcal{V}} d\mathcal{V}\, (\Phi\, \Delta G - G\, \Delta\Phi) = \oint_A d\boldsymbol{A} \cdot \big(\Phi\, \nabla G - G\, \nabla\Phi\big) \tag{3.26}$$

The function $\Phi(\boldsymbol{r})$ is largely arbitrary, but it shall meet the conditions

$$\Phi(\boldsymbol{r}) \begin{cases} \text{twice continuously differentiable} \\ r^2\, \Delta\Phi \overset{r\to\infty}{\longrightarrow} 0 \end{cases} \tag{3.27}$$

For the function $G(\boldsymbol{r})$ and the volume $\mathcal{V}$, we specifically consider

$$G(\boldsymbol{r}) = \frac{1}{r} = \frac{1}{|\boldsymbol{r}|} \quad \text{and} \quad \mathcal{V} = \{\boldsymbol{r} \quad r > \varepsilon\} \tag{3.28}$$

The volume $\mathcal{V}$ is the entire space with the exception of a small sphere at $r = 0$. We now evaluate (3.26) for (3.27) and (3.28). According to (3.1), $\Delta(1/r) = 0$ in the volume $\mathcal{V}$ because the environment of $r = 0$ is excluded from V. The left-hand side of (3.26) therefore results in

$$\int_{\mathcal{V}} d^3r\, \big(\Phi\, \Delta G - G\, \Delta\Phi\big) = -\int_{\mathcal{V}} d^3r\, \frac{\Delta\Phi(\boldsymbol{r})}{r} = -\int d^3r\, \frac{\Delta\Phi(\boldsymbol{r})}{r} + \mathcal{O}(\varepsilon^2) \tag{3.29}$$

Due to the second condition in (3.27), the integral contributions for $r \to \infty$ are finite. Due to $d^3r/r = r\, dr\, d\Omega$ (in spherical coordinates) and the continuity of $\Delta\Phi$, the integral contribution for $r \le \varepsilon$ yield at most a term of the size $\mathcal{O}(\varepsilon^2)$. Therefore, the integration can be extended to the entire space (last step in (3.29)).

The surface A on the right-hand side of (3.26) is the spherical surface $r = \varepsilon$. The surface normal points from the volume $\mathcal{V}$ toward the outside, i.e., $dA = -\boldsymbol{e}_r \, r^2 \, d\Omega$. With $\boldsymbol{e}_r \cdot \boldsymbol{\nabla} = \partial/\partial r$ and $\boldsymbol{\nabla}(1/r) = -\boldsymbol{e}_r/r^2$, we get

$$\int_A d\boldsymbol{A}\cdot\big(\Phi\, \boldsymbol{\nabla} G - G\, \boldsymbol{\nabla}\Phi\big) = -\int d\Omega\ \varepsilon^2 \left(-\frac{\Phi}{\varepsilon^2} - \frac{1}{\varepsilon}\frac{\partial \Phi}{\partial r}\right) = 4\pi\, \Phi(\bar{\boldsymbol{r}}) + \mathcal{O}(\varepsilon) \tag{3.30}$$

In the last step, we used the mean value theorem of integral calculus; $\bar{\boldsymbol{r}}$ is an unknown vector with $|\bar{\boldsymbol{r}}| = \varepsilon$. We insert (3.30) and (3.29) into (3.26) and let ε approach zero:

$$\int d^3r\ \frac{\Delta \Phi(\boldsymbol{r})}{r} = -4\pi\, \Phi(0) \tag{3.31}$$

By repeated partial integration, we turn over the Laplace operator $\Delta = \partial_x^2 + \partial_y^2 + \partial_z^2$ from Φ to $1/r$:

$$\int d^3r\ \Phi(\boldsymbol{r})\, \Delta\, \frac{1}{r} = -4\pi\, \Phi(0) \tag{3.32}$$

The boundary terms (for $r \to \infty$) disappear. From the arbitrariness of $\Phi(\boldsymbol{r})$ follows

$$\boxed{\Delta\, \frac{1}{r} = -\, 4\pi\, \delta(\boldsymbol{r})} \tag{3.33}$$

The replacement $\boldsymbol{r} \to \boldsymbol{r} - \boldsymbol{r}_0$ leads (3.24) (taking into account $\Delta_{\boldsymbol{r}-\boldsymbol{r}_0} = \Delta_{\boldsymbol{r}}$).

Generalization

We now show that (3.33) can be generalized to

$$\boxed{\left(\Delta + k^2\right) \frac{\exp(\pm \mathrm{i} k r)}{r} = -4\pi\, \delta(\boldsymbol{r})} \tag{3.34}$$

With (C.4), it is easy to verify that this is correct for $r \neq 0$:

$$\left(\Delta + k^2\right) \frac{\exp(\pm \mathrm{i} k r)}{r} = \left(\frac{1}{r}\frac{\partial^2}{\partial r^2}\, r + k^2\right) \frac{\exp(\pm \mathrm{i} k r)}{r} \overset{(r\neq 0)}{=} 0 \tag{3.35}$$

We now multiply the left-hand side of (3.34) by an arbitrary test function $f(\boldsymbol{r})$ and integrate over the entire space:

$$\int d^3r\ f(\boldsymbol{r})\ \left(\Delta + k^2\right)\ \frac{\exp(\pm ikr)}{r}$$

$$\stackrel{(3.35)}{=} \int_{r\,\leq\,\varepsilon} d^3r\ f(\boldsymbol{r})\ \left(\Delta + k^2\right)\ \frac{\exp(\pm ikr)}{r}$$

$$= \int_{r\,\leq\,\varepsilon} d^3r\ f(\boldsymbol{r})\ \left(\Delta + k^2\right) \left(\frac{1}{r} \pm \mathrm{i}k - \frac{1}{2}\,k^2 r \pm \cdots\right)$$

$$= \int_{r\,\leq\,\varepsilon} d^3r\ f(\boldsymbol{r})\ \Delta\ \frac{1}{r} + \mathcal{O}(\varepsilon^2) = -4\pi\, f(0) + \mathcal{O}(\varepsilon^2) \qquad (3.36)$$

From the arbitrariness of $f(\boldsymbol{r})$ and with $\varepsilon \to 0$ follows (3.34). The substitution $r \to |\boldsymbol{r} - \boldsymbol{r}'|$ leas to the more general form

$$\left(\Delta + k^2\right)\ \frac{\exp(\pm \mathrm{i}k\,|\boldsymbol{r} - \boldsymbol{r}'|)}{|\boldsymbol{r} - \boldsymbol{r}'|} = -4\pi\,\delta(\boldsymbol{r} - \boldsymbol{r}') \qquad (3.37)$$

Decomposition Theorem for Vector Fields

In this section, we show that every vector field $\boldsymbol{V}(\boldsymbol{r})$ can be expressed by its sources and vortices (provided that $\boldsymbol{V}(\boldsymbol{r})$ decreases sufficiently fast for $r \to \infty$). This decomposition theorem for vector fields is also called *Helmholtz decomposition* or *Helmholtz theorem* or *fundamental theorem of vector calculus*. With (3.24), we can write

$$\boldsymbol{V}(\boldsymbol{r}_0) = \int d^3r\ \boldsymbol{V}(\boldsymbol{r})\,\delta(\boldsymbol{r} - \boldsymbol{r}_0) = -\frac{1}{4\pi}\int d^3r\ \boldsymbol{V}(\boldsymbol{r})\ \Delta\ \frac{1}{|\boldsymbol{r} - \boldsymbol{r}_0|} \qquad (3.38)$$

$$= -\frac{1}{4\pi}\int d^3r\ \frac{\Delta \boldsymbol{V}(\boldsymbol{r})}{|\boldsymbol{r} - \boldsymbol{r}_0|} = \frac{1}{4\pi}\int d^3r\ \frac{\mathrm{curl\,curl}\,\boldsymbol{V}(\boldsymbol{r}) - \mathrm{grad\,div}\,\boldsymbol{V}(\boldsymbol{r})}{|\boldsymbol{r} - \boldsymbol{r}_0|}$$

First, by a double partial integration, the Laplace operator $\Delta = \partial_x^2 + \partial_y^2 + \partial_z^2$ is turned over to $\boldsymbol{V}$. Subsequently, (2.29), $\Delta \boldsymbol{V} = \mathrm{grad}\,(\mathrm{div}\,\boldsymbol{V}) - \mathrm{curl}\,(\mathrm{curl}\,\boldsymbol{V})$ is used. For the first term on the r.h.s. of (3.38), we specify the next steps:

$$\int d^3r\ \frac{\mathrm{curl\,curl}\,\boldsymbol{V}(\boldsymbol{r})}{|\boldsymbol{r} - \boldsymbol{r}_0|} = \int d^3r\ \frac{1}{|\boldsymbol{r} - \boldsymbol{r}_0|}\ \boldsymbol{\nabla} \times \mathrm{curl}\,\boldsymbol{V}(\boldsymbol{r})$$

$$\stackrel{\text{p.i.}}{=} \int d^3r\ \left(\mathrm{curl}\,\boldsymbol{V}(\boldsymbol{r})\right) \times \boldsymbol{\nabla}\ \frac{1}{|\boldsymbol{r} - \boldsymbol{r}_0|}$$

$$= -\int d^3r \,\big(\text{curl}\, \boldsymbol{V}(\boldsymbol{r})\big) \times \boldsymbol{\nabla}_0 \,\frac{1}{|\boldsymbol{r}-\boldsymbol{r}_0|}$$

$$= \int d^3r \; \boldsymbol{\nabla}_0 \times \frac{\text{curl}\, \boldsymbol{V}(\boldsymbol{r})}{|\boldsymbol{r}-\boldsymbol{r}_0|} = \boldsymbol{\nabla}_0 \times \int d^3r \;\frac{\text{curl}\, \boldsymbol{V}(\boldsymbol{r})}{|\boldsymbol{r}-\boldsymbol{r}_0|} \qquad (3.39)$$

In the step denoted by (p.i.), two minus signs occur: one from the partial integration itself and another by interchanging of the sequence in the vector product. When applied to $1/|\boldsymbol{r}-\boldsymbol{r}_0|$, the substitution $\boldsymbol{\nabla} \to -\boldsymbol{\nabla}_0$ can be made; $\boldsymbol{\nabla}$ acts on $\boldsymbol{r}$ and $\boldsymbol{\nabla}_0$ on $\boldsymbol{r}_0$. The derivative with $\boldsymbol{\nabla}_0$ can now be written in front of the integral because $\boldsymbol{r}_0$ is only a parameter. For the second term on the r.h.s. of (3.38), the calculation proceeds in the same way. After renaming the variables, the result is

$$\boxed{\boldsymbol{V}(\boldsymbol{r}) = \frac{1}{4\pi}\,\text{curl} \int d^3r' \;\frac{\text{curl}\, \boldsymbol{V}(\boldsymbol{r}')}{|\boldsymbol{r}-\boldsymbol{r}'|} - \frac{1}{4\pi}\,\text{grad} \int d^3r' \;\frac{\text{div}\, \boldsymbol{V}(\boldsymbol{r}')}{|\boldsymbol{r}-\boldsymbol{r}'|}} \qquad (3.40)$$

The operator curl and grad in front of the integrals act on the variable $\boldsymbol{r}$.

The derivation of (3.40) is valid under the restriction that fields drop off sufficiently fast for $r \to \infty$; thus, in particular, the integrals in (3.38)–(3.40) must be defined. For localized sources and vortices, (3.40) yields $|\boldsymbol{V}| \le \text{const.}/r^2$ for $r \to \infty$. This is a sufficient condition for the existence of the integrals.

The important result (3.40) implies in particular the following:

1. Every vector field $\boldsymbol{V}(\boldsymbol{r})$ is determined by its *vortices* curl $\boldsymbol{V}$ and its *sources* div $\boldsymbol{V}$.
   As an example, we consider electrostatics (Part II) with the field equations curl $\boldsymbol{E} = 0$ and div $\boldsymbol{E} = 4\pi\varrho$. It then follows from (3.40) how the electric field $\boldsymbol{E}$ is determined by the charge distribution $\varrho(\boldsymbol{r})$.
2. Every vector field $\boldsymbol{V}(\boldsymbol{r})$ can be written as the sum of a vortex field and of a gradient field, i.e.,

$$\boldsymbol{V}(\boldsymbol{r}) = \text{curl}\, \boldsymbol{W}(\boldsymbol{r}) + \text{grad}\, \Phi(\boldsymbol{r}) \qquad (3.41)$$

   Equation (3.40) shows how the fields $\boldsymbol{W}(\boldsymbol{r})$ and $\Phi(\boldsymbol{r})$ result from $\boldsymbol{V}(\boldsymbol{r})$.
3. In

$$\text{curl}\, \boldsymbol{V}(\boldsymbol{r}) = 0 \;\longleftrightarrow\; \boldsymbol{V}(\boldsymbol{r}) = \text{grad}\, \Phi(\boldsymbol{r}) \qquad (3.42)$$

the step from left to right means a vortex-free field can be represented as a gradient field. This follows directly from (3.40). The opposite direction follows from (2.27).

4. In

$$\operatorname{div} \boldsymbol{V}(\boldsymbol{r}) = 0 \quad \longleftrightarrow \quad \boldsymbol{V}(\boldsymbol{r}) = \operatorname{curl} \boldsymbol{W}(\boldsymbol{r}) \tag{3.43}$$

the conclusion from left to right means a source-free field can be represented as a vortex field. This follows directly from (3.40). The opposite direction follows from (2.25).

If a field has no curl, it is said to be irrotational. If a field has no divergence, it is said to be solenoidal (like a magnetic field). The terminus solenoid used for a cylindrical coil.

Exercises

3.1 *Delta function as a sequence of functions*

Show

$$\lim_{\ell\to 0} \int_{-\infty}^{\infty} dx\; d_\ell^{(1)}(x - x_0)\, f(x) = f(x_0) \quad \text{with}$$

$$d_\ell^{(1)}(x) = \frac{1}{\sqrt{2\pi}\,\ell}\, \exp\left(-\frac{x^2}{2\ell^2}\right)$$

To do this, assume that $f(x)$ can be expanded into a Taylor series around x_0. Also show

$$\lim_{\ell\to 0} \int_{-\infty}^{\infty} dx\; d_\ell^{(2)}(x - x_0)\, f(x) = f(x_0) \quad \text{with} \quad d_\ell^{(2)}(x) = \frac{\sin(\pi x/\ell)}{2\,\sin(\pi x/2)}$$

Use a suitable integration variable.

3.2 *Integral representation of the δ function*

Compare the function

$$g(x) = \int_{-\infty}^{\infty} dk\; \exp\left(-\frac{\ell^2 k^2}{2}\right)\, \exp(ikx)$$

with $d_\ell^{(1)}(x)$ from Exercise 3.1. Derive from this an integral representation for the δ-function.

3.3 *Representation of the delta function as a sum*

Justify that the δ-function in the interval $-1, 1$ can be represented by the sum

$$\delta(x) = \frac{1}{2} \lim_{N\to\infty} \sum_{n=-N}^{N} \exp\left(\mathrm{i}\,\pi n x\right) = \frac{1}{2} \sum_{n=-\infty}^{\infty} \exp\left(\mathrm{i}\,\pi n x\right) \tag{3.44}$$

Convert the finite sum to the form $d_\ell^{(2)}(x)$ from Exercise 3.1 and perform the limit $N \to \infty$.

3.4 *Delta function of a function*

The function $h(x)$ has a single simple zero at x_0. Justify the relation

$$\delta\big(h(x)\big) = \frac{1}{|h'(x_0)|}\,\delta(x - x_0)$$

Chapter 4

Lorentz Tensors

The most important formulas for the Lorentz transformation, for the definition of Lorentz tensors and their differentiation are compiled. Basic knowledge of the special theory of relativity, in particular the derivation and meaning of the Lorentz transformation, are assumed. For this, I refer to Part IX of my Mechanics [1].

The occurring structures are needed in Chapter 18. They are comparable to those presented in Chapter 2.

Lorentz Transformation

The orthogonal transformations discussed in Chapter 2 mediate between different Cartesian coordinate systems of the three-dimensional space. The now discussed Lorentz transformations (LT) mediate between different inertial systems (IS). An inertial frame is a reference frame that is not accelerated with respect to the fixed star sky. An IS is an inertial frame with coordinates, mostly the Cartesian coordinates x, y and z and a time coordinate t. Here t is the time displayed by a clock that is at rest in IS. The coordinates x, y and z correspond to lengths of resting rules.

The space-time coordinates are denoted by x^α,

$$(x^\alpha) = \left(x^0, x^1, x^2, x^3\right) = (ct, x, y, z) \tag{4.1}$$

where c is the speed of light. Greek indices always run from 0 to 3, and Latin indices from 1 to 3. A set of specific values of the space-time coordinates defines an *event*. Such an event has the coordinate values x^α in IS and the values x'^α in IS′ (Figure 4.1). The Lorentz transformation

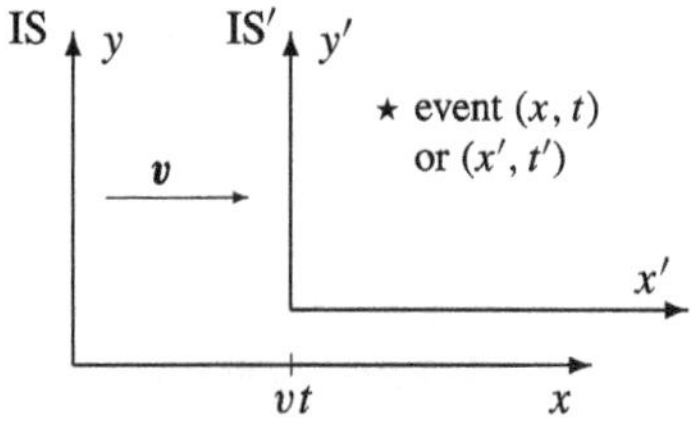

Figure 4.1 An event (⋆) has the coordinates x, t in the inertial system IS and the coordinates x', t' in the system IS′. The system IS′ moves relative to IS with the constant velocity $\boldsymbol{v}$. The coordinate values are connected by a Lorentz transformation.

between these coordinate values is linear:

$$x'^{\alpha} = \Lambda^{\alpha}_{\ \beta}\, x^{\beta} + b^{\alpha} \quad \text{(Lorentz transformation)} \tag{4.2}$$

As a summation convention, we sum over equal indices, one of which is at the top and the other at the bottom; the summation sign is not shown explicitly. The decisive condition for the coefficients $\Lambda^{\alpha}_{\ \beta}$ is the invariance of the four-dimensional line element ds. For two infinitesimally neighboring events, this interval is

$$ds^2 = \eta_{\alpha\beta}\, dx^{\alpha}\, dx^{\beta} = \eta_{\alpha\beta}\, dx'^{\alpha}\, dx'^{\beta} \tag{4.3}$$

The quantities $\eta_{\alpha\beta}$ and $\eta^{\alpha\beta}$ are given by the number assignment

$$\eta = (\eta_{\alpha\beta}) = (\eta^{\alpha\beta}) = \begin{pmatrix} 1 & 0 & 0 & 0 \\ 0 & -1 & 0 & 0 \\ 0 & 0 & -1 & 0 \\ 0 & 0 & 0 & -1 \end{pmatrix} \tag{4.4}$$

They are therefore independent of the IS. The space with the line element (4.3) is called *Minkowski space*.

If the two events are connected by a light signal, then $ds = 0$. Therefore, (4.3) guarantees the constancy of the speed of light. From (4.2) follows $dx'^{\beta} = \Lambda^{\beta}_{\ \alpha}\, dx^{\alpha}$. We insert this into (4.3) and obtain

$$\Lambda^{\alpha}_{\ \gamma}\, \Lambda^{\beta}_{\ \delta}\, \eta_{\alpha\beta} = \eta_{\gamma\delta} \quad \text{or} \quad \Lambda^{\mathrm{T}}\, \eta\, \Lambda = \eta \tag{4.5}$$

This condition can be written as a matrix multiplication with the matrix $\Lambda = (\Lambda^{\beta}_{\ \delta})$; here β is the row index and δ is the column index. The condition $\Lambda^{\mathrm{T}}\, \eta\, \Lambda = \eta$ corresponds to the condition $\alpha^{\mathrm{T}}\alpha = 1$ for orthogonal transformations. The orthogonal transformations are included as a subgroup in the Lorentz transformations because they leave ds^2 invariant, too.

We now consider the following special case: IS and IS′ have have the same spatial-temporal origin ($b^\alpha = 0$), parallel Cartesian coordinate axes and a relative motion in the x-direction. Then, $y' = y$ and $z' = z$ and (4.2) becomes the *special Lorentz transformation* with

$$\Lambda = (\Lambda_\alpha^\beta) = \begin{pmatrix} \Lambda_0^0 & \Lambda_1^0 & 0 & 0 \\ \Lambda_0^1 & \Lambda_1^1 & 0 & 0 \\ 0 & 0 & 1 & 0 \\ 0 & 0 & 0 & 1 \end{pmatrix}, \qquad b^\alpha = 0 \tag{4.6}$$

Equations (4.5) determines the remaining Λ_α^β except for a parameter ψ:

$$\begin{pmatrix} \Lambda_0^0 & \Lambda_1^0 \\ \Lambda_0^1 & \Lambda_1^1 \end{pmatrix} = \begin{pmatrix} \cosh\psi & -\sinh\psi \\ -\sinh\psi & \cosh\psi \end{pmatrix} \tag{4.7}$$

If IS′ moves relative to IS with the speed v, the origin of IS′ is, on the one hand, $x = vt$, and, on the other hand, $x' = 0$. The LT reads $x' = \Lambda_0^1 ct + \Lambda_1^1 x$. From this follows the relationship between the *rapidity* ψ and the velocity v:

$$\tanh\psi = \frac{v}{c}, \qquad \psi = \operatorname{artanh}\frac{v}{c} \tag{4.8}$$

We write down the special LT, (4.2) with (4.6)–(4.8), explicitly:

$$x' = \frac{x - vt}{\sqrt{1 - v^2/c^2}}, \quad y' = y, \quad z' = z, \quad ct' = \frac{ct - xv/c}{\sqrt{1 - v^2/c^2}} \tag{4.9}$$

Tensor Definition

A quantity V^α which transforms like the coordinates x^α

$$V'^\beta = \Lambda_\alpha^\beta \, V^\alpha \tag{4.10}$$

is called *Lorentz tensor* 1st rank. Examples for Lorentz vectors are dx^α and the 4-velocity $u^\alpha = c\, dx^\alpha / ds$.

A *Lorentz tensor* of Nth rank is an N-fold indexed quantity, which transforms as

$$T'^{\alpha_1 \ldots \alpha_N} = \Lambda^{\alpha_1}_{\beta_1} \ldots \Lambda^{\alpha_N}_{\beta_N} T^{\beta_1 \ldots \beta_N} \tag{4.11}$$

Common are also the designations four-tensor or 4-tensor. "Tensor" always refers to the entirety of the indexed quantities. A *Lorentz scalar* or 0th rank tensor is a non-indexed quantity that is invariant under LT. Examples of Lorentz scalars are the proper length of a rod and the proper time interval $d\tau = ds/c$ of a clock.

The use of upper indices in (4.1) is a convention. We assign to every vector V^α a corresponding one with lower indices:

$$V_\beta = \eta_{\beta\alpha} V^\alpha \tag{4.12}$$

We multiply (4.12) by $\eta^{\gamma\beta}$ and use $\eta^{\gamma\beta} \eta_{\beta\alpha} = \delta^\gamma_\alpha$. This results in the reverse relation

$$V^\gamma = \eta^{\gamma\beta} V_\beta \tag{4.13}$$

The introduction of the quantities with lower and upper indices serves to simplify the notation, for example, $ds^2 = dx_\alpha dx^\alpha$. For the coordinates with lower indices, we get

$$(x_\alpha) = (x_0, x_1, x_2, x_3) = (ct, -x, -y, -z) \tag{4.14}$$

This assignment is made accordingly for all tensors, for example

$$T_{\alpha\beta} = \eta_{\alpha\alpha'} \eta_{\beta\beta'} T^{\alpha'\beta'}, \quad T^\alpha{}_\beta = \eta_{\beta\beta'} T^{\alpha\beta'}, \quad T_\alpha{}^\beta = \eta_{\alpha\alpha'} T^{\alpha'\beta} \tag{4.15}$$

The $T^{\alpha\beta\ldots}$ are called *contravariant* components of the tensor, and the $V_{\alpha\beta\ldots}$ *covariant* components. We usually dispense with this detailed designation and call the indexed quantity itself as a tensor, or as a covariant or contravariant tensor. The term "covariant" also has another meaning, namely, "form invariant".

We refer to the $T^\alpha{}_\beta$ as mixed components. With them, attention must be paid to the order of the indices because

$$T^\alpha{}_\beta = \eta_{\beta\gamma} T^{\alpha\gamma} \overset{\text{(in general)}}{\neq} \eta_{\beta\gamma} T^{\gamma\alpha} = T_\beta{}^\alpha \tag{4.16}$$

For a symmetric tensor, we have $T^{\alpha\gamma} = T^{\gamma\alpha}$; in this case, the indices can also be written on top of each other $T^\alpha{}_\beta = T_\beta{}^\alpha = T^\alpha_\beta$.

The indices of the transformation matrix Λ^{α}_{β} can be written on top of each other because this quantity is not a tensor. In order to fulfill the tensor definition, Λ would first of all had to be defined in IS and IS′; Λ is, however, a quantity that is associated with the transition between IS and IS′.

We specify the transformation of the covariant tensors:

$$V'_{\alpha} \overset{(4.12)}{=} \eta_{\alpha\beta}\, V'^{\beta} \overset{(4.10)}{=} \eta_{\alpha\beta}\, \Lambda^{\beta}_{\gamma}\, V^{\gamma} \overset{(4.13)}{=} \eta_{\alpha\beta}\, \Lambda^{\beta}_{\gamma}\, \eta^{\gamma\delta}\, V_{\delta} = \overline{\Lambda}{}^{\delta}_{\alpha}\, V_{\delta} \tag{4.17}$$

In the first step, it was taken into account that the $\eta_{\alpha\beta}$ are defined as numbers independently of the IS. In the last step, we introduced the matrix $\overline{\Lambda}$:

$$\overline{\Lambda}{}^{\delta}_{\alpha} = \eta_{\alpha\beta}\, \Lambda^{\beta}_{\gamma}\, \eta^{\gamma\delta} \tag{4.18}$$

We multiply this by $\Lambda^{\alpha}_{\epsilon}$:

$$\overline{\Lambda}{}^{\delta}_{\alpha}\, \Lambda^{\alpha}_{\epsilon} = \eta_{\alpha\beta}\, \Lambda^{\beta}_{\gamma}\, \eta^{\gamma\delta}\, \Lambda^{\alpha}_{\epsilon} \overset{(4.5)}{=} \eta^{\gamma\delta}\, \eta_{\gamma\epsilon} = \delta^{\delta}_{\epsilon} \tag{4.19}$$

Correspondingly, $\Lambda^{\delta}_{\alpha}\, \overline{\Lambda}{}^{\alpha}_{\epsilon} = \delta^{\delta}_{\epsilon}$. The reverse transformations follow from (4.19)

$$V^{\gamma} = \delta^{\gamma}_{\beta}\, V^{\beta} = \overline{\Lambda}{}^{\gamma}_{\alpha}\, \Lambda^{\alpha}_{\beta}\, V^{\beta} = \overline{\Lambda}{}^{\gamma}_{\alpha}\, V'^{\alpha} \tag{4.20}$$

$$V_{\gamma} = \delta^{\beta}_{\gamma}\, V_{\beta} = \overline{\Lambda}{}^{\beta}_{\alpha}\, \Lambda^{\alpha}_{\gamma}\, V_{\beta} = \Lambda^{\alpha}_{\gamma}\, V'_{\alpha} \tag{4.21}$$

We summarize the following: Contravariant vectors are transformed with Λ^{α}_{β}, covariant vectors with $\overline{\Lambda}{}^{\alpha}_{\beta}$.

The calculation rules for the tensors in the three-dimensional space (Chapter 2) can be transferred to the Lorentz tensors. If S and T are tensors, then the following applies:

1. Addition: $a\, S^{\alpha_1 \ldots \alpha_N} + b\, T^{\alpha_1 \ldots \alpha_N}$ is a tensor of rank N; here a and b are numbers.
2. Multiplication: $S^{\alpha_1 \ldots \alpha_N}\, T^{\beta_1 \ldots \beta_M}$ is a tensor of rank $N + M$.
3. Contraction: $\eta_{\beta\gamma}\, S^{\alpha_1 \ldots \beta \ldots \gamma \ldots \alpha_N} = S^{\alpha_1 \ldots \beta \ldots}{}_{\beta}{}^{\ldots \alpha_N}$ is a tensor of the $(N-2)$th rank. In particular, $ds^2 = dx^{\alpha}\, dx_{\alpha}$ and $S^{\alpha}\, T_{\alpha}$ Lorentz scalars.
4. Tensor equations: If $S^{\alpha} = U^{\alpha\beta}\, T_{\beta}$ applies in every IS, then $U^{\alpha\beta}$ is a 2nd rank tensor.

Minkowski and Levi-Civita tensor

By (4.4), $\eta = (\eta_{\alpha\beta}) = (\eta^{\alpha\beta})$ is defined as a constant matrix. Nevertheless, the indexed quantities $\eta^{\alpha\beta}$ and $\eta_{\alpha\beta}$ can be considered as tensors, too, and transformed accordingly,

$$\eta'_{\alpha\beta} \overset{(4.17)}{=} \overline{\Lambda}^{\,\gamma}_{\alpha}\, \overline{\Lambda}^{\,\delta}_{\beta}\, \eta_{\gamma\delta} \overset{(4.5)}{=} \overline{\Lambda}^{\,\gamma}_{\alpha}\, \overline{\Lambda}^{\,\delta}_{\beta}\, \Lambda^{\mu}_{\gamma}\, \Lambda^{\nu}_{\delta}\, \eta_{\mu\nu} \overset{(4.19)}{=} \eta_{\alpha\beta} \tag{4.22}$$

The tensor η is called *Minkowski tensor*. Due to

$$\eta^{\alpha}{}_{\beta} \overset{(4.15)}{=} \eta^{\alpha\gamma}\, \eta_{\gamma\beta} = \delta^{\alpha}_{\beta} \tag{4.23}$$

the Kronecker symbol δ^{α}_{β} is a 4-tensor, too. Since η is symmetric, the indices here can be written on top of each other, $\eta^{\alpha}{}_{\beta} = \eta_{\beta}{}^{\alpha} = \eta^{\alpha}_{\beta} = \delta^{\alpha}_{\beta}$.

Another constant quantity, which meets the tensor properties, is the totally antisymmetric tensor:

$$\epsilon^{\alpha\beta\gamma\delta} = \begin{cases} +1 & \text{if } (\alpha,\beta,\gamma,\delta) \text{ even permutation of } (0,1,2,3) \\ -1 & \text{if } (\alpha,\beta,\gamma,\delta) \text{ odd permutation of } (0,1,2,3) \\ 0 & \text{other} \end{cases} \tag{4.24}$$

This quantity is also called *Levi-Civita tensor*. If on the right-hand side of (4.11) there is an additional factor $\det\Lambda$ is, then the quantity is called pseudotensor. (From (4.5) follows $\det\Lambda = \pm 1$. The case $\det\Lambda = -1$ is, for example, obtained for the reflection $x \to -x$.) The Levi-Civita tensor is a pseudotensor:

$$\epsilon'^{\alpha\beta\gamma\delta} = \big(\det\Lambda\big)\, \Lambda^{\alpha}_{\alpha'}\, \Lambda^{\beta}_{\beta'}\, \Lambda^{\gamma}_{\gamma'}\, \Lambda^{\delta}_{\delta'}\, \epsilon^{\alpha'\beta'\gamma'\delta'} = \big(\det\Lambda\big)^2\, \epsilon^{\alpha\beta\gamma\delta} = \epsilon^{\alpha\beta\gamma\delta} \tag{4.25}$$

The covariant components of the Levi-Civita tensor are given by

$$\epsilon_{\alpha\beta\gamma\delta} = \eta_{\alpha\alpha'}\, \eta_{\beta\beta'}\, \eta_{\gamma\gamma'}\, \eta_{\delta\delta'}\, \epsilon^{\alpha'\beta'\gamma'\delta'} = -\,\epsilon^{\alpha\beta\gamma\delta} \tag{4.26}$$

Tensor Fields

We extend the tensor definition to tensor fields. The functions $S(x)$, $V^{\alpha}(x)$ and $T^{\alpha\beta}(x)$ are each a scalar, vector or tensor field if

$$S'(x') = S(x) \tag{4.27}$$

$$V'^{\alpha}(x') = \Lambda^{\alpha}_{\beta}\, V^{\beta}(x) \tag{4.28}$$

$$T'^{\alpha\beta}(x') = \Lambda^{\alpha}_{\gamma}\, \Lambda^{\beta}_{\delta}\, T^{\gamma\delta}(x) \tag{4.29}$$

Here, the arguments are transformed, too, i.e., $x' = (x'^{\alpha}) = (\Lambda^{\alpha}_{\ \beta}\, x^{\beta})$ in the argument.

Tensor fields can be differentiated with respect to the arguments. The partial derivative $\partial/\partial x^{\alpha}$ transforms like a covariant vector. From (4.20) follows

$$\frac{\partial x^{\beta}}{\partial x'^{\alpha}} = \overline{\Lambda}_{\alpha}^{\ \beta} \tag{4.30}$$

This results in

$$\frac{\partial}{\partial x'^{\alpha}} = \frac{\partial x^{\beta}}{\partial x'^{\alpha}}\,\frac{\partial}{\partial x^{\beta}} = \overline{\Lambda}_{\alpha}^{\ \beta}\,\frac{\partial}{\partial x^{\beta}} \tag{4.31}$$

Thus,

$$\partial_{\alpha} = \frac{\partial}{\partial x^{\alpha}} \tag{4.32}$$

transforms like a covariant vector. Accordingly,

$$\partial^{\alpha} = \frac{\partial}{\partial x_{\alpha}} \tag{4.33}$$

is a contravariant vector. From the vector property of ∂^{α} and ∂_{α} it follows that the d'Alembert operator

$$\Box = \partial^{\alpha}\partial_{\alpha} = \eta^{\alpha\beta}\partial_{\alpha}\partial_{\beta} = \frac{1}{c^2}\,\frac{\partial^2}{\partial t^2} - \Delta \tag{4.34}$$

is a Lorentz scalar.

Covariance

The formal transformation properties facilitate the utilization of symmetries. The symmetries considered in Chapters 2 and 4 are the isotropy of space (equivalence of differently oriented IS) and the relativity principle (equivalence of differently moving IS).

Due to the isotropy of space, fundamental laws cannot depend on the orientation of a coordinate system S (we refer to the three-dimensional space). They must have the same form in a rotated S′, i.e., they must be covariant under orthogonal transformations. The laws must therefore have the form of tensor equations (with 3-tensors). This has already been discussed in the section on covariance in Chapter 2.

Due to the relativity principle, fundamental laws must not depend on the relative velocity of an IS. They must therefore have the same form in an IS′ moving relative to IS, i.e., they must be covariant under Lorentz transformations. The laws must therefore have the form of Lorentz tensor equations. Examples are the inhomogeneous Maxwell equations

$$\partial_\alpha F^{\alpha\beta} = \frac{4\pi}{c}\, j^\beta \tag{4.35}$$

from Chapter 18. This relation is covariant because ∂_α, $F^{\alpha\beta}$ and j^β are all Lorentz tensors. A vector notation (as in three-dimensional, for example, $\boldsymbol{L} = \widehat{\Theta} \cdot \boldsymbol{\omega}$ instead of $L_i = \sum_j \Theta_{ij}\, \omega_j$) is also possible; however, we do not use it.

The covariance requirement restricts the form of the physical laws significantly. This facilitates the establishment of such laws.

Exercises

4.1 *Lorentz tensor of second rank*

The relationship $V^\alpha = T^{\alpha\beta}\, W_\beta$ shall be valid in every inertial system. It shall be known that V^α and W^α are Lorentz vectors. Prove that then $T^{\alpha\beta}$ is a Lorentz tensor.

PART II
Electrostatics

Chapter 5

Coulomb Law

We present the basic properties of the Coulomb force, i.e., the electrostatic interaction. The charge and the electric field are defined as measuring quantities.

Common matter consists of atomic nuclei and electrons. The *Coulomb force* between the electrons and atomic nuclei determines the various forms of matter and most of the observable phenomena (such as physical properties of a solid body or chemical reactions). The nuclear (strong) interaction is responsible for the structure of the atomic nuclei; as long as the atomic nuclei are stable, it does not make itself noticeable. The gravitational force is known from everyday life. Due to its relative smallness, it usually plays no role in the organization of common matter.

The Coulomb force can be compared to the gravitational force: It has the same distance dependence but is stronger by many orders of magnitude. Just as the gravitational force is proportional to the mass of the particles involved, the Coulomb force is proportional to a quantity (property of the particles), which is called *charge*. The Coulomb interaction can be investigated by scattering experiments; the most famous experiment is the Rutherford scattering experiment. Macroscopic bodies can also achieve "charge" by means of suitable preparation so that they exert Coulomb forces on each other. Matter consisting of neutral atoms or molecules is uncharged; the positive and negative charges add up to zero.

The charge q of a body is a physical observable. Its meaning is specified step by step in the following. This leads to the definition of charge as an observable.

First of all, we assume that the extension of the charged particles under consideration is small compared to their distances. Then, we may assume that the charges are concentrated in one point. We call this concept point charge; It corresponds to the point mass in mechanics. A point charge is characterized by its position vector $\boldsymbol{r}$ and its charge q.

We compile a series of findings that can be generalized from many experiments. For this, we refer to two point charges q_1 and q_2 at rest:

1. The Coulomb force is a central force. This means that it acts in the direction of the line connecting two charges.
2. The Coulomb force is subject to the principle of counteraction (Newton's 3rd axiom), i.e., the forces on the charges are oppositely equal in size, $\boldsymbol{F}_2 = -\boldsymbol{F}_1$.
3. The Coulomb force is proportional to the product of the charges q_1 and q_2. The charges can be positive or negative. Charges with the same sign repel each other, while charges with different signs attract each other.
4. The Coulomb force is inversely proportional to the square of the distance $r_{12} = |\boldsymbol{r}_1 - \boldsymbol{r}_2|$; here $\boldsymbol{r}_1$ and $\boldsymbol{r}_2$ are the position vectors of the two point charges. Together with point 1–3 results in the law of force:

$$\boxed{\boldsymbol{F}_1 = k\, q_1 q_2\, \frac{\boldsymbol{r}_1 - \boldsymbol{r}_2}{|\boldsymbol{r}_1 - \boldsymbol{r}_2|^3}} \tag{5.1}$$

   The constant k depends on the definition of the unit for the charges q_1 and q_2. The repulsion of equal charges implies $k > 0$.
5. The superposition principle applies to the Coulomb forces. The resulting force on a charge is the sum of the Coulomb forces between this charge and all other charges.
6. The law of conservation applies to the charges: The sum of the charges of a closed system is conserved. The charges of elementary particles are invariant properties of these particles.
7. In nature, charges are quantized. The positive charge of a proton is denoted by $q = e$, and the negative charge of an electron by $q = -e$. For charged macroscopic bodies, one finds in general generally $|q| \gg e$ so that quantization plays no role. plays a role.

The *Coulomb law* specified by these points is the basis of *electrostatics*. Electrostatics is limited to charges at rest or stationary charge distributions.

We consider a proton and an electron at a distance of the Bohr radius $r_{12} = a_B \approx 0.53\,\text{Å}$. Coulomb and gravitational forces act between these particles:

$$F_{e-p} = \begin{cases} 8 \cdot 10^{-8}\,\text{N} & \text{(Coulomb force)} \\ 4 \cdot 10^{-47}\,\text{N} & \text{(gravitational force)} \end{cases} \tag{5.2}$$

The Coulomb force is about 39 orders of magnitude stronger; this ratio is independent of the distance. At a distance of $r_{12} = 5\,\text{fm} = 5 \cdot 10^{-15}\,\text{m}$, two colliding protons repel each other with a Coulomb force of about 10 newton. At the distance $r_{12} \lesssim 1\,\text{fm}$, the Coulomb force is overcome by the approximately 100 times stronger, short-range nucleon-nucleon interaction. This *strong* interaction is effectively attractive and leads to the binding of nucleons (Z protons and N neutrons) in an atomic nucleus. For large nuclei, however, the Coulomb force becomes more important because it acts over the entire nucleus and is proportional to Z^2. This leads eventually to an instability of larger nuclei. Thus, uranium and transuranium elements are radioactive; they decay by α-decay or spontaneous fission.

The points 1–7 listed are to be understood as a generalization of experimental findings. The superposition of the forces and the $1/r^2$ dependence can only be verified in a finite number of experiments. Therefore, there could also be deviations from Coulomb's law, for example, in the case of very strong forces or very large or very small distances.

Validity Range

The $1/r^2$-dependence of the Coulomb force is verified experimentally over a wide range. In the Rutherford experiment, helium nuclei (α-particles) are scattered by gold nuclei. From the angular distribution of the scattered particles (Rutherford cross section), the r-dependence of the scattering potential can be deduced. This experiment confirms the validity of the Coulomb law in the range $10^{-12} \ldots 10^{-11}$ cm. Other experiments (such as high energy electron-positron scattering) test the Coulomb force at even smaller distances. For such experiments, however, a quantum mechanical description is necessary. For very small distances (and correspondingly high energies), new particles are generated in such an experiment; a force law such as (5.1) cannot describe such processes.

The correct quantum mechanical description (e.g., the calculation of the cross section for electron–positron scattering) is carried out within the framework of quantum electrodynamics. If the quantum effects are small, they can be described by corrections to the power law (5.1). For the Coulomb interaction in the hydrogen atom, one obtains here corrections of size $\alpha = e^2/\hbar c \approx 1/137$ (i.e., in the percentage range) for distances $\hbar/m_{\rm e}c \approx 4 \cdot 10^{-11}$ cm; here $m_{\rm e}$ is is the electron mass.

For every day distances (centimeters, meters), the Coulomb law can be confirmed by laboratory experiments with macroscopic charged bodies. For very large distances, one may consider the magnetic field of planets. The spatial dependence of these fields is linked to the $1/r^2$ dependence of (5.1). Up to distances of about $R = 10^5$ km, no deviations were found. Deviations for large r were expected if the photon (energy quantum of the electromagnetic wave) had a small mass $m_\gamma \neq 0$. In this case, deviations would occur then for distances of the size $R \sim \hbar/m_\gamma c$. All experimental results are, however, compatible with $m_\gamma = 0$.

Measurement of the Charge

We place any two charges q_1 and q_2 one after the other in the vicinity of a third charge. The forces F_1 and F_2 act on these charges. The measurement of the ratio F_1/F_2 is a measurement of the charge ratio q_1/q_2. If one now defines arbitrarily a certain charge as the unit charge, then the charge is defined as a measuring quantity (observable).

To select the unit of charge, we consider the following possibilities:

(i) It would be obvious to define the charge quantum (e.g., the charge e of a proton or positron) as the charge "1" or as "1 unit of charge = 1 CU". Then, the constant k in (5.1) had to be determined experimentally. It had the dimension $k = {\rm N\,m^2/(CU)^2}$.

Research on electricity led to the Maxwell equations before the quantization of charge was found. Therefore, such a determination is not customary. The following points, (ii) and (iii), introduce the unit coulomb (C). In this unit, the charge quantum becomes

$$e = 1.602 \cdot 10^{-19}\ {\rm C} \tag{5.3}$$

(ii) The charge unit could be defined by (5.1). But historically it is defined otherwise.

The charge 1 coulomb = 1 C (a long time ago, Cb was used instead of C) is defined by the amount of silver (1.118 mg) deposited in a silver nitrate solution. With such a definition, the constant k in (5.1) can be determined experimentally,

$$k \approx 9 \cdot 10^9 \, \frac{\mathrm{N\,m^2}}{\mathrm{C^2}} \tag{5.4}$$

This value implies that 1 coulomb is a very large charge for common matter. Two charges with 1 coulomb at a distance of 1 meter will exert a tremendous force of the size of 10^{10} newton on each other. Due to the strength of the Coulomb force, matter organizes itself in such a way that the positive charges (a total of approximately 10^6 coulombs in one mole of carbon) are compensated by about the same number of negative ones. For currents (charge per time), on the other hand, 1 ampere = 1 A = 1 C/s is a rather common quantity.

(iii) One could define the unit coulomb by prescribing the value of the force that two charges of strength 1 C at a distance of 1 m exert on each other. Practically, one could use (5.4) with an equal sign as the definition of the unit C. Such a definition of the coulomb has the advantage that it is approximatively compatible with the older definition according to point (ii). In fact, one follows a similar way, which is described in the following.

A definition of the unit of charge is equivalent to that of the current unit:

$$1\,\mathrm{A} = 1\,\mathrm{ampere} = 1\,\frac{\mathrm{coulomb}}{\mathrm{second}} = 1\,\frac{\mathrm{C}}{\mathrm{s}} \tag{5.5}$$

The same current $I = \Delta q / \Delta t$ shall flow through two parallel, infinitely long wires at a distance of d. On each wire section of length Δl acts a magnetic force ΔF:

$$\frac{\Delta F}{\Delta l} = k\,\frac{2I^2}{c^2 d} \tag{5.6}$$

Here c is the speed of light ($c \approx 3 \cdot 10^8$ m/s). The magnetic forces are discussed in Chapter 13. The relationship between electric and magnetic forces follows from the relativistic structure of the ED (Part IV). This relationship requires that the same constant k occurs in (5.1) and (5.6).

The same current I flows through two parallel wires at a distance of $d = 1$ m. The current strength $I = 1\,\text{A} = 1\,\text{C/s}$ is defined such that the force per length is $\Delta F/\Delta l = 2 \cdot 10^{-7}$ N/m is The numerical value is chosen so that the resulting coulomb is almost equal to that of the older definition (ii). According to (5.6), the new definition specifies the constant k is fixed:

$$k \stackrel{\text{def}}{=} 10^{-7}\,\frac{\text{N}c^2}{\text{A}^2} \approx 9 \cdot 10^9\,\frac{\text{Nm}^2}{\text{C}^2} \quad \text{(SI system)} \tag{5.7}$$

The measuring system with the unit ampere as the fourth basic unit is called "Practical MKSA system" (MKSA stands for meter, kilogram, second and ampere) or Système International d'Unités (SI for short or the SI system). The designation SI is the modern one, and this unit system is extended to other observables, too.

In this SI system, the force constant k is usually expressed by $1/(4\pi\varepsilon_0)$, where ε_0 is called "dielectric constant of the vacuum".

(iv) On the basis of (5.1), it appears natural to define the unit of charge by

$$k \stackrel{\text{def}}{=} 1 \quad \text{(Gaussian unit system)} \tag{5.8}$$

The unit of charge is then defined as the charge that exerts a force of 1 newton on a charge of the same size at a distance of 1 meter. From historical reasons, it is customary at this point to use $\text{cm} = 10^{-2}$ m as the length unit and $\text{dyne} = \text{g cm/s}^2 = 10^{-5}$ N as the force unit. The unit of charge is called esu (for electrostatic unit). The following results from (5.1) and (5.8):

$$1\,\text{dyne} = \frac{(1\,\text{esu})^2}{\text{cm}^2} \quad \text{or} \quad \text{esu} = \frac{\text{cm}^{3/2}\,\text{g}^{1/2}}{\text{s}} \tag{5.9}$$

In this unit, the quantity of the elementary charge is

$$e \approx 4.803 \cdot 10^{-10}\,\text{esu} \tag{5.10}$$

By comparing this with (5.3), it can be seen how many coulombs are contained in 1 esu, (A.5). Alternatively, the electrostatic unit (esu) is also called statcoulomb (statC).

Choice of the unit system

The choice of measurement system is a question of purpose, convenience and convention. The most important reasons for choosing one of the two systems discussed are as follows:

- SI or MKSA system:
  (1) In many countries, this is the legal or official system. (2) Experimental quantities are mostly given in this system. All technical applications refer to the SI. Introductory physicists often use the SI.
- Gaussian system:
  (1) This system is particularly suitable for displaying of the relativistic structure of the ED. (2) In the scientific physics literature, the Gaussian system is mostly used.

Due to the factual usages, the restriction to a single system is not possible. For a theory course, the decisive argument is that in the Gaussian system the relativistic structure of electrodynamics ED is more transparent. Therefore,

- in this book, the **Gaussian system of units** is used as a matter of principle.

However, we also use the SI for practical estimations. The units of the SI system are listed in the Appendix. The Appendix also displays the most important formulas of electrodynamics in the SI.

In the Gaussian system, the electric and magnetic fields, $\boldsymbol{E}$ and $\boldsymbol{B}$, are measured in the same units. This makes sense for a number of reasons:

– In an inertial system (IS), there shall be a static charge distribution but no currents. Then, $\boldsymbol{E} \neq 0$ but $\boldsymbol{B} = 0$. In an other inertial system IS$'$ moving relative to it, the charge distribution appears as a current distribution, i.e., $\boldsymbol{B}' \neq 0$. Whether there is an electric or a magnetic field is present is therefore (partly) dependent on the observer; the transition IS $\longleftrightarrow$ IS$'$ transforms the fields into each other. In the Gaussian system, this is immediately recognizable as a relativistic effect; it is recognized by the factor v/c in $B' \sim (v/c)\,E$).

- For electromagnetic waves, one finds $|\boldsymbol{E}| = |\boldsymbol{B}|$ in the Gaussian system. Such waves are a fully relativistic phenomenon. For a charge moving with velocity v in the field of the wave, the magnetic force on it is by a factor v/c weaker than the electric one.
- If the point charges considered in (5.1) move with the velocities v_1 and v_2, this results in a magnetic interaction which is weaker by the factor $v_1 v_2/c^2$ than the Coulomb force.

These remarks refer to effects that will be dealt with in detail later; in this respect, they cannot be fully understood at this point. The points listed are merely intended to make it plausible that, from a theoretical point of view, there are good reasons in favor of the Gaussian system. The main reason is that relativistic effects can be recognized by factors v/c.

Electric Field

We consider N charges $q_1, \ldots, q_N$, which are rest at the positions $\boldsymbol{r}_1, \ldots, \boldsymbol{r}_N$. According to point 5 of the experimental findings (summarized above), a further charge q at $\boldsymbol{r}$ feels the force

$$\boldsymbol{F}(\boldsymbol{r}) = \sum_{i=1}^{N} q\, q_i\, \frac{\boldsymbol{r} - \boldsymbol{r}_i}{|\boldsymbol{r} - \boldsymbol{r}_i|^3} = q\,\boldsymbol{E}(\boldsymbol{r}) \tag{5.11}$$

The force field $\boldsymbol{F}(\boldsymbol{r})$ is a vector field. The ratio $\boldsymbol{F}/q$ defines the *electric field* $\boldsymbol{E}(\boldsymbol{r})$,

$$\boxed{\boldsymbol{E}(\boldsymbol{r}) = \frac{\boldsymbol{F}(\boldsymbol{r})}{q}} \tag{5.12}$$

The quantity $\boldsymbol{E}$ is also called *electric field strength*; however, we usually speak of the electric field.

Definition (5.12) for the electric fields applies generally, independent of the configuration (N point charges here) which causes the electrostatic force $\boldsymbol{F}$. It defines the electric field as *measurable quantity* or observable. The charge q shall be so small that it does not significantly affect the force field caused by the other charges. We exclude, for example, that the charge q causes displacements of the other charges or that there are metal bodies that lead to induced charges. Formally, this can be explained by the supplement $q \to 0$ on the right-hand side of (5.12). This is meant when one speaks of a "test charge" q.

The electric field of N point charges can be read from (5.11):

$$\boldsymbol{E}(\boldsymbol{r}) = \sum_{i=1}^{N} q_i \, \frac{\boldsymbol{r} - \boldsymbol{r}_i}{|\boldsymbol{r} - \boldsymbol{r}_i|^3} \tag{5.13}$$

This is equal to the sum of the electric fields of the individual charges. From the superposition principle of the forces follows that for the electric field. Since we are restricting ourselves here to charges at rest, the field is static. The field $\boldsymbol{E}(\boldsymbol{r})$ of a point charge is sketched in Figure 5.1, and the field of two point charges in Figure 6.1.

Charge densities

We introduce the *charge density* $\varrho(\boldsymbol{r})$:

$$\varrho(\boldsymbol{r}) = \frac{\text{charge}}{\text{volume}} = \begin{cases} \varrho_{\text{at}} = \lim\limits_{\Delta V \to 0} \dfrac{\Delta q}{\Delta V} \\[2ex] \langle \varrho_{\text{at}} \rangle = \dfrac{\Delta q}{\Delta V} \end{cases} \tag{5.14}$$

Here ΔV is a small volume at $\boldsymbol{r}$ containing the charge Δq. As indicated in (5.14), the term charge density is used with different meanings. If ΔV goes toward zero, the atomic (or microscopic) charge density ϱ_{at} is obtained. In matter, the density ϱ_{at} varies on the scale of a few fermi because the positive charges are concentrated in the atomic nuclei. In practice, one may rather consider in the mean charge distribution $\langle \varrho_{\text{at}} \rangle$, which is calculated from

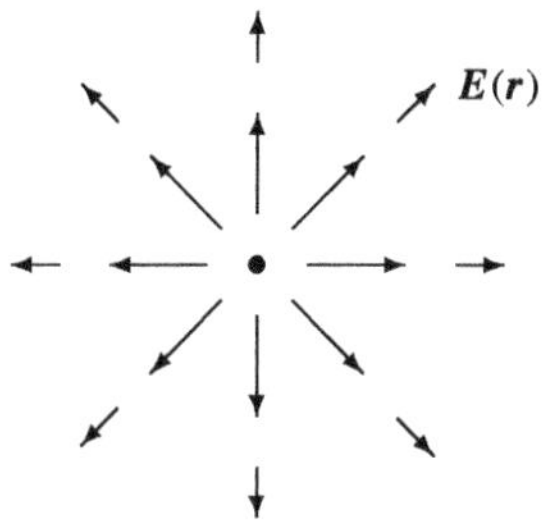

Figure 5.1 Electric field $\boldsymbol{E}(\boldsymbol{r})$ of a point charge.

ϱ_{at} by averaging over *finite* volumes ΔV of suitable size (for example, $\Delta V = (100\,\text{Å})^3$).

These concepts (microscopic versus macroscopic) apply also to the mass density (mass per volume). The atomic mass density reflects the concentration of the mass in the atomic nuclei. If, on the other hand, on says that water has the density $\varrho_{\text{mat}} = 1\,\text{g/cm}^3$, one refers to the average mass density.

The charge density of N point charges can be expressed by δ-functions (3.23):

$$\varrho(\boldsymbol{r}) = \sum_{i=1}^{N} q_i\,\delta(\boldsymbol{r} - \boldsymbol{r}_i) \tag{5.15}$$

This corresponds to the microscopic charge density ϱ_{at}, if we identify the electrons and atomic nuclei with the point charges.

A *continuous*, spatially limited charge density can be represented by N point charges (Figure 5.2). To do this, we divide the range with $\varrho \neq 0$ into N partial volumes ΔV_i:

$$\begin{aligned}\varrho(\boldsymbol{r}) &= \int d^3r'\,\varrho(\boldsymbol{r}')\,\delta(\boldsymbol{r} - \boldsymbol{r}') = \sum_{i=1}^{N} \int_{\Delta V_i} d^3r'\,\varrho(\boldsymbol{r}')\,\delta(\boldsymbol{r} - \boldsymbol{r}') \\ &\approx \sum_{i=1}^{N} \left(\int_{\Delta V_i} d^3r'\,\varrho(\boldsymbol{r}') \right) \delta(\boldsymbol{r} - \boldsymbol{r}_i) = \sum_{i=1}^{N} q_i\,\delta(\boldsymbol{r} - \boldsymbol{r}_i) \end{aligned} \tag{5.16}$$

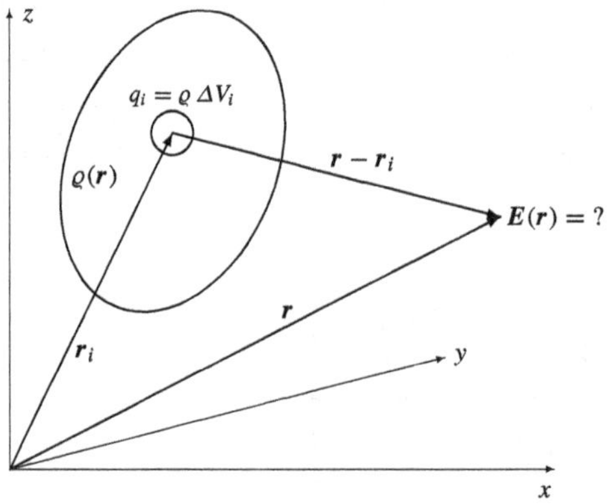

Figure 5.2 We consider continuous charge distribution $\varrho(\boldsymbol{r})$ (indicated by the elliptic contour). This distribution may be decomposed into small individual charge elements q_i at $\boldsymbol{r}_i$. The field $\boldsymbol{E}(\boldsymbol{r})$ is then the sum of the fields of all point charges q_i.

As an approximation, the charges in $\varDelta V_i$ are each replaced by a point charge q_i at $\boldsymbol{r}_i$. Thereby, $\boldsymbol{r}_i$ is a point within $\varDelta V_i$ (for example, the center of charge) and q_i is the total charge in $\varDelta V_i$. For $\varDelta V_i \to 0$, the error of this approximation becomes arbitrarily small.

After replacing the continuous charge distribution by N point charges, the expression (5.13) can be used for the electric field:

$$\begin{aligned}\boldsymbol{E}(\boldsymbol{r}) &= \sum_{i=1}^{N} q_i \,\frac{\boldsymbol{r}-\boldsymbol{r}_i}{|\boldsymbol{r}-\boldsymbol{r}_i|^3} = \sum_{i=1}^{N}\left(\int_{\varDelta V_i} d^3r'\,\varrho(\boldsymbol{r}')\right)\frac{\boldsymbol{r}-\boldsymbol{r}_i}{|\boldsymbol{r}-\boldsymbol{r}_i|^3}\\ &= \sum_{i=1}^{N}\varDelta V_i\,\varrho(\bar{\boldsymbol{r}}_i)\,\frac{\boldsymbol{r}-\boldsymbol{r}_i}{|\boldsymbol{r}-\boldsymbol{r}_i|^3} \approx \int d^3r'\,\varrho(\boldsymbol{r}')\,\frac{\boldsymbol{r}-\boldsymbol{r}'}{|\boldsymbol{r}-\boldsymbol{r}'|^3}\end{aligned} \tag{5.17}$$

This step is illustrated in Figure 5.2. First, the q_i introduced in (5.16). Then with the mean value theorem of the integral calculus for the integration over the partial volumes is used (implying $\bar{\boldsymbol{r}}_i \in \varDelta V_i$). The resulting sum is then a approximation for the integral (last expression). For $\varDelta V_i \to 0$ (and $N \to \infty$), the errors of this derivation (5.17) go to zero.

At a certain point, the derivation of (5.17) must be examined more closely: The continuous charge distribution was represented by many point charges. However, the concept of point charge requires that the actual expansion of the charge (here given by $\varDelta V_i$) is small compared to the distance $|\boldsymbol{r}-\boldsymbol{r}_i|$. This condition is not fulfilled for the charge contribution $d^3r'\,\varrho(\boldsymbol{r}')$ at $\boldsymbol{r}' = \boldsymbol{r}$. Therefore, we first exclude an environment $U_\ell = \{\boldsymbol{r}'\ |\boldsymbol{x}| = |\boldsymbol{r}-\boldsymbol{r}'| \le \ell\}$ from the integration range in (5.17). For fixed ℓ, we use then $|\boldsymbol{r}-\boldsymbol{r}_i| \ge \ell \gg (\varDelta V_i)^{1/3}$ so that the replacement by point charges for $\varDelta V_i \to 0$ is justified. In a second step, we show that the contribution of the excluded environment U_ℓ to (5.17) is negligibly small:

$$\left|\int_{U_\ell} d^3r'\,\varrho(\boldsymbol{r}')\,\frac{\boldsymbol{r}-\boldsymbol{r}'}{|\boldsymbol{r}-\boldsymbol{r}'|^3}\right| = \big|\varrho(\bar{\boldsymbol{r}})\big|\int_{x\,\le\,\ell} d^3x\,\frac{1}{x^2} = 4\pi\,\big|\varrho(\bar{\boldsymbol{r}})\big|\,\ell \overset{\ell\to 0}{\longrightarrow} 0 \tag{5.18}$$

We have used the mean value theorem of integral calculus and the continuity of $\varrho(\boldsymbol{r})$. The value for ℓ can be chosen to be arbitrarily small. This means that the contribution of the environment U_ℓ can be neglected. Consequently, there is no longer a restriction of the integration range in (5.17).

The derivation of (5.17) assumes a continuous charge distribution. Nevertheless, (5.17) can also be used for point charges: The insertion of $\varrho = \sum q_i\,\delta(\boldsymbol{r} - \boldsymbol{r}_i)$ leads to the correct result (5.13). This means that

$$\boxed{\boldsymbol{E}(\boldsymbol{r}) = \int d^3r'\,\varrho(\boldsymbol{r}')\,\frac{\boldsymbol{r}-\boldsymbol{r}'}{|\boldsymbol{r}-\boldsymbol{r}'|^3}} \qquad (5.19)$$

the valid relationship between any charge density and the electric field. This formula is a generalization of the Coulomb law (5.1). To illustrate this, we consider the Coulomb force on a test charge caused by N point charges or by a continuous charge distribution:

$$\boldsymbol{F} = q\,\boldsymbol{E}(\boldsymbol{r}) = \begin{cases} q\displaystyle\sum_{i=1}^{N} q_i\,\frac{\boldsymbol{r}-\boldsymbol{r}_i}{|\boldsymbol{r}-\boldsymbol{r}_i|^3} \\[2ex] q\displaystyle\int dq'\,\frac{\boldsymbol{r}-\boldsymbol{r}'}{|\boldsymbol{r}-\boldsymbol{r}'|^3} \end{cases} \quad \text{(Coulomb force)} \qquad (5.20)$$

Taking into account the superposition of the forces, the first line follows from the Coulomb force (5.1). The second line is an obvious generalization of the first line. With $dq' = \varrho(\boldsymbol{r}')\,d^3r'$, the second line is equivalent to (5.19). Thus, Equation (5.19) is a generalized form of Coulomb's law.

We also introduce the force density on a given charge distribution,

$$\boldsymbol{f}(\boldsymbol{r}) = \frac{\Delta\boldsymbol{F}}{\Delta V} = \frac{\Delta q}{\Delta V}\,\boldsymbol{E}(\boldsymbol{r}) = \varrho(\boldsymbol{r})\,\boldsymbol{E}(\boldsymbol{r}) \qquad (5.21)$$

Here ΔV is a volume element at $\boldsymbol{r}$ and Δq is the charge contained therein.

Exercises

5.1 *Charge density for spherical shell and circular disk*

A spherical shell and a circular disk (both infinitesimal thin, and with radius R) are homogeneously charged (total charge q). Determine the charge density for both cases (using δ- and Θ-functions).

Chapter 6

Field Equations

In Equation (5.19), the electric field $\boldsymbol{E}(\boldsymbol{r})$ was defined as a functional of the charge density $\varrho(\boldsymbol{r})$. We introduce the electrostatic potential, establish the field equations of electrostatics and determine the field energy.

With the help of

$$\operatorname{grad} \frac{1}{|\boldsymbol{r}-\boldsymbol{r}'|} = -\frac{\boldsymbol{r}-\boldsymbol{r}'}{|\boldsymbol{r}-\boldsymbol{r}'|^3} \tag{6.1}$$

we reshape (5.19):

$$\boldsymbol{E}(\boldsymbol{r}) = \int d^3r' \, \varrho(\boldsymbol{r}') \, \frac{\boldsymbol{r}-\boldsymbol{r}'}{|\boldsymbol{r}-\boldsymbol{r}'|^3} = -\operatorname{grad} \int d^3r' \, \frac{\varrho(\boldsymbol{r}')}{|\boldsymbol{r}-\boldsymbol{r}'|} = -\operatorname{grad} \Phi(\boldsymbol{r}) \tag{6.2}$$

In the last step, the *scalar* or *electrostatic potential* Φ was introduced:

$$\boxed{\Phi(\boldsymbol{r}) = \int d^3r' \, \frac{\varrho(\boldsymbol{r}')}{|\boldsymbol{r}-\boldsymbol{r}'|} + \text{const.} = \int d^3r' \, \frac{\varrho(\boldsymbol{r}')}{|\boldsymbol{r}-\boldsymbol{r}'|}} \tag{6.3}$$

The constant disappears when calculating the observable field $\boldsymbol{E}$; it has therefore of no physical significance. It is usually set equal to zero.

The electric field, like any vector field (Chapter 3), is determined by its sources and vortices. The wanted field equations consist of the specification of these sources $\operatorname{div} \boldsymbol{E}$ and vortices $\operatorname{curl} \boldsymbol{E}$. We apply the Laplace operator to the integral in (6.3). Due to

$$\delta \, \frac{1}{|\boldsymbol{r}-\boldsymbol{r}_0|} = -4\pi \, \delta(\boldsymbol{r}-\boldsymbol{r}_0) \tag{6.4}$$

this results in $-4\pi\varrho$. This means that $\operatorname{div}\boldsymbol{E} = -\Delta\Phi = 4\pi\varrho$. From (2.27) and (6.2) follows $\operatorname{curl}\boldsymbol{E} = \operatorname{curl}\operatorname{grad}\Phi = 0$. The sources and vortices of the electric field are thus

$$\begin{array}{ll} \operatorname{div}\boldsymbol{E}(\boldsymbol{r}) = 4\pi\varrho(\boldsymbol{r}) & \text{field equations} \\ \operatorname{curl}\boldsymbol{E}(\boldsymbol{r}) = 0 & \text{of electrostatics} \end{array} \tag{6.5}$$

The first equation is also called the inhomogeneous field equation (because of the source term on the right-hand side); the second equation is the homogeneous field equation. The field equations are differential equations that determine the fields locally; they refer to a specific point point $\boldsymbol{r}$ and its surroundings. In contrast to this (5.19) is an integral statement.

Alternatively, the basic equations (6.5) can be formulated for the electrostatic potential:

$$\Delta\Phi(\boldsymbol{r}) = -4\pi\varrho(\boldsymbol{r}), \quad \boldsymbol{E}(\boldsymbol{r}) = -\operatorname{grad}\Phi(\boldsymbol{r}) \tag{6.6}$$

The first equation is called *Poisson equation*, and the second equation links the potential to the electric field. The Poisson equation describes the local relationship between the potential Φ and the charge density ϱ. For $\varrho = 0$, the Poisson equation becomes the *Laplace equation* $\Delta\Phi = 0$. The formulation with $\Phi(\boldsymbol{r})$ has the advantage that only a single function $\Phi(\boldsymbol{r})$ has to be considered (or determined) instead of the three functions $\boldsymbol{E}(\boldsymbol{r})$.

A basic problem of electrostatics is to calculate the field $\Phi(\boldsymbol{r})$ or $\boldsymbol{E}(\boldsymbol{r})$ for a given charge distribution $\varrho(\boldsymbol{r})$. The formal solution is given by (5.19) or (6.3). If metal surfaces are present (Chapters 7 and 8), the specification of charges is partially replaced by the specification of boundary conditions.

The following applies for a point charge

$$\varrho(\boldsymbol{r}) = q\,\delta(\boldsymbol{r}-\boldsymbol{r}_0) \quad \stackrel{(6.3)}{\longrightarrow} \quad \Phi(\boldsymbol{r}) = \frac{q}{|\boldsymbol{r}-\boldsymbol{r}_0|} \tag{6.7}$$

This is the physical pendent to the mathematical statement (6.4). For N point charges, q_i at the positions $\boldsymbol{r}_i$ (6.3) leads to

$$\Phi(\boldsymbol{r}) = \sum_{i=1}^{N} \frac{q_i}{|\boldsymbol{r}-\boldsymbol{r}_i|} \tag{6.8}$$

Field Lines

Equipotential surfaces are spatial areas of all points with the same value of $\Phi(\boldsymbol{r})$:

$$\Phi(\boldsymbol{r}) = \text{const.} \quad \text{(equipotential surface)} \tag{6.9}$$

The surface normal points in the direction of grad $\Phi = -\boldsymbol{E}$. The lines for which the vectors $\boldsymbol{E}$ at each point are tangents are called *field lines*. A curve $\boldsymbol{r} = \boldsymbol{r}(\lambda)$ is a field line if

$$\frac{d\boldsymbol{r}(\lambda)}{d\lambda} \times \boldsymbol{E}(\boldsymbol{r}(\lambda)) = 0 \quad \text{(field lines } \boldsymbol{r}(\lambda)\text{)} \tag{6.10}$$

The derivative of $\boldsymbol{r}(\lambda)$ with respect to the parameter λ results in a tangent vector to the curve; this vector is parallel to the field for each λ.

Equipotential surfaces and field lines are used to graphically illustrate a field configuration (Figure 6.1).

Integral forms

For a qualitative discussion and a sketch the field lines, the following integral forms of the field equations are useful. They are closely linked to the coordinate-independent and descriptive definitions (Chapter 1) of the respective vector operations.

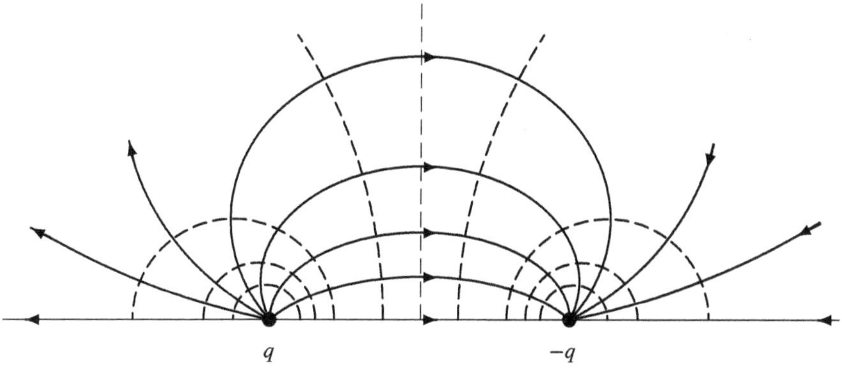

Figure 6.1 Field lines (solid lines) and equipotential surfaces (dashed lines) for two point charges, q and $-q$. A three-dimensional picture is obtained by a rotation around the connecting axis of the two charges.

We apply Gauss' theorem (1.25) to the inhomogeneous field equation:

$$\oint_A d\boldsymbol{A} \cdot \boldsymbol{E} = 4\pi \int_V d^3r \; \varrho(\boldsymbol{r}) = 4\pi \, Q_V \quad \text{(Gauss law)} \tag{6.11}$$

Here $A = A(V)$ is the area that limits the volume V. This statement is called *Gauss law*: The surface integral over $\boldsymbol{E}$ equals to the enclosed charge Q_V (times 4π).

For $Q_V > 0$, more field lines run out of the volume V than into it. In particular, a positive point charge generates star-like field lines; for a negative point, the star shape is inverted (Figure 6.1). For stronger fields, the field lines are denser.

The field lines of a point charge (Figure 6.1) as well as of a charged sphere (Figure 6.3) point radially outward. The difference between the two cases can be illustrated by the equipotential surface for equidistant Φ-values. For the sphere, you then obtain a finite number of spheres. For the point charge, one gets an infinite number of spheres that become increasingly denser for $r \to 0$.

We apply Stokes' theorem to the homogeneous field equation:

$$\oint_C d\boldsymbol{r} \cdot \boldsymbol{E} = 0 \tag{6.12}$$

In particular, this means that in electrostatics *there are no closed field lines.*

Field of a Homogeneously Charged Sphere

We determine the field $\boldsymbol{E}(\boldsymbol{r})$ of a homogeneously charged sphere (Figure 6.2). The charge density is given by

$$\varrho(\boldsymbol{r}) = \varrho(r) = \begin{cases} \varrho_0 & (r \le R) \\ 0 & (r > R) \end{cases} \tag{6.13}$$

with $\varrho_0 = $ const. To calculate the electric field, the following options are available:

1. Gauss' law
2. Solution of the field equation $\Delta\Phi = -4\pi\,\varrho$
3. Evaluation of the integral (6.3).

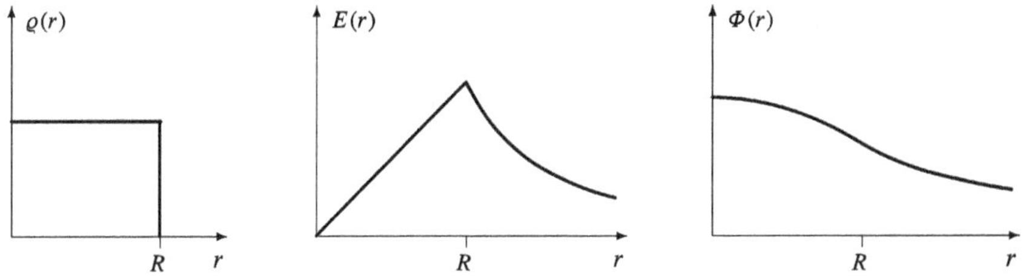

Figure 6.2 Charge distribution $\varrho(r)$, field strength $E(r)$ and potential $\Phi(r)$ of a homogeneously charged sphere.

Alternative to 2, the field equation $\operatorname{div}\boldsymbol{E} = 4\pi\varrho$ can be solved. Alternative to 3, the integral (5.19) can be evaluated. In the following, we discuss the solution in the first two ways; the third way is left to Exercise 6.2.

We use spherical coordinates r, θ and ϕ. Due to the spherical symmetry of the charge distribution (6.13), the potential $\Phi(\boldsymbol{r}) = \Phi(r, \theta, \phi)$ cannot depend on the angles. This means

$$\Phi = \Phi(r), \quad \boldsymbol{E} = -\operatorname{grad}\Phi(r) = -\Phi'(r)\,\boldsymbol{e}_r = E(r)\,\boldsymbol{e}_r. \tag{6.14}$$

We first solve the problem with the help of Gauss' law. We evaluate (6.11) for a spherical surface A of radius r:

$$4\pi r^2 E(r) = 4\pi \int_0^r 4\pi r'^2\, dr'\, \varrho(r') = \begin{cases} 4\pi Q\,\dfrac{r^3}{R^3} & (r < R) \\[2ex] 4\pi Q & (r > R) \end{cases} \tag{6.15}$$

Here $Q = (4\pi/3)\,\varrho_0\, R^3$ is the total charge of the sphere. This results in

$$E(r) = \begin{cases} \dfrac{Q\,r}{R^3} & (r < R) \\[2ex] \dfrac{Q}{r^2} & (r > R) \end{cases} \tag{6.16}$$

From $\Phi'(r) = -E(r)$, we can determine the potential by a simple integration:

$$\Phi(r) = \begin{cases} \dfrac{Q}{R}\left(\dfrac{3}{2} - \dfrac{r^2}{2R^2}\right) & (r < R) \\[2ex] \dfrac{Q}{r} & (r > R) \end{cases} \tag{6.17}$$

The integration constant was chosen so that $\Phi(\infty) = 0$; this determination is arbitrary and usual. The results (6.16) and (6.17) are sketched in Figure 6.2. In the outer region, the charge distribution (6.13) appears like a point charge localized at $r = 0$. From (6.15), it follows that this applies for any spherically symmetric charge distribution.

Figure 6.3 shows the field lines and the equipotential lines of the homogeneously charged sphere.

We now calculate the field of the homogeneously charged sphere from the Poisson equation (6.6). In doing so, we learn some typical strategies (symmetry, integration constants, boundary and continuity conditions) for solving field equations.

We use spherical coordinates. The symmetry of the problem implies (6.14). For this potential, the Poisson equation becomes

$$\Delta\,\Phi(r) = \frac{1}{r^2}\,\frac{d}{dr}\left(r^2\,\frac{d\Phi}{dr}\right) = -4\pi\,\varrho(r) \qquad (6.18)$$

Since Φ only depends on the coordinate r, the Poisson equation is reduced to an ordinary differential equation. The right side is a constant for $r > R$ as well as for $r < R$. We integrate the differential equation separately in

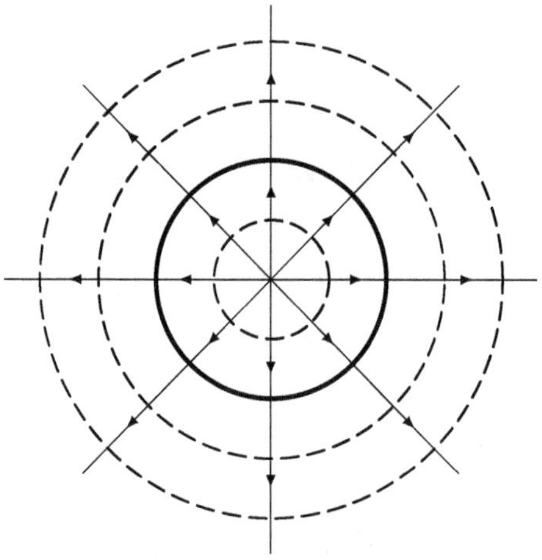

Figure 6.3 Field lines (radial) and equipotential surfaces (circular) of a homogeneously charged sphere. The equipotential surfaces are spheres. The electric field lines run radially outward.

these two ranges:

$$r > R \quad \left(r^2 \Phi'\right)' = 0, \qquad\qquad \Phi = -\frac{C_1}{r} + C_2$$
$$r < R \quad \left(r^2 \Phi'\right)' = -4\pi \varrho_0 r^2, \quad \Phi = -\frac{2\pi}{3} \varrho_0 r^2 - \frac{C_3}{r} + C_4 \tag{6.19}$$

Arbitrarily, we set $\Phi(\infty) = 0$, i.e., $C_2 = 0$. The term $-C_3/r$ leads to $\Delta(C_3/r) = -4\pi C_3 \delta(\boldsymbol{r})$ and corresponds to a point charge at $r = 0$. It is contained in our general solution because for the transition from (6.18) to (6.19), we multiplied both sides of the differential equation by r^2. Since there is no such point charge, the integration constant C_3 must be set equal to zero. Therefore,

$$\Phi(r) = \begin{cases} -\dfrac{C_1}{r} & (r > R) \\[2ex] C_4 - \dfrac{Q\, r^2}{2 R^3} & (r < R) \end{cases} \tag{6.20}$$

We still have to connect these two parts of the solution. With the right-hand side of (6.18), the left-hand side must also have a jump at $r = R$. If $(r^2 \Phi')'$ has a jump, then $r^2 \Phi'$ has a kink (see also Figure 3.2). Thus, Φ' and Φ are continuous for $r = R$, i.e.,

$$\Phi(R + \varepsilon) = \Phi(R - \varepsilon) \quad \text{and} \quad \Phi'(R + \varepsilon) = \Phi'(R - \varepsilon) \quad \text{for } \varepsilon \to 0 \tag{6.21}$$

Here we use (6.20):

$$-\frac{C_1}{R} = C_4 - \frac{Q}{2R}, \quad \frac{C_1}{R^2} = -\frac{Q}{R^2} \tag{6.22}$$

From this follows $C_1 = -Q$ and $C_4 = 3Q/2R$. With these constants, the result (6.20) coincides with (6.17).

We have proceeded from (6.18), i.e., from a differential equation of second order. The solution must therefore contain two integration constants, which are determined by two boundary conditions. Since we set up the solution in two ranges (0 to R and R to ∞) separately, we have obtained four integration constants. This requires four boundary conditions: $\Phi(0)$ is finite (no point charge at $r = 0$), $\Phi(r)$ and $\Phi'(r)$ are continuous at $r = R$ (charge distribution has a jump), and $\Phi(\infty) = 0$ (arbitrary).

Electrostatic Energy

We calculate the energy of the electrostatic field. The result is evaluated for the homogeneously charged sphere.

To create a point charge q in the field $\boldsymbol{E} = -\text{grad}\,\Phi$ of $\boldsymbol{r}_1$ to $\boldsymbol{r}_2$, the work

$$W_{12} = -\int_1^2 d\boldsymbol{r} \cdot \boldsymbol{F} = -q \int_1^2 d\boldsymbol{r} \cdot \boldsymbol{E} = q\left(\Phi(\boldsymbol{r}_2) - \Phi(\boldsymbol{r}_1)\right) \tag{6.23}$$

must be performed. The potential difference $\Phi(\boldsymbol{r}_2) - \Phi(\boldsymbol{r}_1)$ is called *voltage*. The work is equal to the product of charge and voltage. It does not depend on the path between $\boldsymbol{r}_1$ and $\boldsymbol{r}_2$; this also follows from curl $\boldsymbol{E} = 0$ and (1.28). Due to (6.23), potential differences are defined as measurements quantities (observables).

The definition of the electric field by (5.12) and the relation $\boldsymbol{E} = -\text{grad}\,\Phi$ agree in the Gauss and SI system. This means that (6.23) applies in both unit systems. The work is measured in joules (or erg). In the Gaussian system, the unit of the potential Φ (or the voltage) is $\Phi = \text{erg/esu}$ (with $\text{erg} = \text{g}\,\text{cm}^2/\text{s}^2$ and esu of (5.9)). In the SI system, the unit $\Phi = \text{J/C} = \text{V}$ is given the designation *volt*.

Work has the dimension of energy. The quantity

$$W(\boldsymbol{r}) = q\,\Phi(\boldsymbol{r}) \tag{6.24}$$

is the *potential energy* of the charge q in the electrostatic field $\boldsymbol{E}$. Along equipotential surfaces, charges can be shifted without any work. A charge is accelerated in the direction of the field lines; it moves along a field line as long as it is sufficiently slow.

In (6.24), the charge q itself does not contribute to the potential Φ. As a generalization of (6.24), we consider charge distribution $\varrho(\boldsymbol{r})$ in an external field Φ_{ext}; in Φ_{ext}, the contribution of ϱ itself shall not be included. Each charge element $dq = \varrho\, d^3r$ then yields a contribution of the form (6.24). The summation leads to

$$W = \int d^3r\; \varrho(\boldsymbol{r})\; \Phi_{\text{ext}}(\boldsymbol{r}) \tag{6.25}$$

In the following, we calculate which energy a continuous charge distribution has in its *own* field.

We first determine the electrostatic energy of N point charges. To do this, we consider $i - 1$ point charges q_j, which are positioned at $\boldsymbol{r}_j$ $(j = 1, ..., i - 1)$. The potential energy W_i of another point charge q_i in the field of existing charges follows from (6.24) and (6.8):

$$W_i(\boldsymbol{r}_i) = q_i \sum_{j=1}^{i-1} \frac{q_j}{|\boldsymbol{r}_i - \boldsymbol{r}_j|} \tag{6.26}$$

This is equal to the work required to move the charge q_i from infinity to $\boldsymbol{r}_i$. We now imagine that we successively move the charges $i = 1, 2, \ldots, N$ from infinity to the positions $\boldsymbol{r}_i$. The work to be done is equal to the potential energy W of the system of N point charges, i.e.,

$$W = \sum_{i=2}^{N} W_i(\boldsymbol{r}_i) = \sum_{i=2}^{N} \sum_{j=1}^{i-1} \frac{q_i\, q_j}{|\boldsymbol{r}_i - \boldsymbol{r}_j|} = \frac{1}{2} \sum_{i,j,\, i\neq j}^{N} \frac{q_i\, q_j}{|\boldsymbol{r}_i - \boldsymbol{r}_j|} \tag{6.27}$$

To calculate the potential energy for a continuous charge distribution $\varrho(\boldsymbol{r})$, we replace, as in (5.16), the charge distribution by N partial charges $\Delta q_i = \varrho(\boldsymbol{r}_i)\, \Delta V_i$. With $N \to \infty$ and $\Delta V_i \to 0$, this division can be made arbitrarily fine. Then, (6.27) becomes

$$W = \frac{1}{2} \int d^3r \int d^3r' \, \frac{\varrho(\boldsymbol{r})\, \varrho(\boldsymbol{r}')}{|\boldsymbol{r} - \boldsymbol{r}'|} \qquad (\varrho \text{ continuous}) \tag{6.28}$$

For transition from (6.27) to (6.28), first an range around $\boldsymbol{r} = \boldsymbol{r}'$ must be excluded; this corresponds to the restriction $i \neq j$ in (6.27). If the charge distributions are continuous, this range gives a negligibly small contribution as in (5.18).

The restriction "ϱ continuous" must not be omitted in (6.28). For point charges, the result would contain the infinitely large energy of a point charge in its own field; then the result would be undefined. This self-energy can be further investigated by calculating the finite energy of a homogeneously charged sphere and then by letting the radius of this sphere going toward zero.

With (6.3), (6.28) becomes

$$W = \frac{1}{2} \int d^3r \; \varrho(\boldsymbol{r})\; \Phi(\boldsymbol{r}) \tag{6.29}$$

In contrast to (6.25), Φ is the potential caused by ϱ and not an external potential; this difference leads to the factor 1/2.

We insert the Poisson equation into (6.29) and perform a partial integration:

$$W = -\frac{1}{8\pi}\int d^3r \; \Phi(\boldsymbol{r}) \, \Delta\Phi(\boldsymbol{r}) = \frac{1}{8\pi}\int d^3r \; \left|\boldsymbol{E}(\boldsymbol{r})\right|^2 \tag{6.30}$$

For finite charge distributions, Φ and $\boldsymbol{E}$ approach zero sufficiently quickly for $r \to \infty$ so that there are no boundary terms from partial integration. Equation (6.30) implies that

$$w(\boldsymbol{r}) = \frac{1}{8\pi}\left|\boldsymbol{E}(\boldsymbol{r})\right|^2 \quad \text{energy density in electrostatics} \tag{6.31}$$

can be interpreted as the energy density of the electric field.

Homogeneously charged sphere

We calculate the electrostatic energy of the homogeneously charged sphere. From (6.31) and (6.16), we obtain

$$w(\boldsymbol{r}) = \frac{Q^2}{8\pi} \cdot \begin{cases} \dfrac{r^2}{R^6} & (r < R) \\[2ex] \dfrac{1}{r^4} & (r > R) \end{cases} \tag{6.32}$$

and

$$W = \int_0^\infty 4\pi r^2 \, dr \; w(r) = \frac{3}{5}\,\frac{Q^2}{R} \tag{6.33}$$

If the radius of the sphere approaches zero (with unchanged total), the electrostatic energy W diverges. This corresponds to the infinitely large self-energy of a point charge. In this respect, the point charge is an unrealistic theoretical construct. However, just like the point mass of mechanics, it is a useful idealization if the finite size of a charge distribution does not matter (if it is small compared to all other relevant lengths).

We use the homogeneously charged sphere as a model for the electron and set the electrostatic energy equal to the rest energy:

$$W = m_\mathrm{e} c^2 = \frac{3}{5}\frac{e^2}{R_\mathrm{e}} \tag{6.34}$$

This results in the "classical electron radius"

$$R_\mathrm{e} = \frac{3}{5}\frac{e^2}{m_\mathrm{e} c^2} \approx 1.7\ \mathrm{fm} = 1.7 \cdot 10^{-13}\ \mathrm{cm} \tag{6.35}$$

For these and other numerical estimations, you should remember

$$\frac{e^2}{\mathrm{fm}} \approx 1.44\,\mathrm{MeV}, \qquad \frac{e^2}{\text{Å}} \approx 14.4\,\mathrm{eV}, \qquad m_\mathrm{e} c^2 \approx 0.5\,\mathrm{MeV} \tag{6.36}$$

Here e is the elementary charge (5.10) in the Gaussian unit system. The quantities in (6.36) are energies. It is usual and often practical to specify energies in electron volt (eV). In eV, "e" is the elementary charge e_SI in SI units:

$$\mathrm{eV} = e_\mathrm{SI}\ \mathrm{volt} \approx 1.6 \cdot 10^{-19}\,\mathrm{C\,V} = 1.6 \cdot 10^{-19}\ \mathrm{J} \tag{6.37}$$

The homogeneously charged sphere is, of course, not a realistic model of the electron. First of all, there are no forces that hold the charge distribution together. In addition, even at significantly larger distances, quantum effects become important. If the position of an electron is restricted $\hbar/m_\mathrm{e} c = 4 \cdot 10^{-11}$ cm, then — due to the uncertainty relation — it has relativistic momenta and the question of particle creation arises. In this charge sphere model, also the spin and the associated magnetic moment of the electron are missing.

Exercises

6.1 *Gauss' law: Point charge in a sphere*

Verify Gauss' law for a point charge inside a sphere. The sphere has the radius R, and the point charge has the distance a from the center. Use spherical coordinates.

6.2 *Homogeneously charged sphere*

Determine the electrostatic potential

$$\Phi(\boldsymbol{r}) = \int d^3r' \, \frac{\varrho(\boldsymbol{r}')}{|\boldsymbol{r}-\boldsymbol{r}'|}$$

for a homogeneously charged sphere (charge q, radius R). To do this, let $\boldsymbol{r}$ point into the z-direction and carry out the integration in spherical coordinates. Calculate the electric field $\boldsymbol{E}(\boldsymbol{r})$.

6.3 *Homogeneously charged circular cylinder*

Determine the electric field of a homogeneously charged infinitely long circular cylinder (radius R, length L, charge/length $= q/L$, $L \to \infty$). Solve problem (i) with the help of Gauss law and (ii) using the Poisson equation. Note that the potential does not vanish at infinity.

6.4 *Electrostatic potential of the hydrogen atom*

The electrostatic potential in a hydrogen atom in the ground state is of the form

$$\Phi = \frac{e}{r}\left(1+\frac{r}{a_{\mathrm{B}}}\right)\exp\left(-\frac{2r}{a_{\mathrm{B}}}\right) \tag{6.38}$$

where e is the elementary charge and $a_{\mathrm{B}} = 0.53$ Å is the Bohr radius.

Determine the electric field $\boldsymbol{E}(\boldsymbol{r})$ and the charge density $\varrho(\boldsymbol{r})$. Using Gauss' law, calculate the charge $q(R)$ that is inside a sphere of radius R. Sketch $q(R)$ and interpret the result.

6.5 *NaCl crystal*

Calculate the electrostatic interaction energy of a lattice ion in a one-dimensional NaCl crystal.

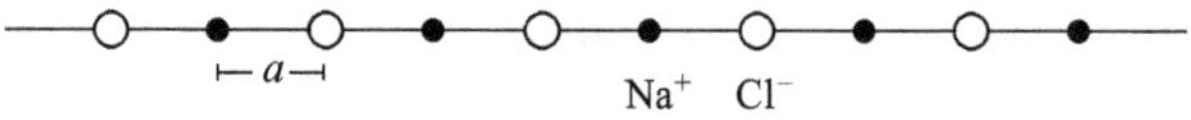

6.6 *Parallel charged wires*

Calculate and sketch the equipotential surfaces and field lines of two parallel, infinitely long, thin wires at a distance of $2a$, whose charge per length is equal to q/L and $-q/L$, respectively. First, consider a single wire and calculate its electric field using Gauss' law. Superpose the potentials and fields of the two wires.

Note: The differential equation for the field lines can be solved by using an integrating factor. This results in an orthogonal circular mesh.

6.7 *Homogeneously charged thin rod*

The charge density of a thin, homogeneously charged rod (charge q, length $2a$) is

$$\varrho(\boldsymbol{r}) = \frac{q}{2a}\,\delta(x)\,\delta(y)\,\Theta(a - |z|) \tag{6.39}$$

Evaluate the integral formula for the potential in cylindrical coordinates. Calculate and sketch the equipotential surfaces and field lines. The differential equation for the field lines can be solved with the help of an integrating factor.

Chapter 7

Boundary Value Problems

We consider electrostatic boundary value problems of the following type: A volume V is enclosed by metal surfaces, and inside V the charge distribution is given. The electrostatic potential Φ inside V shall be determined. An example of such a problem is sketched in Figure 7.1. We investigate the existence and the uniqueness of the solution Φ of the boundary value problem. We discuss also the numerical solution of such a problem.

The charges inside volume V induced surface charges on the metal. Since these surface charges are initially unknown, the potential Φ cannot be determined using (6.3). In a metal, however, $\Phi = \text{const.}$ holds. This provides a *boundary condition* for the determination of $\Phi(\boldsymbol{r})$ in V. This boundary condition replaces the specification of the (unknown) charges on the metal.

Boundary Condition for Metal

In a metal, there are electrons that can approximatively move freely. In the following, we consider the static equilibrium which adjusts itself in the long term. In this case, the following applies to the field acting in the metal:

$$\boldsymbol{E} = 0 \quad \text{(effective field in the metal)} \tag{7.1}$$

For $\boldsymbol{E} \neq 0$, there are forces on the mobile charges, i.e., time dependent processes (non-static case). If charges are brought into the vicinity of a

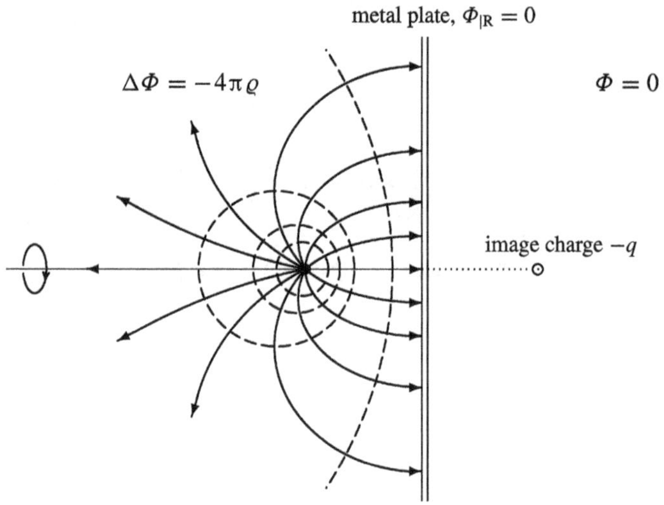

Figure 7.1 A point charge in front of an earthed (grounded), flat metal plate presents a simple boundary value problem. The point charge induces charges on the plate. The resulting field lines are shown in the figure (solid lines with arrows) and equipotential surfaces (dashed lines) are sketched. A rotation around the symmetry axis (straight line through the charge and the image charge) results in a three-dimensional image. In Chapter 8, the field is actually calculated using an image charge.

metal piece, then the mobile charges are shifted until $\boldsymbol{E} = 0$ applies in the metal.

The microscopic fields (in particular the fields of the atomic nuclei and electrons) do not disappear. The relation between these microscopic fields and the effective fields considered here is dealt with in Part VI.

From (7.1) follows

$$\Phi(\boldsymbol{r}) = \text{const.} \quad \text{(in metal)} \tag{7.2}$$

Figure 7.2 shows the boundary surface between a metal body 1 and the vacuum 2. The potential Φ is to be determined inside the volume V (i.e., in a vacuum); in V, a given charge distribution is admitted.

With the help of Gauss' and Stokes' theorem, we derive the boundary condition for the surface R between the vacuum and the metal. From two small line elements of length ℓ, which are parallel to the boundary surface, we form the rectangular surface $\varDelta A$ (Figure 7.2). The distance between the

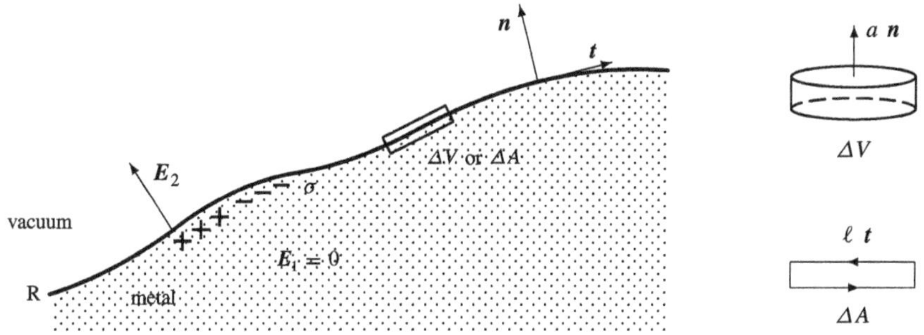

Figure 7.2 A vacuum region limited by a metal. Inside the metal, the mean electric field disappears, $\boldsymbol{E}_1 = 0$. At the boundary of the metal, a surface charge σ may be present. To derive the boundary condition for the field in a vacuum, a volume element ΔV or, alternatively, a surface element ΔA is considered, the shape of which is sketched on the right.

two line elements can be chosen (arbitrarily) small. Using $\operatorname{curl}\boldsymbol{E} = 0$ and Stokes' theorem, we get

$$0 = \int_{\Delta A} d\boldsymbol{A} \cdot \operatorname{curl}\boldsymbol{E} = \oint d\boldsymbol{r} \cdot \boldsymbol{E} = \ell\, \boldsymbol{t} \cdot (\boldsymbol{E}_2 - \boldsymbol{E}_1) \tag{7.3}$$

Here $\boldsymbol{E}_1$ denotes the field in the metal, and $\boldsymbol{E}_2$ the one in a vacuum. Since $\boldsymbol{E}_1 = 0$, the result is

$$\boldsymbol{t} \cdot \boldsymbol{E}_2(\boldsymbol{r})\big|_{\mathrm{R}} = 0 \tag{7.4}$$

Here $\boldsymbol{t}$ is an arbitrary tangent vector of the boundary surface R.

From two small surface elements a, which are parallel to the surface, we form a volume element ΔV (Figure 7.2). The distance between the two surface elements can be chosen (arbitrarily) small. We apply Gauss' law to this volume element with the surface area $A = A(\Delta V)$:

$$\oint_{A(\Delta V)} d\boldsymbol{A} \cdot \boldsymbol{E} = a\boldsymbol{n} \cdot (\boldsymbol{E}_2 - \boldsymbol{E}_1) = 4\pi \int_{\Delta V} d^3r\, \varrho = 4\pi q \tag{7.5}$$

Here, $\boldsymbol{n}$ is the normal vector of the boundary surface, which points into the vacuum, and q is the charge confined to ΔV. Since we can make the distance between the two surfaces arbitrarily small without changing (7.5), only the charges localized on the surface contribute to q. We introduce the surface charge $\sigma(\boldsymbol{r}) = q/a =$ charge per surface and take into account $\boldsymbol{E}_1 = 0$. Thus, (7.5) becomes

$$\boldsymbol{n} \cdot \boldsymbol{E}_2(\boldsymbol{r})\big|_{\mathrm{R}} = 4\pi\sigma(\boldsymbol{r}) \tag{7.6}$$

The mobile charges of the metal shift until (7.4) is satisfied. The surface charges cannot move in the direction of the surface normal.

The following boundary conditions are given by (7.4) and (7.6) for the vacuum boundary R:

- The tangential component of the electric field disappears at the boundary R.
- The normal component of the electric field at the boundary R equals 4π times the surface charge of the metal.

Setup of the Boundary Value Problem

After discussing the physical boundary conditions, we formulate the boundary value problem. A coherent volume V is enclosed by one or more separate metal surfaces which we number by the index i. The boundary R of V then consists of the borders R_i of the individual metal bodies. In V, the potential field $\Phi(\boldsymbol{r})$ is sought. The boundary conditions result from (7.4) and (7.6):

$$\Phi\big|_{\mathrm{R}_i} = \Phi_i = \text{const.} \quad \text{and} \quad \frac{\partial \Phi}{\partial n}\bigg|_{\mathrm{R}} = -4\pi\sigma(\boldsymbol{r}) \tag{7.7}$$

Here $\partial/\partial n$ stands for $\boldsymbol{n} \cdot \boldsymbol{\nabla}$. Metal bodies that are separated from each other can have different potential values Φ_i.

Specific potential values at the boundary are called Dirichlet boundary condition:

$$\Phi\big|_{\mathrm{R}} = \Phi_0(\boldsymbol{r}) \quad \text{(Dirichlet boundary condition)} \tag{7.8}$$

This condition is somewhat more general than the first part of (7.7) because we do no longer require piecewise constant potential values. Alternatively, the normal component of the potential gradient might be given. This is called the Neumann boundary condition:

$$\frac{\partial \Phi}{\partial n}\bigg|_{\mathrm{R}} = -4\pi\sigma(\boldsymbol{r}) \quad \text{(Neumann boundary condition)} \tag{7.9}$$

In general, the surface charges on the metal are not known. Given are instead the potential values Φ_i on the individual metal bodies, i.e., a special form of the Dirichlet boundary condition. For the general discussion of this chapter, however, this specialization plays no role. We investigate therefore the potential problem with a Dirichlet boundary condition:

$$\boxed{\begin{array}{ll} \Delta\,\Phi(\boldsymbol{r}) = -4\pi\,\varrho(\boldsymbol{r}) & \text{in } V \\ \Phi(\boldsymbol{r}) = \Phi_0(\boldsymbol{r}) & \text{on R} \end{array}} \qquad \begin{array}{l} \text{boundary value problem:} \\ \varrho(\boldsymbol{r}),\ \Phi_0(\boldsymbol{r}) \text{ given,} \\ \Phi(\boldsymbol{r}) \text{ searched for} \end{array} \tag{7.10}$$

We see in the following sections that the problem formulated in this way has a unique solution. From this solution, the surface charges can then be calculated:

$$\sigma(\boldsymbol{r}) = -\frac{1}{4\pi}\,\frac{\partial\Phi}{\partial n}\bigg|_{\text{R}} \qquad \text{(surface charge)} \tag{7.11}$$

Particular and homogeneous solution

The solution of (7.10) can be set up as the sum of a particular and a homogeneous solution. It is easy to check that

$$\Phi(\boldsymbol{r}) = \Phi_{\text{part}}(\boldsymbol{r}) + \Phi_{\text{hom}}(\boldsymbol{r}) = \int_V d^3r'\,\frac{\varrho(\boldsymbol{r}')}{|\boldsymbol{r}-\boldsymbol{r}'|} + \Phi_{\text{hom}}(\boldsymbol{r}) \tag{7.12}$$

solves the Poisson equation (7.10). Here, Φ_{hom} is an arbitrary solution of $\Delta\,\Phi_{\text{hom}} = 0$. From the boundary condition in (7.10) follows

$$\Phi_{\text{hom}}\big|_{\text{R}} = \left(\Phi_0 - \Phi_{\text{part}}\right)\Big|_{\text{R}} \tag{7.13}$$

We consider two examples for the split (7.12):

(i) There are no metal surface. The volume V is the entire space, and the boundary condition is $\Phi(\infty) = 0$. Then, Φ_{part} is the known solution (6.3), and Φ_{hom} disappears. In this case, the boundary condition (i.e., $\Phi(\infty) = 0$) is often not stated explicitly.

(ii) There is a plane metal surface, Figure 7.1, on which $\Phi|_{\text{R}} = 0$ holds. The volume V is the subspace defined by $x < 0$. In this case, $\Phi_{\text{hom}}(\boldsymbol{r})$ is equal to the potential of an image charge (Chapter 8).

Numerical Solution: Existence of the Solution

Electrostatic problems can be solved analytically in simple cases (for example, for spherical symmetry or other simple geometries). A number of standard methods are presented in Chapter 8. In this section, we outline the numerical treatment of the problem (7.10). This is a general solution method.

The volume V in which the Poisson equation is to be solved may be covered with discrete grid points. For the sake of simplicity, we consider an equidistant Cartesian grid with the points

$$\boldsymbol{r}_{n_1 n_2 n_3} = \sum_{i=1}^{3} d\, n_i\, \boldsymbol{e}_i \tag{7.14}$$

Here, d is the distance between neighboring points. The values of n_i are integers; their range is given by the volume V (Figure 7.3).

We are looking for the potential $\Phi(\boldsymbol{r}_{n_1 n_2 n_3}) = \Phi(n_1, n_2, n_3)$ at the grid points in the interior of V (without the boundary points). Given the charge density $\varrho(\boldsymbol{r}_{n_1 n_2 n_3}) = \varrho(n_1, n_2, n_3)$ at these points and the potential $\Phi|_{\mathrm{R}} = \Phi_0$ at the boundary points.

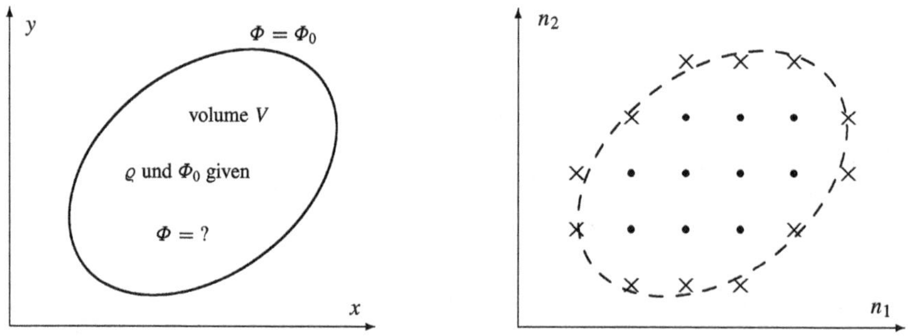

Figure 7.3 In order to solve numerically the boundary value problem shown on the left, a grid is placed over the volume V. At the boundary of the grid ($\times$), the potential values are specified, and at the inner points ($\bullet$), they are unknown. The differential equation $\Delta\Phi = -4\pi\varrho$ results in a linear equation system for the unknown potential values. In a realistic calculation, the grid spacing must be much smaller than the length expansion of V.

The derivatives are now approximated by difference quotients, for example,

$$\frac{\partial \Phi(n_1 + 1/2, n_2, n_3)}{\partial x} \approx \frac{\Phi(n_1 + 1, n_2, n_3) - \Phi(n_1, n_2, n_3)}{d} \tag{7.15}$$

$$\begin{aligned} \frac{\partial^2 \Phi(n_1, n_2, n_3)}{\partial x^2} & \\ \approx \frac{1}{d}\left(\frac{\partial \Phi(n_1 + 1/2, n_2, n_3)}{\partial x} - \frac{\partial \Phi(n_1 - 1/2, n_2, n_3)}{\partial x}\right) & \\ \approx \frac{\Phi(n_1 + 1, n_2, n_3) + \Phi(n_1 - 1, n_2, n_3) - 2\,\Phi(n_1, n_2, n_3)}{d^2} & \end{aligned} \tag{7.16}$$

The auxiliary points $n_1 \pm 1/2$ occur in the intermediate step only. The Poisson equation thus becomes

$$\begin{aligned} \Delta \Phi = \frac{1}{d^2}\Big(&\Phi(n_1 + 1, n_2, n_3) + \Phi(n_1 - 1, n_2, n_3) + \Phi(n_1, n_2 + 1, n_3) \\ &+ \Phi(n_1, n_2 - 1, n_3) + \Phi(n_1, n_2, n_3 + 1) + \Phi(n_1, n_2, n_3 - 1) \\ &- 6\,\Phi(n_1, n_2, n_3)\Big) = -4\pi\,\varrho(n_1, n_2, n_3) \end{aligned} \tag{7.17}$$

For $d \to 0$, the error of this approximation goes to zero. At the same time, the boundary can be better and better approximated by points on the grid.

Equation (7.17) applies to all grid points (n_1, n_2, n_3) *in the interior* of V. Let the number of these points be N. Thus, (7.17) is a linear, inhomogeneous system of equations for N unknowns $\Phi(n_1, n_2, n_3)$. The inhomogeneities are the potential at the boundary points and the charge density in the interior of V. Such a system of equations can be be written in the form $Ax = b$, where the column vector x contains the unknown potential values, the vector b contains the given values for the charge density and the potential boundary values, and the $N \times N$-matrix A contains the coefficients of the system of equations. The unique solution of the system of equations is $x = A^{-1}b$ (the uniqueness of the solution will be shown in the next section). Here we assume $\det A \neq 0$, which is fulfilled for a sensible choice of the grid points. (For a given numerical accuracy, there is a lower limit for the grid spacing; if you choose the grid too narrow, then the equations become *numerically* linearly dependent, and the matrix A cannot be inverted).

The Neumann boundary condition results in the following modifications: First of all, (7.17) again applies for the points in the interior of V. The derivative at the boundary is approximated by the corresponding difference quotient. For M boundary points results in (7.9) M additional linear equations. At the same time, the number of unknowns increases from N to $N + M$ since the value of Φ on R is unknown. This results again in a linear, inhomogeneous system of equations with $N + M$ equations for $N + M$ unknowns. In this case, you can immediately recognize from the system of equations that with Φ is also $\Phi +$const. solution; the solution is therefore not unique.

The procedure outlined does not necessarily represent the optimal strategy for a numerical solution. For example, a choice of grid points adapted to the problem could improve the accuracy of the solution or the speed on the computer (for example, by using the "finite element method"). In addition, there are various possibilities to approximate derivatives by differences, for example, by using more than the nearest neighboring points. The Laplace equation could also initially be approximated by a equivalent variation method (Exercise 7.5). Before the practical calculation on the computer, you should be aware of various numerical methods and existing programs. A recommended book with programs is *Numerical Recipes* by W. H. Press *et al.* (second edition, Cambridge University Press, 1992).

Uniqueness of the Solution

We show that the solution of (7.10) is unique. For this purpose, we consider any two solutions $\Phi_1(\boldsymbol{r})$ and $\Phi_2(\boldsymbol{r})$ of (7.10). For the difference

$$\Psi(\boldsymbol{r}) = \Phi_1(\boldsymbol{r}) - \Phi_2(\boldsymbol{r}) \tag{7.18}$$

then follows from (7.10)

$$\Delta\Psi = 0 \quad \text{in} V \quad \text{and} \quad \Psi\big|_{\text{R}} = 0 \tag{7.19}$$

We write the first Green's theorem (1.30) for $G = \Psi$ and $\Phi = \Psi$:

$$\int_V d^3r \left(\Psi\,\Delta\Psi + \nabla\Psi\cdot\nabla\Psi\right) = \oint_{\text{R}} dF\;\Psi\,\frac{\partial\Psi}{\partial n} \tag{7.20}$$

Since $\Psi|_\mathrm{R} = 0$ the right-hand side disappears. This also applies if Neumann's boundary condition (7.9) is used instead of that by Dirichlet, which for the difference Ψ of the solutions yields

$$\left.\frac{\partial \Psi}{\partial n}\right|_\mathrm{R} = 0 \quad \text{(Neumann)} \tag{7.21}$$

On the left-hand side of (7.20), we set $\Delta\Psi = 0$. This gives us from (7.20)

$$\int_V d^3r\ \left(\nabla\Psi\right)^2 = 0 \quad \longrightarrow \quad \nabla\Psi = 0 \quad \longrightarrow \quad \Psi(\boldsymbol{r}) = \text{ const.} \tag{7.22}$$

This allows two possible solutions $\Phi_1(\boldsymbol{r})$ and $\Phi_2(\boldsymbol{r})$ which can only differ by one constant. For Dirichlet's boundary condition, this constant must be zero because of $\Psi|_\mathrm{R} = 0$; the solution is therefore unique. For Neumann's boundary condition, the solution $\Phi(\boldsymbol{r})$ is only fixed up to such an insignificant constant.

Faraday cage

A Faraday cage is an arbitrarily shaped, closed metal surface. There are no charges inside the enclosed volume V. Then, the potential Φ must fulfill the following conditions:

$$\Delta\Phi = 0 \quad \text{in } V \quad \text{ and } \quad \Phi\big|_\mathrm{R} = \Phi_0 = \text{const.} \tag{7.23}$$

Obviously, now

$$\Phi(\boldsymbol{r}) \equiv \Phi_0 = \text{const. in } V \tag{7.24}$$

is a solution. Since the solution is unique, this is already the solution we are looking for. From $\Phi(\boldsymbol{r}) \equiv$ const. follows that the field $\boldsymbol{E}$ disappears inside the closed metal surface:

$$\boldsymbol{E} = 0 \quad \text{inside Faraday cage} \tag{7.25}$$

Arbitrary external electric fields are shielded by a closed metal cage. In practice, the cage surface can be replaced by a metal grid; to do this, the mesh spacing must be small compared to the length on which the external fields change. In electrostatics, we limit ourselves to static fields. However, it should be mentioned that for frequencies that are not too high, the shielding by a closed metal surface also works.

Exercises

7.1 *Poisson equation on one- and two-dimensional grid*

Formulate the Poisson equation for $\Phi(x)$ on a one-dimensional and for $\Phi(x, y)$ on a two-dimensional equidistant grid with grid spacing d.

7.2 *Poisson's equation on the grid: Hollow metal cube*

A cubic cavity (edge length L) has metal walls. The cube is now cut into two halves by a plane. The cutting plane shall be parallel to two side areas and in the middle of them. The potential $\Phi = \Phi_0$ is given to one half, and on the other half, $\Phi = 0$ applies. The two metal halves are isolated from each other.

Calculate the potential inside this arrangement numerically using a grid spacing $d = L/3$. Imagine now a smaller grid spacing like for example $d = L/100$. Point out where the numerical solution will deviate significantly from the true solution (in spite of the small grid spacing).

7.3 *Volume I enclosed by metal plates*

The volume

$$V_{\mathrm{I}} = \left\{\boldsymbol{r}: 0 \le x \le a,\ 0 \le y \le b,\ -\infty \le z \le \infty\right\}$$

is enclosed by metal plates. The two plates at $x = 0$ and $x = a$ are earthed, and the other two at $y = 0$ and $y = b$ have the potential Φ_0. Due to the translational symmetry in the z-direction, the problem can be reduced to two dimensions: $\Phi(\boldsymbol{r}) = \Phi(x, y)$. Solve the Laplace equation in the interior of the volume using an separation approach. What is the general solution? Determine the constants of this solution so that the boundary conditions are fulfilled.

7.4 *Volume II enclosed by metal plates*

The volume

$$V_{\mathrm{II}} = \left\{\boldsymbol{r}: 0 \le x \le a,\ 0 \le y \le \infty,\ -\infty \le z \le \infty\right\}$$

is enclosed by metal plates. The two side plates at $x = 0$ and $x = a$ are earthed; the base plate at $y = 0$ has the potential Φ_0. Due to the translational symmetry in the z-direction, the problem can be reduced to two dimensions: $\Phi(\boldsymbol{r}) = X(x)\,Y(y)$. Solve the Laplace equation in the interior of the volume by a separation approach and give the general solution. Determine the constants of this solution so that the boundary conditions are fulfilled.

Show that the potential can also be written in the form

$$\Phi(x, y) = \frac{2\,\Phi_0}{\pi}\ \arctan\left[\frac{\sin(\pi\, x/a)}{\sinh(\pi\, y/a)}\right] \tag{7.26}$$

It is sufficient to show that this potential solves the posed boundary value problem (because the solution of the problem is unique). Calculate and sketch the equipotential surfaces and the field lines.

7.5 *Variation principle for the field energy*

The potential $\Phi(\boldsymbol{r})$ is given at the boundary R of the volume V. In V, the field satisfies the variation principle,

$$W\Phi = \frac{1}{8\pi}\int_V d^3r\ \big(\text{grad}\,\Phi(\boldsymbol{r})\big)^2 = \text{minimal}$$

From this, derive a differential equation for $\Phi(\boldsymbol{r})$.

Explanation: The electrostatic field adjusts itself such that the field energy is minimal. The field energy $W\Phi$ is a functional of Φ. The minimality condition implies $\delta W = W\Phi + \delta\Phi - W\Phi = 0$, where $\delta\Phi(\boldsymbol{r})$ is a small variation that disappears at the boundary.

Chapter 8

Applications

We discuss some methods for analytically solving the electrostatic boundary value problem. We introduce the capacitor and calculate the capacitance of a spherical capacitor. The field equations for the vortex-free flow of an ideal fluid are introduced and compared with those of electrostatics.

Image Charge

Figure 8.1 shows a point charge in front of a flat metal plate. The distance of the charge to the plate is a. The plate has the potential $\Phi = 0$; the plate is also said to be earthed. We are looking for the electrostatic potential and the induced charge on of the metal plate. This problem can be solved with the method of image charge (also called mirror charge or mirror image charge).

The metal plate divides the space into two halves: $x > 0$ and $x < 0$. In the right subspace, the solution is

$$\Phi(\boldsymbol{r}) = 0 \quad \text{for } x > 0 \tag{8.1}$$

This satisfies the equation $\Delta\Phi = 0$ and the boundary conditions $\Phi = 0$ at $x = 0$. This solution also applies if the space $x > 0$ is filled by metal or if the metal plate has a finite thickness. This part of the solution is of no particular interest.

We want to know the field in the range $x < 0$. For this, we are looking for a solution of the equations

$$\Delta\,\Phi(\boldsymbol{r}) = -4\pi q\;\delta(\boldsymbol{r} + a\,\boldsymbol{e}_x) \quad \text{for } x < 0, \quad \Phi\big|_{\mathrm{R}} = \Phi(x = 0, y, z) = 0 \tag{8.2}$$

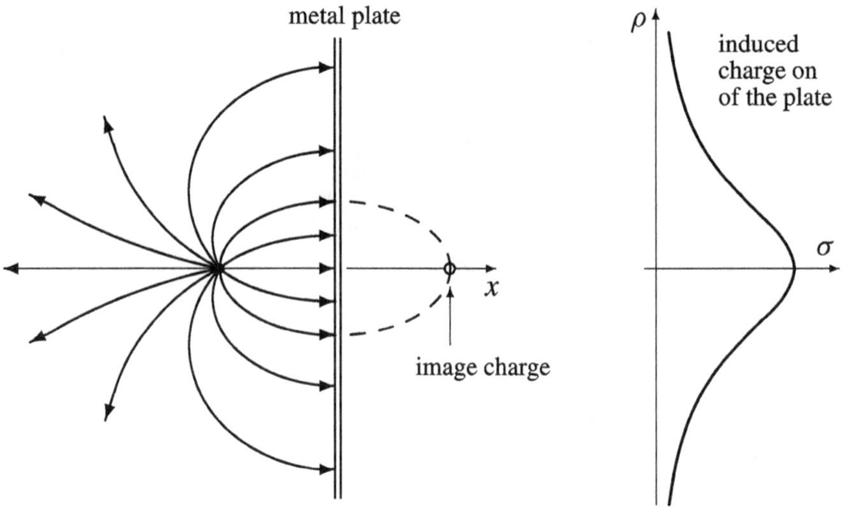

Figure 8.1 The problem "Point charge q and earthed metal plate" is replaced by the problem "Point charge q and image charge $-q$". To the left of the metal plate, both problems result in the same field; the schematic figure shows some field lines. The field strength $\boldsymbol{E}$ at the plate determines the induced charges. Their distribution on the metal plate is sketched on the right.

This is a boundary value problem of the form (7.10). We consider first another problem, namely, a charge q at $-a\boldsymbol{e}_x$ and a mirror charge of the opposite magnitude at $a\boldsymbol{e}_x$:

$$\Delta\Phi(r) = -4\pi q\left(\delta(\boldsymbol{r}+a\boldsymbol{e}_x) - \delta(\boldsymbol{r}-a\boldsymbol{e}_x)\right) \tag{8.3}$$

The field lines for the two charges have already been sketched in Figure 6.1. The solution of (8.3) is known:

$$\Phi(\boldsymbol{r}) = q\left(\frac{1}{|\boldsymbol{r}+a\boldsymbol{e}_x|} - \frac{1}{|\boldsymbol{r}-a\boldsymbol{e}_x|}\right) \tag{8.4}$$

It is easy to check that this $\Phi(\boldsymbol{r})$ also solves (8.2). We have thus found the following solution to the problem "Point charge and metal plate": For $x > 0$, $\Phi = 0$, and for $x < 0$, Equation (8.4) holds. The second term in (8.4) solves the equation $\Delta\Phi = 0$ in the range $x < 0$. It therefore corresponds to the term Φ_{hom} in (7.12). This term is chosen in such a way that the boundary condition is fulfilled.

In general, the image charge method consists of finding charges outside of V in such a way that, together with the charges in V, they produce a potential that satisfies the boundary condition. This image charge method can also be applied to other simple geometries, such as the problems "Charge distribution and plate", "Charge and two perpendicular plates" or "Point charge and sphere" (Exercises 8.1 and 8.2).

Induced Charge

The electric field can be calculated from (8.4):

$$\boldsymbol{E}(\boldsymbol{r}) = -\operatorname{grad}\Phi = q\left(\frac{\boldsymbol{r}+a\boldsymbol{e}_x}{|\boldsymbol{r}+a\boldsymbol{e}_x|^3} - \frac{\boldsymbol{r}-a\boldsymbol{e}_x}{|\boldsymbol{r}-a\boldsymbol{e}_x|^3}\right) \quad (x < 0) \tag{8.5}$$

From (7.11), we obtain the surfaces charge σ on the metal plate:

$$\sigma(y, z) = \frac{-E_x(x=0, y, z)}{4\pi} = \frac{-qa}{2\pi\,(a^2+y^2+z^2)^{3/2}} = \frac{-qa}{2\pi\,(a^2+\rho^2)^{3/2}} \tag{8.6}$$

This induced surface charge is maximum at $(y, z) = (0, 0)$; it decreases with increasing distance $\rho^2 = y^2+z^2$ form the original charge (Figure 8.1, right). The total charge on the metal plate is

$$q_{\text{induced}} = 2\pi\int_0^\infty d\rho\;\rho\;\sigma(\rho) = -aq\int_0^\infty d\rho\;\frac{\rho}{(a^2+\rho^2)^{3/2}} = -q \tag{8.7}$$

These charges are called *induced charges.* In general, these charges are induced by an external field. The external field acts on the freely moving charges of the metal (the electrons in the conduction band of the metal lattice). In the problem under consideration, the surface charges are induced by the field by the point charge in front of the plate. If this point charge is continuously increased, charge flows away from the plate via the earthing (which ensures $\Phi|_{\text{R}} = 0$).

The induced charge on the metal plate and the point charge exert forces on each other. The force on a point charge q is $\boldsymbol{F} = q\boldsymbol{E}'$, where $\boldsymbol{E}'$ is the field of all charges other than the point charge itself. The total field $\boldsymbol{E}$ is given in the range $x <$ is given by (8.5). Therefore, $\boldsymbol{E}'$ is equal to the field of the image charge. In the range $x < 0$, this is also the field of the induced charges. This means that the force acting on the point charge is

$$F = \frac{q^2}{4a^2}\, e_x \tag{8.8}$$

An oppositely equal force acts on the plate.

Green's Function

Let a volume V be bounded by grounded metal bodies. In V, a charge distribution ϱ is given. The problem to be solved is then

$$\Delta\Phi(\boldsymbol{r}) = -4\pi\,\varrho(\boldsymbol{r}) \quad \text{in } V \quad \text{and} \quad \Phi(\boldsymbol{r})\big|_{\mathrm{R}} = 0 \tag{8.9}$$

We first consider a point charge at $\boldsymbol{r}_0$ instead of a continuous charge distribution:

$$\Delta\Phi_0(\boldsymbol{r}) = -4\pi\, q\,\delta(\boldsymbol{r}-\boldsymbol{r}_0) \quad \text{in } V \quad \text{and} \quad \Phi_0(\boldsymbol{r})\big|_{\mathrm{R}} = 0 \tag{8.10}$$

From the solution Φ_0 of this problem, we can construct the solution Φ of the more general problem (8.9). The solution Φ_0 for the charge $q = 1$ is denoted as *Green's function* $G(\boldsymbol{r}, \boldsymbol{r}_0)$:

$$\Delta G(\boldsymbol{r}, \boldsymbol{r}_0) = -4\pi\,\delta(\boldsymbol{r}-\boldsymbol{r}_0) \quad \text{in } V \quad \text{and} \quad G(\boldsymbol{r}, \boldsymbol{r}_0)\big|_{r\in\mathrm{R}} = 0 \tag{8.11}$$

The solution of the original problem (8.9) can now be written with the help of this Green's function,

$$\Phi(\boldsymbol{r}) = \int d^3r'\; G(\boldsymbol{r}, \boldsymbol{r}')\,\varrho(\boldsymbol{r}') \tag{8.12}$$

It is easy to check with the help of (8.11) that this is indeed a solution of (8.9).

Green's function G depends on the shape of the boundary R. We consider two simple examples:

1. Let there be no metal plate; the boundary condition is $\Phi(\infty) = 0$. According to (3.24), the solution of (8.11) is

   $$G(\boldsymbol{r}, \boldsymbol{r}_0) = \frac{1}{|\boldsymbol{r}-\boldsymbol{r}_0|} \tag{8.13}$$

   The integral (8.12) becomes

   $$\Phi(\boldsymbol{r}) = \int d^3r'\; \frac{\varrho(\boldsymbol{r}')}{|\boldsymbol{r}-\boldsymbol{r}'|} \tag{8.14}$$

   This is the well-known solution (6.3).

2. Let there be a grounded metal plate $x = 0$; the boundary condition is $\Phi = 0$ for $x = 0$. The solution (8.4) for a point charge gives Green's function of this problem

$$G(\boldsymbol{r}, \boldsymbol{r}_0) = \frac{1}{|\boldsymbol{r} - \boldsymbol{r}_0|} - \frac{1}{|\boldsymbol{r} - \boldsymbol{r}_1|} \quad \text{with} \quad \boldsymbol{r}_1 = \boldsymbol{r}_0 - 2\,(\boldsymbol{r}_0 \cdot \boldsymbol{e}_x)\,\boldsymbol{e}_x \tag{8.15}$$

The integral (8.12) thus becomes

$$\Phi(\boldsymbol{r}) = \int d^3r_0\ \varrho(\boldsymbol{r}_0) \left(\frac{1}{|\boldsymbol{r} - \boldsymbol{r}_0|} - \frac{1}{|\boldsymbol{r} - \boldsymbol{r}_1|} \right) = \int d^3r'\ \frac{\varrho(\boldsymbol{r}') + \varrho_{\text{IC}}(\boldsymbol{r}')}{|\boldsymbol{r} - \boldsymbol{r}'|} \tag{8.16}$$

This solution can be described as follows: The given charge distribution ϱ is decomposed into individual charge elements $dq = \varrho\, d^3r$. To each charge element dq, an image charge dq_{IC} is assigned. This results in an image charge distribution $\varrho_{\text{IC}}(\boldsymbol{r})$. Compared to ϱ, ϱ_{IC}, the image distribution has the opposite sign and is mirrored at the plane $x = 0$.

Capacitor

We consider a system of N metal bodies with the surfaces R_i. Charges are only allowed on the metal bodies. Then, the boundary value problem reads

$$\Delta\,\Phi(\boldsymbol{r}) = 0, \quad \Phi\big|_{\mathrm{R}_i} = \Phi_i \quad (i = 1, \ldots, N), \quad \Phi(\infty) = 0 \tag{8.17}$$

The shape and position of the metal bodies, i.e., the "geometry of the capacitor problem", are considered fixed (in space and time). The field configuration is then determined by the N potential values $\Phi_1, \ldots, \Phi_N$.

If $\Phi(\boldsymbol{r})$ is solution of (8.17), then $\alpha\,\Phi(\boldsymbol{r})$ is the solution of the problem with the potential values $\alpha\,\Phi_i$ on the metal bodies. For $\alpha\,\Phi_i$, also the fields $\boldsymbol{E}(\boldsymbol{r})$, the surface charges and the charges Q_i obtain the factor α. Therefore, the relation between the Q_i and the Φ_j must be *linear*:

$$Q_i = \sum_{j=1}^{N} C_{ij}\,\Phi_j \tag{8.18}$$

The coefficients C_{ij} depend only on the given geometry. The field configuration is also determined by the charges Q_i; the Φ_j can therefore also be regarded as functions of the Q_j. Therefore, relation (8.18) can be resolved for the Φ_i; the inverse of the matrix $C = (C_{ij})$ exists.

A *capacitor* in the narrower sense is an arrangement of two metal bodies (Figure 8.2), which carry opposite charges of the same size, i.e.,

$$N = 2, \quad Q_1 = -Q_2 = Q \quad \text{(capacitor)} \tag{8.19}$$

Simple examples are the plate capacitor and the spherical capacitor, Figure 8.3: For two conductors, in general, two potential or two charge values must be specified. Due to the restriction $Q_1 = -Q_2 = Q$, the field is defined by a single value, for example, by the charge aQ. The potential difference between the two metal bodies is referred to as the *voltage* U:

$$U = \int_1^2 d\boldsymbol{r} \cdot \boldsymbol{E} = \Phi_1 - \Phi_2 \tag{8.20}$$

If we multiply the potential $\Phi(\boldsymbol{r})$ by a number α, this implies $Q \to \alpha\, Q$ and $U \to \alpha\, U$. It follows that Q and U depend linearly on each other,

$$Q = \text{const.} \cdot U = C\,U \tag{8.21}$$

This is a special case of (8.18). The quantity C describes the receptivity of the arrangement for charges and is therefore called *capacitance*

$$C = \frac{Q}{U} \quad \text{(capacitance)} \tag{8.22}$$

The quantity defined in this way is positive because $Q = Q_1 > 0$ implies $U = \Phi_1 - \Phi_2 > 0$. Definition (8.22) can also be used for a single conductor 1; then the voltage U equals $\Phi_1 - \Phi(\infty) = \Phi_1$.

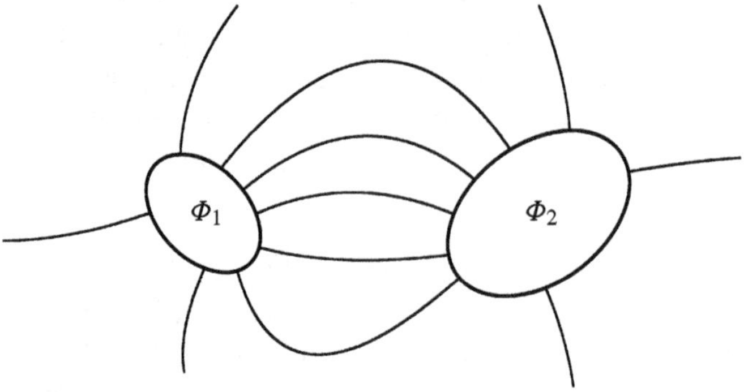

Figure 8.2 Schematic field lines between two metal bodies with Φ_1 = const. and Φ_2 = const. The field lines condense between the bodies; such an arrangement is called capacitor.

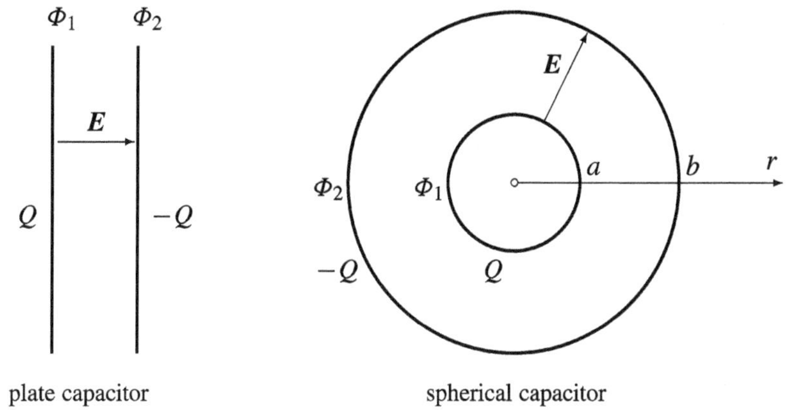

Figure 8.3 A capacitor in the narrower sense means two metal bodies with the charges $Q_1 = Q$ and $Q_2 = -Q$. The voltage is the potential difference $U = \Phi_1 - \Phi_2$ between the bodies. It determines the capacitance $C = Q/U$ of the capacitor. For simple geometries (such as plate and spherical capacitors), the field of the capacitor and its capacitance C can be easily calculated.

We determine the capacitance C of a capacitor consisting of two concentric metal spheres. Sphere 1 has the radius a, and sphere 2 has the radius $b > a$, Figure 8.3. From the spherical symmetry follows $\Phi(\boldsymbol{r}) = \Phi(r)$ and therefore $\boldsymbol{E}(\boldsymbol{r}) = E(r)\,\boldsymbol{e}_r$. We apply Gauss' law to a sphere with different radii r and obtain

$$E(r) = \begin{cases} 0 & (r < a) \\ Q/r^2 & (a < r < b) \\ 0 & (r > b) \end{cases} \tag{8.23}$$

With $\Phi(\infty) = 0$, we get

$$\Phi(r) = \begin{cases} Q/a - Q/b & (r < a) \\ Q/r - Q/b & (a < r < b) \\ 0 & (r > b) \end{cases} \tag{8.24}$$

From this, we can read off the potential values on the spheres:

$$\Phi_1 = \frac{Q}{a} - \frac{Q}{b}, \quad \Phi_2 = 0 \tag{8.25}$$

The capacitance (8.22) of the spherical capacitor is

$$C = \frac{Q}{\Phi_1 - \Phi_2} = \frac{a\,b}{b-a} \tag{8.26}$$

The capacitance has the dimension of a length, $[C] = \mathrm{cm}$; such a designation can be found on capacitors in old radios. Today, the capacitance of real capacitors is usually specified in the SI system:

$$C = \frac{Q}{U} = \frac{\mathrm{C}}{\mathrm{V}} = \mathrm{F} = \text{farad} \quad \text{(SI system)} \tag{8.27}$$

A capacitance of $C = 1\,\mathrm{cm}$ in the Gaussian unit system corresponds approximately to 1 picofarad ($\mathrm{pF} = 10^{-12}\,\mathrm{F}$) in the SI system.

From (6.29), we also obtain the potential energy W of a charged capacitor,

$$W = \frac{1}{2}\int d^3r\; \varrho(\boldsymbol{r})\, \Phi(\boldsymbol{r}) = \frac{1}{2}\left(Q_1\,\Phi_1 + Q_2\,\Phi_2\right)$$

$$\overset{(8.19,8.20)}{=} \frac{Q\,U}{2} = \frac{Q^2}{2\,C} = \frac{C\,U^2}{2} \tag{8.28}$$

This is equal to the field energy $W = \int d^3r\, \boldsymbol{E}^2/8\pi$.

Differentiable Complex Functions

In this section, we focus on two-dimensional problems, where the potential Φ depends on only two Cartesian coordinates coordinates:

$$\Phi(\boldsymbol{r}) = \Phi(x, y) \tag{8.29}$$

A number of such problems can be solved with the help of differentiable complex functions.

A *complex* function $f(z)$ can be expressed in the form

$$f(z) = f(x + \mathrm{i}\,y) = u(x, y) + \mathrm{i}\,v(x, y) \tag{8.30}$$

In this section, z is not used for the third Cartesian coordinate but for $z = x + \mathrm{i}\,y$. The complex function consists of two real functions $u(x, y)$ and $v(x, y)$, which depend on the real variables x and y.

We now consider *differentiable* complex functions. For complex functions $f(z)$, differentiability is a strong restriction because the limit values of the difference quotient $(f(z) - f(z_0))/(z - z_0)$ for $z \to z_0$ must exist and coincide for an any arbitrary approach z_0. We specifically consider the approximations parallel to the x- or y-axis:

$$\frac{df}{dz} = \lim_{\Delta z \to 0} \frac{\Delta f}{\Delta z} = \begin{cases} u_x + \mathrm{i}\, v_x & (\Delta z = \Delta x) \\ -\mathrm{i}\, u_y + v_y & (\Delta z = \mathrm{i}\, \Delta y) \end{cases} \tag{8.31}$$

We use the notation $u_x = \partial u / \partial x$. In order for the two limits to match, we must require

$$u_x = v_y \quad \text{and} \quad u_y = -v_x \tag{8.32}$$

These *Cauchy–Riemann differential equations* are a necessary condition for the existence of the derivative of $f(z)$; they therefore follow from the assumed differentiability. By further differentiation and permutation of the partial derivatives (u and v are twice differentiable), we obtain

$$\Delta u = u_{xx} + u_{yy} = 0 \quad \text{and} \quad \Delta v = v_{xx} + v_{yy} = 0 \tag{8.33}$$

Thus, every (twice) differentiable complex function $f(z)$ provides a solution to Laplace's equation; the function $f(z)$ is also called a complex potential. From

$$\operatorname{grad} u \cdot \operatorname{grad} v = u_x\, v_x + u_y\, v_y \overset{(8.32)}{=} 0 \tag{8.34}$$

it follows that the lines u = const. and v = const. are perpendicular to each other. For the equipotential lines v = const., the associated field lines are given by u = const. and vice versa. By shifting in the direction of the third Cartesian coordinate, from which potential (8.29) does not depend, the equipotential lines become equipotential surfaces.

A differentiable complex function is also called an analytic function or holomorphic function. Such an analytic function is indeed differentiable arbitrarily often.

As an example, we consider the complex potential

$$f(z) = z + \frac{1}{z} = u(x, y) + \mathrm{i}\, v(x, y) \tag{8.35}$$

With the exception of $z = 0$, the function $f(z)$ can be differentiated any number of times. The field and equipotential lines (Figure 8.4) are given by

$$u = x + \frac{x}{x^2 + y^2} = \text{const.}, \quad v = y - \frac{y}{x^2 + y^2} = \text{const.} \tag{8.36}$$

The equipotential line $v = 0$ consists of the circle $x^2 + y^2 = 1$ and the x-axis ($y = 0$). The circle can be interpreted as the contour of a circular cylinder. At a large distance from the cylinder, $v \approx y =$ const. represent the equipotential lines of a homogeneous electric field. Therefore, the complex potential (8.35) provides the solution to the problem "conducting circular cylinder in an homogeneous field" (Figure 8.4).

A complex function $f = u(x, y) + \mathrm{i}\, v(x, y)$ assigns each point of the x-y-plane to a point of the u-v-plane. The function can therefore be described as a mapping of the x-y-plane into the u-v-plane. For *differentiable* functions, this mapping is *conformal*, i.e., angles (but not distances) are preserved. Conformal mappings can be used to transform certain contours into others.

The method of conformal mapping has only been presented very briefly here. Due to the availability of powerful computers, it has become less important. However, it may be useful, when combining numerical methods with analytical solutions.

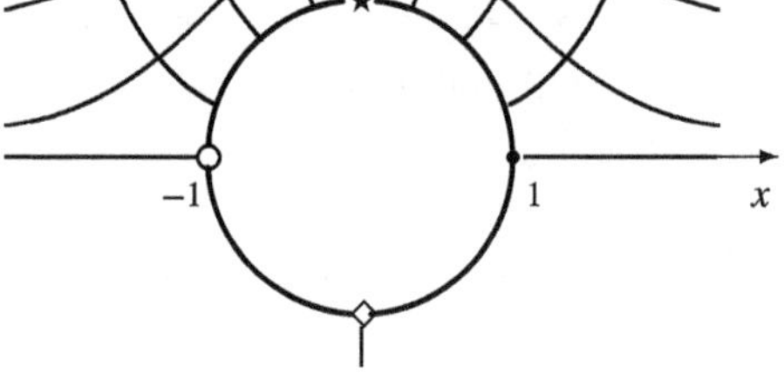

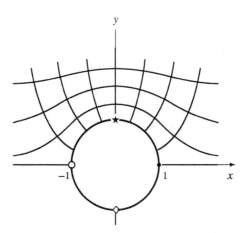

Figure 8.4 For the function $f(z) = z + 1/z = u(x, y) + \mathrm{i}\, v(x, y)$, the lines $v(x, y) =$ const. and $u(x, y) =$ const. are sketched. The lines $v =$ const. are asymptotically horizontal and nestle together at smaller distance to the circle. They describe the equipotential lines of the problem "Conducting circular cylinder in the homogeneous field". The perpendicular lines $u =$ const. are then the field lines (starting at the circular contour an going upwards).

Potential Flow

The flow of a liquid or a gas can be characterized by a velocity field $\boldsymbol{v}(\boldsymbol{r}, t)$. In the stationary case, $\boldsymbol{v} = \boldsymbol{v}(\boldsymbol{r})$ applies. For sufficiently small velocities, the flow is vortex-free:

$$\text{curl}\, \boldsymbol{v}(\boldsymbol{r}) = 0 \tag{8.37}$$

The liquid shall be incompressible, i.e., its mass density is constant, $\varrho_{\text{mat}} = \text{const}$. From the continuity equation $\dot{\varrho}_{\text{mat}} + \text{div}\,(\varrho_{\text{mat}}\, \boldsymbol{v}) = 0$ then follows in the stationary case

$$\text{div}\, \boldsymbol{v}(\boldsymbol{r}) = 0 \tag{8.38}$$

The field equations (8.37) and (8.38) have the same form as those of the electrostatics in the source-free case. An inhomogeneity in (8.38) would correspond to a source, i.e., an inflow. The terms “sources” and “vortex” of the vector field (formally defined by divergence and curl) find an illustrative interpretation in the liquid flow.

Due to (8.37), we can use a velocity potential $\Phi(\boldsymbol{r})$,

$$\boldsymbol{v} = \text{grad}\, \Phi(\boldsymbol{r}) \tag{8.39}$$

Thus, (8.38) becomes the Laplace equation $\Delta \Phi = 0$. We consider the flow around contours, such as an airfoil profile. Such a profile means a boundary R for the volume V in which Φ is to be determined. At the boundary R, the normal component v_{normal} of the velocity must be zero. Thus, we obtain a potential problem with a special Neumann's boundary condition:

$$\Delta \Phi(\boldsymbol{r}) = 0 \quad \text{in } V, \qquad \left.\frac{\partial \Phi}{\partial n}\right|_{\text{R}} = 0 \tag{8.40}$$

The solution of a potential problem of electrostatics can be interpreted as solution of a flow problem, too. Figure 8.4 shows the vortex-free flow around the circular cylinder with the profile $x^2 + y^2 = 1$. The equipotential lines can be interpreted as the *streamlines*, along which liquid elements move. (If, on the other hand, the field lines in Figure 8.4 as streamlines, then the surface of the cylinder must be a corresponding source of the fluid.)

Exercises

8.1 *Point charge in front of earthed metal plates*

The volume

$$V = \{\boldsymbol{r}: 0 \le x \le \infty,\ 0 \le y \le \infty,\ -\infty \le z \le \infty\}$$

is enclosed by earthed metal plates at $x = 0$ and $y = 0$. In V, there shall be a point charge q.

Determine the potential $\Phi(\boldsymbol{r})$ in V (with the help of image charges). Calculate the surface charge density and the total charge on the plates. What force acts on the point charge?

8.2 *Point charge in front of metal sphere*

Outside a grounded, conductive hollow sphere (radius R, center $\boldsymbol{r} = 0$), there is a point charge q_1 at $\boldsymbol{r}_1$. Calculate the potential in the inner and outer space of the sphere. To do this, use a suitable image charge q_2 at $\boldsymbol{r}_2$. Calculate the charge density and the total charge on the spherical surface. What force acts between the point charge and the sphere? What changes if the charge is inside the sphere?

What is the solution if the spherical surface has a finite potential value $\Phi_0 = \Phi(R) - \Phi(\infty) \neq 0$?

8.3 *Spherical capacitor*

Two concentric metal spheres with the radii R_1 and R_2 have the potential values Φ_1 and Φ_2 (assume $\Phi(\infty) = 0$). Determine the potential $\Phi(r)$ in the entire space. Which charges Q_1 and Q_2 are on the spheres?

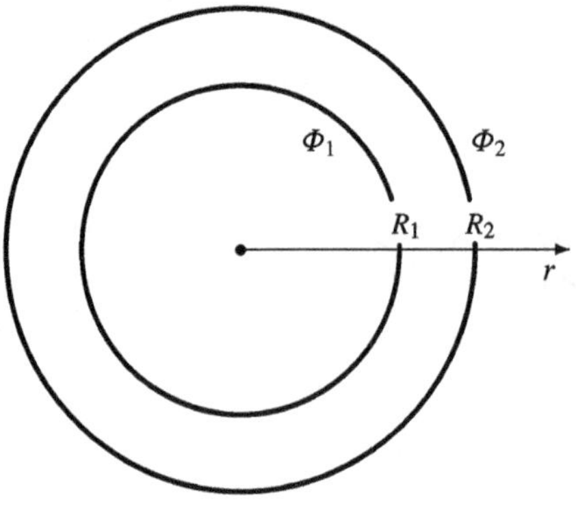

Specialize the result to a spherical capacitor with $Q_1 = -Q_2 = Q$ and $U = \Phi_1 - \Phi_2$, and determine its capacitance $C = Q/U$.

What capacitance is obtained for a single metal sphere?

Consider the special case $d = R_2 - R_1 \ll R_1$ and compare the capacitance with that of a plate capacitor ($C_{\text{plate}} = A_{\text{C}}/(4\pi d)$ without boundary effects).

Show that the capacitance of 1 cm in the unit Gaussian system corresponds approximately to a picofarad in the SI system.

8.4 *Plate capacitor on the grid*

Determine the field of a plate capacitor numerically on the grid:

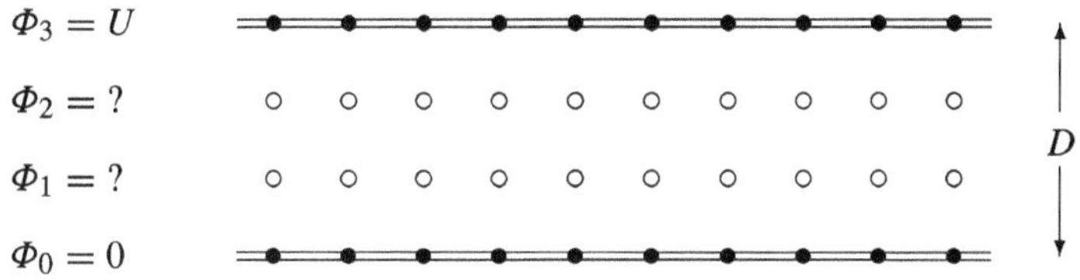

Let the capacitor extend infinitely far parallel to the plates. On the boundary points (•) of the grid, the potential values are given, and at the inner points (◦), the potential shall be determined. Calculate the numerical solution for a grid spacing $d = D/N$ (the figure shows $d = D/3$), and compare it with the exact solution.

8.5 *Complex potential*

Calculate and sketch the equipotential and field lines of the complex potential

$$f(z) = \ln \frac{z-a}{z+a}$$

Give a physical interpretation.

8.6 *Potential flow around a moving sphere*

A sphere (mass M, radius R) moves with the constant velocity $\boldsymbol{v}_0$ through an incompressible liquid (density ϱ_0). The vector $\boldsymbol{r}_0 = \boldsymbol{v}_0\, t$ points to the center of the sphere. A vortex-free flow can be achieved by a velocity field

$\boldsymbol{v}(\boldsymbol{r}) = \text{grad}\,\Phi(\boldsymbol{r})$. Show that the approach

$$\Phi(\boldsymbol{r}, t) = \boldsymbol{c} \cdot \nabla \frac{1}{r'} \quad \text{with } \boldsymbol{r}' = \boldsymbol{r} - \boldsymbol{v}_0\, t$$

fulfills the following conditions:

$$\Delta\,\Phi(\boldsymbol{r}, t) = 0 \quad \text{in } V \quad \text{and} \quad \boldsymbol{v}' \cdot \boldsymbol{e}_{r'} = 0 \quad \text{on R}$$

Here, V is the volume outside the sphere and R is the radius of the sphere. The second condition refers to the rest system of the sphere (indicated by the dashes). Seen from the sphere, the normal component of the velocity on the boundary disappears; this determines the constant vector $\boldsymbol{c}$.

The kinetic energy associated with the motion is of the form $E = E_{\text{sphere}} + E_{\text{liquid}} = M_{\text{eff}}\, \boldsymbol{v}_0^2/2$ and consists of the contribution of the sphere $E_{\text{sphere}} = M\, \boldsymbol{v}_0^2/2$ *and* the fluid. Calculate the effective mass M_{eff} of the sphere.

8.7 *Heat conduction equation*

For the temperature field $T(\boldsymbol{r}, t)$, the following applies

$$\frac{\partial T(\boldsymbol{r}, t)}{\partial t} = \kappa'\, \Delta T(\boldsymbol{r}, t)$$

with a material-dependent constant κ'. In the static case, this reduces to the Laplace equation $\Delta T(\boldsymbol{r}) = 0$.

On two parallel, infinitely extended plates at a distance D, the temperatures T_1 and T_2 are given. The medium between the plates is homogeneous. Specify the temperature profile between the plates in the static case.

Chapter 9

Legendre Polynomials

The Laplace operator is used in many equations in physics; for example, in the Schrödinger equation, in the diffusion equation, in classical wave equations or in the equation for vortex-free flow. The solution of the Laplace equation

$$\Delta \Phi(\boldsymbol{r}) = 0 \quad \text{(Laplace equation)} \tag{9.1}$$

is therefore an important example for solving partial differential equations. We investigate the solution of the Laplace equation for spherical coordinates. This results in the Legendre polynomials (this chapter) and the spherical harmonics (Chapter 11).

Complete, Orthogonal Set of Functions

In connection with the solution of (9.1), we introduce functions (Legendre polynomials, spherical harmonics) which form a *complete orthonormal set of functions* (CONS). We first summarize some essential properties of CONS.

We consider functions $g(x), h(x), \ldots$ which are defined in the interval (a, b). We assign a number to each pair of such these functions:

$$(g, h) = (h, g) = \int_a^b dx \; g(x)\, h(x) \tag{9.2}$$

This quantity is called *scalar product*. For a Cartesian coordinate, the following one might use u$(a, b) = (-\infty, +\infty)$, for the angular coordinate $x = \phi$, on the other hand, we had $(a, b) = (0, 2\pi)$. Furthermore, we allow that x is the abbreviation for several variables, for example, for three Cartesian coordinates.

If $(g, h) = 0$, we call the functions g and h *orthogonal*. If $(g, g) = 1$, the function g is called *normalized*. A countable set (S) of functions

$$\{f_n\} = (f_1, f_2, f_3, \ldots) \tag{9.3}$$

is *orthonormalized* (ON) if

$$(f_n, f_{n'}) = \delta_{nn'} \tag{9.4}$$

In the following, we assume that (9.3) is an ONS. A set of functions is called *complete* (C) if all continuous functions defined in (a, b) can be represented by

$$g(x) = \sum_{n=1}^{\infty} a_n\, f_n(x) \tag{9.5}$$

More precisely, this means that $\int dx\, |g(x) - \sum a_n\, f_n(x)|$ can be made arbitrarily small. We multiply (9.5) by $f_m(x)$, integrate from a to b and use (9.4). This gives us the expansion coefficients

$$a_m = (g, f_m) = (f_m, g) \tag{9.6}$$

By inserting (9.6) into (9.5), we obtain

$$g(x) = \sum_{n=1}^{\infty} (g, f_n)\, f_n(x) = \int_a^b dx'\, g(x') \sum_{n=1}^{\infty} f_n(x')\, f_n(x) \tag{9.7}$$

Since this applies for any function $g(x)$, we get

$$\sum_{n=1}^{\infty} f_n(x)\, f_n(x') = \delta(x - x') \quad \text{(completeness relation)} \tag{9.8}$$

From this follows (9.5) for any g, i.e., the completeness (C) of the set of functions.

An example of the relationships discussed here is the Fourier series. Any function $g(x)$ that is defined in the interval $(a, b) = (0, L)$ can be expanded into the CONS

$$\{f_n\} = \sqrt{\frac{2}{L}}\,\left(\frac{1}{\sqrt{2}},\ \sin\frac{2\pi x}{L},\ \cos\frac{2\pi x}{L},\ \sin\frac{4\pi x}{L},\right.$$
$$\left.\cos\frac{4\pi x}{L},\ \sin\frac{6\pi x}{L},\ \ldots\right) \tag{9.9}$$

If $g(x) = g(x + L)$ applies, then the representation $g(x) = \sum a_n f_n(x)$ is also valid outside the interval (a, b). Mostly, the Fourier expansion is considered for such periodic functions.

A related example is the CONS

$$\{f_n\} = \sqrt{\frac{2}{L}} \left(\sin\frac{\pi x}{L},\ \sin\frac{2\pi x}{L},\ \sin\frac{3\pi x}{L},\ \ldots \right) \tag{9.10}$$

for all functions $g(x)$ that are defined in the interval $(a, b) = (0, L)$ and for which $g(0) = g(L) = 0$ applies. Functions (9.10) may describe the natural oscillations of a cavity resonator, Figure 21.2. In quantum mechanics, (9.10) are the normalized eigenfunctions of the infinitely high potential well, Chapter 11 in [3].

The mathematical structures shown are those of a normalized, complete vector space (also called Banach space). Table 9.1 displays the analogies to the common three-dimensional vector space.

Table 9.1 The structures discussed in the first section of this chapter are those of a orthonormalized, complete vector space. The table contrasts the corresponding designations and relations for vectors. Instead of three dimensions of the common vector space, the function space has an infinite number of dimensions.

Vectors	Designation	Functions
$\boldsymbol{r}$	Vector	$g(x)$
$\{\boldsymbol{e}_n\}$	Basis	$\{f_n(x)\}$
$(\boldsymbol{e}_n \cdot \boldsymbol{e}_{n'}) = \delta_{nn'}$	Orthonormalization	$(f_n, f_{n'}) = \delta_{nn'}$
$\boldsymbol{r} = \sum_{n=1}^{3} a_n\, \boldsymbol{e}_n$	Expansion	$g(x) = \sum_{n=1}^{\infty} a_n\, f_n(x)$
$a_n = (\boldsymbol{e}_n \cdot \boldsymbol{r})$	Expansion coefficients	$a_n = (f_n, g)$
$\boldsymbol{r} = \begin{pmatrix} a_1 \\ a_2 \\ a_3 \end{pmatrix}$	Representation by column vector	$g(x) = \begin{pmatrix} a_1 \\ a_2 \\ a_3 \\ \vdots \end{pmatrix}$
$\sum \boldsymbol{e}_n \circ \boldsymbol{e}_n = 1$	Completeness	$\sum f_n(x)\, f_n(x') = \delta(x - x')$

Legendre Polynomials

We write the Laplace equation $\Delta\Phi = 0$ in spherical coordinates r, θ and ϕ:

$$\frac{1}{r}\frac{\partial^2}{\partial r^2}\left(r\,\Phi\right)+\frac{1}{r^2\,\sin\theta}\,\frac{\partial}{\partial\theta}\left(\sin\theta\,\frac{\partial\Phi}{\partial\theta}\right)+\frac{1}{r^2\,\sin^2\theta}\,\frac{\partial^2\Phi}{\partial\phi^2}=0 \qquad (9.11)$$

The separation approach

$$\Phi(r,\theta,\phi)=\frac{U(r)}{r}\,P(\cos\theta)\,Q(\phi) \qquad (9.12)$$

initially restricts the form of the solutions. However, the general solution can be constructed as a linear combination of the solutions that we obtain with (9.12). For other coordinates (such as Cartesian or cylindrical coordinates), a separation approach leads to the general solution, too.

Extracting a factor $1/r$ in (9.12) and using the argument $\cos\theta$ instead of θ serve for later simplifications. We insert (9.12) into (9.11):

$$\frac{PQ}{r}\,\frac{d^2U}{dr^2}+UQ\,\frac{1}{r^3\sin\theta}\,\frac{d}{d\theta}\left(\sin\theta\,\frac{dP}{d\theta}\right)+\frac{UP}{r^3\,\sin^2\theta}\,\frac{d^2Q}{d\phi^2}=0 \qquad (9.13)$$

We multiply this by $r^3\sin^2\theta$ and divide by UPQ:

$$\frac{1}{Q}\,\frac{d^2Q}{d\phi^2}=-r^2\sin^2\theta\,\frac{1}{U}\,\frac{d^2U}{dr^2}-\frac{\sin\theta}{P}\,\frac{d}{d\theta}\left(\sin\theta\,\frac{dP}{d\theta}\right) \qquad (9.14)$$

The left side does not depend on r or θ, and the right side does not depend on ϕ. Therefore, both sides must be equal to a coordinate-independent number. This number is called *separation constant*; we denote it by $-m^2$. For the left-hand side, we obtain

$$Q''+m^2\,Q=0, \quad \text{also} \quad Q(\phi)=Q_m(\phi)=\exp(\mathrm{i}m\phi) \qquad (9.15)$$

The point in space (r,θ,ϕ) is identical to $(r,\theta,\phi+2\pi)$. The function $\Phi(r,\theta,\phi)$ must therefore have the same value for both arguments, i.e.,

$$Q_m(\phi+2\pi)=Q_m(\phi) \quad\longrightarrow\quad m=0,+1,-1,+2,-2,\ldots \qquad (9.16)$$

It follows that m is an integer. The second independent solution of the differential equation in (9.15) is $\exp(-\mathrm{i}m\phi)$; take into account this solution by allowing positive and negative m values. Equivalent to the two solutions $\exp(\pm im\phi)$ are $\sin(m\phi)$ and $\cos(m\phi)$.

We now set the right-hand side of (9.14) equal to $-m^2$ and divide by $\sin^2\theta$. This results in

$$\frac{r^2}{U}\,\frac{d^2U}{dr^2} = -\frac{1}{P\,\sin\theta}\,\frac{d}{d\theta}\left(\sin\theta\,\frac{dP}{d\theta}\right) + \frac{m^2}{\sin^2\theta} \tag{9.17}$$

The left-hand side is independent of θ and the right-hand side is independent of r. So both sides must be equal to a new separation constant λ. From this, we get two ordinary differential equations for $U(r)$ and $P(\cos\theta)$. We write the differential equation for $P(x) = P(\cos\theta)$ using the variable x:

$$x = \cos\theta, \qquad \frac{d}{dx} = -\,\frac{1}{\sin\theta}\,\frac{d}{d\theta} \tag{9.18}$$

From $\Delta\Phi = 0$ with the approach $\Phi = UPQ/r$, we have obtained the following equations for U, P and Q:

$$\frac{d^2U}{dr^2} - \frac{\lambda}{r^2}\,U(r) = 0 \tag{9.19}$$

$$\frac{d}{dx}\left(\left(1-x^2\right)\frac{dP}{dx}\right) + \left(\lambda - \frac{m^2}{1-x^2}\right)P(x) = 0 \tag{9.20}$$

$$Q(\phi) = \exp(im\phi), \qquad m = 0, \pm 1, \pm 2, \pm 3, \ldots \tag{9.21}$$

In this chapter, we restrict ourselves to the case where the solution does not depend on ϕ, i.e., on

$$m = 0 \quad \text{(cylindrical symmetry)} \tag{9.22}$$

The solutions with $m \neq 0$ are given in Chapter 11. For $m = 0$, the potential Φ is cylindrically symmetric. Then, (9.20) becomes

$$\left(1-x^2\right)P'' - 2x\,P' + \lambda\,P = 0 \quad \begin{array}{l}\text{(differential equation of}\\ \text{the Legendre polynomials)}\end{array} \tag{9.23}$$

As a solution to this differential equation, we obtain the Legendre polynomials.

Recursion formula

Since the coefficients of the differential equation (9.23) only contain powers of x, it makes sense to look for the solution in the form of a power series:

$$P(x) = \sum_{k=0}^{\infty} a_k \, x^k \quad (-1 \le x \le 1) \tag{9.24}$$

The x-range follows from $x = \cos\theta$. We insert (9.24) into (9.23):

$$\sum_{k=0}^{\infty} \Big(a_{k+2} \, (k+2)(k+1) - a_k \, (k-1)k - 2k \, a_k + \lambda \, a_k \Big) \, x^k = 0 \tag{9.25}$$

This equation is fulfilled for all x, only if the coefficient of each x^k vanishes, i.e.,

$$a_{k+2} = \frac{k(k+1) - \lambda}{(k+2)(k+1)} \; a_k \tag{9.26}$$

Choosing some values for a_0 and a_1, this *recursion formula* determines all other coefficients:

$$\begin{aligned} &\text{choosing } a_0 \longrightarrow a_2, \; a_4, \; a_6, \; \ldots \\ &\text{choosing } a_1 \longrightarrow a_3, \; a_5, \; a_7, \; \ldots \end{aligned} \tag{9.27}$$

For very large k, we obtain from (9.26)

$$\frac{a_{k+2}}{a_k} \overset{k\to\infty}{\longrightarrow} 1 \quad \text{also} \quad a_k \overset{k\to\infty}{\longrightarrow} \text{const.} \tag{9.28}$$

For $a_k \to$ const., the power series (9.24) contains for large n the contribution $P = \cdots + \text{const.} \cdot (x^n + x^{n+2} + x^{n+4} + \cdots)$. If const. $\neq 0$, this sum diverges for $x = 1$. This singularity can only be avoided *if the power series breaks off*. The sequence defined by (9.27) and (9.26) terminates when either the start value (a_0 or a_1) equals zero, or if the numerator in (9.26) disappears for a certain k, i.e., when $\lambda = k\,(k+1)$. Since each of the two sequences (9.27) would lead to a singularity, both sequences must break off. This is only possible if

$$\lambda = l\,(l+1), \quad l = \begin{cases} 0, 2, 4, \ldots & \text{and} \quad a_1 = 0 \\ 1, 3, 5, \ldots & \text{and} \quad a_0 = 0 \end{cases} \tag{9.29}$$

For even l, the first sequence in (9.27) breaks off, and for odd, the second. The other sequence must be made to disappear by the choice $a_1 = 0$ or

$a_0 = 0$. After that, only only even or only odd coefficients survive. The solution is therefore of the form

$$P_l(x) = \sum_{k=l,l-2,l-4,\ldots} a_k\, x^k \tag{9.30}$$

The lowest occurring coefficient (a_0 or a_1) can be chosen arbitrarily because the differential equation (9.23) does not specify the normalization of the solution. As a convention, we set

$$P_l(1) = 1 \tag{9.31}$$

As an example, let us construct $P_3(x)$. For odd l, $a_0 = 0$ must be chosen; from this follows $a_2 = a_4 = \cdots = 0$. The coefficient a_1 is initially arbitrary ($a_1 \neq 0$). From the recursion formula follows

$$\begin{aligned} a_3 &= \frac{1(1+1) - 3(3+1)}{(1+2)(1+1)}\, a_1 = -\frac{5}{3}\, a_1, \\ a_5 &= \frac{3(3+1) - 3(3+1)}{(3+2)(3+1)}\, a_3 = 0 \end{aligned} \tag{9.32}$$

and $a_5 = a_7 = \cdots = 0$. This gives us

$$P_3(x) = a_1 \left(x - \frac{5}{3}\, x^3\right) \tag{9.33}$$

From (9.31) then follows $a_1 = -3/2$.

A second-order differential equation like (9.23) has two independent solutions. We have obtained only one solution for a given λ. This is because the second independent solution is irregular (at $x = 1$). In our derivation, we have excluded such solutions. The irregular solutions not specified here are called Legendre polynomials of the second kind. Such solutions may occur in physical problems when the point $x = 1$ (i.e., the z-axis) does not belong to the range in which $\Delta\Phi = 0$ has to be solved.

Explicit expressions

The polynomials defined by (9.30) with (9.26) are called *Legendre polynomials* $P_l(x)$. They can be represented in the following closed form:

$$P_l(x) = \frac{1}{2^l\, l}\, \frac{d^l}{dx^l}\, \left(x^2 - 1\right)^l \quad \text{Legendre polynomials} \tag{9.34}$$

Obviously, this yields a polynomial of degree l which contains only even or only odd powers. In Exercise 9.3, it is shown that the so-defined $P_l(x)$ satisfies the differential equation (9.23).

The lowest polynomials are

$$\begin{aligned} P_0 = 1, \quad P_1 = x, \quad P_2 = \frac{1}{2}\left(3x^2 - 1\right) \\ P_3 = \frac{1}{2}\left(5x^3 - 3x\right), \quad P_4 = \frac{1}{8}\left(35\,x^4 - 30\,x^2 + 3\right) \end{aligned} \tag{9.35}$$

Orthogonality

We prove that the Legendre polynomials are orthogonal in the interval $-1, 1$, i.e., that the scalar product $(P_l, P_{l'})$ disappears for $l \neq l'$. To do this, we write (9.23) in the form of an eigenvalue equation of the differential operator D_{op}:

$$D_{\text{op}}\, P_l = -l\,(l+1)\, P_l \quad \text{with} \quad D_{\text{op}} = \frac{d}{dx}\left(\left(1 - x^2\right) \frac{d}{dx}\right) \tag{9.36}$$

We form the scalar product of this equation with $P_{l'}$ and also write on the corresponding equation with interchanged l and l':

$$\begin{aligned} (P_{l'},\, D_{\text{op}}\, P_l) &= -l\,(l+1)\,(P_{l'},\, P_l) \\ (P_l,\, D_{\text{op}}\, P_{l'}) &= -l'(l'+1)\,(P_l,\, P_{l'}) \end{aligned} \tag{9.37}$$

In the scalar product, the differential operator D_{op} can be shifted to the other side by partial integration. Therefore, the left-hand sides in (9.37) are equal. From this follows

$$\left[\, l'(l'+1) - l\,(l+1) \,\right] \left(P_l,\, P_{l'}\right) = 0 \tag{9.38}$$

For $l \neq l'$, this implies $(P_l, P_{l'}) = 0$; the polynomials are orthogonal. Exercise 9.4 shows $(P_l, P_l) = 2/(2l+1)$. This means that

$$(P_l,\, P_{l'}) = \int_{-1}^{1} dx\ P_l(x)\, P_{l'}(x) = \frac{2}{2l+1}\, \delta_{ll'} \tag{9.39}$$

Completeness

The functions

$$\left\{ \sqrt{\frac{2l+1}{2}}\ P_l(x) \right\} \quad \text{(CONS of Legendre polynomials)} \tag{9.40}$$

form a complete orthonormalized set. The orthonormalization follows from (9.39). The completeness is plausible: Each power x^n can be written as a linear combination of the $P_l(x)$ with $l \leq n$. This means that these functions are complete: Any function $f(x)$ whose Taylor series converges in the interval $(-1, 1)$ can be expressed as a linear superposition these functions.

Functions $f(\theta)$ that are defined in the range $0 \leq \theta \leq \pi$ can be expanded into the Legendre polynomials:

$$f(\theta) = \sum_{l=0}^{\infty} a_l \, P_l(\cos\theta) \tag{9.41}$$

Multiplying this equation by $P_n(\cos\theta)$, integrating over θ and using (9.39) yields the expansion coefficients

$$a_n = \frac{2n+1}{2} \int_{-1}^{+1} d(\cos\theta) \; f(\theta) \; P_n(\cos\theta) \tag{9.42}$$

Inserting these coefficients back into (9.41), the completeness relation can be read off:

$$\sum_{l=0}^{\infty} \frac{2l+1}{2} \, P_l(x) \, P_l(x') = \delta(x - x') = \delta(\cos\theta - \cos\theta') \tag{9.43}$$

Exercises

9.1 *Completeness relation for sine functions*

Consider the orthonormalized set of functions,

$$\left\{ \sqrt{\frac{2}{L}} \, \sin\left(\frac{n\pi x}{L}\right) \right\}, \qquad n = 1, 2, 3 \ldots \qquad \text{and} \qquad x \in 0, L$$

Using (3.44), verify the completeness relation

$$\frac{2}{L} \sum_{0}^{\infty} \sin\left(\frac{n\pi x}{L}\right) \sin\left(\frac{n\pi x'}{L}\right) = \delta(x - x')$$

9.2 *Legendre polynomials from the recursion formula*

Calculate the Legendre polynomials $P_4(x)$ and $P_5(x)$ from the recursion formula

$$a_{k+2} = \frac{k(k+1) - l(l+1)}{(k+1)(k+2)} \, a_k$$

taking into account the normalization $P_l(1) = 1$.

9.3 *Legendre differential equation*

Show that the Legendre polynomials

$$P_l(x) = \frac{1}{2^l l} \frac{d^l}{dx^l} \left(x^2 - 1\right)^l \tag{9.44}$$

fulfill the differential equation $(x^2 - 1)\, P_l'' + 2\,x\, P_l' - l\,(l+1)\, P_l = 0$. To do this, differentiate $(l+1)$ times the equation

$$\left(x^2 - 1\right) \frac{d}{dx} \left(x^2 - 1\right)^l = 2\,l\,x \left(x^2 - 1\right)^l$$

9.4 *Normalization of the Legendre polynomials*

Show $(P_l, P_l) = 2/(2\,l + 1)$ with the help of (9.44) and partial integration. Note: The following applies $\int_{-1}^{1} dx\, (1 - x^2)^l = 2\,(2^l\, l)^2/(2\,l + 1)$

9.5 *Laplace equation in Cartesian and cylindrical coordinates*

Consider a separation approach for the Laplace equation $\Delta\,\Phi = 0$ in Cartesian and cylindrical coordinates. Determine the separation constants and the elementary solutions of the resulting ordinary differential equations.

Chapter 10

Cylindrically Symmetric Problems

With the help of the Legendre polynomials, we give the general cylindrically symmetric solution of the Laplace equation. Thereby a number of examples (point charge, homogeneously charged ring, conducting sphere in a homogeneous field) are presented.

General Cylindrically Symmetric Solution

We use the spherical coordinates r, θ and ϕ. In a cylindrically symmetric problem, the electrostatic potential Φ does not depend on ϕ. From the Laplace equation

$$\Delta\,\Phi(r,\theta) = 0 \quad \text{with} \quad \Phi(r,\theta) = \frac{U(r)}{r}\;P(\cos\theta) \tag{10.1}$$

follow (9.19) and (9.20) with $m = 0$. The differential equation (9.20) is solved by the Legendre polynomials

$$P = P_l(\cos\theta) \quad (l = 0, 1, 2, \ldots) \tag{10.2}$$

The remaining radial equation (9.19) is

$$\frac{r^2}{U}\,\frac{d^2U}{dr^2} = l\,(l+1) \tag{10.3}$$

Since the solution depends on l, we denote it by U_l. One notices that

$$U_l(r) = a_l\, r^{l+1} + \frac{b_l}{r^l} \tag{10.4}$$

is an solution. Since this solution contains two independent constants, it is the general solution of the differential equation of the second order.

Using the approach (10.1), we have obtained solutions of the form $\Phi = P_l\, U_l/r$. Since the Laplace equation is linear in Φ, every linear combination of this is also a solution (superposition principle):

$$\Phi(r,\theta) = \sum_{l=0}^{\infty} \left(a_l\, r^l + \frac{b_l}{r^{l+1}} \right) P_l(\cos\theta) \qquad \text{general cylindrically symmetric solution of the Laplace equation} \tag{10.5}$$

We argue that this is the *general* cylindrically symmetric solution: Due to the completeness of the Legendre polynomials, any cylindrically symmetric function can be expanded into these polynomials:

$$\Phi(r,\theta) = \sum_{l=0}^{\infty} \frac{A_l(r)}{r}\, P_l(\cos\theta) \tag{10.6}$$

We apply the Laplace operator to this, taking (9.20) into account,

$$\sum_{l=0}^{\infty} \left(\frac{d^2}{dr^2} - \frac{l\,(l+1)}{r^2} \right) A_l(r)\; P_l(\cos\theta) = 0 \tag{10.7}$$

We multiply this by P_n, integrate over θ and use the orthogonality of the Legendre polynomials. This gives us

$$\frac{d^2 A_n}{dr^2} - \frac{n(n+1)}{r^2}\, A_n(r) = 0 \tag{10.8}$$

The general solution of this differential equation is $A_n = a_n\, r^{n+1} + b_n\, r^{-n}$. Thus, (10.6) becomes (10.5). The argument (10.6)–(10.8) is an alternative derivation of (10.5), which does not contain any restrictions on Φ (such as the separation approach). However, it requires knowledge of the P_l and its properties (completeness and orthogonality) is a prerequisite.

Partial, linear differential equations usually have infinitely many solutions, $f_1, f_2, f_3, \ldots$. The general solution is then a linear combination $f = \sum a_n f_n$. The structures occurring here are therefore found in many areas.

The Laplace equation is solved by (10.5) with arbitrary coefficients a_l and b_l (this statement is somewhat restricted in the following paragraph). The coefficients are determined by the boundary conditions. In general, a solution is not determined by the differential equation alone but by the differential equation *and* the boundary conditions (and/or initial conditions if time appears as a variable).

If we calculate $\Delta\Phi$ for the expression (10.5), the terms with b_l lead to δ-functions and their derivatives at $r = 0$. Physically, this corresponds to point like charge distributions at $r = 0$. If $r = 0$ belongs to the volume V in which $\Delta\Phi = 0$ applies, then there are no such charges so that $b_l = 0$. If the volume extends to $r = \infty$, the terms with a_l and $l \geq 1$ lead to singularities. Then, $\Phi(\infty) = $ const. requires $a_l = 0$ for $l \geq 1$. In concrete applications, the Laplace equation is always valid only in partial ranges. If, namely, $\Delta\Phi = 0$ applies everywhere, then $\Phi \equiv 0$ (for $\Phi(\infty) = 0$) is the unambiguous and uninteresting solution.

For $\theta = 0$, one gets $P_l(1) = 1$. Thus, (10.5) becomes

$$\Phi(r, 0) = \sum_{l=0}^{\infty}\left(a_l\, r^l + \frac{b_l}{r^{l+1}}\right) = \sum_{l=0}^{\infty}\left(a_l\, z^l + \frac{b_l}{z^{l+1}}\right) \tag{10.9}$$

In cylindrically symmetric problems, the potential on the z-axis, i.e., $\Phi(r, 0)$, can often be found easily. Then, Equation (10.9) suffices to determine the coefficients a_l and b_l of the general solution.

Point Charge

For the example of a point charge at $\boldsymbol{r}_0$,

$$\varrho(\boldsymbol{r}) = q\,\delta(\boldsymbol{r} - \boldsymbol{r}_0) \tag{10.10}$$

we show how the well-known solution $\Phi = q/|\boldsymbol{r} - \boldsymbol{r}_0|$ can be written in the form (10.5). To do this, we choose the coordinate system so that the z-axis is parallel to $\boldsymbol{r}_0$ (Figure 10.1). The distance $|\boldsymbol{r} - \boldsymbol{r}_0|$ can be expressed by r_0, r and θ,

$$\Phi(r, \theta) = \frac{q}{|\boldsymbol{r} - \boldsymbol{r}_0|} = \frac{q}{\sqrt{r^2 + r_0^2 - 2\, r\, r_0 \cos\theta}} \tag{10.11}$$

The potential $\Phi(r, \theta)$ is cylindrically symmetric. The Laplace equation holds everywhere except for an infinitesimal range at $\boldsymbol{r}_0$. Then, $\Phi(r, \theta)$ is of the form (10.5). In the coordinates used, we omit the position of the point charge by restricting ourselves either to $r < r_0$ or to $r > r_0$. This means

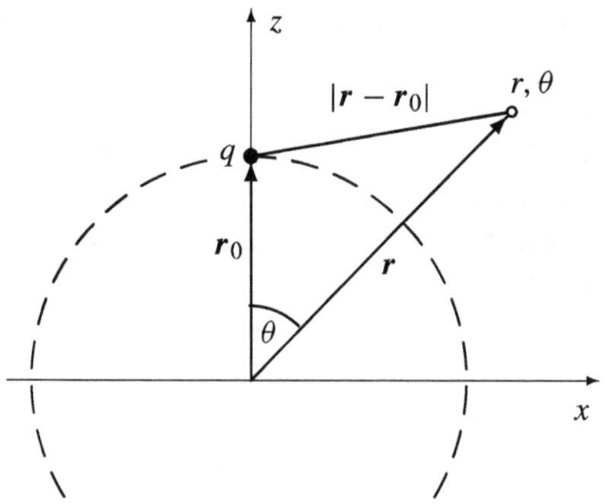

Figure 10.1 A charge q at $\boldsymbol{r}_0 = r_0\, \boldsymbol{e}_z$ generates at $\boldsymbol{r} = (r, \theta)$ the potential $\Phi(r,\theta) = q/|\boldsymbol{r} - \boldsymbol{r}_0|$. The expansion of $\Phi(r,\theta)$ into Legendre polynomials is set up separately for the ranges inside and outside the sphere $r = r_0$ (drawn as a dashed circle).

that the Laplace equation applies in two separate ranges V_1 and V_2:

$$\Delta\,\Phi(\boldsymbol{r}) = 0 \quad \text{for} \quad \begin{cases} V_1 = \{\boldsymbol{r}\ r < r_0\} \\ V_2 = \{\boldsymbol{r}\ r > r_0\} \end{cases} \tag{10.12}$$

For each of these ranges, we may use an expansion of the form (10.5). We determine the coefficients a_l and b_l of these expansions.

We expand the potential (10.11) on the z-axis and compare it with (10.9). At first, the following applies:

$$\big|\boldsymbol{r} - \boldsymbol{r}_0\big|_{\theta\,=\,0} = \sqrt{(r - r_0)^2} = \begin{cases} r_0 - r & (r < r_0) \\ r - r_0 & (r > r_0) \end{cases} \tag{10.13}$$

This gives us for the potential on the z-axis

$$\Phi(r,0) = \begin{cases} \dfrac{q}{r_0 - r} = \dfrac{q}{r_0}\,\dfrac{1}{1 - r/r_0} = \dfrac{q}{r_0}\displaystyle\sum_{l=0}^{\infty}\frac{r^l}{r_0^l} & (r < r_0) \\[2ex] \dfrac{q}{r - r_0} = \dfrac{q}{r}\,\dfrac{1}{1 - r_0/r} = \dfrac{q}{r}\displaystyle\sum_{l=0}^{\infty}\frac{r_0^l}{r^l} & (r > r_0) \end{cases} \tag{10.14}$$

The comparison with (10.9) results in

$$\begin{aligned} a_l &= q/r_0^{l+1}, \quad & b_l &= 0 & (r < r_0) \\ a_l &= 0, & b_l &= q\, r_0^l & (r > r_0) \end{aligned} \tag{10.15}$$

With these coefficients, (10.5) becomes

$$\Phi(r,\theta) = \frac{q}{|\boldsymbol{r}-\boldsymbol{r}_0|} = \begin{cases} \displaystyle\sum_{l=0}^{\infty} q\,\frac{r^l}{r_0^{\,l+1}}\; P_l(\cos\theta) & (r < r_0) \\ \displaystyle\sum_{l=0}^{\infty} q\,\frac{r_0^{\,l}}{r^{\,l+1}}\; P_l(\cos\theta) & (r > r_0) \end{cases} \tag{10.16}$$

With the definitions

$$r_> = \max\,(r, r_0) \quad \text{and} \quad r_< = \min\,(r, r_0) \tag{10.17}$$

we can summarize both expansions in the form:

$$\Phi(r,\theta) = \frac{q}{|\boldsymbol{r}-\boldsymbol{r}_0|} = q\sum_{l=0}^{\infty} \frac{r_<^{\,l}}{r_>^{\,l+1}}\; P_l(\cos\theta) \tag{10.18}$$

Generating function of the Legendre polynomials

The distance $|\boldsymbol{r}-\boldsymbol{r}_0|$ is determined by r, r_0 and $\theta = \sphericalangle(\boldsymbol{r},\boldsymbol{r}_0)$, Figure 10.1. The result (10.18) can therefore be used independently of the orientation of the coordinate system (previously $\boldsymbol{r}_0 \parallel \boldsymbol{e}_z$) for any two vectors $\boldsymbol{r}$ and $\boldsymbol{r}'$:

$$\boxed{\frac{1}{|\boldsymbol{r}-\boldsymbol{r}'|} = \frac{1}{\sqrt{r^2 + r'^2 - 2\,r\,r'\cos\theta}} = \sum_{l=0}^{\infty} \frac{r_<^{\,l}}{r_>^{\,l+1}}\; P_l(\cos\theta)} \tag{10.19}$$

Here, $\theta = \sphericalangle(\boldsymbol{r},\boldsymbol{r}')$, $r_> = \max\,(r, r')$ and $r_< = \min\,(r, r')$. In the intermediate expression, the cosine theorem $|\boldsymbol{r}-\boldsymbol{r}'| = \sqrt{r^2 + r'^2 - 2\,r\,r'\cos\theta}$ is used for the triangle drawn in Figure 10.1 (with $\boldsymbol{r}_0 = \boldsymbol{r}'$).

We multiply (10.19) by r, introduce $t = r'/r$ and consider the case $t < 1$:

$$\frac{1}{\sqrt{1 + t^2 - 2\,t\cos\theta}} = \sum_{l=0}^{\infty} c_l(\theta)\, t^l = \sum_{l=0}^{\infty} P_l(\cos\theta)\, t^l \quad (t < 1) \tag{10.20}$$

The second expression is the general form of a expansion into powers of t. As can be seen, the expansion coefficients c_l are equal to the Legendre polynomials P_l. The expansion of the left side of (10.20) to powers of t thus generates the Legendre polynomials. The left-hand side of (10.20) is therefore called *generating function* of the Legendre polynomials.

Homogeneously Charged Ring

We consider a charged circular ring with the radius a, Figure 10.2. Let the circle lie parallel to the x-y-plane and have the center $(x, y, z) = (0, 0, b)$. The charge q lies uniformly on the circumference $2\pi a$. Then, the charge density is

$$\varrho(\boldsymbol{r}) = \frac{q}{2\pi a}\,\delta(\rho - a)\,\delta(z - b), \quad \rho^2 = x^2 + y^2 \tag{10.21}$$

We first calculate the potential $\Phi(r, 0)$ for a point $\boldsymbol{r} = r\,\boldsymbol{e}_z$ on the z-axis. In Figure 10.2, right, the distance d to any point of the circular ring is shown. For the drawn triangle, the cosine theorem reads

$$d = \sqrt{r^2 + r_0^2 - 2\,r\,r_0\,\cos\alpha}, \quad r_0^2 = b^2 + a^2 \tag{10.22}$$

Each charge element dq gives the contribution dq/d to the potential. On the z-axis, the potential is therefore

$$\Phi(r, 0) = \int \frac{dq}{d} = \frac{q}{d} = \frac{q}{\sqrt{r^2 + r_0^2 - 2\,r\,r_0\,\cos\alpha}} \tag{10.23}$$

In $d = |\boldsymbol{r} - \boldsymbol{r}_0|$, the vectors $\boldsymbol{r}$ and $\boldsymbol{r}_0$ enclose the angle α with each other. Using (10.19), we obtain

$$\Phi(r, 0) = \frac{q}{|\boldsymbol{r} - \boldsymbol{r}_0|} = q \sum_{l=0}^{\infty} \frac{r_<^{\,l}}{r_>^{\,l+1}}\,P_l(\cos\alpha) \tag{10.24}$$

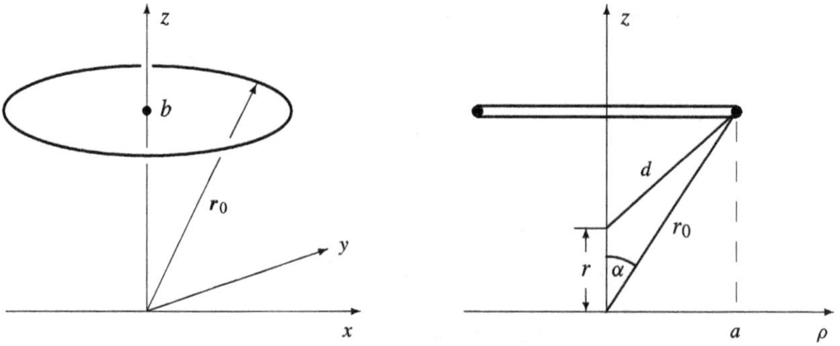

Figure 10.2 A homogeneously charged circular ring is shown on the left in perspective and on the right in a side view. All charge elements of the ring have the same distance d from a given point on the z-axis. Therefore, the potential at this point is equal to $\Phi = q/d$, where q is the total charge of the ring.

where $r_> = \max(r, r_0)$ and $r_< = \min(r, r_0)$. These are again two separate expansions; one applies to the range $r > r_0$ and the other to the range $r < r_0$.

From the comparison of (10.24) with (10.9), we can determine the coefficients a_l and b_l. We insert these coefficients into (10.5):

$$\Phi(r, \theta) = q \sum_{l=0}^{\infty} \frac{r_<^l}{r_>^{l+1}} \; P_l(\cos\alpha)\; P_l(\cos\theta) \tag{10.25}$$

The $P_l(\cos\alpha)$ are part of the coefficients; the $P_l(\cos\theta)$ indicate the angular dependence. The result is a expansion into powers of r/r_0 or r_0/r; it is useless at $r = r_0$ or close to this radius. For large distances, the leading terms are

$$\Phi(r, \theta) \overset{r\to\infty}{\longrightarrow} \frac{q}{r}\left(1 + \frac{r_0 \cos\alpha}{r}\; \cos\theta + \mathcal{O}\big(r_0^2/r^2\big)\right) \approx \frac{q}{r} \tag{10.26}$$

At a very large distance, the charge distribution therefore acts like a point charge at $r = 0$. The next term is classified as a dipole field in Chapter 12.

Conducting Sphere in Homogeneous Field

We consider a conducting sphere in a static electric field that is homogeneous at large distances. The field could be generated by the plates of a capacitor, Figure 10.3.

The sphere (radius R) represents a Faraday cage; the inside solution is $\Phi(\boldsymbol{r}) \equiv \Phi_0 = \text{const}$. We are looking for the potential outside the sphere. The potential must satisfy the Laplace equation

$$\Delta\Phi = 0 \quad \text{in } V = \{\boldsymbol{r} : |\boldsymbol{r}| > R\} \tag{10.27}$$

and the boundary conditions:

$$\Phi(R, \theta) = \Phi_0 \tag{10.28}$$

An external homogeneous field is given by $\boldsymbol{E} = E_0\, \boldsymbol{e}_z$. Far away from the sphere, this is the undisturbed field. For the potential, this means

$$\Phi(r, \theta) \overset{r\to\infty}{\longrightarrow} -E_0\, z + \text{const.} = -E_0\, r\, \cos\theta + \Phi_1 \tag{10.29}$$

In contrast to the examples considered so far, the potential does not approach a constant or zero für $r \to \infty$. This is due to the idealization

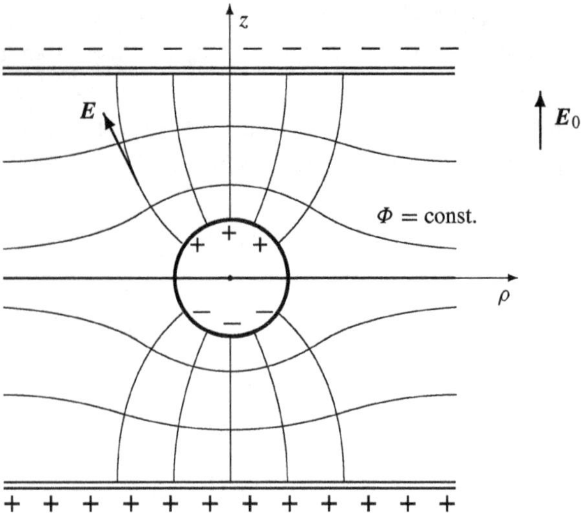

Figure 10.3 A plate capacitor generates a homogeneous electric field $\boldsymbol{E}_0$. This field induces charges on the metal surface of a sphere. The resulting electrostatic field $\boldsymbol{E}$ is sketched schematically.

"homogeneous field", i.e., to the assumption of an infinitely extended, constant field.

Due to the cylindrical symmetry, the potential is of the form (10.5). The coefficients a_l and b_l are determined by the boundary conditions (10.28) and by (10.29). From (10.28) follows

$$\Phi(R,\theta) = \sum_{l=0}^{\infty}\left(a_l\, R^l + \frac{b_l}{R^{l+1}}\right) P_l(\cos\theta) = \Phi_0 = \Phi_0\, P_0(\cos\theta) \quad (10.30)$$

Here, $P_0(\cos\theta) = 1$ was used. We multiply (10.30) by P_n, integrate over θ and use the orthogonality of the Legendre polynomials. From this we obtain

$$\begin{aligned} n = 0 \qquad & a_0 + b_0/R = \Phi_0, \qquad \text{also} \quad b_0 = R\,(\Phi_0 - a_0) \\ n \neq 0 \quad & a_n\, R^n + b_n/R^{n+1} = 0, \quad \text{thus} \quad b_n = -a_n\, R^{2n+1} \end{aligned} \quad (10.31)$$

The boundary condition (10.29) results in

$$\sum_{l=0}^{\infty}\left(a_l\, r^l + \frac{b_l}{r^{l+1}}\right) P_l(\cos\theta) \stackrel{r\to\infty}{\longrightarrow} -E_0\, r\, P_1(\cos\theta) + \Phi_1 \quad (10.32)$$

Here, $P_1(\cos\theta) = \cos\theta$ was used. We form the scalar product with P_n again and get

$$\begin{aligned}
&n = 0 \quad a_0 + b_0/r \overset{r\to\infty}{\longrightarrow} \Phi_1, && \text{also} \quad a_0 = \Phi_1 \\
&n = 1 \quad a_1\, r + b_1/r^2 \overset{r\to\infty}{\longrightarrow} -E_0\, r, && \text{also} \quad a_1 = -E_0 \\
&n > 1 \quad a_n\, r^n + b_n/r^{n+1} \overset{r\to\infty}{\longrightarrow} 0, && \text{also} \quad a_2 = a_3 = \cdots = 0
\end{aligned} \tag{10.33}$$

Thus, (10.31) becomes

$$b_0 = R\,(\Phi_0 - \Phi_1), \quad b_1 = E_0\, R^3, \quad b_2 = b_3 = \cdots = 0 \tag{10.34}$$

We insert the coefficients a_l and b_l into (10.5) and obtain

$$\Phi(r,\theta) = \Phi_1 + (\Phi_0 - \Phi_1)\,\frac{R}{r} - E_0\left(r - \frac{R^3}{r^2}\right)\cos\theta \quad (r \ge R) \tag{10.35}$$

The second term is the field of a possible total charge Q of the sphere,

$$Q = (\Phi_0 - \Phi_1)\, R \tag{10.36}$$

This field is without special interest. We therefore simplify the following discussion by assuming $Q = 0$; this is equivalent to the boundary condition $\Phi_1 = \Phi_0$. Thus, (10.35) becomes

$$\Phi(r,\theta) = \Phi_0 - E_0\, r\cos\theta \,+\, E_0\,\frac{R^3}{r^2}\,\cos\theta \tag{10.37}$$

The first term is an insignificant constant. The second term describes the homogeneous electric field applied from the outside, see (10.29). The third term is generated by the charges on the sphere.

We discuss the course of the equipotential surface $\Phi = \text{const}$. It follows from (10.37) that the equipotential surfaces $\Phi = \Phi_0$ consist of the sphere $r = R$ and the x-y-plane ($\theta = \pi/2$). In Figure 10.3, the sphere is shown as a circle and the x-y-plane as ρ-axis. The equipotential surfaces $\Phi = \Phi_0 \pm \delta\Phi$ (with small $\delta\Phi$) are then close above or below it. For $r \gg R$, the third term in (10.37) is small; thus, the equipotential surfaces are approximately the planes $z = \text{const}$.

We calculate the surface charge σ on the sphere:

$$\sigma = -\frac{1}{4\pi}\,\frac{\partial\Phi}{\partial n}\bigg|_{\mathrm{R}} = -\frac{1}{4\pi}\,\frac{\partial\Phi}{\partial r}\bigg|_{r=R} = \frac{3}{4\pi}\,E_0\,\cos\theta \tag{10.38}$$

These are induced charges that are caused (induced) by the external field. The external field exerts forces on the charges in the metal; this causes the free-moving charges (the electrons in the conduction band of the metal lattice). Specifically, the mobile electrons tend to move in the direction of the positively charged capacitor plate (Figure 10.3). As a result, the lower part of the sphere is negatively charged (excess of electrons), while the upper part is positively charged (deficit of electrons). The sum of these charges (10.38) is zero.

The charge density of the surface charge is given by

$$\varrho(r,\theta) = \sigma(\theta)\,\delta(r-R) \quad \text{with} \quad \sigma(\theta) = \frac{3}{4\pi}\,E_0\,\cos\theta \tag{10.39}$$

The total potential can be expressed in the form

$$\Phi(\boldsymbol{r}) = \Phi_0 + \Phi_{\text{hom}} + \Phi_\sigma \tag{10.40}$$

Here, $\Phi_{\text{hom}} = -E_0\,r\,\cos\theta$ is the field of charges on the capacitor plates (Figure 10.3), and Φ_σ is the field of the surface charges on the sphere. We compare (10.40) with the result calculated above ($\Phi(\boldsymbol{r}) \equiv \Phi_0$ for $r \le R$ and (10.37) for $r \ge R$) and obtain

$$\Phi_\sigma(r,\theta) = \begin{cases} E_0\,r\,\cos\theta = -\,\Phi_{\text{hom}} & (r \le R) \\ E_0\,(R^3/r^2)\cos\theta = \Phi_{\text{dip}} & (r \ge R) \end{cases} \tag{10.41}$$

The surface charges arrange themselves in such a way that the inner area $r \le R$ is field free; the metal surface completely shields the external field. This applies to any closed metal surface (Faraday cage, Chapter 7).

In the outer region $r \ge R$, the surface charges result in a *dipole field* $\Phi_\sigma = \Phi_{\text{dip}}$. Multipole fields (such as dipole field and others) are introduced in Chapter 12. This field can be written in the form

$$\Phi_{\text{dip}} = E_0\,\frac{R^3}{r^2}\,\cos\theta = \frac{\boldsymbol{p}\cdot\boldsymbol{r}}{r^3} \tag{10.42}$$

The *dipole moment* $\boldsymbol{p}$ of the sphere is proportional to the applied field:

$$\boldsymbol{p} = E_0\,R^3\,\boldsymbol{e}_z \tag{10.43}$$

The ratio $\alpha_{\text{e}} = p/E_0$ is called (electric) *polarizability*. For the conductive sphere, the following applies:

$$\alpha_{\text{e}} = R^3 \tag{10.44}$$

Exercises

10.1 *Homogeneously charged circular ring*

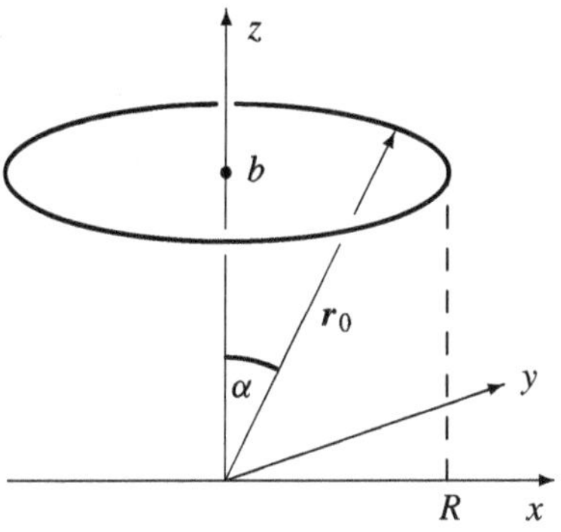

In cylindrical coordinates, the charge density of a circular ring (radius R, charge q) is

$$\varrho(\boldsymbol{r}) = \frac{q}{2\pi R}\,\delta(\rho - R)\,\delta(z - b)$$

Calculate the potential Φ on the z-axis using the integral formula (6.3). Express the result by the distance $|r\boldsymbol{e}_z - \boldsymbol{r}_0|$. Determine the potential $\Phi(r, \theta)$ in the whole space.

10.2 *Two parallel circular rings*

Two parallel circular rings (both with radius R) are homogeneously charged with q and $-q$. The circles are parallel to the x-y plane and have their centers at $(x, y, z) = (0, 0, b)$ and $(0, 0, -b)$, respectively. Calculate the electrostatic potential. Show that for $r \to \infty$ the leading contribution has the form of a dipole field.

10.3 *Homogeneously charged circular disk*

A circular disk (infinitesimal thin, radius R) is homogeneously charged (total charge q). Specify the charge density and calculate the potential on the symmetry axis. Then determine the electrostatic potential for the following cases:

(i) $r \ll R$ up to order r^2 and (ii) $r \gg R$ up to order $1/r^3$

10.4 *Homogeneously charged rotational ellipsoid*

A homogeneously charged rotational ellipsoid with the semi axes $a = b > c$ carries the total charge q. Use cylindrical coordinates (c-axis equals z-axis) and calculate the potential $\Phi(\rho, z)$ first on the z-axis. Determine from this the first two leading terms of the potential for large distances. Show that the potential in the interior is of the form

$$\Phi(\rho, z) = A + B\,z^2 + C\,\rho^2 \quad \text{(interior)} \tag{10.45}$$

and calculate the constants A, B and C. Start with the previously determined potential $\Phi(0, z)$ and with the Poisson equation. Calculate the field energy of the charge distribution.

Chapter 11

Spherical Harmonics

We present the general solution of the Laplace equation in spherical coordinates. For this purpose, the spherical harmonics are introduced. Their most important properties (orthonormalization, completeness, addition theorem) are discussed.

General Solution of the Laplace Equation

The Laplace equation

$$\Delta\,\Phi(r,\theta,\phi)=0 \quad \text{with} \quad \Phi(r,\theta,\phi)=\frac{U(r)}{r}\,P(\cos\theta)\,Q(\phi) \tag{11.1}$$

was reduced to equations (9.19)–(9.21):

$$\frac{d^2U}{dr^2}-\frac{\lambda}{r^2}\,U(r)=0 \tag{11.2}$$

$$\frac{d}{dx}\left(\left(1-x^2\right)\frac{dP}{dx}\right)+\left(\lambda-\frac{m^2}{1-x^2}\right)P(x)=0 \tag{11.3}$$

$$Q_m(\phi)=\exp(\mathrm{i}m\phi), \qquad m=0,\pm1,\pm2,\pm3\,\ldots \tag{11.4}$$

In Chapter 9, we had restricted ourselves to $m=0$ (cylindrical symmetry), and for $P(x)$, we obtained the Legendre polynomials $P_l(x)$. Now we let drop the restriction $m=0$.

For $m\neq0$, the differential equation (11.3) can be solved similarly as in Chapter 9: First, we use $P(x)=(1-x^2)^{m/2}\,T(x)$. The differentiation of the prefactor leads to terms that compensate the term $m^2/(1-x^2)$ in (11.3), see Exercise 11.1. What remains is a differential equation that only contains

T'', T', T and powers of x. This equation can be solved analogously to (9.24)–(9.30) by a power series approach. The convergence of the power series requires

$$\lambda = l(l+1), \quad l = 0, 1, 2, 3, \ldots, \quad m = 0, \pm 1, \pm 2, \ldots \pm l \tag{11.5}$$

The solutions are called *associated Legendre polynomials* P_l^m; they depend on l and m. They can be written in the closed form

$$\boxed{P_l^m(x) = \frac{(-)^m}{2^l\, l}\,\left(1-x^2\right)^{m/2}\,\frac{d^{\,l+m}}{dx^{l+m}}\,\left(x^2-1\right)^l} \tag{11.6}$$

The number m may adopt the values $0, \pm 1, \ldots, \pm l$. An explicit and simple construction of the $P_l^m(x)$ results from the operator method of quantum mechanics (Chapter 36 in [3]).

The dependence of the solution of $x = \cos\theta$ is given by (11.6) and that of ϕ by (11.4). We combine both parts to the *spherical harmonics* Y_{lm}:

$$\boxed{Y_{lm}(\theta,\phi) = \sqrt{\frac{2l+1}{4\pi}\,\frac{(l-m)}{(l+m)}}\;P_l^m(\cos\theta)\,\exp(im\phi)} \tag{11.7}$$

The prefactors in (11.6) and (11.7) are convention.[1] The spherical harmonics satisfy the differential equation

$$\left[\frac{1}{\sin\theta}\frac{\partial}{\partial\theta}\left(\sin\theta\,\frac{\partial}{\partial\theta}\right) + \frac{1}{\sin^2\theta}\,\frac{\partial^2}{\partial\phi^2}\right] Y_{lm}(\theta,\phi) = -l\,(l+1)\;Y_{lm}(\theta,\phi) \tag{11.8}$$

If one performs the ϕ-differentiation and takes into account $x = \cos\theta$, then one obtains (11.3) with $P = P_l^m$. Multiplied by $1/r^2$, the left-hand side of (11.8) is the angular part of the Laplace operator. It results in the term $-l\,(l+1)/r^2$, which then appears in the radial equation (11.2).

The two independent solutions of the radial equation (11.2) are r^{l+1} and r^{-l}. Thus, (11.1) leads to the solutions $r^l\,Y_{lm}$ and $r^{-l-1}\,Y_{lm}$. Since the

[1] We use the same conventions as Jackson [6]. Deviating from this prefactor $(-)^m$ in (11.6) is occasionally shifted to (11.7).

Laplace equation is linear, every linear combination is also a solution:

$$\Phi(r,\theta,\phi) = \sum_{l=0}^{\infty}\sum_{m=-l}^{m=+l}\left(a_{lm}\, r^l + \frac{b_{lm}}{r^{l+1}}\right) Y_{lm}(\theta,\phi) \qquad \text{general solution of } \Delta\Phi = 0 \tag{11.9}$$

The generality of this solution is based on the completeness of the spherical harmonics; we return to this in (11.24). The coefficients a_{lm} and b_{lm} can be chosen such that the potential Φ is real.

As in (10.5), the terms with b_{lm} in $\Delta\Phi$ lead to δ-functions and derivatives of δ-functions at $r = 0$. If $r = 0$ belongs to the volume V in which $\Delta\Phi = 0$ applies, these coefficients must vanish $b_{lm} = 0$. If the volume extends to $r = \infty$, the terms with a_{lm} and $l \geq 1$ lead to singularities. For $\Phi(\infty) = \text{const.}$, we must require $a_{lm} = 0$ for $l \geq 1$. A homogeneous electric field (for $r \to \infty$ an unrealistic idealization), however, has a coefficient $a_{10} \neq 0$, see (10.33).

In contrast to expansion (10.5), equation self does not require any symmetry of the problem. The expansion into spherical harmonics is particularly useful if the sum can be approximated by a few leading terms. In particular, this sum self leads to the multipole expansion (Chapter 12).

Properties of the Spherical Harmonics

We discuss some properties of the associated Legendre polynomials and the spherical harmonics.

For $m = 0$, the associated Legendre polynomials (11.6) reduce to the Legendre polynomials (9.34):

$$P_l^0(x) = P_l(x), \qquad Y_{l0}(\theta,\phi) = \sqrt{\frac{2l+1}{4\pi}}\; P_l(\cos\theta) \tag{11.10}$$

For $m = l$, the derivatives d^{2l}/dx^{2l} in (11.6) must be applied to $x^{2l} + \cdots x^{2l-2} + \cdots$. This results in a factor $(2l)$ so that

$$P_l^l(x) = (-)^l\,\frac{(2l)}{2^l\, l}\left(1-x^2\right)^{l/2} = (-)^l\,\frac{(2l)}{2^l\, l}\,\sin^l\theta \tag{11.11}$$

We note $\sqrt{1-x^2} = |\sin\theta|$; in the range $0 \leq \theta \leq \pi$ we may set $|\sin\theta| = \sin\theta$.

If m is positive, i.e., for $m = |m|$, (11.6) contains the prefactor $(\sin\theta)^{|m|}$. It is associated with the $(l+|m|)$th derivative of a polynomial of degree $2l$, i.e., with a polynomial of degree $l-|m|$:

$$P_l^m(x) = (\sin\theta)^{|m|} \cdot \text{polynomial}^{(l-|m|)}\big(\cos\theta\big) \tag{11.12}$$

For negative m, i.e., for $m=-|m|$, effectively only $l-|m|$ derivatives act on $(1-x^2)^l$ so that a factor $(1-x^2)^{|m|}$ survives; together with the prefactor $(1-x^2)^{-|m|/2}$ one obtains again form (11.12). In fact, the following applies:

$$P_l^{-m}(x) = (-)^m \,\frac{(l-m)}{(l+m)}\, P_l^m(x) \tag{11.13}$$

From this follows

$$Y_{lm}^*(\theta,\phi) = (-)^m \; Y_{l,-m}(\theta,\phi) \tag{11.14}$$

For the sake of clarity, we separate the two indices of the spherical function with a comma.

Orthonormalization

The differential equation (11.3) can be written in the form

$$D_{\text{op}}\, P_l^m = -l\,(l+1)\, P_l^m \quad \text{with} \quad D_{\text{op}} = \frac{d}{dx}\left(\left(1-x^2\right)\frac{d}{dx}\right) - \frac{m^2}{1-x^2} \tag{11.15}$$

Then, the orthogonality of the P_l^m respect to the index l can be proven as in (9.36)–(9.38); hereby m is fixed. Together with the value for $(P_l^m,\, P_l^m)$ we obtain

$$(P_l^m,\, P_{l'}^m) = \int_{-1}^{1} dx\; P_l^m(x)\, P_{l'}^m(x) \;=\; \frac{2}{2l+1}\,\frac{(l+m)}{(l-m)}\;\delta_{ll'} \tag{11.16}$$

The complex solutions $\exp(\pm i m\phi)$ are equivalent to $\sin(m\phi)$ and $\cos(m\phi)$. The cosine functions are orthogonal to each other:

$$\big(\cos(m\phi),\, \cos(m'\phi)\big) = \int_0^{2\pi} d\phi\; \cos(m\phi)\,\cos(m'\phi) = \pi\,\delta_{mm'} \tag{11.17}$$

The same applies to the sine functions; in addition, the cosine and sine functions are mutually orthogonal. To ensure the orthogonality for the

$Q_m = \exp(\mathrm{i}m\phi)$, we generalize definition (9.2) of the scalar product for complex functions:

$$(g, h) = (h, g)^* = \int_a^b dx\; g^*(x)\, h(x) \tag{11.18}$$

For real functions, this agrees with the original definition (9.2). With the generalized definition (11.18), we obtain

$$(Q_m, Q_{m'}) = \int_0^{2\pi} d\phi\, \left[\exp(\mathrm{i}m\phi)\right]^* \exp(\mathrm{i}m'\phi) = 2\pi\, \delta_{mm'} \tag{11.19}$$

The orthogonality of the spherical harmonics follows from (11.19) and (11.16):

$$\left(Y_{l'm'}, Y_{lm}\right) = \int_0^{2\pi} d\phi \int_{-1}^{1} d(\cos\theta)\;\, Y^*_{l'm'}(\theta, \phi)\; Y_{lm}(\theta, \phi) = \delta_{ll'}\, \delta_{mm'} \tag{11.20}$$

The prefactor in (11.7) is chosen such that the spherical harmonics are normalized. Occasionally, the angles θ and ϕ are combined to the solid angle $\Omega = (\theta, \phi)$. Then, (11.20) can also be expressed in the form

$$\left(Y_{l'm'}, Y_{lm}\right) = \int d\Omega\;\, Y^*_{l'm'}(\Omega)\; Y_{lm}(\Omega) = \delta_{ll'}\, \delta_{mm'} \tag{11.21}$$

where $d\Omega = d\phi\, d(\cos\theta)$.

Completeness

We consider integrable functions $f(\theta, \phi)$ that are defined in the range $0 \le \theta \le \pi$ and $0 \le \phi \le 2\pi$. Such function may be expanded into spherical harmonics:

$$f(\theta, \phi) = \sum_{l=0}^{\infty} \sum_{m=-l}^{+l} C_{lm}\; Y_{lm}(\theta, \phi) \tag{11.22}$$

We multiply by $Y^*_{l'm'}$, integrate over the angles and take the orthonormalization into account:

$$C_{lm} = \int_0^{2\pi} d\phi \int_{-1}^{1} d(\cos\theta)\;\, f(\theta, \phi)\; Y^*_{lm}(\theta, \phi) \tag{11.23}$$

If we put the expansion coefficients C_{lm} back into (11.22), we obtain the formal expression for the completeness of the spherical harmonics:

$$\sum_{l=0}^{\infty} \sum_{m=-l}^{+l} Y_{lm}^{*}(\theta', \phi')\, Y_{lm}(\theta, \phi) = \delta(\cos\theta - \cos\theta')\, \delta(\phi - \phi') \quad (11.24)$$

The properties shown here correspond to those discussed in Chapter 9, where the scalar product was modified according to (11.18). The infinite-dimensional, complete vector space with the scalar product (11.18) is called *Hilbert space*.

From the completeness of the spherical harmonics, it follows that self is a general solution of the Laplace equation. Due to this completeness, any potential $\Phi(r, \theta, \phi)$ can be written as the $\sum_{lm} A_{lm}(r)\, Y_{lm}/r$. Inserting this sum in the Laplace equation $\Delta\Phi = 0$ results in equation (11.2) for $A_{lm}(r)$. The general solution of this ordinary differential equation is $A_{lm} = a_{lm}\, r^{l+1} + b_{lm}/r^{l}$. Thus, one obtains (11.9) again. This solution procedure does not contain any restrictions (such as a separation approach).

Explicit expressions

The lowest associated Legendre polynomials are

$$P_0^0 = 1, \quad P_1^0 = x, \quad P_1^1 = -\sqrt{1-x^2}, \quad P_1^{-1} = \frac{1}{2}\sqrt{1-x^2} \quad (11.25)$$

We also give the spherical harmonics for $l = 0$ and $l = 1$,

$$Y_{00} = \frac{1}{\sqrt{4\pi}}, \quad Y_{10} = \sqrt{\frac{3}{4\pi}}\, \cos\theta, \quad Y_{1,\pm 1} = \mp\sqrt{\frac{3}{8\pi}}\, \sin\theta\, \exp(\pm \mathrm{i}\phi) \quad (11.26)$$

and for $l = 2$:

$$Y_{20} = \sqrt{\frac{5}{16\,\pi}}\, \left(3\cos^2\theta - 1\right)$$

$$Y_{2,\pm 1} = \mp\sqrt{\frac{15}{8\pi}}\, \cos\theta\, \sin\theta\, \exp(\pm \mathrm{i}\phi)$$

$$Y_{2,\pm 2} = \sqrt{\frac{15}{32\,\pi}}\, \sin^2\theta\, \exp(\pm 2\mathrm{i}\phi) \quad (11.27)$$

The quantities $r^l\ Y_{lm}$ can be expressed by Cartesian coordinates:

$$\begin{aligned} r\ Y_{10} &= \sqrt{3/4\pi}\ z \\ r\ Y_{1,\pm 1} &= \sqrt{3/8\pi}\ (\mp x - \mathrm{i}\, y) \\ r^2\ Y_{20} &= \sqrt{5/16\pi}\ \left(2z^2 - x^2 - y^2\right) \\ r^2\ Y_{2,\pm 1} &= \sqrt{15/8\pi}\ (\mp x - \mathrm{i}\, y)\, z \\ r^2\ Y_{2,\pm 2} &= \sqrt{15/32\pi}\ (x \pm \mathrm{i}\, y)^2 \end{aligned} \tag{11.28}$$

One sees immediately that applying the Laplace operator $\Delta = \partial_x^2 + \partial_y^2 + \partial_z^2$ to these expressions results in zero.

Addition Theorem

We want to expand $1/|\boldsymbol{r} - \boldsymbol{r}'|$ into spherical harmonics. The result yields addition theorem for spherical harmonics.

Any two position vectors are represented in spherical coordinates by $\boldsymbol{r} = (r, \theta, \phi)$ and $\boldsymbol{r}' = (r', \theta', \phi')$ as shown. Then the scalar product

$$\begin{aligned} \boldsymbol{r} \cdot \boldsymbol{r}' &= r\, r' \cos\varphi = x\, x' + y\, y' + z\, z' \\ &= r\, r' \big(\sin\theta \sin\theta' \cos(\phi' - \phi) + \cos\theta \cos\theta' \big) \end{aligned} \tag{11.29}$$

the distance $|\boldsymbol{r} - \boldsymbol{r}'|$ and

$$\frac{1}{|\boldsymbol{r} - \boldsymbol{r}'|} = \frac{1}{\sqrt{r^2 + r'^2 - 2\boldsymbol{r} \cdot \boldsymbol{r}'}} \tag{11.30}$$

Functions of r, θ, ϕ and r', θ', ϕ'. Figure 11.1 shows that these quantities are also functions of r, r' and the intermediate angle φ. Correspondingly, (11.30) can be calculated into Legendre polynomials $P_l(\cos\varphi)$ or into the spherical harmonics $Y_{lm}(\theta, \phi)$ and $Y_{lm}(\theta', \phi')$ can be expanded as

$$\begin{aligned} \frac{1}{|\boldsymbol{r} - \boldsymbol{r}'|} &= \sum_{l=0}^{\infty} \frac{r_<^l}{r_>^{l+1}}\ P_l(\cos\varphi) \\ &= \sum_{l,l',m,m'} A_{ll'mm'}(r, r')\ Y^*_{l'm'}(\theta', \phi')\ Y_{lm}(\theta, \phi) \end{aligned} \tag{11.31}$$

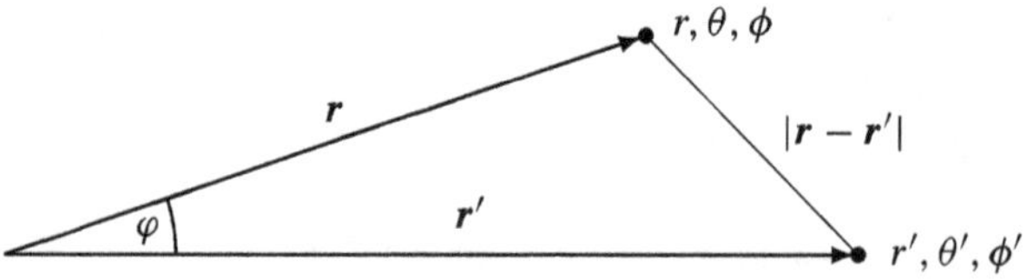

Figure 11.1 The distance $|\boldsymbol{r}-\boldsymbol{r}'|$ is a function of r and r' and of the angles θ, ϕ and θ', ϕ'. It can be expanded into the spherical harmonics depending on these angles. Alternatively, the distance can be understood as a function of r and r' and the angle φ. It can therefore be expanded into the Legendre polynomials $P_l(\cos\varphi)$. Comparing both expansions yields the addition theorem of the spherical harmonics.

The expansion into the Legendre polynomials is known from (10.19). With $\{Y_{lm}\}$, $\{Y^*_{lm}\}$ is also a CONS; we can therefore use the θ', ϕ'-dependence also into the conjugate complex spherical harmonics.

We apply the Laplace operator to (11.31), where we use (11.8) for the angle differentiation of the spherical harmonics:

$$\Delta \frac{1}{|\boldsymbol{r}-\boldsymbol{r}'|} = \sum_{l,l',m,m'} \left(\frac{1}{r}\frac{d^2}{dr^2}\,r - \frac{l\,(l+1)}{r^2}\right) A_{ll'mm'}(r,r')\; Y^*_{l'm'}(\Omega')\; Y_{lm}(\Omega) \tag{11.32}$$

The left-hand side is equal to $-4\pi\,\delta(\boldsymbol{r}-\boldsymbol{r}')$. We express this delta function in spherical coordinates r, θ and ϕ and use the completeness relation (11.24) for the spherical harmonics:

$$\begin{aligned}\Delta \frac{1}{|\boldsymbol{r}-\boldsymbol{r}'|} &= -\frac{4\pi}{r^2}\,\delta(r-r')\;\delta(\cos\theta-\cos\theta')\;\delta(\phi-\phi')\\ &= -\frac{4\pi}{r^2}\,\delta(r-r')\sum_{l,m} Y^*_{lm}(\theta',\phi')\;Y_{lm}(\theta,\phi)\end{aligned} \tag{11.33}$$

Since the spherical harmonics are linearly independent, the coefficients in the expansions (11.32) and (11.33) must coincide. The comparison results in

$$A_{ll'mm'}(r,r') = A_{lm}(r,r')\;\delta_{ll'}\;\delta_{mm'} \tag{11.34}$$

and

$$\left(\frac{1}{r}\frac{d^2}{dr^2}\,r - \frac{l\,(l+1)}{r^2}\right) A_{lm}(r,r') = -\frac{4\pi}{r^2}\,\delta(r-r') \tag{11.35}$$

For $r \neq r'$, the differential equation has the solutions r^l and r^{-l-1}. These solutions do not depend on m, i.e., $A_{lm} = A_l$. Since (11.31) has a finite value for $r = 0$ and $r = \infty$, only the solutions $A_l = a_l\, r^l$ for $r < r'$ and $A_l = b_l/r^{l+1}$ for $r > r'$ in question. According to (11.35), the second derivative of $A_l(r)$ at r' results in a δ function; then the first has a jump and A_l itself is continuous. The continuity at $r = r'$ results in $a_l\, r'^l = b_l/r'^{l+1}$, i.e.,

$$A_{lm}(r, r') = A_l(r, r') = \begin{cases} a_l\, r^l & (r < r') \\ a_l\, r'^{2l+1}/r^{l+1} & (r > r') \end{cases} \tag{11.36}$$

We multiply (11.35) by r and integrate both sides of $r = r' - \varepsilon$ to $r = r' + \varepsilon$. The right-hand side results in $-4\pi/r'$. On the left-hand side, the contribution of $l\,(l+1)/r^2$-term for $\varepsilon \to 0$. The first term $\int dr\, (r\, A_l)''$ results in $(r\, A_l)'_{r'+\varepsilon} - (r\, A_l)'_{r'-\varepsilon}$. This gives us

$$a_l\, r'^{2l+1} \frac{-l}{r'^{l+1}} - a_l\, (l+1)\, r'^l = -\frac{4\pi}{r'} \tag{11.37}$$

From this follows $a_l = (4\pi/(2l+1))/r'^{l+1}$. With this and with (11.36), (11.34) becomes

$$A_{ll'mm'}(r, r') = A_l(r, r')\, \delta_{ll'}\, \delta_{mm'} = \frac{4\pi}{2l+1} \frac{r_<^l}{r_>^{l+1}}\, \delta_{ll'}\, \delta_{mm'} \tag{11.38}$$

Here, as usual, $r_> = \max(r, r')$ and $r_< = \min(r, r')$. We insert (11.38) into (11.31):

$$\boxed{\frac{1}{|\boldsymbol{r} - \boldsymbol{r}'|} = 4\pi \sum_{l=0}^{\infty} \sum_{m=-l}^{+l} \frac{1}{2l+1} \frac{r_<^l}{r_>^{l+1}}\, Y_{lm}^*(\theta', \phi')\, Y_{lm}(\theta, \phi)} \tag{11.39}$$

The comparison with the expansion into Legendre polynomials (first expansion in (11.31)) results in the *addition theorem* for spherical functions:

$$P_l(\cos\varphi) = \frac{4\pi}{2l+1} \sum_{m=-l}^{+l} Y_{lm}^*(\theta', \phi')\, Y_{lm}(\theta, \phi) \tag{11.40}$$

where φ is the angle between the angles formed by θ, ϕ and θ', ϕ' given directions.

Exercises

11.1 *Associated Legendre polynomials*

Use the approach $P(x) = (1 - x^2)^{m/2}\, T(x)$ in the differential equation

$$\left((1 - x^2)\, P'(x)\right)' + \left[\lambda - \frac{m^2}{1 - x^2}\right] P(x) = 0$$

for the associated Legendre polynomials. What is the resulting differential equation for $T(x)$? Solve this equation using a power series approach.

11.2 *Expansion of the scalar product into spherical harmonics*

Set the Cartesian components of the two position vectors $\boldsymbol{r}$ and $\boldsymbol{r}'$ by spherical harmonics. Use this to calculate the scalar product $\boldsymbol{r} \cdot \boldsymbol{r}'$. Verify the result using the addition theorem for spherical harmonics.

11.3 *Spherical shell with given potential*

The potential is given on a spherical shell (radius R):

$$\Phi(R, \theta, \phi) = \Phi_0 \sin\theta \cos\phi$$

There are no charges in the ranges $r > R$ and $r < R$. For $r \to \infty$, the electric field $\boldsymbol{E} = E_0\, \boldsymbol{e}_z$. Determine the potential $\Phi(r, \theta, \phi)$ inside and outside the sphere.

Chapter 12

Multipole Expansion

We consider a static, localized charge distribution:

$$\varrho(\boldsymbol{r}) = \varrho(r, \theta, \phi) = \begin{cases} \text{arbitrary} & (r < R_0) \\ 0 & (r > R_0) \end{cases} \tag{12.1}$$

In the range $r > R_0$, the electrostatic potential Φ can be expended into powers of R_0/r. This multipole expansion is derived and discussed in the following.

We start from (6.3) and use expansion (11.39) for the range $r > R_0$:

$$\begin{aligned} \Phi(\boldsymbol{r}) &= \int d^3r' \, \frac{\varrho(\boldsymbol{r}')}{|\boldsymbol{r} - \boldsymbol{r}'|} \\ &= \int d^3r' \, \varrho(\boldsymbol{r}') \sum_{l,m} \frac{4\pi}{2l+1} \, \frac{r'^l}{r^{l+1}} \, Y_{lm}^*(\theta', \phi') \, Y_{lm}(\theta, \phi) \end{aligned} \tag{12.2}$$

The vectors $\boldsymbol{r}$, $\boldsymbol{r}'$ and the distance $|\boldsymbol{r} - \boldsymbol{r}'|$ are sketched in Figure 12.1. Due to (12.1), only values with $r' \le R_0$ contribute to the integral. Due to $r > R_0$, $r_> = r$ and $r_< = r'$ were set in (11.39). With the *spherical multipole moments*

$$\boxed{q_{lm} = \sqrt{\frac{4\pi}{2l+1}} \int d^3r' \, \varrho(\boldsymbol{r}') \, r'^l \, Y_{lm}^*(\theta', \phi') \quad \text{multipole moment}} \tag{12.3}$$

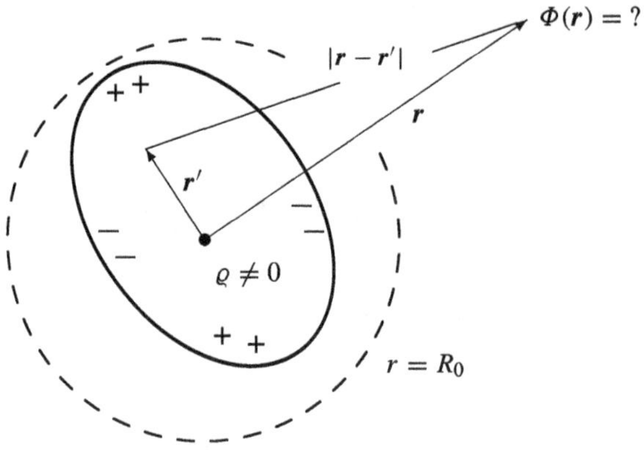

Figure 12.1 A charge distribution (here elliptically bounded) lies within the sphere $r = R_0$. In the range $r > R_0$, the potential $\Phi(\boldsymbol{r})$ is expanded into powers of R_0/r. The expansion coefficient at $1/r^{l+1}$ is proportional to the spherical function $Y_{lm}(\theta, \phi)$ and to the multipole moment q_{lm}.

(12.2) becomes

$$\Phi(\boldsymbol{r}) = \sum_{l=0}^{\infty} \sum_{m=-l}^{+l} \sqrt{\frac{4\pi}{2l+1}} \frac{q_{lm}}{r^{l+1}} Y_{lm}(\theta, \phi) \quad (r > R_0) \tag{12.4}$$

In the range $r > R_0$, Φ is solution of the Laplace equation and therefore of the form (11.9). Due to $q_{lm} \propto R_0^l$, (12.4) is a expansion into powers of R_0/r. This systematic expansion is called *multipole expansion.* Occasionally, the q_{lm} are also defined without the prefactor $4\pi/(2l+1)^{1/2}$, for example, in [6] or in the 2nd edition of this book.

The first terms of the multipole expansion (12.4) are

$$\Phi(\boldsymbol{r}) = \frac{q_{00}}{r} + \frac{q_{10}}{r^2} \cos\theta \mp \frac{1}{\sqrt{2}} \frac{q_{1,\pm 1}}{r^2} \sin\theta \exp(\pm i\phi) \pm \cdots \tag{12.5}$$

If the distance r is sufficiently large, even the lowest non-vanishing terms can already provide a good approximation for $\Phi(\boldsymbol{r})$ result. This means a significant simplification since the function $\varrho(r, \theta, \phi)$ is effectively characterized by a few numbers (the lowest multipole moments) can be characterized.

In the cylindrically symmetric case with $\varrho = \varrho(r, \theta)$ applies $q_{lm} = q_{l0}\,\delta_{m0}$ with

$$q_{l0} = \int d^3r\, \varrho(r, \theta)\, r^l\; P_l(\cos\theta) \tag{12.6}$$

Specifically for a homogeneously charged sphere (charge density ϱ_0, radius R), we obtain from this

$$q_{l0} = 2\pi\,\varrho_0 \int_{-1}^{1} d\cos\theta \int_0^R dr\; r^{l+2}\; P_l(\cos\theta) = \frac{4\pi\,\varrho_0\, R^3}{3}\,\delta_{l0} = q\,\delta_{l0} \tag{12.7}$$

We have used the orthonormalization (9.39) of the Legendre polynomials has been utilized. In this case, the expansion (12.5) already stops after the first term and gives the exact result $\Phi = q/r$ with the total charge q.

Cartesian Representation

For the lowest multipole moments ($l \le 2$), one often uses Cartesian coordinates instead of spherical coordinates. For this purpose, a slightly different form of multipole expansion is needed.

The distance $|\boldsymbol{r}-\boldsymbol{r}'| = (x_1-x_1')^2+(x_2-x_2')^2+(x_3-x_3')^{21/2}$ depends on the Cartesian coordinates x_1, x_2, x_3 and x_1', x_2', x_3'. We expand the inverse distance into a Taylor series into powers of x_1', x_2' and x_3':

$$\frac{1}{|\boldsymbol{r}-\boldsymbol{r}'|} = \frac{1}{r} - \sum_{i=1}^{3} x_i' \,\frac{\partial}{\partial x_i}\,\frac{1}{r} + \frac{1}{2}\sum_{i,j=1}^{3} x_i'\,x_j'\;\frac{\partial}{\partial x_i}\,\frac{\partial}{\partial x_j}\,\frac{1}{r} + \cdots \tag{12.8}$$

The occurring derivatives were calculated according to

$$\left[\frac{\partial}{\partial x_i'}\,\frac{1}{|\boldsymbol{r}-\boldsymbol{r}'|}\right]_{r'=0} = -\left[\frac{\partial}{\partial x_i}\,\frac{1}{|\boldsymbol{r}-\boldsymbol{r}'|}\right]_{r'=0} = -\frac{\partial}{\partial x_i}\,\frac{1}{r}. \tag{12.9}$$

Due to $\Delta(1/r) = 0$ in the range $r > R_0$, we can write the last term in (12.8) as follows:

$$\frac{1}{|\boldsymbol{r}-\boldsymbol{r}'|} = \frac{1}{r} - \sum_{i=1}^{3} x_i' \,\frac{\partial}{\partial x_i}\,\frac{1}{r} + \frac{1}{2}\sum_{i,j=1}^{3} \left(x_i'\,x_j' - r'^2\,\frac{\delta_{ij}}{3}\right)\frac{\partial}{\partial x_i}\,\frac{\partial}{\partial x_j}\,\frac{1}{r} + \cdots \tag{12.10}$$

This added term ultimately simplifies the result.

We introduce the following Cartesian multipole moments:

$$q = \int d^3r' \; \varrho(\boldsymbol{r}') \qquad \text{(charge)} \tag{12.11}$$

$$p_i = \int d^3r' \; x_i' \; \varrho(\boldsymbol{r}') \qquad \text{(dipole moment)} \tag{12.12}$$

$$Q_{ij} = \int d^3r' \left(3x_i' x_j' - r'^2 \, \delta_{ij}\right) \varrho(\boldsymbol{r}') \qquad \text{(quadrupole moment)} \tag{12.13}$$

Under orthogonal transformations, q is a scalar, p_i is a vector and Q_{ij} is a second rank tensor.

We multiply (12.10) by $\varrho(\boldsymbol{r}') \, d^3r'$ and carry out perform the integration; this results in the potential $\Phi(\boldsymbol{r})$. For the integrals that occur, we use (12.11)–(12.13):

$$\Phi(\boldsymbol{r}) = \int d^3r' \, \frac{\varrho(\boldsymbol{r}')}{|\boldsymbol{r} - \boldsymbol{r}'|} = \frac{q}{r} - \sum_{i=1}^{3} p_i \, \frac{\partial}{\partial x_i} \frac{1}{r} + \frac{1}{6} \sum_{i,j=1}^{3} Q_{ij} \, \frac{\partial}{\partial x_i} \frac{\partial}{\partial x_j} \frac{1}{r} + \cdots \tag{12.14}$$

We carry out the differentiations and obtain

$$\boxed{\Phi(\boldsymbol{r}) = \frac{q}{r} + \frac{\boldsymbol{p} \cdot \boldsymbol{r}}{r^3} + \frac{1}{2} \sum_{i,j=1}^{3} Q_{ij} \, \frac{x_i \, x_j}{r^5} + \cdots \qquad (r > R_0)} \tag{12.15}$$

Just like (12.4), this is an expansion into powers of R_0/r. The explicitly listed contributions are the *monopole*, the *dipole* and the *quadrupole field*.

To evaluate (12.14), we used $\partial(1/r)/\partial x_i = -x_i/r^3$ and

$$\frac{\partial}{\partial x_i} \frac{\partial}{\partial x_j} \frac{1}{r} = \frac{\partial}{\partial x_i} \frac{\partial}{\partial x_j} \frac{1}{\sqrt{x_1^2 + x_2^2 + x_3^2}} = 3 \, \frac{x_i \, x_j}{r^5} - \frac{\delta_{ij}}{r^3} \tag{12.16}$$

Since the trace of the matrix (Q_{ij}) disappears,

$$\sum_{i=1}^{3} Q_{ii} \overset{(12.13)}{=} 0 \tag{12.17}$$

the last term in (12.16) does not lead to a contribution in (12.15). We have achieved the freedom of the quadrupole tensor by adding the term with δ_{ij} in (12.10). This simplified the result (12.15).

Comparison between spherical and Cartesian multipoles

Expansions (12.4) and (12.15) give successively for large r the terms of order $1/r$, $1/r^2$, $1/r^3$ .... Both expansions are equivalent to each other. By comparing (12.3) with (12.11)–(12.13), it can be seen that the Cartesian multipole moments q, p_i and Q_{ij} are equivalent to the spherical multipole moments q_{00}, q_{1m} and q_{2m}. From (12.3) with (11.14) and (11.28) we obtain

$$q_{00} = \int d^3r' \, \varrho(\boldsymbol{r}') = q \tag{12.18}$$

$$q_{10} = \int d^3r' \, z' \, \varrho(\boldsymbol{r}') = p_3 \tag{12.19}$$

$$q_{1,\pm 1} = \frac{1}{\sqrt{2}} \int d^3r' \, (\mp x' + \mathrm{i}\, y') \, \varrho(\boldsymbol{r}') = \frac{\mp p_1 + \mathrm{i}\, p_2}{\sqrt{2}} \tag{12.20}$$

$$q_{20} = \frac{1}{2} \int d^3r' \, (3\, z'^2 - r'^2) \, \varrho(\boldsymbol{r}') = \frac{Q_{33}}{2} \tag{12.21}$$

For a given l, there are $2l + 1$ components q_{lm}. For $l = 0$ and $l = 1$, there are the same number of Cartesian components, namely, one q and three p_i. In contrast, the five quantities q_{2m} are initially opposed by the $3 \times 3 = 9$ quantities Q_{ij}. Since Q_{ij} is symmetric, only 6 of the 9 quantities are independent. The trace freedom (12.17) reduces this further to five independent quantities.

For higher terms, more and more Cartesian components are used. For example, the Cartesian octupole would already have 3 indices and 27 components. This is an unnaturally complicated representation; in fact, there are only 7 quantities (the seven q_{3m}) which are independent. For higher multipoles, the spherical expansion (12.4) is therefore much simpler and more elegant.

Dependence on the coordinate system

The multipole moments generally depend on the origin and the orientation of the coordinate system (CS) in which they are calculated. The origin is usually placed in the area of the charge distribution so that charge distribution R_0 in (12.1) is as small as possible because then the expansion converges best to the powers of R_0/r.

The charge (multipole q_{00}) is independent of the origin and of the the orientation of CS. In general, the lowest non-vanishing multipole moment is independent of the choice of the origin of the CS.

If the lowest moment is a dipole, then the dipole moment $\boldsymbol{p}$ is independent of the CS and therefore the characteristic of the charge distribution. For example, for a point charge q at $\boldsymbol{r}_0$, the dipole moment $\boldsymbol{p} = q\,\boldsymbol{r}_0$ depends on the CS because a displacement of the CS changes $\boldsymbol{r}_0$. For two point charges, q for $\boldsymbol{r}_1$ and $-q$ for $\boldsymbol{r}_2, \boldsymbol{p} = q\,(\boldsymbol{r}_1 - \boldsymbol{r}_2)$, and, on the other hand, invariant against over shifts of the CS.

The symmetric quadrupole tensor Q_{ij} can be realized by a suitable be brought to diagonal form by a suitable rotation of the CS (principal axis transformation). Due to its trace freedom, it is then characterized by two diagonal elements. With rotational symmetry around one axis, this is further reduced to *one* quantity; a rotationally symmetric charge distribution therefore has *the* quadrupole moment. Of the spherical multipole moments q_{lm}, only $q_{20} = Q_{33}/2$ is not equal to zero.

Point Multipole

The potential (12.4) was calculated for the range $r > R_0$ in which the charge distribution (12.1) disappears. We now consider this potential now without the restriction $r > R_0$. The potential is of the form (11.9) with $a_{lm} = 0$ and $b_{lm} = q_{lm}\sqrt{4\pi/(2l+1)}$. It is therefore the solution of the Laplace equation with the exception of a neighborhood of the point $r = 0$. The charge density $\varrho = -\Delta$ associated with this potential $\Phi/4\pi$ is therefore localized at $r = 0$.

We repeat this train of thoughts for the example of the homogeneously charged sphere, (6.3) and (12.6). The multipole expansion results in $\Phi = q/r$ for $r > R$, where q is the total charge and R is the radius of the sphere. If we now consider the solution $\Phi = q/r$ in the entire space, then $\Delta\Phi = 0$ for $r \neq 0$. From $\Delta(q/r) = -4\pi\varrho$, we get $\varrho = q\,\delta(\boldsymbol{r})$, i.e., the charge density of a point charge at $r = 0$.

The generalization of this consideration leads to *point multipoles*, i.e., point-like charge distributions that generate the corresponding multipole fields. The point dipole is discussed in the following.

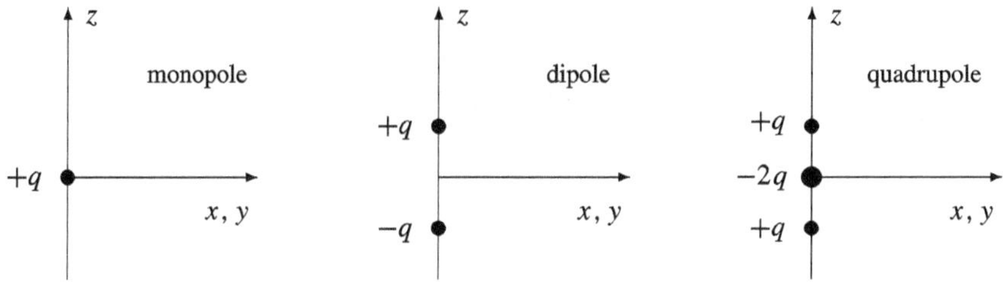

Figure 12.2 A point charge (left) has only one monopole moment; all other multipole moments disappear. In the middle image, you can see a point dipole of strength p is obtained if the distance $2a$ between the two charges approaches zero, and the charges according to $q = p/2a$ are increased. In the configuration shown on the right, the monopole and dipole moments disappear. If the distance a between the charges ($q \propto 1/a^2$) approaches zero, a point quadrupole is obtained.

To introduce the point dipole, we assume two point charges of the strength $q = \pm p/(2a)$ at a distance of $2a$ (Figure 12.2). The charge distribution

$$\varrho(\boldsymbol{r}) = \frac{p}{2a}\Big(\delta(\boldsymbol{r} - a\boldsymbol{e}_z) - \delta(\boldsymbol{r} + a\boldsymbol{e}_z)\Big) \tag{12.22}$$

has the dipole moment $\boldsymbol{p} = p\,\boldsymbol{e}_z$, which is independent of a. We now let a approach zero:

$$\varrho(\boldsymbol{r}) = \lim_{a\to 0} \frac{p}{2a}\Big(\delta(\boldsymbol{r}-a\boldsymbol{e}_z)-\delta(\boldsymbol{r}+a\boldsymbol{e}_z)\Big) = -p\,\frac{\partial}{\partial z}\,\delta(\boldsymbol{r}) = -(\boldsymbol{p}\cdot\boldsymbol{\nabla})\,\delta(\boldsymbol{r}) \tag{12.23}$$

For this we evaluate (6.3):

$$\begin{aligned}\Phi(\boldsymbol{r}) &= -\int d^3r'\,\frac{(\boldsymbol{p}\cdot\boldsymbol{\nabla}')\,\delta(\boldsymbol{r}')}{|\boldsymbol{r}-\boldsymbol{r}'|} \stackrel{\text{p.i.}}{=} \int d^3r'\,\delta(\boldsymbol{r}')\,\big(\boldsymbol{p}\cdot\boldsymbol{\nabla}'\big)\,\frac{1}{|\boldsymbol{r}-\boldsymbol{r}'|}\\ &= -\int d^3r'\,\delta(\boldsymbol{r}')\,(\boldsymbol{p}\cdot\boldsymbol{\nabla})\,\frac{1}{|\boldsymbol{r}-\boldsymbol{r}'|} = -\,(\boldsymbol{p}\cdot\boldsymbol{\nabla})\,\frac{1}{r} = \frac{\boldsymbol{p}\cdot\boldsymbol{r}}{r^3}\end{aligned} \tag{12.24}$$

Here, $\boldsymbol{\nabla}'$ acts on $\boldsymbol{r}'$, and $\boldsymbol{\nabla}$ acts on $\boldsymbol{r}$. The charge density (12.23) is point-like and results in a pure dipole field (12.24). (It can also be easily verified that all other multipole moments disappear.) Thus, (12.23) is the charge distribution of a *point dipole* with the dipole moment $\boldsymbol{p}$.

We calculate the electric field of the point dipole:

$$E(r) = -\operatorname{grad}\left(\frac{p \cdot r}{r^3}\right) = \frac{3\,r\,(p \cdot r) - p\,r^2}{r^5} \quad \text{(point dipole, } r \neq 0) \tag{12.25}$$

The differentiation was calculated according to $\nabla(1/r^3) = -3\,r/r^5$. This is only for $r \neq 0$ permissible because in singular places such derivations can lead to distributions. In Exercise 12.2, the shape of the field at $r = 0$ is examined in more detail. One possibility for the regularization of such singularities is to start from a suitable extended assuming a suitable extended charge distribution. For the dipole field, one could take the charge density (10.39), for example, which has only one dipole moment (i.e., no other multipole moments). In this case, the field is well defined everywhere. One then determines the limiting value of this field if the extent of the charge distribution (at constant dipole moment) approaches zero.

For the charge distribution shown in Figure 12.2, right part, we obtain for $a \to 0$ and $q = Q_{33}/4a^2$ with fixed Q_{33} the charge distribution of a point quadrupole:

$$\varrho(r) = \frac{Q_{33}}{4}\,\delta''(z)\,\delta(y)\,\delta(x) \tag{12.26}$$

Just like a point charge, point multipoles are idealizations of finite charge distributions. According to (12.4) and (12.15), the field outside a charge distribution is a sum of multipole fields; for the exterior region, the charge distribution can therefore be effectively be replaced by points multipoles.

In simple cases, finite charge distributions in the outer region also have a exterior region and have a pure multipole field. For example, the field of the homogeneously charged sphere in the external region is a monopole field; it is identical to that of a point charge. The upper surface charge (10.39) of the polarized metal sphere results in a dipole field in the outer region (10.41).

Energy in the External Field

We calculate the potential energy of a charge distribution in a external electric field $E_{\text{ext}} = -\operatorname{grad} \Phi_{\text{ext}}$. The index "ext" indicates that the field of the charge distribution itself is not contained in $\boldsymbol{E}_{\text{ext}}$ which is included. The charge distribution should be rigid, i.e., it shall not change under the induced of the external field.

The charge distribution sketched in Figure 12.3 has a limited extension:

$$\varrho(\boldsymbol{r}) = \varrho(\boldsymbol{r}_0 + \boldsymbol{r}') = \widetilde{\varrho}(\boldsymbol{r}') = \begin{cases} \text{arbitrary} & (r' < R_0) \\ 0 & (r' > R_0) \end{cases} \tag{12.27}$$

You will choose $\boldsymbol{r}_0$ so that R_0 is as small as possible. With $\widetilde{\varrho}$ we denote the charge distribution in a CS with $\boldsymbol{r}_0$ as the origin.

The external field changes only slightly in the region of the charge distribution. Then, we can assume $\Phi_{\text{ext}}(\boldsymbol{r})$ by a approximate Taylor expansion:

$$\Phi_{\text{ext}}(\boldsymbol{r}) = \Phi_{\text{ext}}(\boldsymbol{r}_0 + \boldsymbol{r}') = \Phi_{\text{ext}}(\boldsymbol{r}_0) + \sum_i \frac{\partial \Phi_{\text{ext}}}{\partial x_i} x_i'$$

$$+ \sum_{i,j} \frac{\partial^2 \Phi_{\text{ext}}}{\partial x_i \, \partial x_j} \frac{x_i' x_j'}{2} + \cdots$$

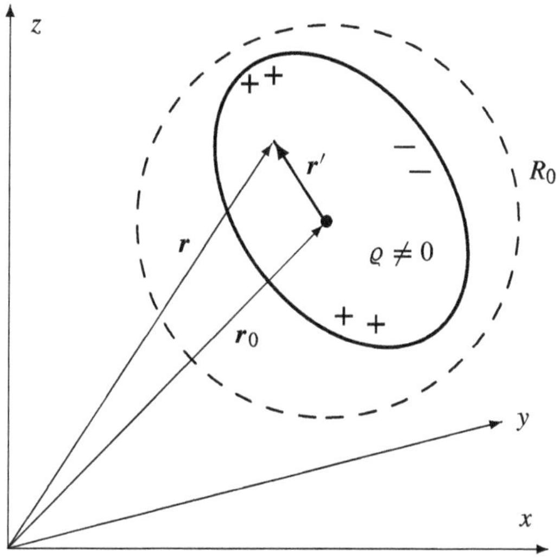

Figure 12.3 To determine the energy of a localized charge distribution in a external field, we expand the field $\Phi(\boldsymbol{r}) = \Phi(\boldsymbol{r}_0 + \boldsymbol{r}')$ by one position $\boldsymbol{r}_0$ within the charge distribution. This effectively expands the charge distribution into multipoles.

$$= \Phi_{\text{ext}}(\boldsymbol{r}_0) + \sum_{i=1}^{3} \frac{\partial \Phi_{\text{ext}}}{\partial x_i}\, x_i'$$

$$+\frac{1}{6} \sum_{i,j=1}^{3} \frac{\partial^2 \Phi_{\text{ext}}}{\partial x_i\, \partial x_j} \left(3 x_i' x_j' - r'^2\, \delta_{ij}\right) + \cdots \qquad (12.28)$$

All derivatives are to be taken at the point $\boldsymbol{r}' = 0$. Since in the range of the charge distribution $\Delta\, \Phi_{\text{ext}} = 0$ applies, we can add the term with δ_{ij}.

We calculate the potential energy of the charge distribution (12.27) in the external field (12.28):

$$W \overset{(6.25)}{=} \int d^3 r\; \varrho(\boldsymbol{r})\; \Phi_{\text{ext}}(\boldsymbol{r}) = \int d^3 r'\; \widetilde{\varrho}(\boldsymbol{r}')\; \Phi_{\text{ext}}(\boldsymbol{r}_0 + \boldsymbol{r}')$$

$$\overset{(12.28)}{=} q\; \Phi_{\text{ext}}(\boldsymbol{r}_0) - \boldsymbol{p} \cdot \boldsymbol{E}_{\text{ext}}(\boldsymbol{r}_0) - \frac{1}{6} \sum_{i,j=1}^{3} Q_{ij}\; \frac{\partial E_{\text{ext},\,j}(\boldsymbol{r}_0)}{\partial x_i} + \cdots \qquad (12.29)$$

where q, p_i and Q_{ij} are the multipole moments calculated with $\widetilde{\varrho}$; i.e., the multipole moments with respect to a CS with the origin at $\boldsymbol{r}_0$.

The force of the external field on the given charge distribution is thus

$$\boldsymbol{F}(\boldsymbol{r}_0) = -\operatorname{grad} W(\boldsymbol{r}_0) = \begin{cases} q\, \boldsymbol{E}_{\text{ext}}(\boldsymbol{r}_0) & \text{(monopole)} \\ (\boldsymbol{p} \cdot \boldsymbol{\nabla})\, \boldsymbol{E}_{\text{ext}}(\boldsymbol{r}_0) & \text{(dipole)} \end{cases} \qquad (12.30)$$

The specified special cases apply if the charge distribution is only a monopole or only one dipole moment. For the dipole, the formula $\boldsymbol{\nabla}(\boldsymbol{p} \cdot \boldsymbol{E}) = (\boldsymbol{p} \cdot \boldsymbol{\nabla})\boldsymbol{E} + (\boldsymbol{E}\, \boldsymbol{\nabla})\boldsymbol{p} + \boldsymbol{p} \times \operatorname{curl} \boldsymbol{E} + \boldsymbol{E} \times \operatorname{curl} \boldsymbol{p}$ is used. Since $\boldsymbol{p}$ does not depend on $\boldsymbol{r}_0$, the derivatives of $\boldsymbol{p}$ do not contribute. Furthermore, $\operatorname{curl} \boldsymbol{E} = 0$.

The potential energy of a dipole

$$W = -\boldsymbol{p} \cdot \boldsymbol{E}_{\text{ext}} = -p\, E_{\text{ext}} \cos\theta \qquad (12.31)$$

is minimal for $\boldsymbol{p} \parallel \boldsymbol{E}$. The torque

$$\boldsymbol{M} = \boldsymbol{p} \times \boldsymbol{E}_{\text{ext}} \qquad (12.32)$$

tries to turn the dipole vector into the energetically preferred position. The following applies: $|\boldsymbol{M}| = |p\, E_{\text{ext}} \sin\theta\,|$.

Exercises

12.1 *Multipoles of the homogeneously charged rod*

The electrostatic potential of a homogeneously charged rod is known from Exercise 6.7. From this, determine the two leading multipole fields of the rod. Check the result by calculating the quadrupole moments directly from the charge density.

12.2 *Singularity of the point dipole field*

An electric point dipole has the potential $\Phi(\boldsymbol{r}) = \boldsymbol{p}\cdot\boldsymbol{r}/r^3$. Calculate from this the charge density $\varrho(\boldsymbol{r})$. Show that the electric field has the form

$$\boldsymbol{E}(\boldsymbol{r}) = \mathrm{P}\,\frac{3\,\boldsymbol{r}\,(\boldsymbol{p}\cdot\boldsymbol{r}) - \boldsymbol{p}\,r^2}{r^5} - \frac{4\pi}{3}\boldsymbol{p}\,\delta(\boldsymbol{r}) \tag{12.33}$$

Here, P denotes the principal value, which is a spherical environment $r \le \epsilon$ during an integration; after integration, the limit $\epsilon \to 0$ is to be taken. Distributions are over the integrals with any differentiable function $f(\boldsymbol{r})$ defined. Therefore, (12.33) is equivalent to

$$\int d^3r\; f(\boldsymbol{r}) \left[\boldsymbol{E}(\boldsymbol{r}) - \mathrm{P}\,\frac{3\,\boldsymbol{r}\,(\boldsymbol{p}\cdot\boldsymbol{r}) - \boldsymbol{p}\,r^2}{r^5} \right] = -\,\frac{4\pi}{3}\; f(0)\,\boldsymbol{p} \tag{12.34}$$

Calculate from the potential $\Phi(\boldsymbol{r}) = \boldsymbol{p}\cdot\boldsymbol{r}/r^3$ first of all the electric field for $r \neq 0$. It follows from the result that the integration in (12.34) applies to a sphere with an (arbitrarily) small radius ϵ which can be restricted.

12.3 *Cartesian and spherical quadrupole components*

Explicitly describe the nine Cartesian components Q_{ij} of the quadrupole tensor explicitly by the spherical components q_{2m}.

12.4 *Quadrupole tensor of rotational ellipsoid and circular cylinder*

Determine the quadrupole tensor of the following homogeneously charged bodies:

(i) Rotational ellipsoid with semi-axes a and b.
(ii) Circular cylinder with length L and radius R.

Choose a suitable coordinate system in each case.

PART III

Magnetostatics

Chapter 13

Magnetic Field

The electric field was defined by the force between charges at rest. The magnetic field is defined by the force between moving charges.

Moving charges generally imply that the problem is time-dependent. In Part III, which begins here, we first study the simpler case of stationary currents in conductors. For this, the magnetic field is time-independent; this branch of electrodynamics is called magnetostatics.

In this chapter, we introduce the current and the current density. We then formulate the law of force for current carrying conductors and thereby define the magnetic field. Subsequently, the magnetic field is written as a functional of the current distribution. This functional is evaluated for a simple case.

Current and Current Density

Under suitable experimental conditions, charges flow through a wire. The amount of charge that passes per time through a cross-sectional area of the wire defines the *current* I,

$$I = \text{current} = \frac{\text{charge}}{\text{time}} = \frac{dq}{dt} \tag{13.1}$$

The charge was introduced in Chapter 5 as a measurement quantity. By (13.1), the current is defined as an observable. In the SI, currents are measured in the unit ampere (A, named after the physicist Ampère).

The *current density* is the current per cross-sectional area[1] $\varDelta a$ of the wire:

$$j = \frac{\text{current}}{\text{area}} = \frac{I}{\varDelta a} \tag{13.2}$$

The position of a thin wire can be described by a curve $\boldsymbol{\ell} = \boldsymbol{\ell}(s)$, where s is a suitable path parameter. The tangent vector $d\boldsymbol{\ell}/d\ell$ points into the direction of the current. Then we define the vector $\boldsymbol{j}$ for the current density by

$$\boldsymbol{j} = \frac{I}{\varDelta a}\,\frac{d\boldsymbol{\ell}}{d\ell} \tag{13.3}$$

The multiplication of the current density with a surface element results in

$$\boldsymbol{j} \cdot d\boldsymbol{a} = \text{current through } d\boldsymbol{a} \tag{13.4}$$

The surface element $d\boldsymbol{a} = d\boldsymbol{a}_{\parallel} + d\boldsymbol{a}_{\perp}$ can be divided into the parts parallel and perpendicular to $\boldsymbol{j}$. The perpendicular part makes no contribution, and for the parallel part, (13.4) is reduced to (13.2).

With $\varDelta a\, d\ell = \varDelta V = d^3r$, we can also write (13.3) as

$$\boldsymbol{j}(\boldsymbol{r}, t)\, d^3r = I\, d\boldsymbol{\ell} \tag{13.5}$$

In this way, an extended current density $\boldsymbol{j} = \boldsymbol{j}(\boldsymbol{r}, t)$ can be decomposed into wire elements with currents. Similarly, in electrostatics, an extended charge distribution was decomposed into individual elements $dq = \varrho\, d^3r$ (Figure 13.1, right).

In magneto statics, we restrict ourselves to time independent current densities:

$$\text{Magneto statics:}\quad \boldsymbol{j}(\boldsymbol{r}, t) = \boldsymbol{j}(\boldsymbol{r}) \tag{13.6}$$

Microscopic definition

In a microscopic and classical treatment, we consider point charges q_i, which have the position vectors $\boldsymbol{r}_i$ and the velocities $\boldsymbol{v}_i = \dot{\boldsymbol{r}}_i(t)$. For this,

---

[1] In general, the letters A and $\boldsymbol{A}$ are used as symbols for areas. As this is in conflict with the common designation of the vector potential (Chapter 14), we use a and $\boldsymbol{a}$ for the area in Parts III–V. All these letters are slanted (italic); units (such as A for ampere) are printed in roman (upright).

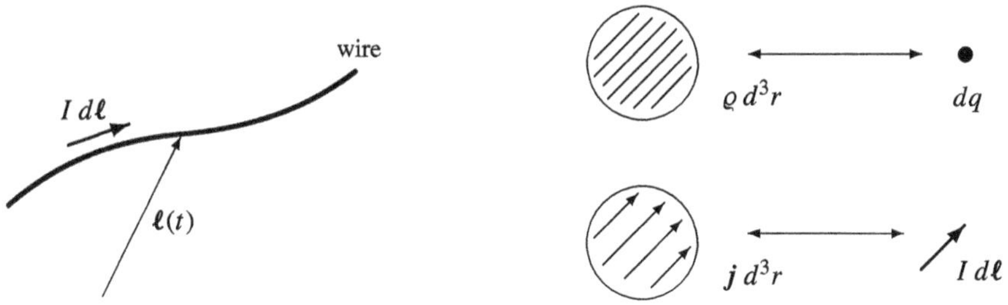

Figure 13.1 The course of a thin wire (left) is described by a curve $\boldsymbol{\ell}(s)$. The current I flows through the wire. Then each line element $d\boldsymbol{\ell}$ of the curve contributes with $I\,d\boldsymbol{\ell}$ to the current density. In a continuous current distribution, the volume element d^3r gives the contribution $\boldsymbol{j}\,d^3r = I\,d\boldsymbol{\ell}$. This is comparable to the decomposition of a charge distribution $\varrho(\boldsymbol{r})$ into the elements $dq = \varrho\, d^3r$ (right).

we obtain the charge and current density

$$\varrho_{\rm at}(\boldsymbol{r},t) = \sum_{i=1}^{N} q_i\,\delta(\boldsymbol{r}-\boldsymbol{r}_i(t)), \quad \boldsymbol{j}_{\rm at}(\boldsymbol{r},t) = \sum_{i=1}^{N} q_i\,\boldsymbol{v}_i\,\delta(\boldsymbol{r}-\boldsymbol{r}_i(t)) \qquad (13.7)$$

As point charges, we imagine the electrons and the atomic nuclei in matter; the index "at" stands for atomic.

We now consider finite volumes ΔV, each of which contains many charges. For this, we define the mean charge and current density:

$$\varrho(\boldsymbol{r},t) = \frac{\Delta q}{\Delta V} = \frac{1}{\Delta V}\sum_{\Delta V} q_i, \quad \boldsymbol{j}(\boldsymbol{r},t) = \frac{1}{\Delta V}\sum_{\Delta V} q_i\,\boldsymbol{v}_i \qquad (13.8)$$

Formally, these quantities result from (13.7) by a spatial averaging over volumes of the size ΔV.

If all charges are equal, $q_i = q$, there is a simple relationship between the charge density and the current density (see Figure 13.2):

$$\boldsymbol{j}(\boldsymbol{r},t) = \frac{q}{\Delta V}\sum_{\Delta V} \boldsymbol{v}_i = \frac{\Delta q}{\Delta V}\,\frac{1}{\Delta N}\sum_{\Delta V} \boldsymbol{v}_i = \varrho(\boldsymbol{r},t)\,\boldsymbol{v}(\boldsymbol{r},t) \qquad (13.9)$$

Here, ΔN is the number and Δq is the total charges in ΔV, and $\boldsymbol{v}(\boldsymbol{r},t)$ is the *mean* velocity of the charges in ΔV.

We discuss the current density for a current carrying metal wire. The metal can be regarded as a crystal lattice of positive ions in which the quasi-free electrons (one per ion) can move around. The average velocity

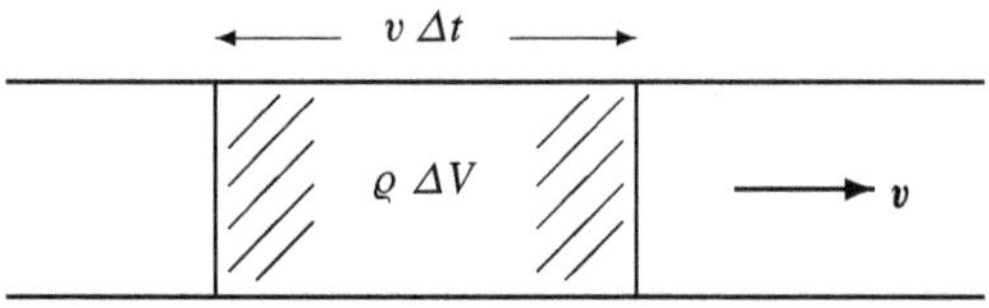

Figure 13.2 In a conductor (cross-sectional area Δa), the charge density ϱ moves with the speed $\boldsymbol{v}$. During the time Δt the charge $\Delta q = \varrho\,\Delta V = \varrho\,\Delta a\,v\,\Delta t$ (shaded region) is transported through a cross-sectional area of the wire. According to (13.1) and (13.2), the current density is then $j = \Delta q/(\Delta t\,\Delta a) = \varrho\,v$.

of the ions and the bound electrons then disappears; they do not contribute to the current density. In contrast, the mobile electrons have an average speed when the voltage is applied. The resulting contribution to the current density can be calculated using (13.9), where ϱ is the charge density of the mobile electrons and $\boldsymbol{v}$ their mean velocity. In the undisturbed matter, the velocities $\boldsymbol{v}_i$ of the individual mobile electrons have statistical directions, so that the mean velocity disappears. An electric field, however, leads to an additional component of all velocities in the direction of the field and thus to an non-vanishing average velocity. This mean velocity is small compared to the velocities $|\boldsymbol{v}_i|$ of the individual electrons (Exercise 13.1).

Continuity Equation

For any process, the sum of the charges in a closed system is constant; the charge is a conserved quantity. The change of the charge $\int d^3r\,\varrho(\boldsymbol{r},t)$ in a volume V per time must therefore be equal to the current through the surface $a = a(V)$ of the volume:

$$\frac{d}{dt}\int_V d^3r\ \varrho(\boldsymbol{r},t) + \oint_{a(V)} d\boldsymbol{a}\cdot\boldsymbol{j}(\boldsymbol{r},t) = 0 \qquad (13.10)$$

If the current density $\boldsymbol{j}$ points outward on average (i.e., in the direction of $d\boldsymbol{a}$), then the second term is positive and charge flows out of V. Accordingly, the charge in V decreases and the first term is negative.

We consider (13.10) for a fixed volume V. Then the time derivative in the first term only acts on $\varrho(\boldsymbol{r},t)$. The second term is transformed using Gauss' theorem. Thus, (13.10) becomes

$$\int_V d^3r\left(\frac{\partial\varrho(\boldsymbol{r},t)}{\partial t} + \operatorname{div}\boldsymbol{j}(\boldsymbol{r},t)\right) = 0 \qquad (13.11)$$

As this applies to arbitrary volumes, the integrand must disappear:

$$\boxed{\frac{\partial \varrho(\boldsymbol{r}, t)}{\partial t} + \operatorname{div} \boldsymbol{j}(\boldsymbol{r}, t) = 0 \quad \text{continuity equation}} \tag{13.12}$$

In short form, this *continuity equation* reads $\dot{\varrho} + \operatorname{div} \boldsymbol{j} = 0$. The continuity equation is the differential expression for the conservation of charge. In the static case, it reduces to

$$\operatorname{div} \boldsymbol{j}(\boldsymbol{r}) = 0 \quad \text{(magnetostatic)} \tag{13.13}$$

Law of Force

In electrostatics, the electric field $\boldsymbol{E}$ was defined by the measurable force $\boldsymbol{F} = q\boldsymbol{E}$ on a test charge q. Similarly, we define the magnetic field $\boldsymbol{B}(\boldsymbol{r})$ by the measurable forces on currents. For this purpose, we consider a thin wire with the current I; the wire is at rest and the current I is constant. In the presence of other current carrying conductors, forces act on the wire. A small line element $d\boldsymbol{\ell}$ of the wire is subjected to a force $d\boldsymbol{F}$, for which one finds the following experimentally:

$$dF \propto I, \quad dF \propto d\ell, \quad d\boldsymbol{F} \perp d\boldsymbol{\ell} \tag{13.14}$$

The force $d\boldsymbol{F}$ can therefore be expressed in the form

$$\boxed{d\boldsymbol{F}(\boldsymbol{r}) = \frac{I}{c}\, d\boldsymbol{\ell} \times \boldsymbol{B}(\boldsymbol{r}) \quad \text{definition of } \boldsymbol{B}} \tag{13.15}$$

With this relation, the vector field $\boldsymbol{B}(\boldsymbol{r})$ is defined as an observable (a measurable quantity) because all other quantities $d\boldsymbol{F}$, $d\boldsymbol{\ell}$ and I are measurable. The definition of $\boldsymbol{B}(\boldsymbol{r})$ depends on the constant c. We choose c as the speed of light

$$c \approx 3 \cdot 10^8 \ \frac{\text{m}}{\text{s}} \quad \text{(speed of light)} \tag{13.16}$$

This choice implies that the fields $\boldsymbol{E}$ and $\boldsymbol{B}$ are measured in the same units; this follows from $I\, d\ell / c = q$. As we will see later (Chapter 20), the choice implies that $\boldsymbol{E}$ and $\boldsymbol{B}$ have the same amplitude in an electromagnetic wave.

We specify the measurement prescription (13.15) for $\boldsymbol{B}$ further in such a way that the current through $d\boldsymbol{\ell}$ does not significantly change field configuration (analogous to the test charge). In addition, for the sake of simplicity, it is assumed that the charge density of the wires disappears, so that no electrostatic forces occur.

The field defined by (13.15) is called *magnetic field,*

$$\boldsymbol{B}(\boldsymbol{r}) = \text{magnetic field} \tag{13.17}$$

In the literature, different designations for $\boldsymbol{B}(\boldsymbol{r})$ are used, in particular *magnetic flux density* or also *magnetic induction.* Since $\boldsymbol{B}$ is the counterpart of the electric field strength $\boldsymbol{E}$, the term "magnetic field strength" would be appropriate. The term "magnetic field strength" is, however, used for a different quantity (for historical reasons). In Chapter 29, we introduce the decomposition $\boldsymbol{B} = \boldsymbol{H} + 4\pi\boldsymbol{M}$ for fields in matter, where $\boldsymbol{H}$ is the "magnetic field strength" and $\boldsymbol{M}$ is the "magnetization". To avoid confusing designations, we will simply speak of the electric field $\boldsymbol{E}$ and the magnetic field $\boldsymbol{B}$.

Once $\boldsymbol{B}$ has been determined as the measurement quantity (observable), it can be determined experimentally which magnetic field is caused by a current carrying wire. At the position $\boldsymbol{r}'$, let there be a wire piece $d\boldsymbol{\ell}$ through which the current I flows. This then causes at the position $\boldsymbol{r}$ a contribution $d\boldsymbol{B}$ with the properties

$$\begin{aligned} &dB \propto I\, d\ell, \quad d\boldsymbol{B} \perp d\boldsymbol{\ell}, \quad d\boldsymbol{B} \perp (\boldsymbol{r}-\boldsymbol{r}') \\ &dB \propto 1/|\boldsymbol{r}-\boldsymbol{r}'|^2, \quad dB \propto \sin\left(\sphericalangle(d\boldsymbol{\ell}, \boldsymbol{r}-\boldsymbol{r}')\right) \end{aligned} \tag{13.18}$$

(Figure 13.3, left). These findings can be summarized by

$$d\boldsymbol{B}(\boldsymbol{r}) = \frac{I}{c}\, d\boldsymbol{\ell} \times \frac{\boldsymbol{r}-\boldsymbol{r}'}{|\boldsymbol{r}-\boldsymbol{r}'|^3} \tag{13.19}$$

Using (13.5) and summing over all contributions yields

$$\boxed{\boldsymbol{B}(\boldsymbol{r}) = \frac{1}{c}\int d^3r'\; \boldsymbol{j}(\boldsymbol{r}') \times \frac{\boldsymbol{r}-\boldsymbol{r}'}{|\boldsymbol{r}-\boldsymbol{r}'|^3}} \tag{13.20}$$

The magnetic field can thus be calculated from a given stationary current distribution.

Table 13.1 shows that the basic concepts of magneto-statics are largely analogous to those of electrostatics. The relations of magnetostatics are

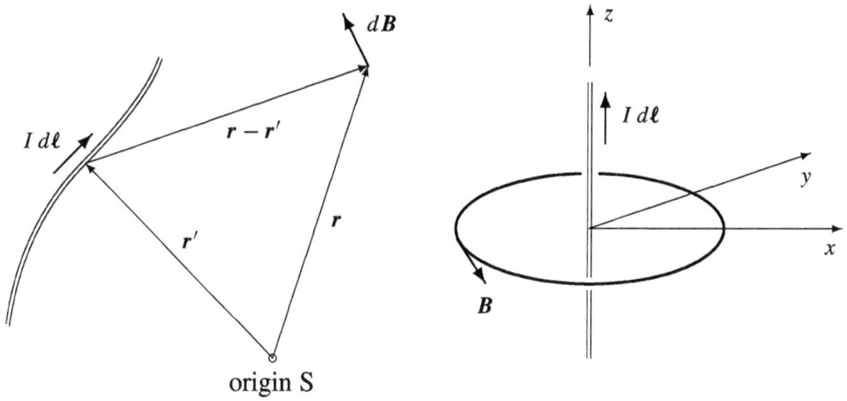

Figure 13.3 The current element $I\,d\boldsymbol{\ell}$ at position $\boldsymbol{r}'$ generates at position $\boldsymbol{r}$ a contribution $d\boldsymbol{B}$ to the magnetic field which is perpendicular to $d\boldsymbol{\ell}$ and $\boldsymbol{r}-\boldsymbol{r}'$ (left). If one adds up all these contributions for an infinitely long, straight wire, the result are circles for the field lines (right).

Table 13.1 Comparison of the relationships of electrostatics and magnetostatics (the last four entries in the right column are explained in Chapter 14.)

Electrostatic	Relation	Magnetostatic
$dq = \varrho\, d^3r$	Charge/current element	$I\,d\boldsymbol{\ell} = \boldsymbol{j}\,d^3r$
$d\boldsymbol{F} = dq\,\boldsymbol{E}$ $\boldsymbol{F} = q\,\boldsymbol{E}$	Field definition	$d\boldsymbol{F} = (I/c)\,d\boldsymbol{\ell}\times\boldsymbol{B}$ $\boldsymbol{F} = q\,(\boldsymbol{v}/c)\times\boldsymbol{B}$
$d\boldsymbol{E}(\boldsymbol{r}) = dq\,\dfrac{\boldsymbol{r}-\boldsymbol{r}'}{\lvert\boldsymbol{r}-\boldsymbol{r}'\rvert^3}$	Force law	$d\boldsymbol{B}(\boldsymbol{r}) = \dfrac{I}{c}\,d\boldsymbol{\ell}\times\dfrac{\boldsymbol{r}-\boldsymbol{r}'}{\lvert\boldsymbol{r}-\boldsymbol{r}'\rvert^3}$
$\operatorname{div}\boldsymbol{E} = 4\pi\varrho$ $\operatorname{curl}\boldsymbol{E} = 0$	Field equations	$\operatorname{curl}\boldsymbol{B} = 4\pi\boldsymbol{j}/c$ $\operatorname{div}\boldsymbol{B} = 0$
$\boldsymbol{E} = -\operatorname{grad}\Phi$	Field expressed by potential	$\boldsymbol{B} = \operatorname{curl}\boldsymbol{A}$
$\Delta\Phi = -4\pi\varrho$	Field equation	$\Delta\boldsymbol{A} = -(4\pi/c)\boldsymbol{j}$
$\Phi(\boldsymbol{r}) = \displaystyle\int d^3r'\,\frac{\varrho(\boldsymbol{r}')}{\lvert\boldsymbol{r}-\boldsymbol{r}'\rvert}$	Potential from source density	$\boldsymbol{A}(\boldsymbol{r}) = \dfrac{1}{c}\displaystyle\int d^3r'\,\frac{\boldsymbol{j}(\boldsymbol{r}')}{\lvert\boldsymbol{r}-\boldsymbol{r}'\rvert}$

somewhat more complicated because the source of the field (the current density) is a vector field, whereas it is a scalar field (the charge density) in electrostatics.

Force on point charge

From (13.5) and (13.15) follows the force density $\boldsymbol{f}$, which a magnetic field $\boldsymbol{B}$ exerts on a current distribution:

$$\boldsymbol{f}(\boldsymbol{r}) = \frac{d\boldsymbol{F}}{dV} = \frac{1}{c}\,\frac{I\,d\boldsymbol{\ell}}{dV} \times \boldsymbol{B} = \frac{1}{c}\,\boldsymbol{j}(\boldsymbol{r}) \times \boldsymbol{B}(\boldsymbol{r}) \tag{13.21}$$

For the current density $\boldsymbol{j}(\boldsymbol{r}) = q\,\boldsymbol{v}\,\delta(\boldsymbol{r} - \boldsymbol{r}_0)$ of a moving point charge, this results in $\boldsymbol{f}(\boldsymbol{r}) = \boldsymbol{F}\,\delta(\boldsymbol{r} - \boldsymbol{r}_0)$. Here,

$$\boxed{\boldsymbol{F} = q\,\frac{\boldsymbol{v}}{c} \times \boldsymbol{B}(\boldsymbol{r}_0)} \tag{13.22}$$

is the force $\boldsymbol{F}$ on a point charge at $\boldsymbol{r}_0$, which moves with the velocity $\boldsymbol{v} = \dot{\boldsymbol{r}}_0$. If $\boldsymbol{v}$ points in the x-direction and $\boldsymbol{B}$ in the y-direction, the force $\boldsymbol{F}$ acts in the z-direction.

The force (13.22) corresponds to the force $\boldsymbol{F} = q\boldsymbol{E}$ which we used for defining the electric field in electrostatics. For the definition of the magnetic field, we did not start from (13.22) because a moving charge causes time-dependent fields.

Magnetic field of a straight wire

We calculate the magnetic field of an infinitely long wire (Figure 13.3, right), through which the current I flows. We place the z-axis onto the wire and use cylindrical coordinates ρ, φ and z. Then the current density reads

$$\boldsymbol{j}(\boldsymbol{r}) = I\,\delta(x)\,\delta(y)\,\boldsymbol{e}_z = I\,\boldsymbol{e}_z\,\frac{\delta(\rho)}{2\pi\,\rho} \tag{13.23}$$

The delta functions are to be chosen such that the integral $\int \boldsymbol{j} \cdot d\boldsymbol{a}$ over a small cross-sectional area of the wire gives the current I; here, $d\boldsymbol{a} = dx\,dy\,\boldsymbol{e}_z = \rho\,d\varphi\,d\rho\,\boldsymbol{e}_z$.

To calculate the magnetic field, we start from (13.20). Due to the cylindrical symmetry, $\boldsymbol{B}(\boldsymbol{r}) = \boldsymbol{B}(\rho, \varphi, z)$ can only depend on ρ. Therefore, it is sufficient to determine the field in the plane $z = 0$ in dependence of $\boldsymbol{r} = \rho\,\boldsymbol{e}_\rho$. The current density $\boldsymbol{j}(\boldsymbol{r}')$ is non-zero only for $\rho' = 0$, so that

$\boldsymbol{r}' = z'\boldsymbol{e}_z$ can be set. This means that $\boldsymbol{r} - \boldsymbol{r}' = \rho\,\boldsymbol{e}_\rho - z'\boldsymbol{e}_z$. From (13.20) with (13.23), we thus obtain

$$\boldsymbol{B}(\rho) = \frac{I}{c}\int_{-\infty}^{\infty} dz' \int_0^{\infty} \rho'\,d\rho' \int_0^{2\pi} d\varphi'\, \frac{\delta(\rho')}{2\pi\,\rho'}\, \boldsymbol{e}_z \times \frac{\rho\,\boldsymbol{e}_\rho - z'\boldsymbol{e}_z}{(\rho^2 + z'^2)^{3/2}}$$
$$= \frac{I}{c}\,\boldsymbol{e}_\varphi \int_{-\infty}^{\infty} dz'\, \frac{\rho}{(\rho^2 + z'^2)^{3/2}} = \frac{2I}{c}\,\frac{\boldsymbol{e}_\varphi}{\rho} \tag{13.24}$$

This result is called *Biot–Savart law*. At all positions, the magnetic field points in the direction of $\boldsymbol{e}_\varphi$. The field lines are therefore circles (Figure 13.3, right). The Biot–Savart law is historically older than the more general formulation (13.20).

Force between two parallel wires

As a simple application of (13.15) and (13.24), we calculate the force per length between two parallel (infinitely long, straight) wires at a distance d. Both wires carry the current I. We pick out a line element $d\boldsymbol{\ell}_1$ of the first wire. The other line elements $d\boldsymbol{\ell}_1'$ of the same wire do not produce a field at this point because the connecting vector between the two line elements is parallel to $d\boldsymbol{\ell}_1'$. Therefore, only the field of the other wire exerts a force. According to (13.24), this field has the size $B_2 = 2I_2/(cd)$ and is perpendicular to the chosen line element. Using (13.15), we obtain

$$\frac{dF_1}{d\ell_1} = \frac{I_1 B_2}{c} = \frac{2I_1 I_2}{c^2}\,\frac{1}{d} = \frac{dF_2}{d\ell_2} \tag{13.25}$$

The forces act parallel to the shortest connection between the wires. If the current flows in the same direction for both wires, the forces are attractive. If the currents flow in opposite directions, the wires repel each other.

Exercises

13.1 *Velocity of the metal electrons*

The kinetic energy of free metal electrons is of the size $E_{\text{kin}} \approx 10\,\text{eV}$. Calculate from this their mean velocity $\overline{v} = \langle \boldsymbol{v}^2 \rangle^{1/2}$.

For each cell of a metal lattice (with volume $\varDelta V = 1\,\text{Å}^3$), there is one free electron. A current $I = 1\,\text{A}$ flows a wire with a cross-sectional area $a = 1\,\text{mm}^2$. Calculate the associated mean drift velocity v_{drift} of the electrons.

Chapter 14

Field Equations

Starting from (13.20), we establish the field equations of magneto-statics. We present integral forms of these equations and treat simple applications (homogeneous wire, infinitely long coil).

We transform (13.20):

$$B(r) = \frac{1}{c}\int d^3r'\; j(r') \times \frac{r - r'}{|r - r'|^3} = -\frac{1}{c}\int d^3r'\; j(r') \times \nabla \frac{1}{|r - r'|}$$

$$= \frac{1}{c}\,\nabla \times \int d^3r'\; \frac{j(r')}{|r - r'|} = \text{curl}\, A(r) \qquad (14.1)$$

The nabla operator only acts on r. In the last step, we have introduced the *vector potential*

$$A(r) = \frac{1}{c}\int d^3r'\; \frac{j(r')}{|r - r'|} + \text{grad}\, \Lambda(r) \qquad (14.2)$$

We may add an arbitrary vortex-free field to $A(r)$ without changing $B(r)$. According to (3.41), such an additional field can be written as a gradient field, i.e., in the form $\text{grad}\, \Lambda$ with arbitrary $\Lambda(r)$. We specifically choose $\Lambda = 0$ so that

$$\boxed{A(r) = \frac{1}{c}\int d^3r'\; \frac{j(r')}{|r - r'|}} \qquad (14.3)$$

This choice of Λ implies

$$\text{div}\, A(r) = 0 \quad \text{(Coulomb gauge)} \qquad (14.4)$$

To see this, one calculates the divergence of the right-hand side of (14.3), uses $\nabla\,|\boldsymbol{r}-\boldsymbol{r}'|^{-1} = -\nabla'\,|\boldsymbol{r}-\boldsymbol{r}'|^{-1}$ and shifts ∇' by partial integration to $\boldsymbol{j}(\boldsymbol{r}')$. Then, the integrand is proportional to $\operatorname{div}\boldsymbol{j}$ and disappears due to (13.13).

The indeterminacy of $\boldsymbol{A}$ in (14.2) can be formulated as follows: The transformation

$$\boldsymbol{A}(\boldsymbol{r}) \quad\longrightarrow\quad \boldsymbol{A}(\boldsymbol{r}) + \operatorname{grad}\Lambda(\boldsymbol{r}) \quad \text{(gauge transformation)} \qquad (14.5)$$

leaves the physical field (defined as an observable) $\boldsymbol{B}(\boldsymbol{r})$ unchanged. Such a transformation is called *gauge transformation*. The invariance against a transformation is a symmetry. Symmetries are often used to simplify problems. Here we used the gauge symmetry to fix $\operatorname{div}\boldsymbol{A}$ according to (14.4). This choice is called *Coulomb gauge*.

We calculate the curl of the magnetic field:

$$\begin{aligned}\operatorname{curl}\boldsymbol{B}(\boldsymbol{r}) &= \operatorname{curl}\operatorname{curl}\boldsymbol{A} \overset{(2.29)}{=} \operatorname{grad}\operatorname{div}\boldsymbol{A} - \Delta\boldsymbol{A} \overset{(14.4)}{=} -\Delta\boldsymbol{A}\\ &\overset{(14.3)}{=} -\frac{1}{c}\int d^3r'\,\boldsymbol{j}(\boldsymbol{r}')\,\Delta\,\frac{1}{|\boldsymbol{r}-\boldsymbol{r}'|} \overset{(3.24)}{=} \frac{4\pi}{c}\,\boldsymbol{j}(\boldsymbol{r}) \qquad (14.6)\end{aligned}$$

The divergence of $\boldsymbol{B}$ disappears because of (2.24), $\operatorname{div}\operatorname{curl}\boldsymbol{A} = 0$. This gives us for the sources and vortices of the magnetic field

$$\boxed{\begin{array}{ll}\operatorname{div}\boldsymbol{B}(\boldsymbol{r}) = 0 & \text{field equations}\\ \operatorname{curl}\boldsymbol{B}(\boldsymbol{r}) = \dfrac{4\pi}{c}\,\boldsymbol{j}(\boldsymbol{r}) & \text{of magnetostatics}\end{array}} \qquad (14.7)$$

These field equations are the basic equations of magnetostatics. The integral (13.20) is the solution of these field equations for a given localized current density distribution.

We may insert $\boldsymbol{B} = \operatorname{curl}\boldsymbol{A}$ into the field equations. The homogeneous field equation is automatically fulfilled. Due to the Coulomb gauge, $\operatorname{curl}\boldsymbol{B} = \operatorname{curl}\operatorname{curl}\boldsymbol{A} = -\Delta\boldsymbol{A}$. This gives us

$$\boxed{\Delta\boldsymbol{A}(\boldsymbol{r}) = -\,\frac{4\pi}{c}\,\boldsymbol{j}(\boldsymbol{r}), \qquad \boldsymbol{B}(\boldsymbol{r}) = \operatorname{curl}\boldsymbol{A}(\boldsymbol{r})} \qquad (14.8)$$

as alternative basic equations of magnetostatics. The integral (14.3) is the solution of these field equation for a given localized current density distribution.

If there were magnetic sources, the field equation $\operatorname{div} \boldsymbol{B} = 0$ had to be replaced by $\operatorname{div} \boldsymbol{B} = 4\pi\,\varrho_{\text{magn}}$. A magnetic point charge (i.e., a magnetic monopole) would have the field $\boldsymbol{B} = q_{\text{magn}}\,\boldsymbol{r}/r^3$.

The quantization of electric charge and angular momentum implies that magnetic monopoles must also be quantized [6]. The magnetic elementary charge q_{magn} would then be larger than the electric one by a factor $\hbar c/e^2 \approx 137$. A magnetic monopole would ionize matter correspondingly strongly. The search for magnetic monopoles has so far been unsuccessful; one can therefore assume that there are no such particles.

Ampère Law

We consider an arbitrary, coherent surface a with the contour C and apply Stokes' theorem to the field equation $\operatorname{curl} \boldsymbol{B} = 4\pi\boldsymbol{j}/c$:

$$\oint_C d\boldsymbol{r} \cdot \boldsymbol{B} = \frac{4\pi}{c} \int_a d\boldsymbol{a} \cdot \boldsymbol{j} = \frac{4\pi}{c}\, I_F \quad \text{(Ampère law)} \tag{14.9}$$

This *Ampère law* states that the line integral $\oint_C d\boldsymbol{r} \cdot \boldsymbol{B}$ along the boundary of the surface a is equal to the current I through the surface area (multiplied by $4\pi/c$). The Ampère law is an integral form of the inhomogeneous field equation and corresponds to the Gauss law of electrostatics. For simple geometries, the Ampère law can be the easiest way to solve a problem.

For a given contour C, there are infinitely many different areas a; for example, for a circle C, a could be the plane circle area or a spherical half-shell. The integral in (14.9) does not depend on the specific shape of a: Two different surface areas a_1 and a_2 with the same contour C can be combined to form a closed surface. Then, integral $\oint d\boldsymbol{a} \cdot \boldsymbol{j}$ over this closed surface can be converted into the volume integral $\int d^3r \operatorname{div} \boldsymbol{j}$ which disappears due to (13.13). Therefore, $\oint d\boldsymbol{a} \cdot \boldsymbol{j} = \int_{a_1} d\boldsymbol{a} \cdot \boldsymbol{j} - \int_{a_2} d\boldsymbol{a} \cdot \boldsymbol{j} = 0$; the surface integrals over a_1 and a_2 have the same value.

Magnetic Flux

We derive an integral formulation for the homogeneous field equation. For this purpose, we define the *magnetic flux* Φ_{m} through a surface a by

$$\Phi_{\text{m}} = \int_a d\boldsymbol{a} \cdot \boldsymbol{B} \quad \text{(magnetic flux)} \tag{14.10}$$

Since $B = \Phi_{\rm m}/\text{area}$, $\boldsymbol{B}$ is also called *magnetic flux density*. From the field equation $\operatorname{div}\boldsymbol{B} = 0$, it follows that the magnetic flux through a closed surface a disappears:

$$\oint_a d\boldsymbol{a} \cdot \boldsymbol{B} = \int_V d^3r \ \operatorname{div}\boldsymbol{B} = 0 \tag{14.11}$$

Here, V is the volume that is enclosed by a. This means that as many field lines go into V as go out again. As a rule, this means that the field lines are closed.[1] According to (14.11), there are no magnetic charges that are the origin of field lines.

The Ampère law and $\oint d\boldsymbol{a} \cdot \boldsymbol{B} = 0$ are integral forms of the field equations. These statements are helpful for sketching the field lines.

Homogeneous Current through Wire

As a simple application of Ampère's law, we consider a cylindrical wire through which a homogenous current I flows. The symmetry axis of the straight, infinitely long wire is the z-axis. Then, the current density is

$$\boldsymbol{j}(\boldsymbol{r}) = \boldsymbol{e}_z \cdot \begin{cases} \dfrac{I}{\pi R^2} & (\rho \le R) \\ 0 & (\rho > R) \end{cases} \tag{14.12}$$

We use cylindrical coordinates ρ, φ and z. From (14.3) and $\boldsymbol{j} \parallel \boldsymbol{e}_z$ follows $\boldsymbol{A} = A(\boldsymbol{r})\,\boldsymbol{e}_z$. This problem is invariant under rotations around the z-axis and under displacements in the z-direction. Due to this cylindrical symmetry, $A = A(\rho, \varphi, z)$ cannot depend on φ and z, therefore

$$\boldsymbol{A}(\boldsymbol{r}) = A(\rho)\,\boldsymbol{e}_z \quad \text{and} \quad \boldsymbol{B}(\boldsymbol{r}) = \operatorname{curl}\boldsymbol{A} = B(\rho)\,\boldsymbol{e}_\varphi \tag{14.13}$$

The field lines are circles (as in Figure 13.3, right).

[1] For example, consider the field line diagram of a finite, current-carrying coil (Figure 14.2 right) or a current-carrying wire circuit (Figure 15.1). However, there are exceptions to this statement, see Luca Zilberti, the misconception of closed magnetic flux lines, *IEEE Magnetics Letters* 8 (2017) 1–5.

As a contour in the Ampère law (14.9) we choose a circle with $\rho =$ const. and $z =$ const. With $d\boldsymbol{r} = \rho\, d\varphi\, \boldsymbol{e}_\varphi$, we get

$$\oint d\boldsymbol{r} \cdot \boldsymbol{B} = 2\pi\rho\, B(\rho) = \frac{4\pi}{c}\, I_F = \frac{4\pi}{c} \cdot \begin{cases} I\,\rho^2/R^2 & (\rho \le R) \\ I & (\rho > R) \end{cases} \tag{14.14}$$

From this follows

$$\boldsymbol{B}(\boldsymbol{r}) = B(\rho)\, \boldsymbol{e}_\varphi = \frac{2I}{c}\, \boldsymbol{e}_\varphi \cdot \begin{cases} \rho/R^2 & (\rho \le R) \\ 1/\rho & (\rho > R) \end{cases} \tag{14.15}$$

The ρ -dependence of the field is depicted in Figure 14.1. For $R \to 0$, we obtain the special case (13.24) of the thin wire.

Infinitely Long Coil

We consider a coil with N_S windings per length ℓ_S on a circular cylinder (Figure 14.2). Such a device is also called solenoid. We again use cylindrical coordinates ρ, φ and z; the z-axis shall be the symmetry axis of the cylinder. Let the coil be wound in such a way that the individual turns are circles with $\rho = R$ and $z =$ const. Then, the current flows in the $\boldsymbol{e}_\varphi$ direction, and current density is

$$\boldsymbol{j}(\boldsymbol{r}) = j(\rho)\, \boldsymbol{e}_\varphi = \frac{N_\mathrm{S}\, I}{\ell_\mathrm{S}}\, \delta(\rho - R)\, \boldsymbol{e}_\varphi \tag{14.16}$$

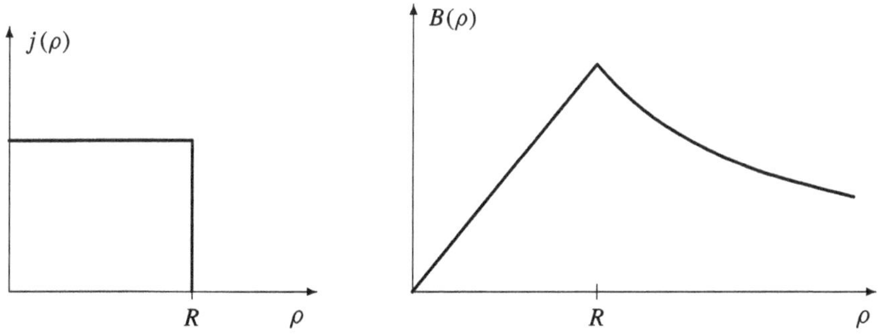

Figure 14.1 Current density $\boldsymbol{j} = j(\rho)\, \boldsymbol{e}_z$ and magnetic field $\boldsymbol{B} = B(\rho)\, \boldsymbol{e}_\varphi$ of a wire transporting a homogeneous current.

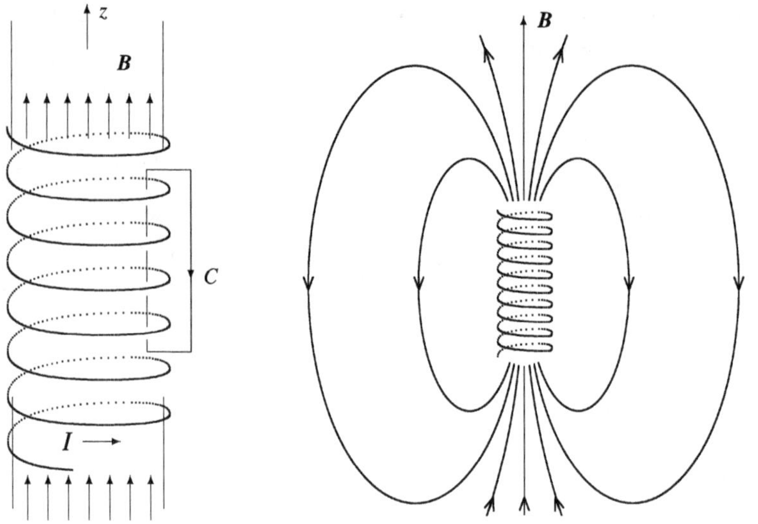

Figure 14.2 On the interior of an infinitely long coil (left), the magnetic field is homogeneous, and on the outside, it disappears. Using the Ampère law for the contour C, the field strength inside the coil is easily obtained. On the right is the field of a finite coil.

From (14.3) and $\boldsymbol{j} \perp \boldsymbol{e}_z$ follows $\boldsymbol{A} \perp \boldsymbol{e}_z$, thus $\boldsymbol{A}(\boldsymbol{r}) = A_\rho\, \boldsymbol{e}_\rho + A_\varphi\, \boldsymbol{e}_\varphi$. Due to the cylindrical symmetry, the components A_ρ and A_φ can only depend on ρ, i.e., $\boldsymbol{A} = A_\rho(\rho)\, \boldsymbol{e}_\rho + A_\varphi(\rho)\, \boldsymbol{e}_\varphi$. For the assumed Coulomb gauge (14.4), $\operatorname{div} \boldsymbol{A} = \rho^{-1} \partial(\rho\, A_\rho)/\partial\rho = 0$. This has the solution $A_\rho(\rho) = \text{const.}/\rho$. Since $\boldsymbol{A}$ must not diverge at $\rho \to 0$, the constant must disappear, i.e., $A_\rho = 0$ and

$$\boldsymbol{A}(\boldsymbol{r}) = A_\varphi(\rho)\, \boldsymbol{e}_\varphi = A(\rho)\, \boldsymbol{e}_\varphi \tag{14.17}$$

Since the direction $\boldsymbol{e}_\varphi$ is coordinate-dependent, one cannot directly conclude from $\boldsymbol{j} \parallel \boldsymbol{e}_\varphi$ and (14.3) to $\boldsymbol{A} \parallel \boldsymbol{e}_\varphi$.

From (14.17) follows for the magnetic field

$$\boldsymbol{B}(\boldsymbol{r}) = \operatorname{curl} \boldsymbol{A} = \frac{1}{\rho}\frac{d}{d\rho}\Big(\rho\, A(\rho)\Big)\, \boldsymbol{e}_z = B(\rho)\, \boldsymbol{e}_z \tag{14.18}$$

Then, $\operatorname{curl} \boldsymbol{B} = -B'(\rho)\, \boldsymbol{e}_\varphi$. Using this and (14.16), the field equation $\operatorname{curl} \boldsymbol{B} = (4\pi/c)\boldsymbol{j}$ becomes

$$-\frac{d}{d\rho}\frac{1}{\rho}\frac{d}{d\rho}\Big(\rho\, A(\rho)\Big) = \frac{4\pi}{c}\frac{N_{\mathrm{S}}\, I}{\ell_{\mathrm{S}}}\, \delta(\rho - R) \tag{14.19}$$

We integrate this once:

$$\frac{1}{\rho}\left(\rho\,A(\rho)\right)' = -\frac{4\pi}{c}\,\frac{N_{\mathrm{S}}\,I}{\ell_{\mathrm{S}}}\,\left(\Theta(\rho - R) + C_1\right) \tag{14.20}$$

A further integration leads to

$$\rho\,A(\rho) = -\frac{4\pi}{c}\,\frac{N_{\mathrm{S}}\,I}{\ell_{\mathrm{S}}}\left(\frac{(\rho^2 - R^2)\,\Theta(\rho - R)}{2} + C_1\,\frac{\rho^2}{2} + C_2\right) \tag{14.21}$$

From (14.3) follows $A(0) = 0$ and $A(\infty) = 0$ since for this all contributions to the integral cancel each other out. Therefore, $C_2 = 0$ and $C_1 = -1$, i.e.,

$$\boldsymbol{A}(\boldsymbol{r}) = A(\rho)\,\boldsymbol{e}_\varphi = \frac{4\pi}{c}\,\frac{N_{\mathrm{S}}\,I}{\ell_{\mathrm{S}}}\,\boldsymbol{e}_\varphi \cdot \begin{cases} \rho/2 & (\rho < R) \\ R^2/(2\rho) & (\rho > R) \end{cases} \tag{14.22}$$

Inside the coil, there is a homogeneous field with a strength $B_0 = (4\pi/c)\,N_{\mathrm{S}}\,I/\ell_{\mathrm{S}}$, on the outside, the field is zero. A coil is the standard device for generating a homogeneous magnetic field, comparable to the plate capacitor in electrostatics.

If we assume $\boldsymbol{B} = 0$ in the exterior and $\boldsymbol{B} = B_0\,\boldsymbol{e}_z$ in the interior, B_0 can be obtained from the Ampère law. The contour C shown in Figure 14.2 leads (14.9) to

$$\oint_C d\boldsymbol{r} \cdot \boldsymbol{B} = \ell_{\mathrm{S}}\,B_0 = \frac{4\pi}{c}\,N_{\mathrm{S}}\,I \tag{14.23}$$

An infinitely long coil is a theoretical model that can be used as an approximation for a long but finite coil. The field (14.23) will be approximately valid for the finite coil, except for the vicinity of the ends. Figure 14.2 sketches the field lines for the finite coil. For the finite coil, all field lines that pass through the inside are closed on the outside of the coil (due to (14.11)). Considering especially the x-y-plane, the flux is

$$\Phi_{\mathrm{m}} = \pi\,R^2\,\frac{4\pi}{c}\,\frac{N_{\mathrm{S}}\,I}{\ell_{\mathrm{S}}} \tag{14.24}$$

through the circular area $\rho < R$. The same flow passes with the opposite sign through the area $\rho > R$. In the outside area, the flow is distributed over a much larger area so that the field $\boldsymbol{B}$ (i.e., the flux density) is correspondingly smaller.

Self-inductance

According to (14.1), the magnetic field $\boldsymbol{B}$ is proportional to the current density. For a given geometry of a wire loop or coil, $\boldsymbol{B}$ is therefore proportional to the current I through the wire:

$$\boldsymbol{B}(\boldsymbol{r}) \propto I \tag{14.25}$$

This also applies to the magnetic flux Φ_{m} through a closed loop:

$$\Phi_{\mathrm{m}} = \int_a d\boldsymbol{a} \cdot \boldsymbol{B}(\boldsymbol{r}) \propto I \tag{14.26}$$

As discussed following (14.9), Φ_{m} does not depend on the specific choice for the area a (for a given contour). For a wire loop, the ratio Φ_{m}/I is therefore a constant. This ratio multiplied by N_{S}/c is defined as the *self-inductance*:

$$L = N_{\mathrm{S}}\, \frac{\Phi_{\mathrm{m}}}{c\, I} \quad \text{(self-inductance)} \tag{14.27}$$

Here, N_{S} is the number of closed wire turns, through which the same magnetic flux passes ($N_{\mathrm{S}} = 1$ for a single wire loop). For the cylindrical coil considered above with the cross-sectional area $A_{\mathrm{S}} = \pi R^2$, the self-inductance becomes

$$L = N_{\mathrm{S}}\, \frac{\Phi_{\mathrm{m}}}{c\, I} = \frac{N_{\mathrm{S}}}{c\, I}\, A_{\mathrm{S}}\, \frac{4\pi}{c}\, \frac{N_{\mathrm{S}}\, I}{\ell_{\mathrm{S}}} = \frac{4\pi}{c^2}\, N_{\mathrm{S}}^2\, \frac{F_{\mathrm{S}}}{\ell_{\mathrm{S}}} \tag{14.28}$$

The unit of inductance is $L = \mathrm{cm}$. In the SI unit system, the factor $4\pi/c^2$ must be replaced by μ_0; the inductance is then measured in Henry (H), $L = \mathrm{Vs/A} = \mathrm{H}$ (Appendix A).

For non-stationary currents, it comes to an induced voltage U at the wire ends of the coil (Chapters 16 and 26). The self-inductance (14.28) is defined such that $U = -L\, dI/dt$ holds. This corresponds to the relation $U = Q/C$ for the capacitor.

Boundary Value Problem

In Chapter 7, we introduced boundary value problems of electrostatics. In such problems, the unknown source distribution at the boundary surfaces is replaced by boundary conditions for the fields. We briefly discuss the analogous case for magnetostatics.

Let the current density be zero within a volume V. Then, $\operatorname{curl} \boldsymbol{B} = 0$ applies, and we can write $\boldsymbol{B}$ as a gradient field:

$$\boldsymbol{B}(\boldsymbol{r}) = \operatorname{grad} \Psi(\boldsymbol{r}) \quad \text{for } \boldsymbol{j}(\boldsymbol{r}) = 0 \tag{14.29}$$

From $\operatorname{div} \boldsymbol{B} = 0$ then follows the boundary value problem

$$\Delta \Psi = 0 \quad \text{in } V, \quad \text{and boundary condition} \tag{14.30}$$

According to Chapter 7, the solution can be fixed by a Dirichlet or a Neumann's boundary condition. The general solution of (14.31) can be constructed as in Chapter 10 or 11.

The specification of the boundary conditions requires a discussion of the polarizability of matter (Part VI) and other phenomena (superconductivity). We limit ourselves here to a short phenomenological description of two special cases. In both cases, we consider a hollow sphere in a homogeneous magnetic field:

1. Let the magnetic polarizability of the material be very high. Then, field lines that hit the sphere continue to run on the spherical shell before they leave the sphere again. As a result, the inner area remains virtually field free.

   The (theoretical) limiting case of infinitely high polarizability leads to the boundary condition $\boldsymbol{B} \cdot \boldsymbol{t} = 0$ or $\Psi|_{\mathrm{R}} = \text{const.}$ In this limiting case, one obtains for a sphere the same field line pattern as in Figure 10.3 for the electric field.
2. The sphere consists of a superconductor. Then, $\boldsymbol{B} \cdot \boldsymbol{n} = 0$ and $(\partial \Psi / \partial n)|_{\mathrm{R}} = 0$ because a normal component of $\boldsymbol{B}$ induces a permanent supercurrent that compensates the magnetic field. Inside the sphere, $\Psi \equiv \text{const.}$ is the solution of the Laplace equation with the boundary condition $(\partial \Psi / \partial n)|_{\mathrm{R}} = 0$. Therefore, the interior of a superconductor is completely field-free. For a sphere, the magnetic field lines run like the equipotential surfaces in Figure 10.3; the field lines are repelled out of the sphere.

A closed superconducting surface is a method for shielding magnetic fields. For electric fields, on the other hand, a closed normal conducting surface suffices (Faraday cage, last section in Chapter 7). Both shielding methods also work for time-dependent fields, as long as they vary sufficiently slow.

Exercises

14.1 *Current flowing through a hollow cylinder*

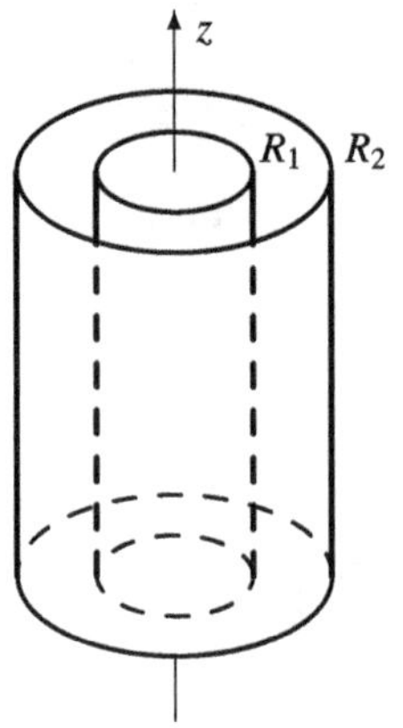

An infinitely long hollow cylinder (inner radius R_1, outer radius R_2) is homogeneously flowed through by the current I. Calculate the magnetic field $\boldsymbol{B}$ using Ampère's law. Sketch $|\boldsymbol{B}|$ as a function of the distance from the symmetry axis.

14.2 *Current carrying wire*

Solve the field equation $\Delta \boldsymbol{A} = -4\pi \boldsymbol{j}/c$ for an infinitely long, cylindrical wire (radius R) through which the current I flows homogeneously. Determine the magnetic field $\boldsymbol{B}$.

14.3 *Cylindrical coil*

For an infinitely long coil, the current density is given in cylindrical coordinates:

$$\boldsymbol{j}(\boldsymbol{r}) = j(\rho)\, \boldsymbol{e}_\rho = \frac{NI}{L}\, \delta(\rho - R)\, \boldsymbol{e}_\rho$$

Calculate the vector potential from the integral formula (14.3). Take into account the symmetry of the problem. Use partial integration and

$$J = \int_0^{2\pi} d\varphi\, \frac{\cos(n\,\varphi)}{1 - 2a\,\cos\varphi + a^2} = \frac{2\pi\, a^n}{1 - a^2} \qquad (|a| < 1,\;\; n = 0, 1, 2, \ldots)$$

Chapter 15

Magnetic Dipole

We consider a localized current distribution:

$$\boldsymbol{j}(\boldsymbol{r}) = \begin{cases} \text{arbitrary} & (r < R_0) \\ 0 & (r > R_0) \end{cases} \tag{15.1}$$

As in Figure 12.1, the field outside the distribution shall be determined. In the range $r > R_0$, the vector potential $\boldsymbol{A}$ can be expanded into powers of R_0/r; this is the multipole expansion known from Chapter 12. We restrict ourselves to the lowest term in this expansion, the magnetic dipole.

We insert

$$\frac{1}{|\boldsymbol{r}-\boldsymbol{r}'|} \overset{(r'<r)}{=} \frac{1}{r} - \sum_{i=1}^{3} x_i' \, \frac{\partial}{\partial x_i} \frac{1}{r} + \cdots = \frac{1}{r} + \frac{\boldsymbol{r}\cdot\boldsymbol{r}'}{r^3} + \cdots \tag{15.2}$$

into Equation (14.3):

$$\begin{aligned} \boldsymbol{A}(\boldsymbol{r}) &= \frac{1}{c}\int d^3r' \, \frac{\boldsymbol{j}(\boldsymbol{r}')}{|\boldsymbol{r}-\boldsymbol{r}'|} \\ &= \frac{1}{c\,r}\int d^3r' \, \boldsymbol{j}(\boldsymbol{r}') + \frac{1}{c\,r^3}\int d^3r' \, (\boldsymbol{r}\cdot\boldsymbol{r}')\, \boldsymbol{j}(\boldsymbol{r}') + \cdots \end{aligned} \tag{15.3}$$

In a spatially limited current distribution, there are in each direction as many positive as negative contributions. Therefore, the following applies:

$$\int d^3r \, \boldsymbol{j}(\boldsymbol{r}) = 0 \tag{15.4}$$

Thus, the first term on the right-hand side of (15.3) vanishes. For the second term, we need to perform some intermediate calculations. From $\operatorname{div} \boldsymbol{j} = 0$ follows $\sum_n \partial_n (x_i\, j_n) = j_i$. Therefore,

$$\int d^3r\; x_k\; j_i(\boldsymbol{r}) = \sum_{n=1}^{3} \int d^3r\; x_k\; \partial_n \big(x_i\; j_n(\boldsymbol{r})\big) \overset{\text{p.i.}}{=} -\int d^3r\; x_i\; j_k(\boldsymbol{r}) \quad (15.5)$$

We multiply both sides by x'_k, sum over k, go to the vector notation and interchange $\boldsymbol{r}' \leftrightarrow \boldsymbol{r}$:

$$\int d^3r'\; (\boldsymbol{r}\cdot\boldsymbol{r}')\, \boldsymbol{j}(\boldsymbol{r}') = -\int d^3r'\; \boldsymbol{r}' \left[\boldsymbol{r}\cdot \boldsymbol{j}(\boldsymbol{r}')\right] \quad (15.6)$$

Using this, we transform the last term in (15.3):

$$\begin{aligned}\int d^3r'\; (\boldsymbol{r}\cdot\boldsymbol{r}')\, \boldsymbol{j}(\boldsymbol{r}') &= \frac{1}{2}\int d^3r'\; \Big((\boldsymbol{r}\cdot\boldsymbol{r}')\, \boldsymbol{j}(\boldsymbol{r}') - \boldsymbol{r}'\left[\boldsymbol{r}\cdot \boldsymbol{j}(\boldsymbol{r}')\right]\Big) \\ &= -\frac{1}{2}\int d^3r'\; \boldsymbol{r} \times \big(\boldsymbol{r}' \times \boldsymbol{j}(\boldsymbol{r}')\big) \\ &= -\frac{\boldsymbol{r}}{2} \times \int d^3r'\; \boldsymbol{r}' \times \boldsymbol{j}(\boldsymbol{r}') \end{aligned} \quad (15.7)$$

The quantity

$$\boxed{\boldsymbol{\mu} = \frac{1}{2c}\int d^3r\; \boldsymbol{r} \times \boldsymbol{j}(\boldsymbol{r}) \qquad \text{magnetic dipole moment}} \quad (15.8)$$

is identified as the *magnetic dipole moment* the current distribution. We insert (15.7) with (15.8) into (15.3):

$$\boldsymbol{A}(\boldsymbol{r}) = \frac{\boldsymbol{\mu} \times \boldsymbol{r}}{r^3} + \cdots \qquad (r > R_0) \quad (15.9)$$

The next terms result from the continuation of the expansion (15.2). They are each smaller by a factor $\mathcal{O}(R_0/r)$. These terms are not taken into account in the following. The magnetic field calculated from (15.6) reads

$$\boldsymbol{B}(\boldsymbol{r}) = \operatorname{curl} \boldsymbol{A} = \frac{3\boldsymbol{r}\,(\boldsymbol{r}\cdot\boldsymbol{\mu}) - \boldsymbol{\mu}\, r^2}{r^5} \qquad (r > R_0) \quad (15.10)$$

This field has the same structure as the electric dipole field (12.25).

Examples of systems with a magnetic dipole moment are the Earth, a current carrying wire loop or coil, a compass needle or elementary particles such as electrons, protons or neutrons. Some of these systems are considered in more detail in the following.

The magnetic field of a coil (Figure 14.2, right) is a dipole field for large distances (large compared to the coil dimensions). If the size of the coil goes to zero, the current carrying coil becomes a magnetic point dipole.

Currents in the Earth's interior cause a magnetic field. The dominant part of this field is a dipole field, i.e. of the form (15.10). The direction of the magnetic axis (direction of $\boldsymbol{\mu}$) deviates by a few degrees from the axis of rotation (magnetic declination). The Earth's magnetic field reverses its polarity approximately every 200 000 years. The simplest and best known measurement of the Earth's field is that by compass; see the sections "Permanent magnet" and "Energy in the external field".

Point Dipole

We consider the vector potential (15.9) without the restriction $r > R_0$. For this we calculate the current density $\boldsymbol{j} = -(c/4\pi)\,\Delta\boldsymbol{A}$: We disregard

$$\boldsymbol{j}(\boldsymbol{r}) = -\frac{c}{4\pi}\,\Delta\,\frac{\boldsymbol{\mu}\times\boldsymbol{r}}{r^3} = \frac{c}{4\pi}\,\Delta\left(\boldsymbol{\mu}\times\boldsymbol{\nabla}\right)\frac{1}{r} = -c\left(\boldsymbol{\mu}\times\boldsymbol{\nabla}\right)\delta(\boldsymbol{r}) \qquad (15.11)$$

Used in (14.3), this current density results in the dipole field $\boldsymbol{A} = \boldsymbol{\mu}\times\boldsymbol{r}/r^3$. In addition, $\boldsymbol{j}$ disappears for $\boldsymbol{r} \neq 0$. It follows from these two statements that (15.11) is the current density of a *magnetic point dipole*.

Wire Loop

We calculate the dipole moment of a current I flowing through a wire circle with the radius R. We use cylindrical coordinates and place the wire loop so that

$$\boldsymbol{j} = I\,\delta(\rho - R)\,\delta(z)\,\boldsymbol{e}_\varphi \qquad (15.12)$$

We use this in (15.8):

$$\boldsymbol{\mu} = \frac{1}{2c}\int_0^\infty \rho\,d\rho \int_0^{2\pi} d\varphi \int_{-\infty}^{\infty} dz\,\left(\rho\,\boldsymbol{e}_\rho + z\,\boldsymbol{e}_z\right)$$

$$\times\,\boldsymbol{e}_\varphi\,I\,\delta(\rho - R)\,\delta(z) = \frac{\pi R^2 I}{c}\,\boldsymbol{e}_z \qquad (15.13)$$

For large distances ($r \gg R$), the magnetic field of the wire loop is given by (15.10). In the range $r \lesssim R$, the actual field (sketched in Figure 15.1) differs significantly from the dipole field.

Gyromagnetic Ratio

We consider a rigid body with the charge density $\varrho(\boldsymbol{r})$. The body rotates with the angular velocity $\boldsymbol{\omega}$ around a fixed axis, Figure 15.2. This generates the velocity field

$$\boldsymbol{v}(\boldsymbol{r}) = \boldsymbol{\omega} \times \boldsymbol{r} \quad \text{(rigid rotation)} \tag{15.14}$$

The resulting current density $\boldsymbol{j} = \varrho\, \boldsymbol{v}$ yields the dipole moment

$$\boldsymbol{\mu} = \frac{1}{2c} \int d^3r\; \varrho(\boldsymbol{r}) \left(\boldsymbol{r} \times \boldsymbol{v}(\boldsymbol{r})\right) \tag{15.15}$$

The body shall have the mass density $\varrho_{\text{m}}(\boldsymbol{r})$. The rotating mass rotates and leads to an angular momentum $\boldsymbol{L}$. To calculate the angular momentum, we

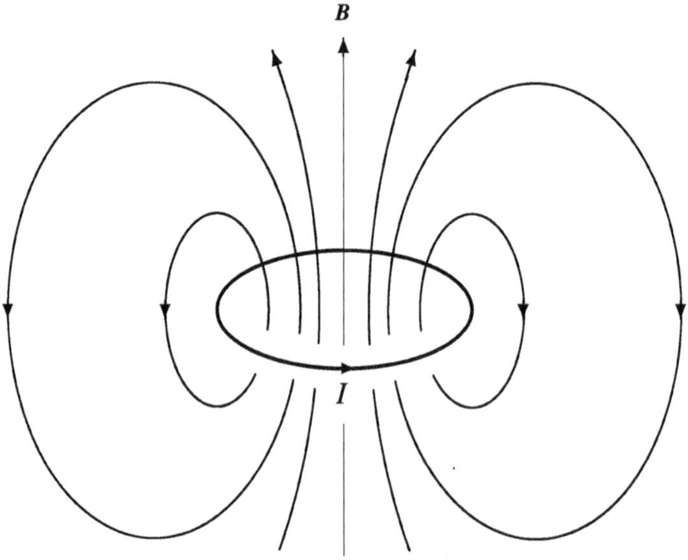

Figure 15.1 Sketch of the magnetic field of a current carrying wire loop. At large distances, the field becomes a dipole field (15.10).

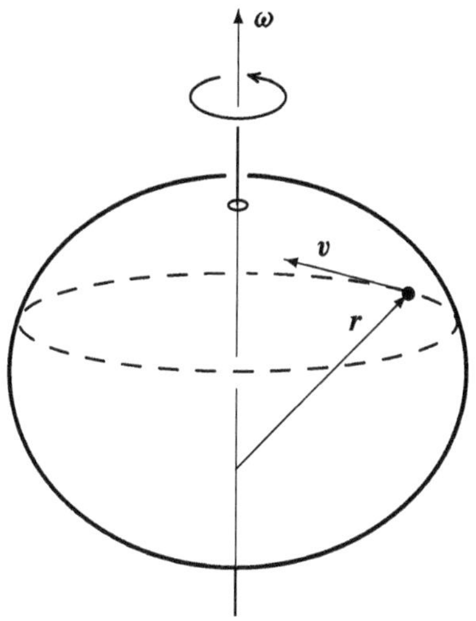

Figure 15.2 The rotation of a rigid body implies the velocity field $\boldsymbol{v}(\boldsymbol{r}) = \boldsymbol{\omega} \times \boldsymbol{r}$; where the vectors ω and $\boldsymbol{r}$ are defined by picture. If the body is massive and charged, then the rotation leads to an angular momentum $\boldsymbol{L}$ and to a magnetic moment $\boldsymbol{\mu}$. The vectors $\boldsymbol{L}$ and $\boldsymbol{\mu}$ are then parallel to each other. The quotient μ/L is called gyromagnetic ratio.

divide the mass density into individual mass points m_ν:

$$\boldsymbol{L} = \sum_\nu m_\nu\, \boldsymbol{r}_\nu \times \boldsymbol{v}_\nu = \int d^3r\ \varrho_{\rm m}(\boldsymbol{r})\,\big(\boldsymbol{r} \times \boldsymbol{v}(\boldsymbol{r})\big) \qquad (15.16)$$

We now introduce the *assumption* that the mass m and the charge q are equally distributed within the body:

$$\frac{\varrho(\boldsymbol{r})}{q} = \frac{\varrho_{\rm m}(\boldsymbol{r})}{m} = f(\boldsymbol{r}) \qquad (15.17)$$

This applies, for example, if the rigid body consists of individual parts which all have the same value $\Delta q/\Delta m$. From (15.15)–(15.17) we obtain

$$\boldsymbol{\mu} = \frac{q}{2mc}\,\boldsymbol{L} \qquad (15.18)$$

The ratio μ/L of magnetic moment to angular momentum is called *gyromagnetic ratio*. Deviations from the assumption (15.17) or relativistic or quantum mechanical effects cause the gyromagnetic ratio to deviate more or less strongly from $q/2mc$. These deviations are expressed by the so-called g-factor:

$$\boldsymbol{\mu} = g\,\frac{q}{2mc}\,\boldsymbol{L} \tag{15.19}$$

We apply this model of a rotating mass and charge distribution to the intrinsic angular momentum, the spin, of an electron. The projection s_z of the quantum mechanical spin can assume the values $\pm\hbar/2$. $s_z = \hbar/2$ has the magnetic moment of the electron (with $q = -e$) in the z-direction and then has the value

$$\mu_{\mathrm{e}} = -\frac{g}{2}\,\mu_{\mathrm{B}}, \qquad \mu_{\mathrm{B}} = \frac{e\hbar}{2m_{\mathrm{e}}c} \tag{15.20}$$

The experimental magnetic moment of the electron is about twice as large ($g \approx 2$) than one would expect in the naive model of a rotating, rigid charge and mass density. It is therefore approximately equal to the *Bohr magneton* μ_{B}. The experimental deviation[1] from the value μ_{B} is

$$\frac{g}{2} = 1.001\,159\,652\,180\,73 \pm 0.000\,000\,000\,000\,000\,28 \tag{15.21}$$

The value $g/2 = 1$ follows from the Dirac equation. In the quantum electrodynamics, corrections to this value are calculated: These corrections are classified by orders of the fine structure constant $\alpha = e^2/(\hbar c) \approx 1/137$. The first-order corrections result in $g/2 = 1 + \alpha/(2\pi) \approx 0.001\,161$ (J. Schwinger, *Physical Review* 73 (1948) 416). In higher order in α, values are obtained that agree with (15.21). The magnetic moment of the electron is a quantity that can be determined with a particular high precision, both experimentally and theoretically.

Protons and neutrons are also spin 1/2 particles. We assume the form $\mu_{\mathrm{p,n}} = (g/2)(e\hbar/2mc)$ for their magnetic moments, where m is the mass of the nucleon, $q_{\mathrm{p}} = e$ and $q_{\mathrm{n,fictitious}} = e$. The experiment yields

$$\frac{g_{\mathrm{p}}}{2} \approx 2.79, \qquad \frac{g_{\mathrm{n}}}{2} \approx -1.91 \tag{15.22}$$

All g-factors are roughly comparable to 1. Thus, the classical idea of a rotating mass and charge density explains at least the order of magnitude of the magnetic moments. This model implies in particular $\mu \propto 1/m$. Thereby, it explains that the magnetic moment of the proton is about 10^3

[1] Hanneke *et al.*, New measurement of the electron magnetic moment and the fine structure constant, arxiv:0801.1134 [physics.atom-ph].

times smaller than that of the electron. Of course, the model of the rotating, rigid mass and charge density is unrealistic in other respects.

The charge of the neutron is actually zero; the value $g_{\rm n} \approx -1.91$ applies to the fictitious value $q_{\rm n,fictitious} = e$. In a classical picture, one could assume that the neutron has a charge density that is positive for small radii but negative for larger radii. For this charge density, $q_{\rm n} = \int d^3r\, \varrho = 0$ must apply. In the integral (15.15), the larger distances are weighted stronger so that a negative g factor is possible.

Permanent Magnet

Currents in the atomic range determine the magnetic properties of various materials (dia-, para- and ferromagnetism, Chapter 32). In a ferromagnetic material such as iron, the spins of unpaired electrons may all align parallel (for not too high temperatures). This means that the magnetic moments of these electrons are aligned and result in a large total dipole moment. The alignment of the spins is stable for not too high temperatures, yielding a permanently magnetic material, i.e., a permanent magnet. Since the magnetic properties are ultimately determined by currents, the laws of magnetostatics apply to such a permanent magnet. For the permanent magnet, the sources of this field (atomic circular currents and their analogs in elementary particles) are usually not treated explicitly.

The magnetic field of a bar magnet is sketched in Figure 15.3. The field of the bar magnet is similar to that of the finite coil (compare Figure 14.2,

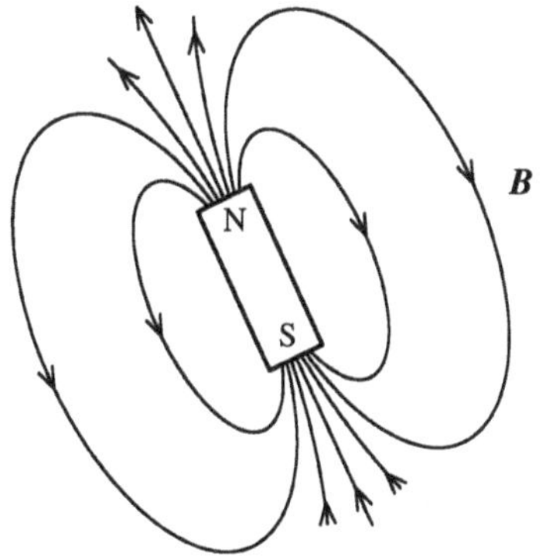

Figure 15.3 Sketch of the field of a bar magnet (permanent magnet). The magnetic field is based on permanent circular currents in the microscopic range.

right and Figure 15.3). The bar consists of ferromagnetic material in which all spins and magnetic moments of the unpaired electrons all aligned in the direction of the bar.

With a suitable bearing, a bar magnet can serve as a compass needle. The end pointing north is called the north pole of the needle. It points roughly to the geographic north pole which is the south pole of the Earth field.

Energy in an External Field

We calculate the potential energy of a current distribution in a external magnetic field $\boldsymbol{B}_{\text{ext}}$. Let the current density be spatially limited:

$$j(\boldsymbol{r}) = j(\boldsymbol{r}_0 + \boldsymbol{r}') = \widetilde{j}(\boldsymbol{r}') = \begin{cases} \text{arbitrary} & (r' < R_0) \\ 0 & (r' > R_0) \end{cases} \tag{15.23}$$

The coordinates were chosen as in Figure 12.3. The current distribution shall not change under the induced of the external field. In the region of the current density, we expand the external magnetic field $\boldsymbol{B}_{\text{ext}}$ into a Taylor series:

$$\boldsymbol{B}_{\text{ext}}(\boldsymbol{r}) = \boldsymbol{B}_{\text{ext}}(\boldsymbol{r}_0 + \boldsymbol{r}') = \boldsymbol{B}_{\text{ext}}(\boldsymbol{r}_0) + (\boldsymbol{r}' \cdot \boldsymbol{\nabla}_0)\, \boldsymbol{B}_{\text{ext}}(\boldsymbol{r}_0) + \cdots \tag{15.24}$$

From the force density $\boldsymbol{f} = \widetilde{\boldsymbol{j}} \times \boldsymbol{B}_{\text{ext}}/c$, (13.21), we obtain the total force $\boldsymbol{F}$ on the current distribution:

$$\begin{aligned} \boldsymbol{F} &= \frac{1}{c} \int d^3r'\; \widetilde{\boldsymbol{j}}(\boldsymbol{r}') \times \boldsymbol{B}_{\text{ext}}(\boldsymbol{r}_0 + \boldsymbol{r}') \\ &= \frac{1}{c} \int d^3r'\; \widetilde{\boldsymbol{j}}(\boldsymbol{r}') \times (\boldsymbol{r}' \cdot \boldsymbol{\nabla}_0)\, \boldsymbol{B}_{\text{ext}}(\boldsymbol{r}_0) + \cdots \\ &= \frac{1}{c} \int d^3r'\; \big((\boldsymbol{\nabla}_0 \cdot \boldsymbol{r}')\, \widetilde{\boldsymbol{j}}(\boldsymbol{r}') \big) \times \boldsymbol{B}_{\text{ext}}(\boldsymbol{r}_0) + \cdots \end{aligned} \tag{15.25}$$

The first term in the expansion (15.24) is omitted here because of (15.4). The terms not shown explicitly are omitted in the following.

In (15.6), $\boldsymbol{r}$ is an arbitrary vector that is independent of the integration. Therefore, this relationship also applies if $\boldsymbol{r}$ is replaced by $\boldsymbol{\nabla}_0$

(acting on $\boldsymbol{r}_0$):

$$\int d^3r' \,(\nabla_0 \cdot \boldsymbol{r}')\,\widetilde{\boldsymbol{j}}(\boldsymbol{r}') = -\int d^3r' \,\boldsymbol{r}' \left(\nabla_0 \cdot \widetilde{\boldsymbol{j}}(\boldsymbol{r}')\right) \tag{15.26}$$

Herewith, we transform (15.25):

$$\begin{aligned}
\boldsymbol{F} &= \frac{1}{2c} \int d^3r' \,\left(\left[\nabla_0 \cdot \boldsymbol{r}'\right]\widetilde{\boldsymbol{j}}(\boldsymbol{r}') - \boldsymbol{r}'\left[\nabla_0 \cdot \widetilde{\boldsymbol{j}}(\boldsymbol{r}')\right]\right) \times \boldsymbol{B}_{\text{ext}}(\boldsymbol{r}_0) \\
&= \frac{1}{2c} \left(\int d^3r' \,\left[\boldsymbol{r}' \times \widetilde{\boldsymbol{j}}(\boldsymbol{r}')\right] \times \nabla_0\right) \times \boldsymbol{B}_{\text{ext}}(\boldsymbol{r}_0) \\
&= \left(\boldsymbol{\mu} \times \nabla_0\right) \times \boldsymbol{B}_{\text{ext}}(\boldsymbol{r}_0) = \nabla_0 \left(\boldsymbol{\mu} \cdot \boldsymbol{B}_{\text{ext}}\right) - \boldsymbol{\mu}\left(\nabla_0 \cdot \boldsymbol{B}_{\text{ext}}\right) \\
&= \nabla_0 \left(\boldsymbol{\mu} \cdot \boldsymbol{B}_{\text{ext}}\right)
\end{aligned} \tag{15.27}$$

The formulas $(\boldsymbol{a} \times \boldsymbol{b}) \times \boldsymbol{c} = \boldsymbol{b}\,(\boldsymbol{a} \cdot \boldsymbol{c}) - \boldsymbol{a}\,(\boldsymbol{b} \cdot \boldsymbol{c})$ and $\nabla_0 \cdot \boldsymbol{B}_{\text{ext}} = 0$ were used. In the result, the operator ∇_0 only acts on $\boldsymbol{B}_{\text{ext}}$ because $\boldsymbol{\mu}$ is a constant. With $\boldsymbol{F} = -\,\text{grad}\, W(\boldsymbol{r})$, we can identify

$$W = -\boldsymbol{\mu} \cdot \boldsymbol{B}_{\text{ext}}(\boldsymbol{r}_0) = -\mu\, B_{\text{ext}} \cos\theta \tag{15.28}$$

as the potential energy of a dipole in the external field. In addition to $\boldsymbol{r}_0$, the potential energy depends on the angle θ between the direction of the dipole and the magnetic field. The field therefore exerts a torque $\boldsymbol{M}$ with the magnitude $M = -\partial W/\partial\theta$. The torque is perpendicular to $\boldsymbol{\mu}$ and $\boldsymbol{B}_{\text{ext}}$, i.e.,

$$\boldsymbol{M} = \boldsymbol{\mu} \times \boldsymbol{B}_{\text{ext}} \tag{15.29}$$

We consider some applications of (15.28) and (15.29):

1. A freely rotating magnetic dipole adjusts itself in equilibrium in such a way that the energy (15.28) is minimal. Then, $\boldsymbol{\mu}$ points in the direction of $\boldsymbol{B}$. This is the principle of the compass.
2. In a known homogeneous field $\boldsymbol{B}_0$, there is a freely rotatable compass needle. Now, an additional magnetic field of unknown strength and perpendicular to $\boldsymbol{B}_0$ is switched on. From the orientation of the needle in the combined field, the unknown field strength can be determined.
3. In a homogeneous magnetic field $\boldsymbol{B} = B\,\boldsymbol{e}_z$, there are electrons. Quantum mechanically, the two spin settings $s_z = \pm\hbar/2$ are possible. The spins can switch their direction by absorption of electromagnetic

radiation. To do this, the frequency of the radiation must fulfill the condition $\hbar\omega = 2\mu_{\mathrm{B}}B$. By measuring the frequency ω, the strength of the magnetic field can be determined.

Exercises

15.1 *Localized current density*

Let the current distribution $\boldsymbol{j}(\boldsymbol{r})$ be spatially limited. Derive

$$\int d^3r\, \boldsymbol{j}(\boldsymbol{r}) = 0$$

from $\operatorname{div} \boldsymbol{j}(\boldsymbol{r}) = 0$. To do this, use $\boldsymbol{j} = (\boldsymbol{j} \cdot \boldsymbol{\nabla})\,\boldsymbol{r}$.

15.2 *Cylindrically symmetric current density*

A cylindrically symmetrical current distribution is given in spherical coordinates:

$$\boldsymbol{j} = j(r, \theta)\, \boldsymbol{e}_\phi \tag{15.30}$$

Show that the vector potential is also of this form and derive an expression for $A(r, \theta)$. Which scalar differential equation for $A(r, \theta)$ follows from $\Delta \boldsymbol{A} = -4\pi \boldsymbol{j}/c$?

Hints: Consider the combination $A_x + \mathrm{i}A_y$. Expand $1/|\boldsymbol{r} - \boldsymbol{r}'|$ in the integral formula for the vector potential into spherical harmonics.

15.3 *Current flowing through a conductor loop*

In a circular wire loop, the current density in cylindrical coordinates is given by

$$\boldsymbol{j} = I\, \delta(\rho - R)\, \delta(z)\, \boldsymbol{e}_\varphi$$

Calculate the vector potential for $\rho \ll R$ and $\rho \gg R$. Show that a dipole field $\boldsymbol{A} = (\boldsymbol{\mu} \times \boldsymbol{r})/r^3$ results for large distances.

Hints: According to Exercise 15.2, the cylindrically symmetrical current density implies a potential with the same symmetry, i.e., $\boldsymbol{A}(\boldsymbol{r}) = A(\rho, z)\, \boldsymbol{e}_\varphi$. In order to determine $A(\rho, z)$, it is sufficient to evaluate the integral formula for A_y with $\varphi = 0$. The resulting integral should only be solver for the considered limiting cases.

15.4 *Helmholtz coils*

Two parallel circular conductor loops carry both the current I (flowing in the same direction). The circles are parallel to the x-y plane, they both have the radius R and their centers are at $(x, y, z) = (0, 0, b)$ and $(0, 0, -b)$.

Determine the vector potential of this arrangement as the superposition of the potentials of the individual wire loops (Exercise 15.3). Expand the vector potential in the vicinity of the coordinate origin up to the order $\mathcal{O}(\rho^3, \rho z^2)$. What relationship must exist between the radius R and the distance $D = 2b$ of the circles, so that the magnetic field in this region is as homogeneous as possible?

15.5 *Rotating homogeneously charged sphere*

A homogeneously charged sphere (radius R, charge q) rotates with the angular velocity $\boldsymbol{\omega}$ around an axis through its center. What is the current density $\boldsymbol{j}$? Determine the vector potential $\boldsymbol{A}$ from the integral formula. Calculate the components of the magnetic field in spherical coordinates.

Hints: Place $\boldsymbol{\omega}$ in the z-direction and calculate $A_x + \mathrm{i}\,A_y$ using the integral formula (14.3). Express $j_x + \mathrm{i}\,j_y$ by Y_{11} and expand $1/|\boldsymbol{r} - \boldsymbol{r}'|$ into spherical harmonics.

15.6 *Surface currents for a homogeneously magnetized sphere*

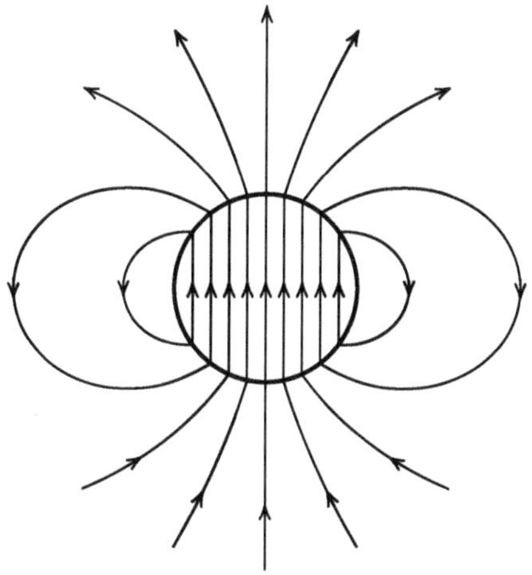

The magnetic field

$$\boldsymbol{B} = \begin{cases} B_0 \boldsymbol{e}_z & (r < R) \\ \dfrac{3\boldsymbol{r}(\boldsymbol{r}\cdot\boldsymbol{\mu}) - \boldsymbol{\mu}\, r^2}{r^5} & (r > R) \end{cases}$$

belongs to a homogeneously magnetized sphere with the dipole moment $\boldsymbol{\mu} = \mu\, \boldsymbol{e}_z$. The equations $\operatorname{div} \boldsymbol{B} = 0$ and $\operatorname{curl} \boldsymbol{B} = 0$ apply in the ranges $r < R$ and $r > R$, respectively. The sources of the field are therefore surface currents.

Due to the cylindrical symmetry, the surface currents are of the form

$$\boldsymbol{j} = \frac{I(\theta)}{\pi R}\, \delta(r - R)\, \boldsymbol{e}_\phi$$

Determine the current $I(\theta)$ and the magnetic moment μ. Derive the following conditions from the field equations:

$$B_r(R + \epsilon) - B_r(R - \epsilon) = 0 \qquad (15.31)$$

$$B_\theta(R + \epsilon) - B_\theta(R - \epsilon) = \frac{4\pi}{c}\, \frac{I(\theta)}{\pi R} \qquad (15.32)$$

15.7 *Small permanent magnet*

A small permanent magnet (dipole moment $\boldsymbol{\mu}$) is positioned at $\boldsymbol{d} = d\, \boldsymbol{e}_x$. It can rotate freely within the x-y plane. A homogeneous external magnetic field $\boldsymbol{B}_0 = B_0\, \boldsymbol{e}_x$ acts on the magnet.

In which direction does $\boldsymbol{\mu}$ point in equilibrium? In which direction does $\boldsymbol{\mu}$ point in equilibrium if there is in addition a wire with the current density $\boldsymbol{j} = I\, \delta(x)\, \delta(y)\, \boldsymbol{e}_z$?

PART IV

Maxwell Equations: Basics

Chapter 16

Maxwell Equations

In Parts II and III, we dealt with electrostatics and magnetostatics separately. For time-dependent processes, there are couplings between electric and magnetic fields. These coupling are named "Faraday law of induction" and "Maxwell displacement current". The coupling terms can be justified by key experiments. By these terms, the field equations of electrostatics and magnetostatics are generalized to the Maxwell equations. The energy and momentum density of the electromagnetic field are determined.

The field equations of electro- and magnetostatics are

$$\begin{aligned} &\operatorname{div} \boldsymbol{E}(\boldsymbol{r}) = 4\pi\,\varrho(\boldsymbol{r}), && \operatorname{curl} \boldsymbol{E}(\boldsymbol{r}) = 0 \quad \text{(electrostatics)} \\ &\operatorname{curl} \boldsymbol{B}(\boldsymbol{r}) = (4\pi/c)\,\boldsymbol{j}(\boldsymbol{r}), && \operatorname{div} \boldsymbol{B}(\boldsymbol{r}) = 0 \quad \text{(magnetostatics)} \end{aligned} \tag{16.1}$$

From this follows the continuity equation in the form

$$\operatorname{div} \boldsymbol{j}(\boldsymbol{r}) = 0 \tag{16.2}$$

The electric and magnetic fields are defined by the forces that they exert. A point charge with the position $\boldsymbol{r}_0$ and the velocity $\boldsymbol{v} = \dot{\boldsymbol{r}}_0$ is acted upon by the forces (5.11) and (13.22):

$$\boldsymbol{F} = q\,\boldsymbol{E}(\boldsymbol{r}_0) \quad \text{and} \quad \boldsymbol{F} = q\,\frac{\boldsymbol{v}}{c} \times \boldsymbol{B}(\boldsymbol{r}_0) \tag{16.3}$$

It is easy to see that there must be connections between the electric and magnetic field, i.e., that (16.1) is not a generally valid theory: In an inertial system (IS), there is a static charge distribution $\varrho(\boldsymbol{r})$. In an IS$'$ that is moving relative to IS, this distribution appears as a charge *and* current distribution. While there is only an electric field in IS, there is also a magnetic field in IS′. Therefore, the fields transform into each other during the transition between inertial frames; they depend on the observer's point of view.

For a limiting case, we can read off the transformation of the fields from (16.3). In an IS, a particle shall move with a constant velocity $\boldsymbol{v}$ in an electric field $\boldsymbol{E}$ and a magnetic field $\boldsymbol{B}$. According to (16.3), the force $q\,(\boldsymbol{E} + (\boldsymbol{v}/c) \times \boldsymbol{B})$ acts on the particle. We now go to the rest system IS′ of the particle. Due to $\boldsymbol{v}' = 0$, Equation (16.3) yields the force $q\boldsymbol{E}'$, where $\boldsymbol{E}'$ is the electric field in IS′. In the non-relativistic limiting case (neglecting terms $\mathcal{O}(v^2/c^2)$), both are forces are equal. Therefore, the following applies

$$\boldsymbol{E}' = \boldsymbol{E} + \frac{\boldsymbol{v}}{c} \times \boldsymbol{B} \quad (v \ll c) \tag{16.4}$$

for the transformation from IS (with $\boldsymbol{E}$ and $\boldsymbol{B}$) to IS′ (with $\boldsymbol{E}'$).

We now present the generalization of (16.1) for time-dependent fields. For this purpose, we provide all fields with a time argument and add two terms:

$$\begin{aligned} &\operatorname{div}\boldsymbol{E}(\boldsymbol{r},t) = 4\pi\,\varrho(\boldsymbol{r},t), \quad \operatorname{curl}\boldsymbol{E}(\boldsymbol{r},t) + \underbrace{\frac{1}{c}\frac{\partial \boldsymbol{B}(\boldsymbol{r},t)}{\partial t}}_{\text{induction}} = 0 \\ &\operatorname{curl}\boldsymbol{B}(\boldsymbol{r},t) - \underbrace{\frac{1}{c}\frac{\partial \boldsymbol{E}(\boldsymbol{r},t)}{\partial t}}_{\text{displacement current}} = \frac{4\pi}{c}\boldsymbol{j}(\boldsymbol{r},t), \quad \operatorname{div}\boldsymbol{B}(\boldsymbol{r},t) = 0 \end{aligned} \tag{16.5}$$

These *Maxwell equations* were established by Maxwell in 1864. The additional terms designated by *induction* and *displacement current* are explained in the following two sections.

We differentiate the Maxwell equation with $\operatorname{div}\boldsymbol{E}$ with respect to the time, and form the divergence of the equation with $\operatorname{curl}\boldsymbol{B}$. The combination of the two resulting equations yields the continuity equation:

$$\frac{\partial \varrho(\boldsymbol{r},t)}{\partial t} + \operatorname{div}\boldsymbol{j}(\boldsymbol{r},t) = 0 \tag{16.6}$$

The forces (16.3) are summarized by the *Lorentz force* $\boldsymbol{F}_{\mathrm{L}}$:

$$\boldsymbol{F}_{\mathrm{L}} = q\left(\boldsymbol{E}(\boldsymbol{r}_0,t) + \frac{\boldsymbol{v}}{c} \times \boldsymbol{B}(\boldsymbol{r}_0,t)\right) \tag{16.7}$$

Equations (16.5) and (16.7) are the basic equations of of electrodynamics. All relevant statements are derived from them.

Faraday's Law of Induction

We connect the term "induction" in (16.5) with simple key experiments.

We integrate Maxwell's equation $\operatorname{curl}\boldsymbol{E} = -\dot{\boldsymbol{B}}/c$ over a time-independent area $\boldsymbol{a}$ with boundary C and use Stokes' theorem:

$$\oint_C d\boldsymbol{r}\cdot\boldsymbol{E} = -\frac{1}{c}\int_a d\boldsymbol{a}\cdot\frac{\partial\boldsymbol{B}(\boldsymbol{r},t)}{\partial t} \tag{16.8}$$

A given contour C can be spanned by different areas, such as a_1 and a_2. The difference $\int_{a_1} d\boldsymbol{a}\cdot\boldsymbol{B} - \int_{a_2} d\boldsymbol{a}\cdot\boldsymbol{B}$ vanishes because it can be transformed into a closed surface integral $\oint d\boldsymbol{a}\cdot\boldsymbol{B} = \int d^3r\,\operatorname{div}\boldsymbol{B} = 0$. The exact choice of surface area is therefore not relevant.

If the closed contour is cut open, there are two adjacent end points 1 and 2 (Figure 16.1). Between them, it comes to the voltage

$$U = \int_1^2 d\boldsymbol{r}\cdot\boldsymbol{E} = \oint d\boldsymbol{r}\cdot\boldsymbol{E} \tag{16.9}$$

For infinitesimally neighboring points 1 and 2, the line integral is equal to the integral over the closed contour C. The *induced voltage* (or induction voltage) U can be measured by placing a wire loop along the contour C

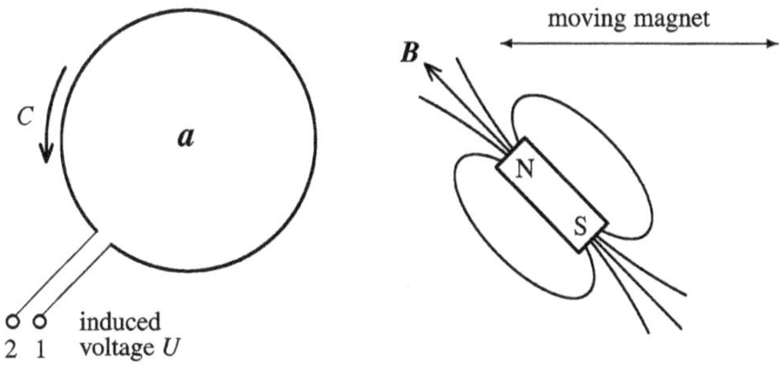

Figure 16.1 A magnet moves relative to a stationary wire loop. Then the time-dependent magnetic field induces a voltage that can be measured at the two wire endings. This effect is described by the Faraday law of induction. This effect is the basis for producing of electricity in a dynamo or generator, and also for the functioning of a transformer or an electrodynamic microphone. For the specified direction of C, the surface normal $\boldsymbol{a}$ points toward the viewer.

(Figure 16.1). Due to the force $q\boldsymbol{E}$, the mobile charges in the wire shift until the mean electric field in the wire disappears (Part VI). The resulting charge accumulations at the endings of the wire then cause the induced voltage.[1] This *induction* of a voltage was observed in 1831 by Faraday. This measurement verifies the equation $\operatorname{curl}\boldsymbol{E} = -\dot{\boldsymbol{B}}/c$.

If a time-dependent magnetic field (such as in Figure 16.1) induces a current in a closed wire loop, then the induced current weakens the external magnetic field. This is called *Lenz's rule*: The induced currents are directed such that their magnetic field weakens the external field (the field of the bar magnet in Figure 16.1). This leads to an explanation of diamagnetism (Chapter 32).

For a time-dependent contour $C(t)$ (with the area $a(t)$), we denote the velocity of the line element $d\boldsymbol{r}$ by $\boldsymbol{v}$ (Figure 16.2); the line element shall carry the charge q. For the moving charge q, the Coulomb force $q\boldsymbol{E}$ must be replaced by the Lorentz force $\boldsymbol{F}_{\rm L} = q(\boldsymbol{E} + (\boldsymbol{v}/c) \times \boldsymbol{B})$. The energy change $q\,dU = q\boldsymbol{E}\cdot d\boldsymbol{r}$ is then replaced by $q\,dU = q(\boldsymbol{E}+(\boldsymbol{v}/c)\times\boldsymbol{B})\cdot d\boldsymbol{r}$,

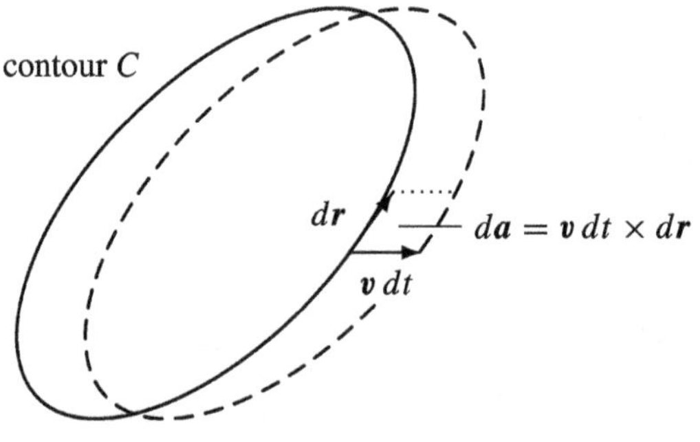

Figure 16.2 The Faraday law of induction (16.12) applies also for a time dependence of the contour $C(t)$. If a selected line element $d\boldsymbol{r}$ moves with the speed $\boldsymbol{v}$, it yields the change $d\boldsymbol{a} = \boldsymbol{v}\,dt \times d\boldsymbol{r}$ of the considered area. For each individual line element $d\boldsymbol{r}$, a local velocity $\boldsymbol{v}$ may be used. This allows then also for continuous deformations of the contour (in contrast to the sketch).

[1]Since the mean electric field in the wire disappears, the electric field strength is shifted into the gap from $2 \to 1$, $U = \oint d\boldsymbol{r}\cdot\boldsymbol{E} = \int_2^1 d\boldsymbol{r}\cdot\boldsymbol{E}_{\rm gap} = -\int_1^2 d\boldsymbol{r}\cdot\boldsymbol{E}_{\rm gap}$. This means that the voltage measured at the gap has the opposite sign.

and (16.9) becomes

$$U = \oint_{C(t)} d\boldsymbol{r} \cdot \left(\boldsymbol{E} + \frac{\boldsymbol{v}}{c} \times \boldsymbol{B} \right) \tag{16.10}$$

This is the induced voltage in a wire loop with the time dependent contour $C(t)$. A time dependence of the contour may, for example, results from the movement of a rigid wire loop.

The induction voltage U can be linked to the change in the magnetic flux $\Phi_{\rm m} = \int_{a(t)} \boldsymbol{a} \cdot \boldsymbol{B}$ inside the contour:

$$\frac{d\Phi_{\rm m}}{dt} = \int \frac{d\boldsymbol{a}}{dt} \cdot \boldsymbol{B} + \int d\boldsymbol{a} \cdot \frac{\partial \boldsymbol{B}(\boldsymbol{r}, t)}{\partial t} \overset{(16.8)}{=} -c \oint d\boldsymbol{r} \cdot \left(\boldsymbol{E} + \frac{\boldsymbol{v}}{c} \times \boldsymbol{B} \right) \tag{16.11}$$

For the transformation of the first integral, we have used $d\boldsymbol{a}/dt = (\boldsymbol{v} \times d\boldsymbol{r})$, which can be seen in Figure 16.2. For the second integral, Equation (16.8) was used.

From (16.10) and (16.11), we can now read the relationship between the induction voltage and the change in magnetic flux:

$$\boxed{U = -\frac{1}{c}\frac{d\Phi_{\rm m}}{dt} = -\frac{1}{c}\frac{d}{dt}\int_{a(t)} d\boldsymbol{a} \cdot \boldsymbol{B}(\boldsymbol{r}, t)} \tag{16.12}$$

This is the general form of the *Faraday law of induction* (also called Faraday's law for short). Faraday's law of induction states the following: The induced voltage is determined by the change in the magnetic flux $\Phi_{\rm m} = \int d\boldsymbol{a} \cdot \boldsymbol{B}$ through the loop.

For (16.12), a time-dependent area $a(t)$ was allowed. Without this time independence, formula (16.12) is reduced to (16.8). Also, the restricted statement (16.8) is referred to as Faraday's law of induction.

One application of induction is the eddy current brake. A conductive, rotating metal disk is exposed at one point to a constant magnetic field that is perpendicular to the disk. The rotation of the disk leads to induction currents (eddy or vortex currents). The finite conductivity of the metal means that energy is converted into heat (31.24). This energy reduces the rotational energy of the disk; this implies a braking effect for the disk.

Maxwell's Displacement Current

We connect the term "displacement current" in (16.5) with simple key experiments.

The Faraday experiments led to the induction term in (16.5); this is the first term for the generalization of (16.1). The second decisive step was made by Maxwell in 1864. Maxwell started from the continuity equation (16.6), which follows from the conservation of charge, see (13.10)–(13.12). Combining the continuity equation with $\operatorname{div} \boldsymbol{E}(\boldsymbol{r}, t) = 4\pi\,\varrho(\boldsymbol{r}, t)$ results in

$$\dot{\varrho} + \operatorname{div}\boldsymbol{j} = \operatorname{div}\left(\frac{1}{4\pi}\,\frac{\partial \boldsymbol{E}}{\partial t} + \boldsymbol{j}\right) = 0 \tag{16.13}$$

This result is not compatible with $\operatorname{curl} \boldsymbol{B}(\boldsymbol{r}, t) = 4\pi \boldsymbol{j}(\boldsymbol{r}, t)/c$ because from this follows $\operatorname{div}\boldsymbol{j}(\boldsymbol{r}, t) = 0$. However, the equations become consistent by adding *Maxwell's displacement current* $\dot{\mathbf{E}}/4\pi$ to $\boldsymbol{j}$ in the inhomogeneous magnetic field equation:

$$\operatorname{curl} \boldsymbol{B}(\boldsymbol{r}, t) - \frac{1}{c}\,\frac{\partial \boldsymbol{E}(\boldsymbol{r}, t)}{\partial t} = \frac{4\pi}{c}\,\boldsymbol{j}(\boldsymbol{r}, t) \tag{16.14}$$

The additional term is enforced by requiring the consistency with the continuity equation. The displacement current can easily be verified experimentally. For this purpose, we consider the integral form of (16.14) in a current-free range:

$$\oint_C d\boldsymbol{r} \cdot \boldsymbol{B} = \frac{1}{c}\,\frac{\partial}{\partial t} \int_a d\boldsymbol{a} \cdot \boldsymbol{E} \quad (\boldsymbol{j} = 0) \tag{16.15}$$

In the same way as the change in the magnetic flux causes an electric field, the change in the electric flux $\Phi_{\mathrm{e}} = \oint d\boldsymbol{a} \cdot \boldsymbol{E}$ causes a magnetic field. For example, short-circuiting a charged plate capacitor leads to closed magnetic field lines (Figure 16.3). This magnetic field could, for example, be observed using a compass needle.

Unified Theory

Maxwell's theory is the standard example of a *unified theory*. This means that phenomena that were initially treated separately can be understood and described within a single theory.

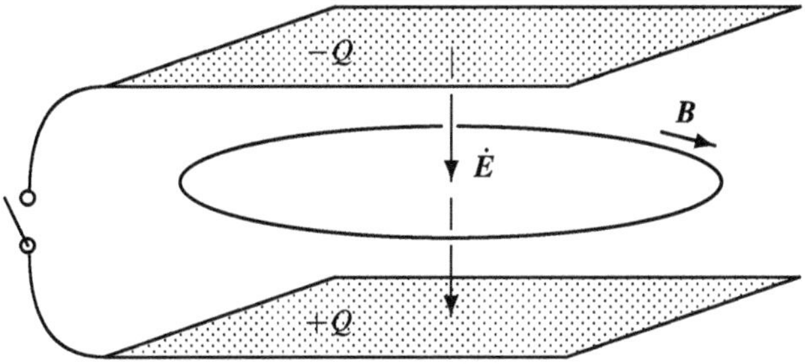

Figure 16.3 When the capacitor is short-circuited, the electric field decays. This time dependence $\dot{\boldsymbol{E}} \neq 0$ causes a magnetic field with closed, circular field lines. This field has the same geometry as a field created by the current density $\boldsymbol{j} \parallel \dot{\boldsymbol{E}}$. Therefore, on the right-hand side of $\operatorname{curl} \boldsymbol{B} = (4\pi/c)\boldsymbol{j}$ is supplemented by Maxwell's displacement current $\dot{\boldsymbol{E}}/c$.

A first example of a unified theory is Newton's gravitation theory.[2] From this theory, Galileo's laws for the free fall and Kepler's laws for the planetary motion can be derived; before Newton there was no connection between these laws. It was Newton's outstanding achievement to explain the fall of an apple and the orbit of Venus as a consequence of the same force and the same law of motion.

Maxwell's equations represent a unified theory because they explain the electric and magnetic phenomena on a common basis. In earlier times, no connection was seen between these phenomena.

In the 1960s, Weinberg, Salam and Glashow developed a unified theory for the electromagnetic and weak interaction. This theory predicted the existence of vector bosons (Z^0 and $W^{\pm}$), which were detected in the 1980s.

It is a goal of physics to explain all interactions, in particular including the strong interaction and the gravitational interaction, within the framework of a unified theory. The Maxwell's theory and its successors (quantum electrodynamics, Weinberg–Salam theory) serve as a template for this goal.

Energy Balance

We determine the energy density of the electromagnetic field. To do this, we investigate the energy balance for a system of fields and N charges

[2] I. Newton, *Philosophiae naturalis principia mathematica*, 1687.

q_i (with the positions $\boldsymbol{r}_i$ and the velocities $\boldsymbol{v}_i$). The electromagnetic field exerts the force (16.7) on the individual particles. This causes changes in the energy E_{mat} of the particles:

$$dE_{\text{mat}} = \sum_{i=1}^{N} \boldsymbol{F}_{\text{L},i} \cdot d\boldsymbol{r}_i = \sum_{i=1}^{N} q_i \left(\boldsymbol{E} + \frac{\boldsymbol{v}_i}{c} \times \boldsymbol{B}\right) \cdot \boldsymbol{v}_i\, dt$$

$$= \sum_{i=1}^{N} q_i\, \boldsymbol{v}_i \cdot \boldsymbol{E}(\boldsymbol{r}_i, t)\, dt \tag{16.16}$$

One might think of electrons in the field of a capacitor or in a particle accelerator. In (16.16), we use the current density $\boldsymbol{j} = \sum q_i\, \boldsymbol{v}_i\, \delta(\boldsymbol{r} - \boldsymbol{r}_i)$:

$$\frac{dE_{\text{mat}}}{dt} = \int d^3r\, \boldsymbol{j}(\boldsymbol{r}, t) \cdot \boldsymbol{E}(\boldsymbol{r}, t) \tag{16.17}$$

With the help of Maxwell's equations, we express $\boldsymbol{j} \cdot \boldsymbol{E}$ by the fields:

$$\boldsymbol{j} \cdot \boldsymbol{E} = \frac{c}{4\pi}\, \boldsymbol{E} \cdot \text{curl}\, \boldsymbol{B} - \frac{1}{4\pi}\, \boldsymbol{E} \cdot \frac{\partial \boldsymbol{E}}{\partial t}$$

$$= -\frac{c}{4\pi}\, \text{div}\left(\boldsymbol{E} \times \boldsymbol{B}\right) + \frac{c}{4\pi}\, \boldsymbol{B} \cdot \text{curl}\, \boldsymbol{E} - \frac{1}{8\pi} \frac{\partial \boldsymbol{E}^2}{\partial t}$$

$$= -\frac{c}{4\pi}\, \text{div}\left(\boldsymbol{E} \times \boldsymbol{B}\right) - \frac{1}{8\pi} \frac{\partial}{\partial t} \left(\boldsymbol{E}^2 + \boldsymbol{B}^2\right) \tag{16.18}$$

We have used $\boldsymbol{\nabla} \cdot (\boldsymbol{E} \times \boldsymbol{B}) = \boldsymbol{B} \cdot (\boldsymbol{\nabla} \times \boldsymbol{E}) - \boldsymbol{E} \cdot (\boldsymbol{\nabla} \times \boldsymbol{B})$. With the abbreviations

$$w_{\text{em}} = w_{\text{em}}(\boldsymbol{r}, t) = \frac{1}{8\pi} \left(\boldsymbol{E}^2 + \boldsymbol{B}^2\right) \quad \text{energy density} \tag{16.19}$$

and

$$\boldsymbol{S} = \boldsymbol{S}(\boldsymbol{r}, t) = \frac{c}{4\pi} \left(\boldsymbol{E} \times \boldsymbol{B}\right) \quad \begin{array}{l}\text{Poynting vector}\\ \text{energy current density}\end{array} \tag{16.20}$$

we write (16.18) in the form

$$\frac{\partial w_{\text{em}}}{\partial t} + \text{div}\, \boldsymbol{S} = -\boldsymbol{j} \cdot \boldsymbol{E} \quad \text{(Poynting theorem)} \tag{16.21}$$

The following discussion shows that w_{em} is the energy density of the electromagnetic field, and $\boldsymbol{S}/c^2$ is the momentum density.

We integrate (16.21) over a time-independent volume using Gauss' theorem:

$$\frac{\partial}{\partial t}\int_V d^3r\; w_{\rm em}(\boldsymbol{r},t) + \oint_{a(V)} d\boldsymbol{a}\cdot\boldsymbol{S}(\boldsymbol{r},t) = -\frac{dE_{\rm mat}}{dt} \qquad (16.22)$$

On the right-hand side, (16.17) was used. With $E_{\rm em} = \int_V d^3r\; w_{\rm em}$ (16.22) becomes

$$\frac{dE_{\rm em}}{dt} + \frac{dE_{\rm mat}}{dt} = -\oint_{a(V)} d\boldsymbol{a}\cdot\boldsymbol{S}(\boldsymbol{r},t) \qquad (16.23)$$

In (16.16), it was summed over all material particles. In $E_{\rm mat} = \int_V d^3r\ldots$, it was assumed that all N particles remain in the volume under consideration.

We now consider a closed system and choose the volume V so that the system lies within V. Then, the integral over the surface $a(V)$ of the volume disappears, and (16.23) becomes

$$E = E_{\rm mat} + E_{\rm em} = \text{const.} \quad \text{(closed system)} \qquad (16.24)$$

The volume V can also be the whole space. For the vanishing of the surface integral, it is then sufficient that the fields fall off at least like $1/r^2$ for $r\to\infty$.

The quantity (16.24) has the dimension of an energy. The energy quantity, which is constant for a closed system, is the total energy of the system. The system consists of fields and particles. It is known that $E_{\rm mat}$ is the energy of the particles. From this follows that $E_{\rm em} = \int d^3r\; w_{\rm em}$ is the energy and $w_{\rm em}$ the energy density of the electromagnetic field. The energy density $w_{\rm em} = (\boldsymbol{E}^2 + \boldsymbol{B}^2)/8\pi$ is a generalization of the energy density known from electrostatics $u = \boldsymbol{E}^2/8\pi$, (6.31).

We now consider (16.23) for a finite volume V. The energy $E_{\rm em}$ of the electromagnetic fields can change due to the fact that (i) energy is transferred to the material particles, or that (ii) electromagnetic energy reaches the surface $a(V)$ of the volume V. This means that the Poynting vector $\boldsymbol{S}$ is the *energy current density* of the electromagnetic field. It has the dimension

$$S = \frac{\text{energy}}{\text{area}\cdot\text{time}} = \frac{\text{power}}{\text{area}} \qquad (16.25)$$

From this follows S/c^2 = momentum/volume, therefore $\boldsymbol{S}/c^2$ is a momentum density. This suggests that $\boldsymbol{S}/c^2$ is the *momentum density* of the electromagnetic field; formally this is confirmed in the following section.

Momentum Balance

The momentum balance is derived in the same way as the energy balance; it is presented only briefly here.[3] The time derivative of the momentum $\boldsymbol{P}_{\text{mat}}$ of a system of N particles follows from Newton's equation of motion:

$$\frac{d\boldsymbol{P}_{\text{mat}}}{dt} = \sum_{i=1}^{N} \frac{d\boldsymbol{p}_i}{dt} = \sum_{i=1}^{N} \boldsymbol{F}_{\text{L},i} = \sum_{i=1}^{N} q_i \left(\boldsymbol{E}(\boldsymbol{r}_i, t) + \frac{\boldsymbol{v}_i}{c} \times \boldsymbol{B}(\boldsymbol{r}_i, t) \right) \qquad (16.26)$$

Here we introduce the charge density $\varrho = \sum q_i\, \delta(\boldsymbol{r} - \boldsymbol{r}_i)$ and the current density $\boldsymbol{j} = \sum q_i\, \boldsymbol{v}_i\, \delta(\boldsymbol{r} - \boldsymbol{r}_i)$:

$$\frac{d\boldsymbol{P}_{\text{mat}}}{dt} = \int d^3r \left(\varrho(\boldsymbol{r}, t)\, \boldsymbol{E}(\boldsymbol{r}, t) + \frac{\boldsymbol{j}(\boldsymbol{r}, t)}{c} \times \boldsymbol{B}(\boldsymbol{r}, t) \right) \qquad (16.27)$$

We transform the integrand again using Maxwell's equations. After a few intermediate calculations[4] this yields

$$\varrho\, \boldsymbol{E} + \frac{\boldsymbol{j} \times \boldsymbol{B}}{c} = -\frac{\partial \boldsymbol{g}_{\text{em}}}{\partial t} + \sum_{i,k=1}^{3} \frac{\partial T_{ik}}{\partial x_k}\, \boldsymbol{e}_i \qquad (16.28)$$

Here, we have introduced the quantities

$$\boxed{\boldsymbol{g}_{\text{em}} = \frac{1}{4\pi c}\, \boldsymbol{E} \times \boldsymbol{B} = \frac{\boldsymbol{S}}{c^2} \quad \text{momentum density}} \qquad (16.29)$$

and

$$T_{ik} = \frac{1}{4\pi} \left(E_i\, E_k + B_i\, B_k - \frac{\delta_{ik}}{2} \left(\boldsymbol{E}^2 + \boldsymbol{B}^2 \right) \right) \qquad (16.30)$$

In the following, $\boldsymbol{g}_{\text{em}}$ is identified as the momentum density of the electromagnetic field. The quantities T_{ik} are referred to as the *Maxwell stress tensor*.

---

[3]This section can be skipped.

[4]These calculations may be found in Chapter 6.8 of [6].

We integrate (16.28) over a volume V that fully encloses the system. The term with $\partial T_{ik}/\partial x_k$ is transformed into an surfaces integral using Gauss' theorem; for the closed system, the fields and thus the T_{ik} vanish at the surface of the volume. Together with (16.27) we obtain

$$\frac{d\boldsymbol{P}_{\text{mat}}}{dt} + \frac{\partial}{\partial t} \int_V d^3r\, \boldsymbol{g}_{\text{em}}(\boldsymbol{r}, t) = \frac{d}{dt} \left(\boldsymbol{P}_{\text{mat}} + \boldsymbol{P}_{\text{em}} \right) = 0 \qquad (16.31)$$

or $\boldsymbol{P}_{\text{mat}} + \boldsymbol{P}_{\text{em}} = \text{const}$. This is the law of conservation of momentum for the closed system. From this follows the interpretation of $\boldsymbol{g}_{\text{em}}$ as the momentum density of the electromagnetic field. From this we may also get the angular momentum density

$$\boldsymbol{l}_{\text{em}} = \boldsymbol{l}_{\text{em}}(\boldsymbol{r}, t) = \boldsymbol{r} \times \boldsymbol{g}_{\text{em}} = \frac{1}{4\pi c}\, \boldsymbol{r} \times \left(\boldsymbol{E} \times \boldsymbol{B} \right) \qquad (16.32)$$

Exercises

16.1 *Induction in a moving rectangular conductor loop*

A rectangular conductor loop (side lengths b_1 and b_2) lies in the x-y-plane and moves with constant, non-relativistic velocity $\boldsymbol{v} = v\, \boldsymbol{e}_x$.

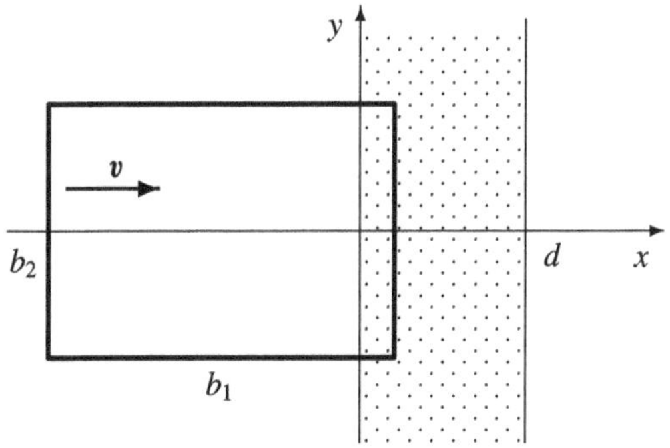

In the range $0 \leq x \leq d < b_1$, there is a constant homogeneous magnetic field $\boldsymbol{B} = B_0\, \boldsymbol{e}_z$ (marked by dots in the sketch).

Calculate the induces voltage $U(t)$ in the conductor loop. Sketch the function $U(t)$.

16.2 *Induction in a moving circular conductor loop*

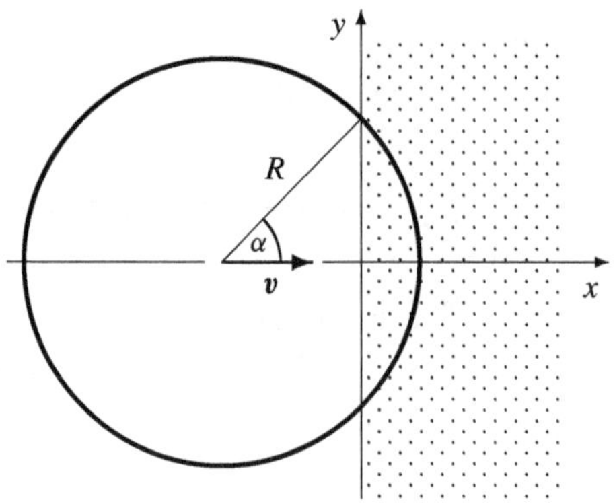

A circular conductor loop moves within the x-y-plane with constant, non-relativistic velocity $\boldsymbol{v} = v\,\boldsymbol{e}_x$. In the range $x > 0$, there is a homogeneous magnetic field $B_0\,\boldsymbol{e}_z$ (marked by dots).

Calculate the induced voltage $U(t)$ in the conductor loop. Sketch the function $U(t)$.

16.3 *Induction in a rotating circular ring*

A conducting circular ring ($z = 0$ and $x^2 + y^2 = r_0^2$) rotates at a constant angular velocity ω around the x-axis. A homogeneous magnetic field $\boldsymbol{B} = B\,\boldsymbol{e}_z$ acts on the ring.

What voltage $U(t)$ is induced in the ring? In the ring's circumference, a small electric bulb (corresponding to a resistance R) is inserted. The bulb then consumes the electrical power $P = U^2/R$. What time-averaged torque M must be exerted on the ring so that the angular velocity remains constant (the mechanical friction shall be neglected)?

16.4 *Magnetic field in a discharging plate capacitor*

A plate capacitor consisting of two parallel circular disks (radius r_0, distance d, $d \ll r_0$, boundary effects are neglected) is slowly discharged via a resistor R. The initial charges on the plates are Q_0 and $-Q_0$.

Determine the charges $\pm Q(t)$ on the plates and the magnetic field $\boldsymbol{B}(t)$.

16.5 *Field angular momentum of a rotating charged sphere*

A homogeneously charged sphere (charge q, radius R) rotates at a constant angular velocity $\boldsymbol{\omega}$ around an axis through the center. Calculate the angular momentum $\boldsymbol{L}_{\text{em}} = \int d^3r \, \boldsymbol{l}_{\text{em}}$ of the electromagnetic field.

Chapter 17

General Solution

We determine the general solution of the Maxwell equations. For this purpose, we first simplify these equations by introducing the potentials Φ and $\boldsymbol{A}$. The general solution consists of the general homogeneous solution (wave solutions) and a particular solution (retarded potentials).

The potentials Φ and $\boldsymbol{A}$ are introduced in such a way that the homogeneous Maxwell equations

$$\operatorname{div} \boldsymbol{B}(\boldsymbol{r}, t) = 0 \tag{17.1}$$

$$\operatorname{curl} \boldsymbol{E}(\boldsymbol{r}, t) + \frac{1}{c}\frac{\partial \boldsymbol{B}(\boldsymbol{r}, t)}{\partial t} = 0 \tag{17.2}$$

are fulfilled. A source-free field, (17.1), can be written as a vortex field:

$$\boldsymbol{B}(\boldsymbol{r}, t) = \operatorname{curl} \boldsymbol{A}(\boldsymbol{r}, t) \tag{17.3}$$

From (17.2) and (17.3) follows

$$\operatorname{curl}\left(\boldsymbol{E} + \frac{1}{c}\frac{\partial \boldsymbol{A}}{\partial t}\right) = 0 \tag{17.4}$$

A vortex-free field can be represented as a gradient field, i.e.,

$$\boldsymbol{E}(\boldsymbol{r}, t) = -\operatorname{grad} \Phi(\boldsymbol{r}, t) - \frac{1}{c}\frac{\partial \boldsymbol{A}(\boldsymbol{r}, t)}{\partial t} \tag{17.5}$$

The six fields $\boldsymbol{E}$ and $\boldsymbol{B}$ can thus be reduced to four fields, namely, the scalar potential Φ and the vector potential $\boldsymbol{A}$. Equation (17.3) is of the same form as in magnetostatics. Compared to electrostatics, however, an additional term appears in (17.5).

The field equations for Φ and $\boldsymbol{A}$ result from the inhomogeneous Maxwell equations. From $\operatorname{div} \boldsymbol{E} = 4\pi\varrho$ follows

$$\Delta\Phi + \frac{1}{c}\frac{\partial(\operatorname{div}\boldsymbol{A})}{\partial t} = -4\pi\varrho \tag{17.6}$$

From $\operatorname{curl} \boldsymbol{B} - \dot{\boldsymbol{E}}/c = 4\pi \boldsymbol{j}/c$ follows

$$\Delta\boldsymbol{A} - \frac{1}{c^2}\frac{\partial^2 \boldsymbol{A}}{\partial t^2} - \operatorname{grad}\left(\operatorname{div}\boldsymbol{A} + \frac{1}{c}\frac{\partial\Phi}{\partial t}\right) = -\frac{4\pi}{c}\boldsymbol{j} \tag{17.7}$$

Here, $\operatorname{curl}\operatorname{rot}\boldsymbol{A} = -\Delta\boldsymbol{A} + \operatorname{grad}\operatorname{div}\boldsymbol{A}$ was used. Equations (17.6) and (17.7) represent four coupled partial differential equations for the four fields Φ and $\boldsymbol{A}$.

Decoupling by Lorenz Gauge

The potentials Φ and $\boldsymbol{A}$ are not uniquely defined by the physical fields $\boldsymbol{E}$ and $\boldsymbol{B}$. The transformation

$$\boldsymbol{A}(\boldsymbol{r},t) \quad\longrightarrow\quad \boldsymbol{A}(\boldsymbol{r},t) + \operatorname{grad}\Lambda(\boldsymbol{r},t) \tag{17.8}$$

does not change the $\boldsymbol{B}$-field; the additional term is the general form of an irrotational field. In order that the $\boldsymbol{E}$-field in (17.5) remains unchanged, too, the scalar potential must be transformed simultaneously:

$$\Phi(\boldsymbol{r},t) \quad\longrightarrow\quad \Phi(\boldsymbol{r},t) - \frac{1}{c}\frac{\partial\Lambda(\boldsymbol{r},t)}{\partial t} \tag{17.9}$$

The transformations (17.8) and (17.9) are called *gauge transformation.* The fields $\boldsymbol{E}$ and $\boldsymbol{B}$ do not change under a gauge transformation with an arbitrary field $\Lambda(\boldsymbol{r},t)$.

Due to the gauge invariance, we can impose a scalar condition on the potentials, which can be fulfilled by a suitable choice of the scalar function $\Lambda(\boldsymbol{r},t)$. As a condition we choose the *Lorenz gauge,*[1]

$$\boxed{\operatorname{div}\boldsymbol{A} + \frac{1}{c}\frac{\partial\Phi}{\partial t} = 0 \quad \text{Lorenz gauge}} \tag{17.10}$$

[1]Named after Ludwig Valentin Lorenz (1829–1891). The Lorentz transformations and related terms (like Lorentz tensors), on the other hand, are named after Hendrik Antoon Lorentz (1853–1928, Nobel Prize 1902). Due to its outstanding importance, H. A. Lorentz is also often credited with the gauge (17.10). One therefore finds occasionally the incorrect spelling "Lorentz gauge".

For arbitrary fields Φ and $\boldsymbol{A}$ were the left side of (17.10) would result in some scalar function $f(\boldsymbol{r}, t) \neq 0$. The transformation (17.8) and (17.9) adds the term $\Delta \Lambda - \partial_t^{\,2} \Lambda / c^2$. One can now choose Λ such that $\Delta \Lambda - \partial_t^{\,2} \Lambda / c^2 = -f$. Then the new potentials fulfill the condition (17.10). In practice, one restricts the potentials from the outset by (17.10).

(17.10), (17.6) and (17.7) become

$$\boxed{\Delta \Phi(\boldsymbol{r}, t) - \frac{1}{c^2} \frac{\partial^2 \Phi(\boldsymbol{r}, t)}{\partial t^2} = -4\pi \varrho(\boldsymbol{r}, t)} \tag{17.11}$$

$$\boxed{\Delta \boldsymbol{A}(\boldsymbol{r}, t) - \frac{1}{c^2} \frac{\partial^2 \boldsymbol{A}(\boldsymbol{r}, t)}{\partial t^2} = -\frac{4\pi}{c} \boldsymbol{j}(\boldsymbol{r}, t)} \tag{17.12}$$

Thereby, we obtained four *decoupled* differential equations for the fields Φ, A_x, A_y and A_z. The Lorenz gauge has led to the decoupling of equations (17.6) and (17.7), i.e., to a significant simplification. Equations (17.11) and (17.12), together with (17.3), (17.5) and (17.10), are equivalent to Maxwell's equations. Due to (17.10), only three of the four fields Φ and $\boldsymbol{A}$ are independent.

Wave equations that are mathematically equivalent to (17.11) or (17.12) also occur in mechanics. For example, density waves (sound waves) are described by an equation of the form (17.11). In this case, the scalar field $\Phi = \varrho_{\mathrm{m}}(\boldsymbol{r}, t) - \varrho_{\mathrm{m},0}$ is the deviation of the mass density ϱ_{m} from its equilibrium value $\varrho_{m,0}$, the constant c is the speed of sound and the right-hand side describes an external force density.

Each component of (17.12) has the same structure as (17.11). We can therefore limit the discussion of the solution to (17.11). The general solution of the differential equation (17.11) is of the form

$$\Phi(\boldsymbol{r}, t) = \Phi_{\mathrm{hom}} + \Phi_{\mathrm{part}} \tag{17.13}$$

Here, Φ_{hom} is the general solution of the homogeneous differential equation, and Φ_{part} is a special (particular) solution of the inhomogeneous differential equation.

Solution of the Homogeneous Equation

We first determine the general solution of the homogeneous equation:

$$\Delta \Phi_{\text{hom}} - \frac{1}{c^2} \frac{\partial^2 \Phi_{\text{hom}}}{\partial t^2} = 0 \tag{17.14}$$

With the separation approach

$$\Phi_{\text{hom}}(\boldsymbol{r}, t) = X(x)\, Y(y)\, Z(z)\, T(t) \tag{17.15}$$

Equation (17.14) becomes

$$\frac{X''}{X} + \frac{Y''}{Y} + \frac{Z''}{Z} - \frac{1}{c^2} \frac{T''}{T} = 0 \tag{17.16}$$

Each individual term must be equal to a constant since the other terms do not depend on the respective coordinate, i.e.,

$$\frac{X''}{X} = -k_x^2, \quad \frac{Y''}{Y} = -k_y^2, \quad \frac{Z''}{Z} = -k_z^2, \quad \frac{T''}{T} = -\omega^2 \tag{17.17}$$

From (17.16), we obtain the following condition for the separation constants:

$$\omega^2 = c^2 \boldsymbol{k}^2 = c^2 \left(k_x^2 + k_y^2 + k_z^2\right) \tag{17.18}$$

We solve the differential equation for $X(x)$:

$$\frac{d^2 X}{dx^2} = -k_x^2\, X(x) \quad \longrightarrow \quad X(x) = \exp(\pm \mathrm{i} k_x x) \tag{17.19}$$

The solutions for $Y(y)$, $Z(z)$ and $T(t)$ are of the same form. To ensure that the solutions do not diverge for $\pm\infty$, the quantities k_x, k_y, k_z and ω must be real. In the spatial part, we use the positive sign in the exponential function but allow negative k values:

$$-\infty < k_x, k_y,\, k_z < \infty \tag{17.20}$$

In the time component, we first consider both signs in $T = \exp(\pm \mathrm{i}\omega t)$. Then we can restrict ourselves on positive frequencies ω:

$$\omega = \omega(k) = c|\boldsymbol{k}| = ck = c\sqrt{k_x^2 + k_y^2 + k_z^2} \tag{17.21}$$

Thus, the separation approach leads to the elementary solutions

$$\Phi_{k_x k_y k_z} = \exp\Big(\mathrm{i}\big[\boldsymbol{k} \cdot \boldsymbol{r} \pm \omega(\boldsymbol{k})\, t\big]\Big) \tag{17.22}$$

which depends on any three real parameters k_x, k_y and k_z. The fourth separation constant ω is fixed by (17.21). Due to the linearity of (17.14), a superposition of the solutions (17.22) with different $\boldsymbol{k}$-values is solution again. Since the potential Φ is real, we take the real part (Re) of this superposition:

$$\boxed{\Phi_{\text{hom}}(\boldsymbol{r}, t) = \text{Re} \int d^3k \left(a_1(\boldsymbol{k}) + \mathrm{i}\, a_2(\boldsymbol{k})\right) \exp\left[\,\mathrm{i}(\boldsymbol{k}\cdot\boldsymbol{r} - \omega t)\right]} \qquad (17.23)$$

Here, $a_1(\boldsymbol{k})$ and $a_2(\boldsymbol{k})$ are real functions. In Exercise 17.1, this form of solution is justified in a somewhat different way.

Since (17.14) is a second-order differential equation in time, the solution is fixed by the initial conditions $\Phi(\boldsymbol{r}, 0) = G(\boldsymbol{r})$ and $\dot{\Phi}(\boldsymbol{r}, 0) = H(\boldsymbol{r})$. The superposition (17.23) contains just as many integration constants (the real functions $a_1(\boldsymbol{k})$ and $a_2(\boldsymbol{k})$), as can be determined by the initial conditions (the real functions $G(\boldsymbol{r})$ and $H(\boldsymbol{r})$). Therefore, (17.23) is the *general solution* of (17.14).

Corresponding solutions can be obtained for the components $A_{i,\,\text{hom}}$ of the vector potential; thereby the condition (17.10) must be fulfilled. The solutions Φ_{hom} and $\boldsymbol{A}_{\text{hom}}$ represent waves which are investigated in more detail in Chapter 20.

Retarded Potentials

For the general solution (17.13), we still need a particular solution, i.e., a special solution of the inhomogeneous equation (17.11). The particular solutions for Φ and $\boldsymbol{A}$, which we derive in the following, are called "retarded potentials".

We perform a Fourier transform of the time dependence of $\Phi(\boldsymbol{r}, t)$ and $\rho(\boldsymbol{r}, t)$:

$$\Phi(\boldsymbol{r}, t) = \frac{1}{\sqrt{2\pi}} \int_{-\infty}^{\infty} d\omega\; \Phi_\omega(\boldsymbol{r}) \exp(-\mathrm{i}\omega t) \qquad (17.24)$$

$$\varrho(\boldsymbol{r}, t) = \frac{1}{\sqrt{2\pi}} \int_{-\infty}^{\infty} d\omega\; \varrho_\omega(\boldsymbol{r}) \exp(-\mathrm{i}\omega t) \qquad (17.25)$$

The sign in the exponential function is a convention; in the reverse transformation, the other sign occurs in each case.

The transformed quantities $\Phi_\omega(\boldsymbol{r})$ and $\varrho_\omega(\boldsymbol{r})$ are functions of $\boldsymbol{r}$ and ω. We insert (17.24) and (17.25) into the wave equation (17.11):

$$\int_{-\infty}^{\infty} d\omega \left(\Delta + \frac{\omega^2}{c^2}\right) \Phi_\omega(\boldsymbol{r}) \exp(-\mathrm{i}\omega t) = -4\pi \int_{-\infty}^{\infty} d\omega\, \varrho_\omega(\boldsymbol{r}) \exp(-\mathrm{i}\omega t) \tag{17.26}$$

The functions $\exp(-\mathrm{i}\omega t)$ are independent of each other for different ω. Therefore,

$$\left(\Delta + \frac{\omega^2}{c^2}\right) \Phi_\omega(\boldsymbol{r}) = -4\pi\, \varrho_\omega(\boldsymbol{r}) \tag{17.27}$$

must apply. We use (3.37) with $k = \omega/c$:

$$\left(\Delta + \frac{\omega^2}{c^2}\right) \frac{\exp(\pm\mathrm{i}\omega|\boldsymbol{r}-\boldsymbol{r}'|/c)}{|\boldsymbol{r}-\boldsymbol{r}'|} = -4\pi\, \delta(\boldsymbol{r}-\boldsymbol{r}') \tag{17.28}$$

This makes it easy to verify that

$$\Phi_\omega(\boldsymbol{r}) = \int d^3r'\, \varrho_\omega(\boldsymbol{r}')\, \frac{\exp(+\mathrm{i}\omega|\boldsymbol{r}-\boldsymbol{r}'|/c)}{|\boldsymbol{r}-\boldsymbol{r}'|} \tag{17.29}$$

is solution of (17.27). As in (17.28), we could also use a minus sign in the exponential function; this leads to a different solution, which is discussed in the following. We substitute (17.29) into (17.24):

$$\begin{aligned}\Phi(\boldsymbol{r},t) &= \frac{1}{\sqrt{2\pi}} \int d^3r' \int_{-\infty}^{\infty} d\omega\, \frac{\varrho_\omega(\boldsymbol{r}')}{|\boldsymbol{r}-\boldsymbol{r}'|}\, \exp\big[-\mathrm{i}\omega \underbrace{(t - |\boldsymbol{r}-\boldsymbol{r}'|/c)}_{=t'}\big] \\ &= \int d^3r'\, \frac{\varrho(\boldsymbol{r}',t')}{|\boldsymbol{r}-\boldsymbol{r}'|}\end{aligned} \tag{17.30}$$

For a static charge distribution $\varrho(\boldsymbol{r},t) = \varrho(\boldsymbol{r})$, this reduces to the integral (6.3) known from electrostatics. However, in the case of time-dependent phenomena, it is not sufficient to include the time t as an additional argument in (6.3). Rather, in the charge distribution, the *earlier* time argument

$$t' = t - \delta t = t - \frac{|\boldsymbol{r}-\boldsymbol{r}'|}{c} \tag{17.31}$$

has to be used. This has the following physical meaning: If the charge distribution changes at a certain point at time t', then the resulting change in the electromagnetic field propagates with the speed of light c. At the distance $|\boldsymbol{r} - \boldsymbol{r}'|$, the field therefore only changes at the *later* time $t = t' + \delta t$. Due to this delayed reaction, the potential solution (17.30) is also called *retarded*; we mark this in the following with an index "ret":

$$\Phi_{\text{ret}}(\boldsymbol{r}, t) = \int d^3r' \; \frac{\varrho(\boldsymbol{r}', t - |\boldsymbol{r} - \boldsymbol{r}'|/c)}{|\boldsymbol{r} - \boldsymbol{r}'|} \tag{17.32}$$

The analogous solution for the vector potential is

$$\boldsymbol{A}_{\text{ret}}(\boldsymbol{r}, t) = \frac{1}{c} \int d^3r' \; \frac{\boldsymbol{j}(\boldsymbol{r}', t - |\boldsymbol{r} - \boldsymbol{r}'|/c)}{|\boldsymbol{r} - \boldsymbol{r}'|} \tag{17.33}$$

Thus, we have found particular solutions of the differential equations (17.11) and (17.12):

$$\Phi_{\text{part}} = \Phi_{\text{ret}}, \quad \boldsymbol{A}_{\text{part}} = \boldsymbol{A}_{\text{ret}} \tag{17.34}$$

In (17.29), we could also have chosen the other sign in the exponential function; because of (17.28), this would also be the solution of (17.27). The other sign leads to the *advanced* solution

$$\Phi_{\text{av}}(\boldsymbol{r}, t) = \int d^3r' \; \frac{\varrho(\boldsymbol{r}', t + |\boldsymbol{r} - \boldsymbol{r}'|/c)}{|\boldsymbol{r} - \boldsymbol{r}'|} \tag{17.35}$$

and to an analogous expression for $\boldsymbol{A}_{\text{av}}$. Each linear combination

$$\Phi_{\text{part}} = a\, \Phi_{\text{ret}} + (1 - a)\, \Phi_{\text{av}} \tag{17.36}$$

is also a particular solution and, together with the homogeneous solution results in the general solution.

For a discussion of the physical significance of the retarded and advanced solution, let us consider the dipole antenna of an FM transmitter. The radiation of the antenna (localized at $\boldsymbol{r}'$) arrives at the radio listener (with a receiving antenna at $\boldsymbol{r}$) only after the time $\delta t = |\boldsymbol{r} - \boldsymbol{r}'|/c$. This radio wave is described by the retarded potential of the oscillating transmitting antenna. The advanced solution, on the other hand, is to be excluded for causality reasons: In the advanced solution, the effect (reception by the radio listener) occurs before the cause (transmission of the radio wave).

For the general solution $\Phi_{\text{general}} = \Phi_{\text{part}} + \Phi_{\text{hom},}$ one can, in principle, use an arbitrary particular solution. Therefore, one could also use the "wrong" solution $\Phi_{\rm part} = \Phi_{\rm av}$ as the particular solution. The experimental boundary conditions (spatial an temporal) would then, however, require the term $\Phi_{\rm hom} = \Phi_{\rm ret} - \Phi_{\rm av}$ in the homogeneous solution. As a result, one would again obtain $\Phi_{\rm ret}$ for the actual solution.

In the following, especially in Chapters 23 and 24, we consider the simple examples for radiating source distributions and calculate the retarded solution.

Exercises

17.1 *Fourier transform of the wave equation*

Solve the homogeneous wave equation

$$\left(\Delta - \frac{1}{c^2}\frac{\partial^2}{\partial t^2}\right)\Phi(\boldsymbol{r},t) = 0$$

by a Fourier transformation of the variables x, y, z and t.

17.2 *Solution of the one-dimensional wave equation*

Show that

$$\Phi(x,t) = \mathrm{Re}\int_{-\infty}^{\infty} dk\; a(k)\,\exp\big(\mathrm{i}\,(kx-\omega t)\big), \quad (\omega = c\,|k|) \tag{17.37}$$

with an arbitrary complex function $a(k)$ solves the one-dimensional wave equation $\left(\partial_x^2 - \partial_t^2/c^2\right)\Phi = 0$. Verify that this homogeneous solution can be written in the following form:

$$\Phi_{\rm hom}(x,t) = f(x-ct) + g(x+ct) \tag{17.38}$$

Here, f and g are real functions. Discuss time evolution of such a solution with $f(x) = f_0\exp(-\gamma\, x^2)$ and $g = 0$.

Chapter 18

Covariance

The Maxwell equations are valid in all inertial frames. Formally this means that their form does not change under Lorentz transformations (LT); they are form invariant or covariant. This is most clearly expressed by a covariant notation. We specify how charge, currents and fields transform under LT. This is the basis for calculating various effects like fields of moving charges or the Doppler effect (Chapters 22 and 23).

The *relativity principle* is the statement that all inertial systems (IS) are equivalent. Equivalent means that the fundamental laws have the same form in all IS. The relativity principle was originally established by Galileo. Until the beginning of last century, Newton's laws were considered to be relativistic laws. The (assumed) validity of Newton's laws in all IS leads to the Galilean transformation between different IS (Chapter 5 in [1]).

The Galilean transformation implies that light moves at different speeds in different IS. If light (or anything else) propagates in IS′ with the velocity c, then the Galilean transformation means

$$\frac{dx'}{dt'} = c \qquad \xrightarrow[\text{Galilean transformation}]{x = x' + v\,t',\ t = t'} \qquad \frac{dx}{dt} = c + v \tag{18.1}$$

Light would therefore propagate in IS with the speed $c + v \neq c$. Maxwell's equations have wave solutions that propagate with the velocity c. According to Galileo's relativity principle, the Maxwell equations are then non-relativistic laws that are valid only in a specific IS. The particular IS should be the one in which the light-bearing medium (ether) rests. This concept appears quite natural if one compares light waves with other waves. For example, one measures *the* speed of sound in the IS in which the sound-carrying medium (such as air) is at rest.

However, experiments have shown that light propagates with the same speed in every IS (Michelson experiment). This is generalized to the statement that Maxwell's equations are relativistic, i.e.,

$$\text{Maxwell's equations apply in every IS} \tag{18.2}$$

The formulation "all IS are equivalent" of the relativity principle is maintained. However, its meaning changes because now the Maxwell's equations (and no longer Newton's laws) are considered as fundamental and relativistic laws. The new *Einstein relativity principle* requires a different transformation between the IS. It assumes the invariance of the interval $c^2dt^2 - d\boldsymbol{r}^2$ (ds is the 4-dimensional line element between two adjacent events). From the invariance of ds^2 follows the constancy of the speed of light:

$$\frac{dx'}{dt'} = c \qquad \xrightarrow[\text{Lorentz transformation}]{c^2dt'^2 - dx'^2 = c^2dt^2 - dx^2} \qquad \frac{dx}{dt} = c \tag{18.3}$$

The linear transformations that leave $ds^2 = c^2dt^2 - d\boldsymbol{r}^2$ invariant are the Lorentz transformations (LT). The relativity principle of Einstein is the physical basis of the theory of special relativity.

The fundamentals of special relativity that are assumed to be known to the extent to which they are usually dealt with in mechanics (e.g., in Part IX of [1]). To clarify the notation, the formal foundations (like the handling of Lorentz tensors) have been summarized in Chapter 4.

Lorentz Invariance of the Charge

We first show that (18.2) implies the Lorentz invariance of the charge. The experimental proof that the charge is independent of the velocity then represents a further verification of the statement (18.2).

The continuity equation (16.6)

$$\frac{\partial \varrho(\boldsymbol{r}, t)}{\partial t} + \operatorname{div} \boldsymbol{j}(\boldsymbol{r}, t) = 0 \tag{18.4}$$

follows from Maxwell's equations (16.5). We combine the time and position coordinates into Lorentz or 4-vectors:

$$\left(x^\alpha\right) = (ct, x, y, z) = (ct, \boldsymbol{r}), \quad \left(\partial_\alpha\right) = \left(\frac{\partial}{\partial x^\alpha}\right) \tag{18.5}$$

The associated contra- and covariant components (Chapter 4) are

$$\left(x_\alpha\right) = (ct, -x, -y, -z) = (ct, -\boldsymbol{r}) \quad \left(\partial^\alpha\right) = \left(\frac{\partial}{\partial x_\alpha}\right) \tag{18.6}$$

We now introduce the following fourfold indexed quantity:

$$\left(j^\alpha\right) = (c\varrho, j_x, j_y, j_z) \tag{18.7}$$

This is initially a definition and not (yet) a statement about the transformation behavior. Using this, Equation (18.4) can be written in the form

$$\partial_\alpha\, j^\alpha(x) = 0 \tag{18.8}$$

It should be remembered (Chapter 4) that we implicitly sum over two equal indices (where one is at the top and one at the bottom). In the argument, x stands for $t, \boldsymbol{r}$ or for x^0, x^1, x^2, x^3. With (18.2), (18.8) also applies in all inertial frames, i.e.,

$$\partial'_\alpha\, j'^\alpha(x') = 0 \tag{18.9}$$

The primed quantities refer to any IS$'$ that moves relative to IS. It follows from (18.9) and (18.8) that $\partial_\alpha\, j^\alpha$ is a Lorentz scalar. Since ∂_α is a 4-vector, it follows that the *current density* (or 4-current density) j^α is a Lorentz or 4-vector:

$$j^\alpha(x) \text{ is a 4-vector field, thus } j'^\alpha(x') = \Lambda^\alpha_{\ \beta}\, j^\beta(x) \tag{18.10}$$

To discuss this transformation of j^α, let us consider the case that there is only a charge density ϱ in IS, i.e., $(j^\alpha) = (c\varrho, 0)$. In an IS$'$ that moves with $\boldsymbol{v}$ relative to IS, the 4-current density is then

$$\left(j'^\alpha\right) = \Lambda(\boldsymbol{v})\,(c\varrho, 0) = \gamma\,(c\varrho, -\varrho\,\boldsymbol{v}) = (c\varrho', -\varrho'\,\boldsymbol{v}) \tag{18.11}$$

Here, $\gamma = 1/\sqrt{1 - v^2/c^2}$. For the charge density, the charge per volume, one gets

$$\varrho' = \frac{dq'}{dV'} = \gamma\,\varrho = \gamma\,\frac{dq}{dV} \tag{18.12}$$

The considered volume element the transformation implies a length contraction in the direction of v, i.e.,

$$dV' = dV\,\sqrt{1 - v^2/c^2} = \frac{dV}{\gamma} \tag{18.13}$$

From the last two equations follows

$$dq = dq' \tag{18.14}$$

Since this applies to every charge element, the charge q is a Lorentz scalar:

$$q = \frac{1}{c}\int d^3r\; j^0(x) = \text{Lorentz scalar} \tag{18.15}$$

In Exercise 18.5, this result is derived more formally. The result means that the charge of a particle is independent of its velocity.

Experimentally, the independence of the charge from the velocity is verified by the neutrality of the hydrogen atom. A proton and an electron have a total charge of zero, regardless of whether the particles are at rest or not. This does not apply to the mass; the rest mass of the hydrogen atom is not the sum of the rest masses of the proton and electron.

The neutrality of the hydrogen atom is proven with high accuracy. If there were changes of the charge of the size $\mathcal{O}(v^2/c^2)$, then a hydrogen atom would have a net charge $|q| \sim 10^{-4}\,e$ because of $v_{\rm e} \sim c/100$ (electron velocity). One mole hydrogen would then have the (tremendously huge) charge $Q \sim 6 \cdot 10^{19}\,e \sim 10\,\mathrm{C}$. The Lorentz invariance of the charge is a consequence of (18.2). Its experimental proof is therefore a verification of (18.2).

The constancy of the charge follows from the continuity equation (18.8) in a double sense: On the one hand, the charge of a closed system does not depend on time, see (13.10)–(13.12). On the other hand, the charge does not depend on its motion (18.14); this follows from the covariance of (18.8).

Covariant Maxwell Equations

We set up the covariant form of Maxwell's equations. In the first step, this is done for the potentials. We start from Equations (17.10)–(17.12):

$$\left(\Delta - \frac{1}{c^2}\frac{\partial^2}{\partial t^2}\right)\Phi(\boldsymbol{r}, t) = -4\pi\varrho(\boldsymbol{r}, t) \tag{18.16}$$

$$\left(\Delta - \frac{1}{c^2}\frac{\partial^2}{\partial t^2}\right)\boldsymbol{A}(\boldsymbol{r},t) = -\frac{4\pi}{c}\,\boldsymbol{j}(\boldsymbol{r},t) \tag{18.17}$$

$$\frac{1}{c}\,\frac{\partial \Phi(\boldsymbol{r},t)}{\partial t} + \nabla \cdot \boldsymbol{A}(\boldsymbol{r},t) = 0 \tag{18.18}$$

We introduce the following fourfold indexed quantity:

$$\left(A^{\alpha}\right) = \left(\Phi,\, A_x,\, A_y,\, A_z\right) \tag{18.19}$$

Initially, this is a definition and not (yet) a statement about the transformation behavior. The equations (18.16) and (18.17) can now be written in the form

$$\boxed{\Box\, A^{\alpha}(x) \;=\; \frac{4\pi}{c}\; j^{\alpha}(x) \qquad \text{covariant Maxwell equations for the potentials}} \tag{18.20}$$

where the d'Alembert operator was used,

$$\Box = \partial_{\beta}\,\partial^{\beta} = \frac{1}{c^2}\,\frac{\partial^2}{\partial t^2} - \Delta \tag{18.21}$$

Maxwell's equations (18.20) apply in all IS. With j^{α}, also the left-hand side must be a Lorentz vector. Since the d'Alembert operator is a Lorentz scalar, it follows for the *4-potential* A^{α} is a Lorentz vector field,

$$A^{\alpha}(x)\ \text{ is a 4-vector field} \tag{18.22}$$

Consequently, $\partial_{\alpha} A^{\alpha}$ is a Lorentz scalar, and the condition

$$\boxed{\partial_{\alpha} A^{\alpha}(x) \;=\; 0 \quad \text{Lorenz gauge}} \tag{18.23}$$

holds in all IS.

The Maxwell equations (18.20) with (18.23) and the continuity equation (18.8) are *covariant equations*, which means they are form-invariant under Lorentz transformations. It should be reminded that "covariant" has the second meaning "lower" for tensor indices (in contrast to contravariant for "upper").

We now establish the covariant form of Maxwell's equations for the measurable fields $\boldsymbol{E}$ and $\boldsymbol{B}$. The connection between these fields and the potentials is given in (17.3) and (17.5):

$$\boldsymbol{E} = -\nabla \Phi - \frac{1}{c}\frac{\partial \boldsymbol{A}}{\partial t}, \qquad \boldsymbol{B} = \nabla \times \boldsymbol{A} \tag{18.24}$$

By

$$F^{\alpha\beta} = \partial^\alpha A^\beta - \partial^\beta A^\alpha \quad \text{(field strength tensor)} \tag{18.25}$$

we define the antisymmetric *electromagnetic tensor* (also electromagnetic field tensor or *field strength tensor*) $F^{\alpha\beta}$. From the definition, it immediately follows that $F^{\alpha\beta}$ is invariant under the gauge transformation (17.8) and (17.9):

$$A^\alpha \quad \longrightarrow \quad A^\alpha - \partial^\alpha \Lambda \quad \text{(gauge transformation)} \tag{18.26}$$

The components of the field strength tensor follow from (18.24) and (18.25):

$$\left(F^{\alpha\beta}\right) = \begin{pmatrix} 0 & -E_x & -E_y & -E_z \\ E_x & 0 & -B_z & B_y \\ E_y & B_z & 0 & -B_x \\ E_z & -B_y & B_x & 0 \end{pmatrix} \tag{18.27}$$

The signs must be treated carefully: $\partial^i = \partial/\partial x_i = -\partial/\partial x^i = -\nabla \cdot \boldsymbol{e}_i$. In the transition to the covariant field strength tensor $F_{\alpha\beta}$, the 00- and ij-components remain the same, while the $0i$- and $i0$-components receive a minus sign:

$$\left(F_{\alpha\beta}\right) \overset{(4.15)}{=} \left(\eta_{\alpha\gamma}\,\eta_{\beta\delta}\,F^{\gamma\delta}\right) = \begin{pmatrix} 0 & E_x & E_y & E_z \\ -E_x & 0 & -B_z & B_y \\ -E_y & B_z & 0 & -B_x \\ -E_z & -B_y & B_x & 0 \end{pmatrix} \tag{18.28}$$

The definition (18.25) implies that $F^{\alpha\beta}$ is a Lorentz tensor of the second rank. From the known transformation behavior of this tensor, the behavior of $\boldsymbol{E}$ and $\boldsymbol{B}$ can be read off. This is done explicitly in Chapter 22.

In (18.20), we may add the term $\partial_\beta\,\partial^\alpha A^\beta = \partial^\alpha\,\partial_\beta\,A^\beta = 0$,

$$\partial_\beta\left(\partial^\beta A^\alpha - \partial^\alpha A^\beta\right) = \frac{4\pi}{c}\,j^\alpha \tag{18.29}$$

These are the four inhomogeneous Maxwell equations for $F^{\alpha\beta}$:

$$\partial_\beta F^{\beta\alpha}(x) = \frac{4\pi}{c}\, j^\alpha(x) \qquad \begin{cases} \alpha = 0 & \operatorname{div} \boldsymbol{E} = 4\pi\varrho \\ \alpha = i & \operatorname{curl} \boldsymbol{B} - \dot{\boldsymbol{E}}/c = 4\pi \boldsymbol{j}/c \end{cases} \tag{18.30}$$

With the help of the totally antisymmetric pseudo tensor $\varepsilon^{\alpha\beta\delta\gamma}$ from (4.24), we define the *dual* field strength tensor

$$\left(\widetilde{F}^{\alpha\beta}\right) = \frac{1}{2}\left(\varepsilon^{\alpha\beta\gamma\delta} F_{\gamma\delta}\right) = \begin{pmatrix} 0 & -B_x & -B_y & -B_z \\ B_x & 0 & E_z & -E_y \\ B_y & -E_z & 0 & E_x \\ B_z & E_y & -E_x & 0 \end{pmatrix} \tag{18.31}$$

The dual field strength tensor is an antisymmetric Lorentz pseudotensor of 2nd rank. The homogeneous Maxwell equations can be expressed by this dual field tensor:

$$\partial_\beta \widetilde{F}^{\beta\alpha} = 0 \qquad \begin{cases} \alpha = 0 & \operatorname{div} \boldsymbol{B} = 0 \\ \alpha = i & \operatorname{curl} \boldsymbol{E} + \dot{\boldsymbol{B}}/c = 0 \end{cases} \tag{18.32}$$

The Maxwell equations for the physical fields $F^{\alpha\beta}$ are thus

$$\boxed{\partial_\beta F^{\beta\alpha} = \frac{4\pi}{c}\, j^\alpha, \quad \partial_\beta \widetilde{F}^{\beta\alpha} = 0 \qquad \text{covariant Maxwell equations}} \tag{18.33}$$

This form has the following advantages over (16.5):

1. The equations have a simpler structure.
2. The structure of the equations reflects the covariance with respect to LT. Of course, this covariance also applies to (16.5), but there it is not obvious.

Compared to (18.33), the formulations (18.20) and (18.23) have the advantage of even greater simplicity because there are only four fields and the field equations are decoupled. On the other hand, (18.20) and (18.23) have the disadvantage that the potentials A^α are no measuring quantities.

Equations (18.33) were more symmetric if the homogeneous equations are replaced by $\partial_\beta \widetilde{F}^{\beta\alpha} = 4\pi\, j^\alpha_{\text{magn}}/c$. This would imply that there are magnetic sources, i.e., in particular magnetic monopoles. Following (14.8), the hitherto unsuccessful search for magnetic monopoles was pointed out.

Relativistic generalization of electrostatics

In view of the covariance of Maxwell's equations, one may ask to what extent Maxwell's equations are just a relativistic generalization of electrostatics. Are the Coulomb law and Einstein's relativity principle sufficient to establish Maxwell's equations?

The field equation of electrostatics

$$\Delta\,\Phi(\boldsymbol{r}) = -4\pi\,\varrho(\boldsymbol{r}) \tag{18.34}$$

may be generalized relativistically as follows. First, the argument $\boldsymbol{r}$ is replaced by $\boldsymbol{r}, t$. This alone, however, would result in a law of distant action because the change of ϱ at one point would imply the simultaneous change of Φ at another point. According to the special relativity, however, effects can propagate at most with the speed of light. This is achieved by the substitution

$$\Delta \;\longrightarrow\; -\Box \tag{18.35}$$

in (18.34). The adequate particular solution is then the retarded potential Φ_{ret}; the retarded potential meets the requirement that the effects propagate at most with the speed of light. Formally, the 3-scalar $\Delta = -\partial_i\,\partial^i$ is generalized to the corresponding 4-scalar $\Box = \partial_\alpha\,\partial^\alpha$.

If one admits IS moving relative to each other, a charge density inevitably also leads to a current density. Therefore, the charge density must be replaced by the 4-current density:

$$\varrho \;\longrightarrow\; \left(j^\alpha\right) = (c\,\varrho, \boldsymbol{j}) \tag{18.36}$$

Accordingly, then Φ on the left-hand side of (18.34) must be replaced by a 4-vector field, i.e., by $(A^\alpha) = (\Phi, \boldsymbol{A})$. In this way, Equation (18.20) may be obtained as a generalization of (18.34).

In this sense, the Maxwell theory can be regarded as a relativistic generalization of electrostatics. The Maxwell theory does, however, not follow unambiguously from the Coulomb law and the covariance requirement. This can be seen in the following counter example. The gravitational force has the same form as the Coulomb force (5.1). Therefore, the field equation of Newton's gravitational theory

$$\Delta\,\Phi_{\text{grav}}(\boldsymbol{r}) = 4\pi\,G\,\varrho_{\text{m}}(\boldsymbol{r}) \tag{18.37}$$

is mathematically equivalent to (18.34). Here, ϱ_{m} is the mass density, G the gravitational constant and Φ_{grav} the gravitational potential. The potential Φ_{grav} determines the force $\boldsymbol{F} = -m \,\mathrm{grad}\, \Phi_{\mathrm{grav}}$ on a test mass m. The relativistic generalization of (18.37) does not lead to Maxwell's equations but to the Einstein's field equations of gravitation. The common structure of (18.34) and (18.37), however, leads to a number of similarities between electromagnetism and gravitation. Thus for example, there are gravitomagnetic effects and gravitational waves.

The field equations of electrostatics and of Newton's gravitational theory are mathematically equivalent. However, the relativistic generalizations lead to different theories. Therefore, Maxwell's equations cannot be derived from electrostatics and the covariance requirement alone; other statements must be added. These are in particular the transformation behavior of the source terms and a simplicity requirement, as will be briefly discuss.

First of all, it is essential that the charge is a Lorentz scalar (only then is j^{α} a 4-vector). In the case of gravity, on the other hand, the mass or energy (as the source of the field) is not a Lorentz scalar, but the 0-component of a Lorentz vector (of the 4-momentum). Furthermore, we have restricted ourselves to a linear (and therefore particularly simple) theory. Whether this leads to a valid theory can only be decided by experiments. Nonlinear terms mean that the field itself is the source of the field. This is the case with gravity, but also in the quantum chromodynamics (quantum theory of the strong interactions). In contrast to this, the electromagnetic field itself is not a source of the field; the photons (the quanta of the field) have no charge. The linearity of (18.34) is exact, while (18.37) is only a linear approximation. Another point that cannot be decided by the formal procedure of relativistic generalization is the possible existence of magnetic charges.

Lorentz Force

We supplement the covariant field equations by the covariant equations of motion. The Lorentz force acts on a point charge,

$$\boldsymbol{F}_{\mathrm{L}} = q\left(\boldsymbol{E} + \frac{\boldsymbol{v}}{c} \times \boldsymbol{B}\right) \tag{18.38}$$

We introduced this force in the framework of electrostatics and magnetostatics. Form the outset, it is therefore not clear whether it is also relativistically correct (in all orders of v/c). We start by assuming that (18.38) applies in the Newtonian limit $v/c \to 0$. In this limit, Newton's 2nd law becomes

$$m\,\frac{d\boldsymbol{v}'}{dt'} = q\boldsymbol{E}' \quad \text{(relativistically correct in IS}') \tag{18.39}$$

Here, IS$'$ is the inertial frame in which the particle has momentarily the velocity $\boldsymbol{v}' = 0$. By a Lorentz transformation, we can obtain from this the relativistic valid equation of motion in any other IS (where the considered particle has the velocity $\boldsymbol{v}$). However, it is easier to start from the covariant form of the relativistic equation of motion

$$m\,\frac{du^\alpha}{d\tau} = f^\alpha \tag{18.40}$$

This equation of motion is known from mechanics (Part IX of [1]). The mass at rest m and the proper time

$$d\tau = dt\,\sqrt{1 - \frac{v^2}{c^2}} \tag{18.41}$$

are Lorentz scalars. The 4-velocity

$$\big(u^\alpha\big) = \left(\frac{dx^\alpha}{d\tau}\right) = \left(\frac{c}{\sqrt{1 - v^2/c^2}},\, \frac{\boldsymbol{v}}{\sqrt{1 - v^2/c^2}}\right) = \gamma\;(c,\,\boldsymbol{v}) \tag{18.42}$$

is a Lorentz vector. This means that the Minkowski force f^α on the right-hand side of (18.40) is a Lorentz vector, too.

We are looking for the Minkowski force f^α that is caused by the electromagnetic field. Due to (18.38) we expect that f^α is linear in the field strength $F^{\alpha\beta}$, linear in the velocity u^α and proportional to the charge q. The simplest approach results in the equations of motion

$$\boxed{m\,\frac{du^\alpha}{d\tau} = \frac{q}{c}\,F^{\alpha\beta}\,u_\beta \qquad \text{covariant equations of motion}} \tag{18.43}$$

The following argumentation shows that this is indeed the correct equation:

1. Equation (18.43) is covariant. If it is correct in one IS, it is valid in all other IS′. It is therefore sufficient to show its validity in a specific IS′.
2. As this special inertial frame, we choose the momentarily comoving IS′. In IS′, we have $v' = 0$, $d\tau = dt'$ and therefore

$$\left(\frac{du'^{\alpha}}{d\tau}\right) = \left(0,\ \frac{d\boldsymbol{v}'}{dt'}\right) \quad \text{(in IS}')\tag{18.44}$$

Since $(u'_{\alpha}) = (c, 0)$ in IS′, the right-hand side of (18.43) becomes

$$\frac{q}{c}\left(F'^{\alpha\beta}u'_{\beta}\right) = \frac{q}{c}\left(F'^{\alpha 0}c\right) = \left(0,\ q\boldsymbol{E}'\right) \quad \text{(in IS}')\tag{18.45}$$

The last two equations show that (18.43) becomes (18.39) in IS′. Therefore, (18.43) is valid in IS′.

We express the relativistic equation of motion (18.43) in terms of the velocity $\boldsymbol{v}$ and the fields $\boldsymbol{E}$ and $\boldsymbol{B}$. For $\alpha = 0$ and $\alpha = i$, this results in

$$\boxed{\frac{d}{dt}\,\frac{mc^2}{\sqrt{1 - v^2/c^2}} = q\boldsymbol{E}\cdot\boldsymbol{v}}\tag{18.46}$$

$$\boxed{\frac{d}{dt}\,\frac{m\boldsymbol{v}}{\sqrt{1 - v^2/c^2}} = q\left(\boldsymbol{E} + \frac{\boldsymbol{v}}{c}\times\boldsymbol{B}\right)}\tag{18.47}$$

The arguments of the fields are $\boldsymbol{r}$ and t. Here, $\boldsymbol{r}$ is the position of the particle and $\boldsymbol{v}(t) = d\boldsymbol{r}/dt$ its velocity.

For $v/c \ll 1$, we obtain from (18.47) the equation

$$m\,\frac{d\boldsymbol{v}}{dt} = q\left(\boldsymbol{E} + \frac{\boldsymbol{v}}{c}\times\boldsymbol{B}\right) + \mathcal{O}\Big(\frac{v^2}{c^2}\Big)\tag{18.48}$$

According to (18.47), the Lorentz force (18.38) is the relativistically correct force. According to (18.48), however, it can also be used in non-relativistic mechanics. Newton's force in the narrower sense is the force on the right-hand side of (18.39). This force differs from the Lorentz force by terms of the order v^2/c^2.

Relativistic energy

For the discussion of (18.46), we assume that a particle (mass m, charge q) moves in an electrostatic field $\boldsymbol{E} = -\operatorname{grad}\Phi(\boldsymbol{r})$. With $d\Phi(\boldsymbol{r}(t))/dt = \operatorname{grad}\Phi \cdot d\boldsymbol{r}/dt = -\boldsymbol{v}\cdot\boldsymbol{E}$ (18.46) becomes

$$\frac{d}{dt}\left(\frac{mc^2}{\sqrt{1-v(t)^2/c^2}} + q\,\Phi(\boldsymbol{r}(t))\right) = 0 \tag{18.49}$$

From this follows $\gamma\, mc^2 + q\,\Phi = \text{const.}$ or more detailed

$$\underbrace{mc^2}_{\text{rest energy}} + \underbrace{mc^2\,(\gamma - 1)}_{\text{kinetic energy}} + \underbrace{q\,\Phi(\boldsymbol{r})}_{\text{potential energy}} = \text{const.} \tag{18.50}$$

We have thus derived a *conserved quantity* from the equations of motion. From the known meaning of $q\,\Phi$ (potential energy), it follows that this is the *energy* of the particle in the potential. The energy of the particle is conserved because the potential is time-independent. The first two terms in (18.50) are together referred to as *relativistic energy* or *energy* of the of the free particle:

$$\boxed{E = \frac{mc^2}{\sqrt{1-v^2/c^2}} = \text{relativistic energy}} \tag{18.51}$$

The fraction E/c is equal to the 0-component of the 4-momentum $p^\alpha = m\,u^\alpha$.

The *kinetic* energy $E_{\text{kin}} = E(v) - E(0) = mc^2(\gamma - 1)$ is the energy that is necessary to bring the particle at rest to the velocity v. The term $E(0) = mc^2$ is the *rest energy* of the particle. The deeper meaning of this relationship is referred to as *equivalence of mass and energy* (Part IX in [1]).

Energy–Momentum Tensor

In this section, a Lorentz tensor 2nd rank is introduced which represents the energy and momentum density of the electromagnetic field. Using this tensor, the conservation laws can be formulated by covariant equations (this section can be skipped).

In Chapter 16, we introduced the energy and momentum density of the electromagnetic field,

$$w_{\rm em} = \frac{1}{8\pi}\left(\boldsymbol{E}^2 + \boldsymbol{B}^2\right), \quad \boldsymbol{g}_{\rm em} = \frac{\boldsymbol{S}}{c^2} = \frac{1}{4\pi c}\,\boldsymbol{E}\times\boldsymbol{B} \tag{18.52}$$

In the discussion of the momentum balance, we have introduced Maxwell's stress tensor T_{ik}, (16.30). In covariant form, these quantities becomes the *energy momentum tensor* $T^{\alpha\beta}$,

$$T^{\alpha\beta} = \frac{1}{4\pi}\left(F^{\alpha}_{\ \gamma}\,F^{\gamma\beta} + \frac{1}{4}\,\eta^{\alpha\beta}\,F_{\gamma\delta}\,F^{\gamma\delta}\right) \tag{18.53}$$

This quantity is a symmetric ($T^{\alpha\beta} = T^{\beta\alpha}$) Lorentz tensor. Using (18.52) and (16.30), this $T^{\alpha\beta}$ becomes

$$T = \left(T^{\alpha\beta}\right) = \begin{pmatrix} w_{\rm em} & c\,\boldsymbol{g}_{\rm em} \\ c\,\boldsymbol{g}_{\rm em} & -T_{ik} \end{pmatrix} \tag{18.54}$$

The tensor elements are the energy and momentum density, $w_{\rm em}$ and $\boldsymbol{g}_{\rm em}$.

We calculate the divergence of this energy–momentum tensor:

$$\begin{aligned} \partial_\alpha\,T^{\alpha\beta} &= \frac{1}{4\pi}\left[\left(\partial^\alpha F_{\alpha\gamma}\right)F^{\gamma\beta} + F_{\alpha\gamma}\left(\partial^\alpha F^{\gamma\beta}\right) + \frac{1}{2}\,F_{\delta\gamma}\,\partial^\beta F^{\delta\gamma}\right] \\ &= -\frac{1}{c}\,F^{\beta\gamma}\,j_\gamma + \frac{1}{8\pi}\,F_{\alpha\gamma}\left(\partial^\alpha F^{\gamma\beta} + \partial^\alpha F^{\gamma\beta} + \partial^\beta F^{\alpha\gamma}\right) \end{aligned} \tag{18.55}$$

The upper and lower positions of two equal indices can be interchanged. For the first term in the top line, the inhomogeneous Maxwell equation was used. The second term was written on twice with a prefactor 1/2. In the third term, an index was renamed.

For further reshaping, we use the homogeneous Maxwell equation

$$\partial^\beta F^{\gamma\delta} + \partial^\delta F^{\beta\gamma} + \partial^\gamma F^{\delta\beta} = 0 \tag{18.56}$$

For three different index values, this form follows from $\varepsilon_{\alpha\beta\gamma\delta}\,\partial^\beta F^{\gamma\delta} = 0$, (18.33) with (18.31). If two indices are equal, (18.56) is valid because of $F^{\alpha\beta} = -F^{\beta\alpha}$. With (18.56), the last two terms in (18.55) give $-\partial^\gamma F^{\beta\alpha}$ or $\partial^\gamma F^{\alpha\beta}$:

$$\partial_\alpha\,T^{\alpha\beta} = -\frac{1}{c}\,F^{\beta\gamma}\,j_\gamma + \frac{1}{8\pi}\,F_{\alpha\gamma}\left(\partial^\alpha F^{\gamma\beta} + \partial^\gamma F^{\alpha\beta}\right) = -\frac{1}{c}\,F^{\beta\gamma}\,j_\gamma \tag{18.57}$$

The term omitted in the last step is antisymmetric in α and γ and cancels when summing over these indices. The result is the differential conservation law for the energy and the momentum, as known from Chapter 16:

$$\partial_\alpha T^{\alpha\beta} = -\frac{1}{c} F^{\beta\gamma} j_\gamma \qquad \begin{cases} \beta = 0 & (16.21) \\ \beta = i & (16.28) \end{cases} \tag{18.58}$$

For the further discussion, we assume that there are no charged particles with which the electromagnetic field can exchange energy or momentum; i.e., $j_\gamma = 0$. Then, $\partial_\alpha T^{\alpha\beta} = 0$ applies. For a closed system, this leads to the conservation of the 4-momentum: We integrate $\partial_\alpha T^{\alpha\beta} = 0$ over a volume that completely encloses the system:

$$0 = \int_V d^3r\, \partial_\alpha T^{\alpha\beta} = \frac{1}{c}\, \partial_t \int_V d^3r\, T^{0\beta} + \int_V d^3r\, \partial_i T^{i\beta} \tag{18.59}$$

The last integral is converted into an integral with Gauss' theorem. For the closed system, the fields on the surface disappear. (If V is the entire space, it is sufficient that the fields for $r \to \infty$ fall off at least as $1/r^2$ fall off.) This means that

$$P^\beta_{\text{em}} = \frac{1}{c} \int_V d^3r\, T^{0\beta} = \text{const.} \quad (\text{closed system, } j^\alpha = 0) \tag{18.60}$$

The 4-momentum $(P^\alpha_{\text{em}}) = (E_{\text{em}}/c, \boldsymbol{P}_{\text{em}})$ contains the energy and momentum of the electromagnetic field. In a closed system without charges, these are conserved quantities.

Exercises

18.1 *Relativistic equations of motion*

Show that (18.46) follows from the equations of motion (18.47).

18.2 *Particles in a constant electric field*

A particle (mass m, charge q) has the initial velocity $v_0\, \boldsymbol{e}_x$. It crosses the homogeneous, constant field $\boldsymbol{E} = E\, \boldsymbol{e}_z$ of a capacitor, which is limited to the range $0 \le x \le L$. Integrate the relativistic equation of motion and calculate the deflection angle α for the case $qEL/(mc) \ll \gamma_0 v_0$, where $1/\gamma_0^2 = 1 - v_0^2/c^2$.

18.3 *Particles in a constant magnetic field*

A particle (mass m, charge q) moves in a constant, homogeneous magnetic field $\boldsymbol{B} = B\,\boldsymbol{e}_z$. Solve the relativistic equations of motion with the initial conditions

$$\big(u^\alpha(\tau = 0)\big) = \big(\gamma_0\, c,\, \gamma_0\, v_0, 0, 0\big), \qquad \gamma_0 = \frac{1}{\sqrt{1 - {v_0}^2/c^2}}$$

18.4 *Homogeneous Maxwell equations*

Show that the homogeneous Maxwell equations (18.32) $\partial_\beta\, \widetilde{F}^{\beta\alpha} = 0$ follow from the definitions $\widetilde{F}^{\alpha\beta} = \varepsilon^{\alpha\beta\gamma\delta}\, F_{\gamma\delta}/2$ and $F^{\alpha\beta} = \partial^\alpha A^\beta - \partial^\beta A^\alpha$.

18.5 *Charge as Lorentz scalar*

The Lorentz vector $j^\alpha(x)$ fulfills the equation $\partial_\alpha\, j^\alpha = 0$. Show that the quantity $q = \int d^3r\; j^0/c$ is a Lorentz scalar. First, convince yourself that q can be written in the form

$$q = \frac{1}{c} \int_{x^0 = \text{const.}} da_\alpha\, j^\alpha \tag{18.61}$$

Here,

$$da_\alpha = \frac{1}{6}\, \varepsilon_{\alpha\beta\gamma\delta}\, da^{\beta\gamma\delta}$$

is a Lorentz vector and $da^{\beta\gamma\delta}$ an antisymmetric tensor, which is defined by the assignments

$$da^{012} = dx^0\, dx^1\, dx^2, \quad da^{102} = -dx^0\, dx^1\, dx^2, \quad \text{and so on}$$

This means that the $da^{\beta\gamma\delta}$ represent three-dimensional "area elements" of the four-dimensional Minkowski space.

Chapter 19

Lagrange Formalism

In mechanics, the Lagrange formalism takes center stage. This is particularly so because writing down the Lagrange function is usually the easiest way to obtain the equations of motion. In electrodynamics, we always have to deal with the same equations of motion, the Maxwell equations, so that this point of view no longer applies. For general investigations, however, a formulation as a variation principle is of interest here, too. This applies, for example, to the connection between symmetries and conservation laws or for the comparison with other field theories. Therefore, in the following, the basic principles of the Lagrange formalism of electrodynamics are briefly presented. This chapter is to be understood as a supplement and can be skipped without loss of continuity.

In the following, the basics of the variation calculus (Part III of [1]) are assumed to be known. In addition, some knowledge of fields in mechanics and their treatment in Lagrange formalism is advantageous (e.g., the introduction to continuum mechanics in Part VIII of [1]).

We start with a simple example from mechanics. A string (mass density ϱ_{m}) is stretched (by a tension force P) between two clamping points (at $x = 0$ and $x = l$); the rest position of the string coincides with the x-axis. The string might be deflected (i.e., displaced) perpendicular to the x-axis by the length $\phi(x, t)$. Then,

$$\mathcal{L}(\phi', \dot{\phi}, \phi, x, t) = \frac{\varrho_{\mathrm{m}}}{2}\,\dot{\phi}^2 - \frac{P}{2}\,\phi'^2 + \phi\, f(x, t) \qquad (19.1)$$

is the *Lagrange density* of the string. Here, $\dot{\phi} = \partial\phi/\partial t$ and $\phi' = \partial\phi/\partial x$. Via an external force density $f(x, t)$, the Lagrange density $\mathcal{L}$ may explicitly depend on x and t. The first term in (19.1) is the kinetic energy density of the string motion, the second is the potential energy due to the clamping and the third is the potential energy density of a deflection in the external force field. A detailed derivation of the Lagrangian density (19.1) can be found in Chapter 30 of [1].

We start with the Hamilton principle:

$$\delta S = \delta \int_{t_1}^{t_2} dt \int_0^l dx\ \mathcal{L}(\phi', \dot{\phi}, \phi, x, t) = 0 \tag{19.2}$$

The variation of $\phi(x,t)$ is restricted by fixed boundary values. From $\delta S = 0$, the equation of motion follows:

$$\frac{\partial^2 \phi}{\partial x^2} - \frac{1}{c^2}\frac{\partial^2 \phi}{\partial t^2} = -\frac{f(x,t)}{P} \tag{19.3}$$

Here, $c = \sqrt{P/\varrho_{\text{m}}}$ is the velocity of the string waves.

We consider now the variational principle for the Maxwell equations:

$$\delta S = \delta \int dt \int d^3r\ \mathcal{L}(\text{fields}) = \frac{1}{c}\,\delta \int d^4x\ \mathcal{L}(\text{fields}) = 0 \tag{19.4}$$

Generally, the Lagrangian density or function is a particularly simple expression of the dynamical variables (trajectories $q_i(t)$ or fields $\phi(x,t)$). Starting from this point of view we determine the Lagrangian density $\mathcal{L} = \mathcal{L}(\text{field})$ for the electromagnetic field.

In a relativistic theory, the action functional $S = \int d^4x\, \mathcal{L}$ and thus $\mathcal{L}$ should be Lorentz scalars. (The weaker condition that δS is a Lorentz scalar would also suffice.) In analogy to Lagrange functions and densities of mechanics, we expect $\mathcal{L}$ to be quadratic in the derivatives of the field, see for example (19.1). For electrodynamics, this means quadratic in $\partial^\beta A^\alpha$ or in the field strength $F^{\alpha\beta}$. The only Lorentz scalar that we can construct from $F^{\alpha\beta} F^{\gamma\delta}$ is $F^{\alpha\beta} F_{\alpha\beta}$ because a contraction $F_\alpha{}^\alpha$ results in zero, and $\varepsilon_{\alpha\beta\gamma\delta}\, F^{\alpha\beta} F^{\gamma\delta}$ were only a pseudo scalar.

In Equation (19.1), the force density $f(x,t)$ may induce deflections of the string and is therefore a source of the field $\phi(x,t)$. In electrodynamics, the current densities j^α are the sources of the field. The simplest scalar for j_α and the field A_α is $j_\alpha A^\alpha$. The quantity $j_\alpha A^\alpha$ is linear in the field (as the last term in (19.1). All these considerations make the approach

$$\boxed{\mathcal{L}\big(\partial^\alpha A^\beta, A^\beta, x\big) = -\frac{1}{16\,\pi}\, F^{\alpha\beta} F_{\alpha\beta} - \frac{1}{c}\, A^\alpha\, j_\alpha(x)} \tag{19.5}$$

plausible. The x-dependence shown in the argument of $\mathcal{L}$ stems from the sources $j_\alpha(x)$; here, x stands for $(t, \boldsymbol{r}) = (x^0, x^1, x^2, x^3)$. A multiplicative constant in $\mathcal{L}$ does not change (19.4). The ratio of the two constants on the r.h.s. in (19.5) is chosen such that (19.4) yields the inhomogeneous Maxwell equations. The fields A^α are considered as dynamic variables; the fields $F^{\alpha\beta}$ are used as abbreviations for $\partial^\alpha A^\beta - \partial^\beta A^\alpha$. The homogeneous Maxwell equations follow directly from $F^{\alpha\beta} = \partial^\alpha A^\beta - \partial^\beta A^\alpha$.

For derivation of the Euler–Lagrange equations, we consider the variations δA^β, where the values at the borders $t = t_1,\ t = t_2$ and $x^i = \pm\infty$ are kept fixed. Then all border terms from the partial integrations vanish and we can transform δS as follows:

$$\begin{aligned}
\delta S &= \frac{1}{c}\int d^4x \left[\mathcal{L}\big(\partial^\alpha(A^\beta + \delta A^\beta), A^\beta + \delta A^\beta, x\big) - \mathcal{L}\big(\partial^\alpha A^\beta, A^\beta, x\big)\right] \\
&= \frac{1}{c}\int d^4x \left[\frac{\partial \mathcal{L}}{\partial(\partial^\alpha A^\beta)}\ \partial^\alpha(\delta A^\beta) + \frac{\partial \mathcal{L}}{\partial A^\beta}\ \delta A^\beta\right] \\
&\overset{\text{p.I.}}{=} \frac{1}{c}\int d^4x \left[-\partial^\alpha \left(\frac{\partial \mathcal{L}}{\partial(\partial^\alpha A^\beta)}\right) + \frac{\partial \mathcal{L}}{\partial A^\beta}\right]\delta A^\beta \ = 0
\end{aligned} \tag{19.6}$$

Since δS must be zero for arbitrary variations δA^β, it follows

$$\frac{\partial}{\partial x_\alpha}\, \frac{\partial \mathcal{L}}{\partial(\partial^\alpha A^\beta)} = \frac{\partial \mathcal{L}}{\partial A^\beta} \tag{19.7}$$

We insert (19.5) into these Euler–Lagrange equations. It should be noted that a derivative with respect to A^β also acts on A_β. Since in each term $\sum_{\alpha\beta} \ldots F^{\alpha\beta}$, the $\partial^\gamma A^\delta$ occurs twice, the derivative of $\mathcal{L}$ with respect to $\partial^\alpha A^\beta$ yields four times $-F_{\alpha\beta}/(16\,\pi)$. As a result, (19.7) becomes

$$\partial^\alpha F_{\alpha\beta} = \frac{4\pi}{c}\ j_\beta \tag{19.8}$$

Coupling between Field and Matter

The coupling term $j^\alpha A_\alpha$ in (19.5) describes the coupling between charged matter (j^α) and the fields (A_α). On the one hand, this coupling implies that the charges are sources of the fields and therefore occur on the right-hand side of (19.8). On the other hand, the coupling term describes the forces that fields exert on charges. These force effects are discussed in this section in more detail.

The force on a charged particle is described by the right-hand side of the equation of motion (18.43),

$$m\,\frac{du^\alpha}{d\tau} = \frac{q}{c}\,F^{\alpha\beta}\,u_\beta \tag{19.9}$$

The corresponding Lagrange function results from that for a free particle, $\mathscr{L}_0 = -mc\,(u^\alpha u_\alpha)^{1/2}$ (Chapter 40 in [1]) and a coupling term:

$$\mathscr{L}\big(u^\beta(\tau), x^\beta(\tau)\big) = -mc\,\sqrt{u^\beta u_\beta} - \frac{q}{c}\,A^\alpha\big(x^\beta(\tau)\big)\,u_\alpha \tag{19.10}$$

Note the use of slightly different symbols for the Lagrange function $\mathscr{L}$ and the Lagrange density $\mathcal{L}$. In (19.10), the dynamic variable is the trajectory $x^\alpha(\tau)$ of the particle (which also determines the velocity $u^\alpha = dx^\alpha/d\tau$ and the momentum $p^\alpha = m u^\alpha$). In contrast to this, the field $A_\alpha(x)$ in (19.10) is considered to be given. In the Hamiltonian principle

$$\delta S = \delta \int_{\tau_1}^{\tau_2} d\tau\;\mathscr{L}\big(u^\beta(\tau), x^\beta(\tau)\big) = 0 \tag{19.11}$$

the trajectory $x^\alpha(\tau)$ is varied with fixed boundary values. This results in the equations of motion

$$\frac{d}{d\tau}\,\frac{\partial \mathscr{L}}{\partial u^\alpha} = \frac{\partial \mathscr{L}}{\partial x^\alpha} \tag{19.12}$$

Using (19.10) and taking $u^\alpha u_\alpha = c^2$ into account, we obtain (19.9). Note that $u^\alpha u_\alpha = c^2$ must not be used in (19.10), because $\mathscr{L}$ is not a physical quantity. Likewise, in classical mechanics, the conservation of energy must not be used in $\mathscr{L}$.

The coupling term in (19.10) is justified by the fact that (19.12) results in the correct equation of motion (19.9). On a first look, it appears different from the coupling term in (19.5). To compare both coupling terms, we start with the 4-current density $j^\alpha(x)$ of a moving point charge:

$$j^\alpha = q\,c\,\delta\big(\boldsymbol{r}-\boldsymbol{r}(t)\big)\,u^\alpha\,\frac{d\tau}{dx^0} = q\,c\int d\tau\;\delta^{(4)}\big(x-x^\beta(\tau)\big)\,u^\alpha(\tau) \qquad (19.13)$$

Here, $(j^\alpha) = \varrho\,(c, \boldsymbol{v})$, $\varrho = q\,\delta(\boldsymbol{r}-\boldsymbol{r}(t))$, $(c, \boldsymbol{v}) = (dx^\alpha/dt) = c\,(dx^\alpha/dx^0)$ and $u^\alpha = dx^\alpha/d\tau$ are used. For the last step, $\int d\tau\;\delta(x^0 - x^0(\tau)) = 1/|dx^0/d\tau|$ (see (3.19)) was used. In the four-dimensional δ-function, x stands for (x^0, x^1, x^2, x^3). We now insert (19.13) into the coupling term in (19.5) and determine its contribution to the action functional $S = \int d^4x\,\mathcal{L}/c$:

$$-\frac{1}{c^2}\int d^4x\;A^\alpha(x)\;j_\alpha(x) \;\overset{(19.13)}{=}\; -\frac{q}{c}\int d\tau\;A^\alpha\big(x^\beta(\tau)\big)\;u_\alpha(\tau) \qquad (19.14)$$

This shows that coupling terms in (19.5) and (19.10) are equivalent to each other.

In the Maxwell equations (19.8), the source distribution $j^\alpha(x)$ is assumed to be given. In the equation of motion (19.9), the field $A_\alpha(x)$ is assumed to be given. In the general case, however, (19.8) and (19.9) represent coupled equations for the fields and the particles. The Lagrange function of the overall system then results from the contribution of the free fields (first term in (19.5)), the contribution of the free particles (first term in (19.10)) and the common coupling term. The variation of $\delta S = 0$ with respect to $A_\alpha(x)$ and $x_\alpha(\tau)$ then results in the coupled equations (19.8) and (19.9).

Instead of $x^\alpha(\tau)$, also $\boldsymbol{r}(t)$ can be used as the dynamic variable; because of $u^\alpha u_\alpha = c^2$, only three of the four functions $x^\alpha(\tau)$ are independent of each other. The associated Lagrange function for the relativistic motion of a particle in the electromagnetic field is then (Chapter 40 in [1])

$$\mathcal{L}\big(\boldsymbol{r}, \boldsymbol{v}, t\big) = -m c^2\sqrt{1-\frac{v(t)^2}{c^2}} - q\,\Phi\big(\boldsymbol{r}, t\big) + \frac{q}{c}\;\boldsymbol{v}\cdot\boldsymbol{A}\big(\boldsymbol{r}, t\big) \qquad (19.15)$$

Here, $\boldsymbol{v}(t) = d\boldsymbol{r}/dt$. From $\delta\int dt\;\mathcal{L} = 0$ follows (18.47).

Exercises

19.1 *Gauge transformation*

Which term is added to the Lagrangian density

$$\mathcal{L} = -\frac{1}{16\pi}\, F^{\alpha\beta}\, F_{\alpha\beta} - \frac{1}{c}\, A^{\alpha}\, j_{\alpha}$$

by the gauge transformation $A^{\alpha} \to A^{\alpha} - \partial^{\alpha}\Lambda$? Show that this term can be written as a divergence $\partial^{\alpha} V_{\alpha}$. What contribution does this term yield in the action functional?

19.2 *Conservation of the 4-momentum*

The Lagrangian density $\mathcal{L}(A^{\alpha}, \partial^{\alpha}A^{\beta}, x^{\gamma}) = -F^{\alpha\beta}F_{\alpha\beta}/(16\pi)$ of the free electromagnetic field does not explicitly depend on x^{γ}; such a dependence would only arise from external sources $j^{\alpha}(x^{\gamma}) \neq 0$. Formally, this temporal and spatial translation invariance can be expressed by $\partial_{\alpha}\mathcal{L} = 0$. Show that from this the conservation law

$$\partial_{\alpha} T^{\alpha\beta} = 0 \quad \text{for} \quad T^{\alpha\beta} = \frac{1}{4\pi}\left[F^{\alpha\gamma}\, F_{\gamma}{}^{\beta} + \frac{1}{4}\,\eta^{\alpha\beta}\, F^{\gamma\delta}\, F_{\gamma\delta}\right]$$

follows, where $T^{\alpha\beta}$ is the energy–momentum tensor. From $\partial_{\alpha} T^{\alpha\beta} = 0$, one obtains the conservation of the 4-momentum in a closed system.

PART V

Maxwell Equations: Applications

Chapter 20

Plane Waves

Part V deals with the most important applications of the Maxwell equations. These applications include in particular electromagnetic waves, the fields of moving charges and the scattering of light by electrons.

In this chapter, we examine the properties of a plane and monochromatic wave. Various types of electromagnetic radiation (radio to γ-rays) are presented. The connection to the quanta of the field, the photons, is discussed. This includes the relation between the polarization of the wave and the spin of photons.

In the source-free case, $j^\alpha = 0$, the Maxwell equations read (18.20, 18.23)

$$\Box A^\alpha = 0 \tag{20.1}$$

$$\partial_\alpha A^\alpha = 0 \tag{20.2}$$

The fields $F^{\alpha\beta} = \partial^\alpha A^\beta - \partial^\beta A^\alpha$ are invariant under the gauge transformation

$$A^\alpha \longleftrightarrow A^{*\alpha} = A^\alpha + \partial^\alpha \Lambda \tag{20.3}$$

It was therefore possible to impose the condition (20.2) on the potentials. We repeat the underlying train of thought. If the potentials $A^{*\alpha}$ do not fulfill the condition (20.2),

$$\partial_\alpha A^{*\alpha} = f(x) \neq 0 \tag{20.4}$$

then we introduce via (20.3) different potentials A^α:

$$\partial_\alpha A^\alpha = f(x) - \Box \Lambda \tag{20.5}$$

We now choose Λ such that (20.2) is fulfilled, i.e.,

$$\Box\, \Lambda(x) = f(x) \tag{20.6}$$

The general solution of this equation is of the form (17.13):

$$\Lambda = \Lambda_{\text{hom}} + \Lambda_{\text{part}} \tag{20.7}$$

The condition (20.2) can be fulfilled by a particular solution Λ_{part}. This means that Λ_{hom} is not fixed. Without violating the condition (20.2), the potentials can therefore be changed once more by the gauge transformation

$$A^{\alpha} \;\longleftrightarrow\; A^{*\alpha} = A^{\alpha} + \partial^{\alpha} \Lambda_{\text{hom}} \tag{20.8}$$

For a given $A^{*\alpha} = A^{*\alpha}_{\text{hom},}$ we choose Λ_{hom} such that $\partial^0 \Lambda_{\text{hom}} = \Phi^*$ yielding

$$A^{\alpha} = A^{*\alpha} - \partial^{\alpha} \Lambda_{\text{hom}} = (0, \boldsymbol{A}) \tag{20.9}$$

or

$$A^0 = \Phi = 0 \quad (\text{gauge for } j^{\alpha} = 0) \tag{20.10}$$

From this and (20.2) it follows that $\operatorname{div} \boldsymbol{A} = 0$, i.e., the Coulomb gauge (also called radiation gauge). Overall, we get

$$\boxed{\left(\Delta - \frac{1}{c^2}\frac{\partial^2}{\partial t^2}\right)\boldsymbol{A}(\boldsymbol{r}, t) = 0, \quad \operatorname{div} \boldsymbol{A}(\boldsymbol{r}, t) = 0, \quad \Phi(\boldsymbol{r}, t) = 0} \tag{20.11}$$

This means that only two of the four fields A^{α} are independent of each other. This corresponds to the two polarization directions of the plane wave.

The general solution of the homogeneous wave equation has been given in Chapter 17. In the following, we focus on the *plane, monochromatic* wave.

Plane Wave Packet

A wave is called *plane* if it depends on one Cartesian coordinate (we choose z) only:

$$\boldsymbol{A}(\boldsymbol{r}, t) = \boldsymbol{A}(z, t) \quad (\text{plane wave}) \tag{20.12}$$

The amplitude of the wave is constant in the planes $z =$ const., this explains the designation "plane". Due to $(\partial_z^{\,2} - \partial_t^{\,2}/c^2)\, f(z \pm ct) = 0$, the fields

$$A_i(z, t) = f_i(z - ct) + g_i(z + ct) \tag{20.13}$$

with arbitrary functions f_i and g_i are solution of Maxwell's equations. These are wave packets that travel with the velocity c in the $\pm z$ direction. From $\operatorname{div} \boldsymbol{A} = 0$, it follows that $A_3' = f_3' + g_3' = 0$, which only allows $A_3 =$ const. Since a constant does not contribute to the fields $\boldsymbol{E}$ and $\boldsymbol{B}$, we can set $A_3 = 0$. This means that $\boldsymbol{A}$ is perpendicular to the direction of propagation. Such a wave is called *transverse*.

As the wave equation is a second-order differential equation in time, it has to be supplemented by two initial conditions, $A_i(z, 0) = F_i(z)$ and $\dot{A}_i(z, 0) = G_i(z)$. The solution (20.13) contains just as many integration constants (the functions f_i and g_i) as can be determined by the initial conditions (the functions F_i and G_i). Therefore, (20.13) is the general solution. The solution (20.13) can also be written in the form (17.23), Exercise 17.2. All the functions considered here (f_i, g_i, F_i, G_i) are real.

With (20.13), the electric and magnetic fields are also functions of the arguments $z \pm ct$. Consider a field configuration at a certain time t_0. At a later time $t_0 + \Delta t$, it is shifted by $\Delta z = \pm c\, \Delta t$ in the z direction, but otherwise it is unchanged. The wave packet $f_i(z - ct)$ in (20.13) thus propagates *without change of its form* in the z-direction with the velocity c; independent of this, the package $g_i(z + ct)$ runs with c in the $(-z)$-direction. In contrast to this, an electromagnetic wave packet in matter or a quantum mechanical wave packet disperses (broadens) over time. Such a *dispersion* of a wave packet is discussed in Chapter 33.

We further explain the meaning of *plane* by comparing the plane wave to a spherical wave. Thereby, we limit ourselves to a fixed frequency $\omega = ck$:

$$\boldsymbol{A}(\boldsymbol{r}, t) = \begin{cases} \operatorname{Re} \boldsymbol{A}_0 \exp\big(\mathrm{i}(\pm kz - \omega t)\big) & \text{(plane wave)} \\[2ex] \operatorname{Re} \dfrac{\boldsymbol{A}_0}{r} \exp\big(\mathrm{i}(\pm kr - \omega t)\big) & \text{(spherical wave)} \end{cases} \tag{20.14}$$

The plane wave equals the elementary solution found in (17.22). The spherical wave solves the wave equation in the range $r \neq 0$, as can be seen from (3.34). The use of the real part (Re) is discussed in the following in more detail.

The waves (20.14) are characterized by their surfaces (or areas) of equal phase, which may be called phase surfaces. For the plane wave, the equal phase condition $\pm kz - \omega t = \text{const.}$ implies that these surfaces are planes; these planes travel with the velocity c into the $\pm z$ direction. For the spherical wave, the equal phase condition $\pm kr - \omega t = \text{const.}$ yields concentric spheres, which run with the velocity c into the $\pm \boldsymbol{e}_r$ direction. As the spheres get larger and large, the amplitude decreases according to $\boldsymbol{A} \propto 1/r$. An outgoing sphere wave (plus sign) can be generated by an oscillating charge distribution at $r = 0$ (Chapter 24). In general, the surfaces of equal phases depend on the sources and on the medium (Chapter 38).

Monochromatic Plane Wave

Waves with only one frequency ω are called *monochromatic*. The time dependence of such waves is of the form $a_1 \cos(\omega t) + a_2 \sin(\omega t)$ with two real amplitudes a_1 and a_2. For many calculations, it is more convenient to use the solution $\exp(-\mathrm{i}\omega t)$ with a complex amplitude $a_1 + \mathrm{i}a_2$:

$$\mathrm{Re}\,(a_1 + \mathrm{i}a_2)\exp(-\mathrm{i}\omega t) = a_1 \cos(\omega t) + a_2 \sin(\omega t) \qquad (20.15)$$

Instead of the real part (Re), one could just as well use the imaginary part (Im). No new solution is obtained if ω is replaced by $-\omega$; we can therefore restrict ourselves to positive frequencies ω.

The fields $\boldsymbol{E}$ and $\boldsymbol{B}$ are real; this is then also true for the vector potential $\boldsymbol{A}$. The restrictions *plane* and *monochromatic* reduce the general solution (17.23) of the wave equation to the elementary solution (17.22). The general expression for a monochromatic plane wave is thus as follows:

$$\boxed{\boldsymbol{A}(\boldsymbol{r}, t) = \mathrm{Re}\,\boldsymbol{A}_0 \exp\big[\,\mathrm{i}(\boldsymbol{k} \cdot \boldsymbol{r} - \omega t)\big]} \qquad (20.16)$$

The wave (20.16) is plane because it only moves in one direction (the of $\boldsymbol{k}$). It is monochromatic because it contains only one frequency. It is the solution of the wave equation in (20.11) if

$$\boxed{\omega = c\,|\boldsymbol{k}| = ck} \qquad (20.17)$$

The frequency ω is positive, and the components k_x, k_y and k_z of the wave vector are real:

$$\omega > 0, \qquad -\infty < k_x, k_y,\ k_z < \infty \tag{20.18}$$

The wave equation (20.11) is solved by (20.16) with an arbitrary, generally complex amplitude $\boldsymbol{A}_0$. The second condition in (20.11) implies

$$\operatorname{div} \boldsymbol{A} = \operatorname{Re}\, \mathrm{i}\boldsymbol{k} \cdot \boldsymbol{A}_0 \exp\big(\mathrm{i}(\boldsymbol{k}\cdot\boldsymbol{r} - \omega t)\big) = 0 \tag{20.19}$$

Since this applies to any position and time, we get $\boldsymbol{k}\cdot\boldsymbol{A}_0 = 0$. We consider the scalar product of (20.16) and $\boldsymbol{k}$; in doing so, we can interchange the real vector $\boldsymbol{k}$ with the Re-sign. From $\boldsymbol{k}\cdot\boldsymbol{A}_0 = 0$ then follows $\boldsymbol{k}\cdot\boldsymbol{A} = 0$ or

$$\boldsymbol{A} \perp \boldsymbol{k} \tag{20.20}$$

We now calculate the electric and magnetic field:

$$\boldsymbol{E}(\boldsymbol{r}, t) = -\frac{1}{c}\frac{\partial \boldsymbol{A}}{\partial t} = \operatorname{Re}\, \mathrm{i}k\boldsymbol{A}_0 \exp\big[\,\mathrm{i}(\boldsymbol{k}\cdot\boldsymbol{r} - \omega t)\big] = \operatorname{Re}\boldsymbol{E}_0 \exp\big[\,\mathrm{i}(\boldsymbol{k}\cdot\boldsymbol{r} - \omega t)\big] \tag{20.21}$$

$$\boldsymbol{B}(\boldsymbol{r}, t) = \nabla\times\boldsymbol{A} = \operatorname{Re}\, \mathrm{i}\boldsymbol{k}\times\boldsymbol{A}_0 \exp\big[\mathrm{i}(\boldsymbol{k}\cdot\boldsymbol{r} - \omega t)\big] = \operatorname{Re}\boldsymbol{B}_0 \exp\big[\mathrm{i}(\boldsymbol{k}\cdot\boldsymbol{r} - \omega t)\big] \tag{20.22}$$

Here, $\Phi = 0$ and $\omega = ck$ were used. In the last expressions, the complex amplitudes $\boldsymbol{E}_0$ and $\boldsymbol{B}_0$ were introduced:

$$\boldsymbol{E}_0 = \mathrm{i}k\boldsymbol{A}_0, \qquad \boldsymbol{B}_0 = \mathrm{i}\boldsymbol{k}\times\boldsymbol{A}_0 \tag{20.23}$$

From (20.21) and (20.22) we get the relations

$$\begin{aligned}\boldsymbol{B}(\boldsymbol{r}, t) &= \operatorname{Re}\, \mathrm{i}\boldsymbol{k}\times\boldsymbol{A}_0 \exp\big[\,\mathrm{i}(\boldsymbol{k}\cdot\boldsymbol{r} - \omega t)\big] \\ &= (\boldsymbol{k}/k)\times\operatorname{Re}\, \mathrm{i}k\boldsymbol{A}_0 \exp\big[\,\mathrm{i}(\boldsymbol{k}\cdot\boldsymbol{r} - \omega t)\big] = (\boldsymbol{k}/k)\times\boldsymbol{E}(\boldsymbol{r}, t)\end{aligned} \tag{20.24}$$

$$\begin{aligned}(\boldsymbol{k}/k)\cdot\boldsymbol{E}(\boldsymbol{r}, t) &= (\boldsymbol{k}/k)\cdot\operatorname{Re}\, \mathrm{i}k\boldsymbol{A}_0 \exp\big[\,\mathrm{i}(\boldsymbol{k}\cdot\boldsymbol{r} - \omega t)\big] \\ &= \operatorname{Re}\, \mathrm{i}\boldsymbol{k}\cdot\boldsymbol{A}_0 \exp\big[\,\mathrm{i}(\boldsymbol{k}\cdot\boldsymbol{r} - \omega t)\big] \overset{(20.19)}{=} 0\end{aligned} \tag{20.25}$$

The real vector $\boldsymbol{k}$ was interchanged with the Re-sign. The amplitudes $\boldsymbol{A}_0$, $\boldsymbol{E}_0$ and $\boldsymbol{B}_0$ are generally complex. For the real physical fields $\boldsymbol{E}$ and $\boldsymbol{B}$, we get from the last two equations:

$$\boxed{\boldsymbol{E}(\boldsymbol{r}, t) \perp \boldsymbol{k}, \qquad \boldsymbol{B}(\boldsymbol{r}, t) \perp \boldsymbol{k}} \tag{20.26}$$

$$\boxed{\big|\boldsymbol{E}(\boldsymbol{r}, t)\big| = \big|\boldsymbol{B}(\boldsymbol{r}, t)\big|, \quad \boldsymbol{E}(\boldsymbol{r}, t) \perp \boldsymbol{B}(\boldsymbol{r}, t)} \tag{20.27}$$

Discussion

The *wave vector* $\boldsymbol{k}$ points into the *propagation direction* of the wave. Consider, for example, an equal phase surface $\boldsymbol{k}\cdot\boldsymbol{r}-\omega t = \text{const}$. This phase surface (plane) shifts with the phase velocity $c = \omega/k$ in the direction of $\boldsymbol{k}/k$. From (20.26) it follows that the energy flux density $\boldsymbol{S} = c\boldsymbol{E} \times \boldsymbol{B}/4\pi$, (16.20), is parallel to $\boldsymbol{k}$. Therefore, $\boldsymbol{k}$ is the direction of energy transport of the wave.

If the amplitude vector and the wave vector are parallel to each other, the wave is called *longitudinal*. Sound waves in gases and liquids are longitudinal, and sound waves in solids can be transverse or longitudinal. In these examples, the wave field is the deviation from the equilibrium density or the displacement from the rest position.

Due to (20.26), electromagnetic waves are *transversal*; the electric and magnetic fields are perpendicular to $\boldsymbol{k}$. In addition, the electric and magnetic fields are perpendicular to each other.

The electric and magnetic fields of the wave have at each point $\boldsymbol{r}, t$ the same absolute value. This means that fields at the same have maxima and zeros at the same points; they are *in phase*.

If one proceeds in the $\boldsymbol{k}$ direction by one wavelength

$$\lambda = \frac{2\pi}{k} \quad \text{(wavelength)} \tag{20.28}$$

then the phase changes by $k\lambda = 2\pi$, and the amplitude $\boldsymbol{A}(\boldsymbol{r}, t)$ has the same value. In particular, the distance between two neighboring maxima of the field $\boldsymbol{A}$ is equal to λ. The wavenumber $k = 2\pi/\lambda$ is equal to 2π times the number of wave crests per length. In addition to the angular (also circular) frequency ω, the frequency ν is used,

$$\nu = \frac{\omega}{2\pi} = \frac{c}{\lambda} \tag{20.29}$$

The frequency is measured in Hertz, and the wavelength in meters:

$$\nu = \frac{1}{\text{s}} = 1\,\text{Hertz} = 1\,\text{Hz}, \qquad \lambda = 1\,\text{m} \tag{20.30}$$

The common designations for waves are summarized in Table 20.1.

Table 20.1 The designations for various properties of an electromagnetic wave. These concepts are also used for other waves (with some exceptions in the third column). For sound waves, the amplitude is the deviation from the equilibrium density (gas, liquid) or the displacement from the equilibrium position (solid body); the speed of sound is $c_S \approx 330\,\mathrm{m/s}$ in air and $c_S \approx 5000\,\mathrm{m/s}$ in iron.

Designation	Symbol	Relations
Amplitude	$\boldsymbol{A}_0$	$\boldsymbol{A}_0 \perp \boldsymbol{k}$
Wave vector	$\boldsymbol{k}$	$\boldsymbol{S} \parallel \boldsymbol{k}$
Wavenumber	k	$k = \lvert\boldsymbol{k}\rvert = \omega/c$
Wavelength	λ	$\lambda = 2\pi/k$
Angular frequency	ω	$\omega = ck$
Frequency	ν	$\nu = \omega/2\pi = c/\lambda$
Phase velocity	c	$c = \omega/k = 3 \cdot 10^8\,\mathrm{m/s}$
Phase	φ	$\varphi = \boldsymbol{k} \cdot \boldsymbol{r} - \omega t$

Example

We consider the special case

$$\boldsymbol{k} = k\,\boldsymbol{e}_z, \quad \boldsymbol{A}_0 = -A_0\,\boldsymbol{e}_x, \quad A_0 = \text{real} \tag{20.31}$$

The fields of this wave

$$E_x = B_y = k\,A_0 \sin(kz - \omega t), \quad E_y = E_z = B_x = B_z = 0 \tag{20.32}$$

are sketched in Figure 20.1. At a given position, the electric and magnetic fields oscillate in phase between the values $\pm k\,A_0$. A wave crest, such as a certain maximum of E_x, travels with the velocity c in the z-direction.

Polarization

We consider a specific propagation direction $\boldsymbol{k}$ of the wave. The orientations of the fields $\boldsymbol{E}$ and $\boldsymbol{B}$ are restricted by $\boldsymbol{E} \perp \boldsymbol{k}$, $\boldsymbol{B} \perp \boldsymbol{k}$ and $\boldsymbol{E} \perp \boldsymbol{B}$. Then $\boldsymbol{E}$ may assume any direction within the planes $\boldsymbol{k} \cdot \boldsymbol{r} = 0$. Under *polarization* we understand the direction of the vector $\boldsymbol{E}$ (or $\boldsymbol{B}$) and its time dependence. The *linear* polarization means a constant direction and *circular* polarization means a rotation of this direction with constant

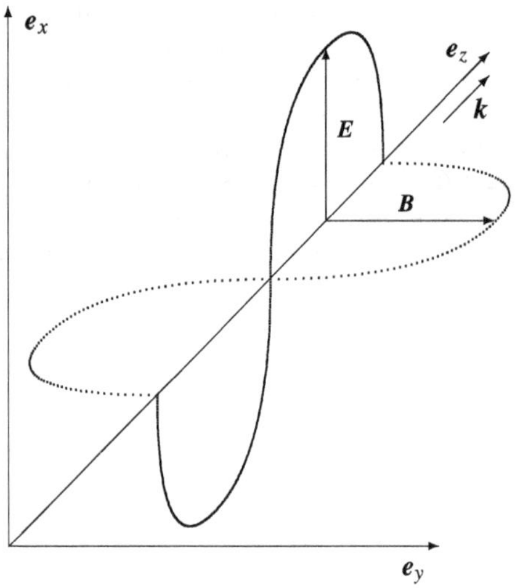

Figure 20.1 Electric field $\boldsymbol{E}(z,t) = E(z,t)\,\boldsymbol{e}_x$ and magnetic field $\boldsymbol{B}(z,t) = B(z,t)\,\boldsymbol{e}_y$ of the linearly polarized wave (20.32) at a certain time. At a given position, the fields oscillate between the values $\pm k\,A_0$. Effectively, the sketched fields shift with the velocity c in the direction of the wave vector $\boldsymbol{k} = k\,\boldsymbol{e}_z$.

angular velocity; this is explained in more detail in the following. Unpolarized light refers to an ensemble of wave packets in which the directions of $\boldsymbol{E}$ are statistically are distributed.

The complex amplitude vector $\boldsymbol{E}_0$ can be decomposed into two real vectors, $\boldsymbol{a}_1$ and $\boldsymbol{a}_2$,

$$\boldsymbol{E}_0 = \boldsymbol{a}_1 + \mathrm{i}\,\boldsymbol{a}_2 = (\boldsymbol{b}_1 + \mathrm{i}\,\boldsymbol{b}_2)\,\exp(-\mathrm{i}\alpha) \tag{20.33}$$

In the last expression, an arbitrary phase factor has been split off. The scalar product $\boldsymbol{E}_0^2 = \boldsymbol{E}_0 \cdot \boldsymbol{E}_0$ is a complex number. We split this number into its magnitude and phase:

$$\boldsymbol{E}_0^2 = \left|\boldsymbol{E}_0^2\right|\,\exp(\mathrm{i}\gamma) = \left(\boldsymbol{b}_1 + \mathrm{i}\,\boldsymbol{b}_2\right)^2\,\exp(-2\mathrm{i}\alpha) \tag{20.34}$$

If we now choose $\alpha = -\gamma/2$, then $(\boldsymbol{b}_1 + \mathrm{i}\,\boldsymbol{b}_2)^2$ is equal to $|\boldsymbol{E}_0^2|$, i.e., real:

$$\left(\boldsymbol{b}_1 + \mathrm{i}\,\boldsymbol{b}_2\right)^2 = \boldsymbol{b}_1^2 - \boldsymbol{b}_2^2 + 2\,\mathrm{i}\,\boldsymbol{b}_1 \cdot \boldsymbol{b}_2 = \text{real} \tag{20.35}$$

Therefore, $\boldsymbol{b}_1 \cdot \boldsymbol{b}_2 = 0$ and $\boldsymbol{b}_1 \perp \boldsymbol{b}_2$. From $\boldsymbol{k} \cdot \boldsymbol{E}_0 = 0$ follows $\boldsymbol{k} \cdot (\boldsymbol{b}_1 + \mathrm{i}\,\boldsymbol{b}_2) = 0$ and thus $\boldsymbol{b}_1 \perp \boldsymbol{k}$ and $\boldsymbol{b}_2 \perp \boldsymbol{k}$. If neither $\boldsymbol{b}_1$ nor $\boldsymbol{b}_2$ are zero, we can use a

Cartesian coordinate system with the base vectors $\boldsymbol{e}_1$, $\boldsymbol{e}_2$ and $\boldsymbol{e}_3$ such that

$$\boldsymbol{e}_1 = \frac{\boldsymbol{b}_1}{b_1}, \quad \boldsymbol{e}_2 = \mp \frac{\boldsymbol{b}_2}{b_2}, \quad \boldsymbol{e}_3 = \frac{\boldsymbol{k}}{k} \tag{20.36}$$

The sign in $\boldsymbol{e}_2$ is set such that a right-handed system results. One of the two vectors $\boldsymbol{b}_1$ or $\boldsymbol{b}_2$ could be zero; then the corresponding basis vector must be chosen perpendicular to the other two.

With (20.36), the electric field (20.21) becomes

$$\begin{aligned} \boldsymbol{E}(\boldsymbol{r}, t) &= \text{Re}\,\big(b_1\,\boldsymbol{e}_1 \mp \mathrm{i} b_2\,\boldsymbol{e}_2\big)\,\exp\big[\,\mathrm{i}(\boldsymbol{k}\cdot\boldsymbol{r} - \omega t - \alpha)\big] \\ &= \boldsymbol{e}_1\,b_1\,\cos(\boldsymbol{k}\cdot\boldsymbol{r} - \omega t - \alpha) \pm \boldsymbol{e}_2\,b_2\,\sin(\boldsymbol{k}\cdot\boldsymbol{r} - \omega t - \alpha) \\ &= E_1(\boldsymbol{r}, t)\,\boldsymbol{e}_1 + E_2(\boldsymbol{r}, t)\,\boldsymbol{e}_2 \end{aligned} \tag{20.37}$$

For the Cartesian components $E_1 = b_1 \cos(\ldots)$ and $E_2 = b_2 \sin(\ldots)$, we obtain

$$\frac{E_1^2}{b_1^2} + \frac{E_2^2}{b_2^2} = 1 \tag{20.38}$$

In the E_1-E_2-plane, the vector $\boldsymbol{E} = (E_1, E_2, 0)$ thus describes an ellipse, Figure 20.2. At a given position $\boldsymbol{r}$, the vector $\boldsymbol{E}$ rotates around the $\boldsymbol{k}$ axis.

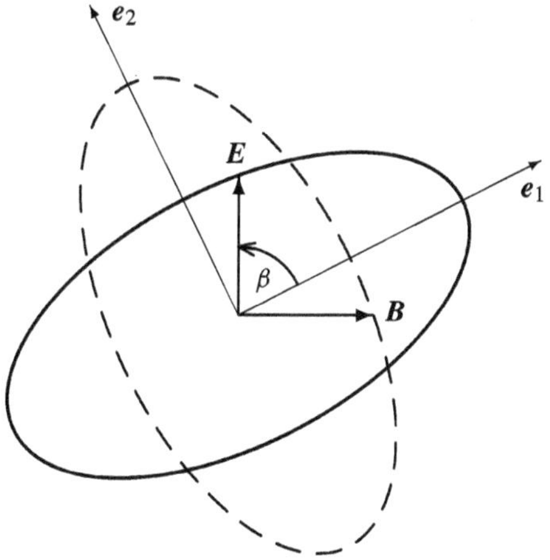

Figure 20.2 At a given position, the field vector $\boldsymbol{E}(\boldsymbol{r}, t)$ of a plane wave rotates with $\beta = \beta(t)$ of (20.39) and describes an ellipse. The magnetic field vector $\boldsymbol{B}$ is perpendicular to $\boldsymbol{E}$; it describes the dashed ellipse. The wave vector is perpendicular on the image plane.

The angle between $\boldsymbol{E}$ and the $\boldsymbol{e}_1$ axis is denoted by β. For $\beta(t)$ we obtain

$$\tan\beta(t) = \frac{E_2}{E_1} = \mp\frac{b_2}{b_1}\ \tan(\omega t + \delta) \tag{20.39}$$

where $\delta = \alpha - \boldsymbol{k}\cdot\boldsymbol{r}$. Depending on the sign, the $\boldsymbol{E}$ vector rotates with the angular frequency ω to the left or to the right. A complete cycle of the ellipse takes place during the time period $T = 2\pi/\omega$.

The $\boldsymbol{B}$-field is given by $\boldsymbol{B}(\boldsymbol{r},t) = (\boldsymbol{k}/k)\times\boldsymbol{E}(\boldsymbol{r},t)$, (20.27). Therefore, the $\boldsymbol{B}$ vector describes an ellipse that is rotated by $\pi/2$ compared to that of $\boldsymbol{E}$.

The derivation of (20.37) was carried out for an arbitrary complex amplitude $\boldsymbol{E}_0$. The result therefore represents the most general case of a plane, monochromatic wave. For a easy representation, the coordinate system was cleverly chosen: First, the $\boldsymbol{e}_3$ axis is placed in the direction of $\boldsymbol{k}$. For any two, coordinate axes that are perpendicular to it, we would get a rotated ellipse instead of (20.38). The choice (20.36) for $\boldsymbol{e}_1$ and $\boldsymbol{e}_2$ causes this ellipse to be in principal axis form.

In the general case, the plane, monochromatic wave is *elliptically polarized*. The position and shape of the ellipse is determined by the vectors $\boldsymbol{b}_1$ and $\boldsymbol{b}_2$. There are two simple special cases: One semi-axis of the ellipse is zero, or both semi-axes are of the same length:

$$\boldsymbol{E}_0\,\exp(\mathrm{i}\alpha) = \boldsymbol{b}_1 + \mathrm{i}\,\boldsymbol{b}_2 = \begin{cases} b\,\boldsymbol{e}_1 \text{ or } b\,\boldsymbol{e}_2 & \text{linearly polarized} \\ b\,(\boldsymbol{e}_1 \pm \mathrm{i}\boldsymbol{e}_2) & \text{circularly polarized} \end{cases} \tag{20.40}$$

The general case can be described by superposition of the two linearly polarized or the two circularly polarized solutions.

Energy and Momentum

We calculate the energy and momentum density of the plane, monochromatic wave ((20.21), (20.22)). At a certain position, these quantities oscillate between zero and a maximum value; we therefore determine the time average. All fields can be expressed in the form

$$a(t) = \mathrm{Re}\,a_0\,\exp(-\mathrm{i}\omega t), \qquad b(t) = \mathrm{Re}\,b_0\,\exp(-\mathrm{i}\omega t) \tag{20.41}$$

with complex a_0 and b_0. The energy and momentum quantities are quadratic in the fields. We therefore examine

$$a(t)\,b(t) = \frac{1}{4}\Big(a_0 \exp(-\mathrm{i}\omega t) + a_0^* \exp(\mathrm{i}\omega t)\Big) \times \Big(b_0 \exp(-\mathrm{i}\omega t) + b_0^* \exp(\mathrm{i}\omega t)\Big) \tag{20.42}$$

By averaging the oscillating components vanish

$$\big\langle a(t)\,b(t)\big\rangle = \frac{1}{4}\,\big(a_0\, b_0^* + a_0^*\, b_0\big) = \frac{1}{2}\,\mathrm{Re}\,\big(a_0\, b_0^*\big) \tag{20.43}$$

The brackets $\langle \ldots \rangle$ stand for the time averaging. We use this result the fields, for example, for $\boldsymbol{E}(\boldsymbol{r}, t) = \boldsymbol{E}_0 \exp(\mathrm{i}\boldsymbol{k}\cdot\boldsymbol{r} - \omega t)$. The time-averaged energy density is then

$$\langle w_{\mathrm{em}} \rangle = \frac{1}{8\pi}\,\big\langle \boldsymbol{E}^2 + \boldsymbol{B}^2 \big\rangle = \frac{1}{16\pi}\,\big(\mathrm{Re}\,\big(\boldsymbol{E}_0\cdot\boldsymbol{E}_0^*\big) + \mathrm{Re}\,\big(\boldsymbol{B}_0\cdot\boldsymbol{B}_0^*\big)\big) = \frac{1}{8\pi}\,\boldsymbol{B}_0\cdot\boldsymbol{B}_0^* = \frac{1}{8\pi}\,\boldsymbol{E}_0\cdot\boldsymbol{E}_0^* \tag{20.44}$$

The result can alternatively be expressed by $\boldsymbol{B}_0$. The time-averaged energy current density is

$$\langle \boldsymbol{S} \rangle = \frac{c}{4\pi}\,\big\langle \boldsymbol{E}\times\boldsymbol{B} \big\rangle = \frac{c}{8\pi}\,\mathrm{Re}\,\big(\boldsymbol{E}_0\times\boldsymbol{B}_0^*\big) = \langle w_{\mathrm{em}} \rangle\, c\,\frac{\boldsymbol{k}}{k} \tag{20.45}$$

The wave transports its energy with the speed c into the direction of the wave vector.

The total energy $\int d^3r\,\langle w_{\mathrm{em}}\rangle$ of the plane, monochromatic wave is infinite. In this respect, this waveform is an unrealistic model. In reality, there are only finite wave packets. Such wave packets can be represented as a superposition (17.23) of plane waves. The properties discussed here, in particular the polarization properties, apply to such wave packets, too.

A finite wave packet (Exercise 20.2) is no longer strictly monochromatic. A limitation to the length l implies the frequency uncertainty

$$\frac{\Delta\nu}{\nu} \gtrsim \frac{\lambda}{l} \qquad \text{(wave packet with extension } l\,) \tag{20.46}$$

For a wave packet with $l \gg \lambda$, the frequency uncertainty can be small. In this case, instead of the wave packet one might use a plane, monochromatic wave, which is easier to handle.

Electromagnetic Spectrum

Table 20.2 assigns electromagnetic waves of different frequencies to various forms of appearance. For large wavelengths, the boundaries often play a major role. The following chapter discusses the modifications for the case that the considered volume is enclosed by conducting boundaries (for example, for cavity waves).

Visible light is only a small part of the spectrum. It lies in the wavelength range

$$\lambda_{\text{visible}} = 4 \ldots 8 \cdot 10^{-7}\,\text{m} \tag{20.47}$$

The lower part corresponds to the colors *violet* and *blue*, and the upper one corresponds to the color *red*. A certain frequency corresponds to a certain color. This explains the term *monochromatic* for the wave (20.16) with fixed frequency.

Atoms can absorb or emit electromagnetic radiation in the visible range of the spectrum (and in neighboring ranges). The wavelengths in (20.47) are much larger than the Bohr radius $a_\text{B} \approx 5 \cdot 10^{-11}$ m, which characterizes

Table 20.2 Electromagnetic waves occur in various appearances. The five decades for radio waves are classified as LF (low frequency, $\lambda = 10\,\text{km} \ldots 1\,\text{km}$), MF (medium frequency $\lambda = 1000\,\text{m} \ldots 100\,\text{m}$), HF (hight frequency $\lambda = 100\,\text{m} \ldots 10\,\text{m}$), VHF (very high frequency $\lambda = 10\,\text{m} \ldots 1\,\text{m}$) and UHF (ultra high frequency, $\lambda = 1\,\text{m} \ldots 10\,\text{cm}$). A microwave oven uses wavelengths that are close to the UHF range. The wavelength range of microwaves goes down to about 0.1 mm; it includes radar with wavelengths $\lambda \sim 3\,\text{cm}$. The range for light can be divided into the sections infrared, visible and ultraviolet. The transition between X-ray and gamma radiation is not sharp. The gamma radiation in the narrower sense refers to the radiation from atomic nuclei. In the cosmic radiation, there are also γ's with much higher frequencies (i.e., energies).

Designation	Frequency ν in Hz
Radio waves	$3 \cdot 10^4 \ldots 3 \cdot 10^9$
Microwaves	$3 \cdot 10^9 \ldots 10^{12}$
Light	$10^{12} \ldots 5 \cdot 10^{17}$
X-rays	$3 \cdot 10^{16} \ldots 3 \cdot 10^{20}$
Gamma rays	$3 \cdot 10^{19} \ldots 3 \cdot 10^{22}$ and higher

the size of the atoms (roughly $\lambda \sim 10^4 a_B$). The emitted radiation occurs in quanta with the energy $\hbar\omega$. The emission of a light quantum takes about the time $\tau \sim 10^{-8}$ s; this results in wave packets of the length $l_c \sim 3$ m and a corresponding frequency uncertainty (20.46). Radiation processes are discussed in Chapter 24.

Monochromatic waves are solutions of Maxwell's equations for arbitrary frequencies in the range $0 < \nu < \infty$. Effectively, there are limitations of the frequency range:

- For $\nu \to 0$, the wavelength λ approaches infinity. A wave packet with a reasonably well-defined frequency ($\Delta\nu \ll \nu$) must therefore be correspondingly large, $l \sim \lambda\,\nu/\Delta\nu \gg \lambda$. At least for experiments on Earth, there are then limitations that lead to boundary conditions and minimum frequencies.

  An example of this is the Schumann resonances, where the boundary of the Earth's surface and the ionosphere lead to discrete frequencies. Their minimum frequency is given by $\nu \approx 8$ Hz, see equation (21.30).
- For $\nu \to \infty$, the particle character of the electromagnetic radiation becomes dominant. This particle character is discussed qualitatively in the rest of this chapter. Quantitatively, such processes are treated with in the context of quantum electrodynamics.

  Transitions in the atomic nucleus max lead to emissions of γ quanta with energies $\hbar\omega$ of a few MeV. For these γ's, the particle character dominates; this cannot be described by classical electromagnetic field.

Photons

Electrodynamics is a classical field theory. It is implicitly assumed that the quantization of the field plays no role. In this section, we outline the relation between the classical field and the field quanta. For practical applications, an elementary quantization of the field is often sufficient.

First, we compare the quantization of a mechanical oscillator with that of the electromagnetic field:

1. The quantization of a one-dimensional oscillator with the frequency ω leads to the energy eigenvalues

$$E_n = \hbar\omega\left(n + \frac{1}{2}\right) \qquad (20.48)$$

where $\hbar$ is the Planck constant. The E_n are the possible energy values of the stationary states of the system. By means of such an oscillator model, for example, the vibrations of an O_2 molecule can be described. Experimentally, one finds in the absorption spectrum (in the infrared) discrete lines which correspond to the energy differences $\hbar\omega$.

2. Electromagnetic radiation consists of individual lumps of energy or quanta of the size

$$E = \hbar\omega \quad \text{(photon)} \tag{20.49}$$

These quanta are called *photons*. Their existence is proven by numerous experiments, for example by the verification of Planck's radiation law, or by the photo effect and the Compton effect.

Standing electromagnetic waves can be excited in a metal cavity (Chapter 21). These excitations are natural oscillations of the system, whose quantization leads to (20.48).

We now consider the *classical limit* of these two points:

1. A mass $m = 1$ g is suspended by a spring and oscillates with the natural frequency $\omega = 1/\text{s}$; the amplitude of the oscillation is $a = 1$ cm. Also this oscillator has the quantum mechanical energy eigenvalues (20.48). The number of oscillation quanta n is, however, so large in this case that the quantization plays no role. For the specified values, the following applies: $n \approx E_{\text{class}}/\hbar\omega \approx 10^{27}$, where $E_{\text{class}} = m a^2 \omega^2/2$. Practically, the energies E_n form a continuum, as is assumed in classical mechanics.
2. The classical limit of the electromagnetic wave is obtained for a laser pulse with an energy $E_{\text{pulse}} = 1$ J. The laser light lies near the red end of the visible spectrum; then the energy of one photon is $\hbar\omega \approx 2\,\text{eV} \approx 3 \cdot 10^{-19}$ J. From this follows the number of photons in one laser pulse:

$$N = \frac{E_{\text{pulse}}}{\hbar\omega} \approx 3 \cdot 10^{18} \text{ photons} \tag{20.50}$$

In the classical limits, the quantization of the energy is irrelevant; it is not taken into account in the associated theory (classical mechanics, classical electrodynamics).

The comparison of laser light with a classical electromagnetic wave requires certain restrictions, which we only hint here. On one hand, laser light is characterized by a so-called coherent state (see Chapter 34 in [3]).

Here the phase of the wave and the number of photons underly the quantum mechanical uncertainty relation. On the other hand, the symmetry of the quantum mechanical wave function (indistinguishability of the photons) is responsible for the particular stability of laser light.

As a model of the photon, we consider a wave packet with the energy $\hbar\omega = \int d^3r \, \langle w_{\rm em}\rangle$. The wave packet has, within the uncertainty (20.46), the wave vector $\boldsymbol{k}$. The momentum density of the electromagnetic field is $\boldsymbol{S}/c^2$. We use this to calculate the momentum of the wave packet:

$$\boldsymbol{p} = \frac{1}{c^2}\int d^3r \, \langle \boldsymbol{S}\rangle \stackrel{(20.45)}{=} \frac{\boldsymbol{k}}{k}\int d^3r \, \frac{\langle w_{\rm em}\rangle}{c} = \frac{\boldsymbol{k}}{k}\,\frac{\hbar\omega}{c} = \hbar\boldsymbol{k} \quad \text{(photon)} \tag{20.51}$$

The experimental proof that photons have indeed this momentum can be provided by the inelastic scattering of photons on electrons (Compton effect). The energy and momentum balance of the particles involved (photon and electron) leads to a characteristic relation between the frequency change of the photon and the scattering angle.

From (20.49) and (20.51) follows

$$E^2 = c^2 p^2 \quad \text{(photon)} \tag{20.52}$$

The comparison with the general relativistic energy–momentum relation $E^2 = m^2c^4 + c^2p^2$ shows that the photons are massless:

$$m = 0 \quad \text{(photon)} \tag{20.53}$$

To set up a quantum mechanical wave equation, one usually starts from the energy–momentum relation and uses the correspondence (or substitution) rules $\boldsymbol{p} \to -\mathrm{i}\hbar\,\boldsymbol{\nabla}$ and $E \to \mathrm{i}\hbar\,\partial_t$. For (20.52), this leads to a wave equation of the form (20.11). The possible spin settings are then described by multi-component wave functions.

Photons are real particles that have an energy $E = \hbar\omega$, a momentum $\boldsymbol{p} = \hbar\boldsymbol{k}$ and the spin 1 (see the following). Just as with material particles there is a wave–particle duality (Chapter 1 of [3]). This means that electromagnetic radiation behaves like a wave (interference effects) but also like a beam of particles (photoelectric effect).

Photons can assume the same wave function in any number. This generally applies to bosons (particles with an integer spin); fermions (particles with half-integer spin), on the other hand, obey the Pauli principle.

If a large number of photons have the same wave function, then the wave function itself (and not only its squared magnitude) is an observable. For example, the electric field emitted by a radio transmitter is measurable. In this case, the quantization of the electromagnetic wave plays no role.

We establish the relation between the polarization of the electromagnetic wave and the spin of photons. To do this, we consider the behavior of A^α under a rotation. We consider a circularly polarized wave with $\boldsymbol{k} = k\boldsymbol{e}_z$. Using $\boldsymbol{E} = -\dot{\boldsymbol{A}}/c = \mathrm{i}k\boldsymbol{A}$ (here without real part formation), the vector potential reads

$$\big(A^\alpha\big) = \big(0, \boldsymbol{E}/\mathrm{i}k\big) \overset{(20.40)}{=} \frac{b}{\mathrm{i}k}\,(0, 1, \pm\mathrm{i}, 0)\,\exp\big[\,\mathrm{i}(kz - \omega t - \alpha)\big] \tag{20.54}$$

A rotation around the z-axis by the angle ϕ_0 is mediated by the transformation matrix

$$\big(\Lambda^\alpha_{\ \beta}\big) = \begin{pmatrix} 1 & 0 & 0 & 0 \\ 0 & \cos\phi_0 & \sin\phi_0 & 0 \\ 0 & -\sin\phi_0 & \cos\phi_0 & 0 \\ 0 & 0 & 0 & 1 \end{pmatrix} \tag{20.55}$$

For the wave (20.54), this transformation results in

$$A'^\alpha = \Lambda^\alpha_{\ \beta}\, A^\beta = \exp(\pm\,\mathrm{i}\phi_0)\, A^\alpha \tag{20.56}$$

In a quantized theory, A^α becomes the wave function of the photons. For rotational symmetry around the z-axis, the wave functions are of the form $\Psi \propto \exp(\mathrm{i}m\phi)$. Here $m\hbar$ is the projection of the angular momentum onto the z-axis and ϕ is the corresponding azimuth angle. When rotating around the angle ϕ_0, $\Psi \propto \exp(\mathrm{i}m\phi)$ obtains the additional factor $\exp(\mathrm{i}m\phi_0)$. A wave function that transforms like (20.56) thus describes a particle with $m = \pm 1$; the projection of the angular momentum onto the z direction has the possible values $\pm\,\hbar$. Therefore, a photon is assigned a spin $\boldsymbol{s}$, for which the projection in the z or momentum direction reads

$$\frac{\boldsymbol{k}}{k}\cdot\boldsymbol{s} = \pm\,\hbar \quad \text{(photon)} \tag{20.57}$$

This means that a photon has the spin of 1. This spin can be parallel or antiparallel to the momentum. These alignments correspond to the two possible circular polarizations.

Quantum mechanically, three spin settings are initially expected for spin 1, $(\boldsymbol{k}/k)\cdot\boldsymbol{s}/\hbar = 0,\ \pm 1$. However, the spin projection zero in the direction of momentum does not occur. For this projection, (20.56) yields $A'^{\alpha} = A^{\alpha}$. This would imply $(A^{\alpha}) \propto (0,0,0,1)$. Such an A^{α} is not possible due to the gauge condition $\operatorname{div}\boldsymbol{A} = 0$.

Two classical wave packets do not scatter at each other: If $\boldsymbol{E}_1(\boldsymbol{r},t)$ and $\boldsymbol{E}_2(\boldsymbol{r},t)$ are solutions of Maxwell's equations, then $\boldsymbol{E} = \boldsymbol{E}_1(\boldsymbol{r},t) + \boldsymbol{E}_2(\boldsymbol{r},t)$ is solution, too; this follows from the linearity of the Maxwell equations. Specifically, the rays from two flashlights can cross each other unhindered. (Actually, the flashlight consists of many wave packets that correspond to individual photons.) Deviating from the classical theory, there is a very small, but measurable cross-section for the scattering of photons by photons. It is caused by the creation and annihilation of virtual electron–positron pairs and can be calculated in quantum electrodynamics.

Exercises

20.1 *Plane electromagnetic wave*

By $\boldsymbol{A} = A(x-ct)\,\boldsymbol{e}_z$ and $\Phi = 0$ a plane electromagnetic wave is defined. Determine the $\boldsymbol{E}$ and $\boldsymbol{B}$ field, the energy density w_{em} and the Poynting vector $\boldsymbol{S}$.

20.2 *One-dimensional wave packet*

The potentials of an electromagnetic wave packet are

$$\boldsymbol{A} = A(x-ct)\,\boldsymbol{e}_z\,, \qquad A(x) = \int_{-\infty}^{\infty} dk\, f(k)\exp(ikx), \qquad \Phi = 0$$

Consider the following cases:

(i) $f(k) = f_0 \exp(-\gamma\, k^2/2)$
(ii) $f(k) = f_0 \exp(\,-\alpha\,|k|)$
(iii) $f(k) = f_0\, \Theta(\kappa - |k|)$

In each case, sketch the shape of the wave packet $A(x-ct)$ and determine its center $\overline{x}$ and its width Δx.

20.3 *Circularly polarized wave packet*

A circularly polarized wave with the components

$$E_x = E_0(x, y) \exp\left[\mathrm{i}(kz - \omega t)\right]$$

$$E_y = \pm\mathrm{i}\, E_0(x, y) \exp\left[\mathrm{i}(kz - \omega t)\right]$$

is limited in x- and y-direction; the amplitude $E_0(x, y)$ is a real and even function in the variables x and y. The wave packet shall extend in these directions over many wavelengths so that

$$\left|\frac{\partial E_0}{\partial x}\right| \ll \frac{E_0}{\lambda} \quad \text{and} \quad \left|\frac{\partial E_0}{\partial y}\right| \ll \frac{E_0}{\lambda}$$

A much weaker z-dependence in $E_0(x, y, z)$ shall limit the extension of the wave packet in the z-direction; this dependence shall not be explicitly considered in the following calculations.

Read off E_z and $\boldsymbol{B}$ from the Maxwell equations (only the leading terms). Calculate the time averaged quantities: Energy density $\langle w_{\mathrm{em}}\rangle$, Poynting vector $\langle \boldsymbol{S}\rangle$ and angular momentum density $\boldsymbol{r} \times \langle \boldsymbol{S}\rangle/c^2$. Set the total energy W equal to $\hbar\omega$, i.e., to the energy of a photon. What is angular momentum of the wave packet?

Chapter 21

Cavity Waves

We investigate wave solutions in a volume which is enclosed by metal walls. Within the volume, the free Maxwell equations apply. On the metal boundaries, charges and currents are induced, which are in general unknown. However, the boundary conditions for the electric and magnetic field can be specified. We consider a cavity resonator and a waveguide.

If a volume is enclosed by a closed metal surfaces only certain natural oscillations of the electromagnetic field are possible; we speak of a cavity resonator. If the volume is unlimited in one direction, waves can propagate in this direction; we call this a waveguide. In both cases, we restrict ourselves to simple geometries of the volume, Figure 21.1.

Due to the mobile charges in the metal, the tangential components of the electric field vanish at the boundary R of the considered volume:

$$\boldsymbol{t} \cdot \boldsymbol{E}(\boldsymbol{r}, t)\big|_{\mathrm{R}} = 0 \tag{21.1}$$

Since $\operatorname{curl} \boldsymbol{E} = -\dot{\boldsymbol{B}}/c = \mathrm{i}\omega \boldsymbol{B}/c$ a periodic magnetic field induces electric fields, which are perpendicular to $\boldsymbol{B}(\boldsymbol{r}, t)$. Due to (21.1), such electric fields are zero in the tangential direction. Therefore, the normal component of $\boldsymbol{B}$ must vanish:

$$\boldsymbol{n} \cdot \boldsymbol{B}(\boldsymbol{r}, t)\big|_{\mathrm{R}} = 0 \tag{21.2}$$

The conditions (21.1) and (21.2) are idealizations; the actual situation in the metal is more complicated and is examined in some detail in Part VI. In Chapter 7, (21.1) for static fields was justified. This might be generalized to time dependent fields as long as these fields do not change too rapidly. For not too high frequencies, the mobile charges can

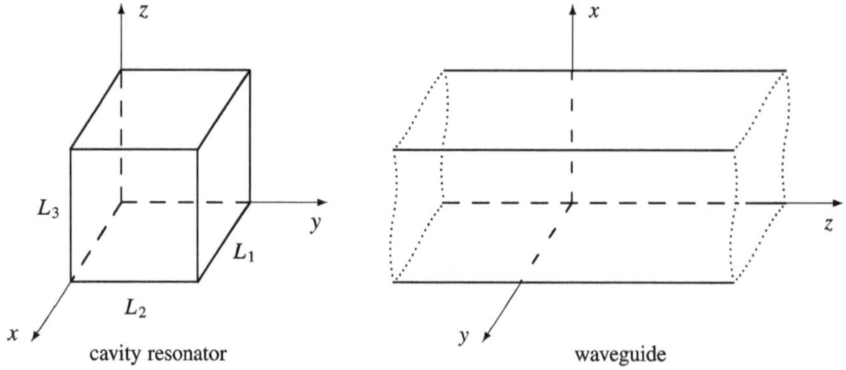

Figure 21.1 The cavity resonator is a closed metal surface, whereas the waveguide is unlimited in one direction. For simplicity, we consider rectangular boundaries.

shift as in the static case. The conditions (21.1) and (21.2) are useful for radio frequencies, but they are inappropriate in the ultraviolet range (where metals become transparent, Chapter 34). In fact, the electromagnetic field penetrates somewhat into the metal also for radio frequencies. This leads to ohmic losses and thus to a damping of the oscillations or waves. Such damping effects are not taken into account here.

There are no charges within the cavity, $\varrho = 0$ and $\boldsymbol{j} = 0$. Then the free Maxwell equations apply:

$$\operatorname{div} \boldsymbol{E} = 0, \quad \operatorname{curl} \boldsymbol{E} = -\dot{\boldsymbol{B}}/c, \quad \operatorname{curl} \boldsymbol{B} = \dot{\boldsymbol{E}}/c, \quad \operatorname{div} \boldsymbol{B} = 0 \qquad (21.3)$$

We use this in $\Delta \boldsymbol{E} = -\operatorname{curl}\operatorname{rot} \boldsymbol{E} + \operatorname{grad}\operatorname{div} \boldsymbol{E}$ and get

$$\boxed{\left(\Delta - \frac{1}{c^2}\frac{\partial^2}{\partial t^2}\right)\boldsymbol{E}(\boldsymbol{r}, t) = 0} \qquad (21.4)$$

The corresponding calculation for $\boldsymbol{B}$ yields

$$\boxed{\left(\Delta - \frac{1}{c^2}\frac{\partial^2}{\partial t^2}\right)\boldsymbol{B}(\boldsymbol{r}, t) = 0} \qquad (21.5)$$

The wave equations (21.4) and (21.5) cannot fully replace Maxwell's equations. According to (21.3), the Cartesian components of $\boldsymbol{E}$ and $\boldsymbol{B}$ are related to each other.

We are now looking for solutions that fulfill the boundary conditions (21.1) and (21.2). Such solutions are determined by the geometry of the cavity. We restrict ourselves to a rectangular cavity and a wave cavity, Figure 21.1. In these cases, a separation approach for Cartesian coordinates leads to the solution.

We start with an separation approach for the x-component of the electric $\boldsymbol{E} = \sum E_i\, \boldsymbol{e}_i$:

$$E_1(x, y, z, t) = X(x)\, Y(y)\, Z(z)\, T(t) \tag{21.6}$$

We insert this into the wave equation (21.4):

$$\frac{X''}{X} + \frac{Y''}{Y} + \frac{Z''}{Z} - \frac{1}{c^2}\frac{T''}{T} = 0 \tag{21.7}$$

Each term must be equal to a constant because the other terms do not depend on the respective coordinate:

$$\frac{X''}{X} = -k_1^2, \quad \frac{Y''}{Y} = -k_2^2, \quad \frac{Z''}{Z} = -k_3^2, \quad \frac{T''}{T} = -\omega^2 \tag{21.8}$$

These separation constants must fulfill (21.7), i.e.,

$$\omega^2 = c^2\left(k_1^2 + k_2^2 + k_3^2\right) \tag{21.9}$$

There are various equivalent forms for the solution:

$$X(x) = \begin{cases} \sin(k_1 x + \alpha_1) & \text{or} \\ \sin(k_1 x),\ \cos(k_1 x) & \text{or} \\ \exp(\pm\, \mathrm{i} k_1 x) \end{cases} \tag{21.10}$$

For X, Y and Z, we use the first form. For T, we use instead $T =$ Re $A\, \exp(-\mathrm{i}\omega t)$ with $\omega > 0$ and a complex amplitude A; this is equivalent to a superposition of $\sin(\omega t)$ and $\cos(\omega t)$, (20.15). An amplitude effectively occurs only once in the product approach (21.6). This means that the solutions are of the form

$$E_1 = \mathrm{Re}\left[C_1\, \sin(k_1 x + \alpha_1)\, \sin(k_2 y + \alpha_2)\, \sin(k_3 z + \alpha_3)\, \exp(-\mathrm{i}\omega t)\right] \tag{21.11}$$

To simplify the notation, we agree with the following:

- The sign Re for the formation of the real part is no longer written on explicitly in the following.

Cavity Resonator

We consider a cuboid as a cavity (Figure 21.1, left), whose walls are made out of metal. The volume of the cavity is given by

$$0 \le x \le L_1, \quad 0 \le y \le L_2, \quad 0 \le z \le L_3 \tag{21.12}$$

The x-component $E_1(x, y, z, t)$ of the electric field is a tangential component at the walls $y = 0$, $y = L_2$, $z = 0$ and $z = L_3$. The boundary condition (21.1) thus becomes

$$E_1(x, y, 0, t) = E_1(x, y, L_3, t) = E_1(x, 0, z, t) = E_1(x, L_2, z, t) = 0 \tag{21.13}$$

For $E_1 \propto Z(z) = \sin(k_3 z + \alpha_3)$, this implies $\alpha_3 = 0$ and $\sin(k_3 L_3) = 0$, and therefore

$$k_3 = \frac{n\,\pi}{L_3} \quad \text{with} \quad n = 0, 1, 2, 3, \ldots \tag{21.14}$$

Negative n values only result in a sign of the amplitude and can be left away. In this chapter we consider n itself (instead of k_3) as the *wavenumber*. For $E_1 \propto Y(y) = \sin(k_2 y + \alpha_2)$ and (21.13) we get analogously $\alpha_2 = 0$ and $k_2 = m\,\pi/L_2$. The boundary condition does not restrict $X(x)$ in E_1. This gives us

$$E_1 \;=\; C_1\, X(x)\, \sin\left(\frac{m\,\pi\, y}{L_2}\right)\, \sin\left(\frac{n\,\pi\, z}{L_3}\right)\, \exp(-\mathrm{i}\omega t) \tag{21.15}$$

The same procedure yields the y and z components

$$E_2 = C_2\, Y(y)\, \sin\left(\frac{l\,\pi\, x}{L_1}\right)\, \sin\left(\frac{n'\,\pi\, z}{L_3}\right)\, \exp(-\mathrm{i}\omega' t) \tag{21.16}$$

$$E_3 = C_3\, Z(z)\, \sin\left(\frac{l'\,\pi\, x}{L_1}\right)\, \sin\left(\frac{m'\,\pi\, y}{L_2}\right)\, \exp(-\mathrm{i}\omega'' t) \tag{21.17}$$

The possible values of n, n', l, l', m and m' are $0, 1, 2, 3, \ldots$. Since in (21.4) the components of $\boldsymbol{E}$ are decoupled, this results in initially independent separation constants, which are indicated by dashes. In addition to

(21.4) and (21.5), however the Maxwell equations apply. From (21.3) we obtain

$$\begin{aligned}\operatorname{div}\boldsymbol{E} = \; & C_1\, X'(x)\, \sin\left(\frac{m\,\pi\,y}{L_2}\right)\, \sin\left(\frac{n\,\pi\,z}{L_3}\right)\, \exp(-\mathrm{i}\omega t) \\ & + C_2\, Y'(y)\, \sin\left(\frac{l\,\pi\,x}{L_1}\right)\, \sin\left(\frac{n'\,\pi\,z}{L_3}\right)\, \exp(-\mathrm{i}\omega' t) \\ & + C_3\, Z'(z)\, \sin\left(\frac{l'\,\pi\,x}{L_1}\right)\, \sin\left(\frac{m'\,\pi\,y}{L_2}\right)\, \exp(-\mathrm{i}\omega'' t) \; = \; 0\end{aligned} \tag{21.18}$$

This equation can only be fulfilled at all times and positions if

$$\omega = \omega' = \omega'', \quad l = l', \quad m = m', \quad n = n' \tag{21.19}$$

and

$$X'(x) \propto \sin\left(\frac{l\,\pi\,x}{L_1}\right), \quad Y'(y) \propto \sin\left(\frac{m\,\pi\,y}{L_2}\right), \quad Z'(z) \propto \sin\left(\frac{n\,\pi\,z}{L_3}\right) \tag{21.20}$$

The integration yields $X(x) \propto \cos(l\,\pi x/L_1)$. Since the amplitude C_1 is not yet fixed in (21.15), we can set $X(x) = \cos(l\,\pi x/L_1)$. With $Y(y)$ and $Z(z)$ we proceed accordingly. This turns (21.15)–(21.17) into

$$E_1 = C_1\, \cos\left(\frac{l\,\pi\,x}{L_1}\right)\, \sin\left(\frac{m\,\pi\,y}{L_2}\right)\, \sin\left(\frac{n\,\pi\,z}{L_3}\right)\, \exp(-\mathrm{i}\omega t) \tag{21.21}$$

$$E_2 = C_2\, \sin\left(\frac{l\,\pi\,x}{L_1}\right)\, \cos\left(\frac{m\,\pi\,y}{L_2}\right)\, \sin\left(\frac{n\,\pi\,z}{L_3}\right)\, \exp(-\mathrm{i}\omega t) \tag{21.22}$$

$$E_3 = C_3\, \sin\left(\frac{l\,\pi\,x}{L_1}\right)\, \sin\left(\frac{m\,\pi\,y}{L_2}\right)\, \cos\left(\frac{n\,\pi\,z}{L_3}\right)\, \exp(-\mathrm{i}\omega t) \tag{21.23}$$

The wavenumbers l, m and n can adopt the values 0, 1, 2, ... Only one wavenumber may be zero; otherwise we would obtain $\boldsymbol{E} = 0$ and therefore also $\boldsymbol{B} = 0$. According to (21.9), the wavenumbers fix the frequency

$$\omega^2 = \left(\omega_{lmn}\right)^2 = c^2\pi^2\left(\frac{l^2}{L_1^2} + \frac{m^2}{L_2^2} + \frac{n^2}{L_3^2}\right) \quad (l, m, n = 0, 1, 2, \ldots) \tag{21.24}$$

For (21.21)–(21.23), $\operatorname{div}\boldsymbol{E} = 0$ becomes

$$C_1\,\frac{l\,\pi}{L_1} + C_2\,\frac{m\,\pi}{L_2} + C_3\,\frac{n\,\pi}{L_3} = 0 \tag{21.25}$$

For given wavenumbers l, m and n two amplitudes can be freely chosen. This corresponds to the two possible polarization directions of the free wave (Chapter 20).

For the periodic fields under consideration, $\partial_t\boldsymbol{B} = -\mathrm{i}\omega\boldsymbol{B}$ applies. All component of the magnetic field follow from this, $\operatorname{curl}\boldsymbol{E} = -\dot{\boldsymbol{B}}/c$ and (21.21)–(21.23),

$$B_3 = -\frac{\mathrm{i}c}{\omega}\left(\frac{\partial E_2}{\partial x} - \frac{\partial E_1}{\partial y}\right) = -\frac{\mathrm{i}c}{\omega}\left(\frac{C_2\,l\,\pi}{L_1} - \frac{C_1\,m\,\pi}{L_2}\right)\cos\left(\frac{l\,\pi\,x}{L_1}\right)$$
$$\times\cos\left(\frac{m\,\pi\,y}{L_2}\right)\sin\left(\frac{n\,\pi\,z}{L_3}\right)\exp(-\mathrm{i}\omega t) \tag{21.26}$$

$$B_1 = -\frac{\mathrm{i}c}{\omega}\left(\frac{\partial E_3}{\partial y} - \frac{\partial E_2}{\partial z}\right), \quad B_2 = -\frac{\mathrm{i}c}{\omega}\left(\frac{\partial E_1}{\partial z} - \frac{\partial E_3}{\partial x}\right) \tag{21.27}$$

The boundary condition (21.2) is fulfilled:

$$B_3(x, y, 0) = B_3(x, y, L_3) = 0 \tag{21.28}$$

In (21.21)–(21.23), (21.26) and (21.27), eventually the real part of the right-hand side has to be taken. Equation (21.24) contains only real quantities. In (21.25), both the real and the imaginary part must be zero.

It is easy to check that the solution found satisfies all Maxwell's equations and the boundary conditions: The boundary conditions are satisfied by the sine function and the restriction to integers for l, m and n. Due to (21.24), the fields $\boldsymbol{E}$ and $\boldsymbol{B}$ are solutions of the wave equations (21.4) and (21.5).

Discussion of the solutions

The free waves investigated in Chapter 20 may occur with arbitrary frequencies ω. The boundary conditions of the cavity, on the other hand, only allow specific *natural frequencies* or *eigenfrequencies* ω_{lmn}, (21.24). The corresponding solutions for $\boldsymbol{E}$ and $\boldsymbol{B}$ are *standing waves*. This means that the position of the wave crests and troughs is constant in time (Figure 21.2). The number of nodes (points at which the field is zero) is $l - 1$, $m - 1$ and $n - 1$; the two boundary nodes are not counted. Each direction can be assigned a wavelength (equal to the distance between two crests):

$$\lambda_1 = \frac{2L_1}{l}, \quad \lambda_2 = \frac{2L_2}{m}, \quad \lambda_3 = \frac{2L_3}{n} \tag{21.29}$$

For example, for $n = 2$ (middle part in Figure 21.2) just one wavelength fits into the cavity.

The solution (21.21)–(21.23) is an undamped, harmonic oscillation. For a given n, l and m, it is a *natural oscillation* (or *eigenmode*) of the cavity. A natural oscillation can be excited with arbitrary strength; the strength is given by the amplitudes C_i. An example of an eigenmode is shown in Figure 21.2). The natural vibrations or eigenmodes of the cavity can be compared with those of a clamped string.

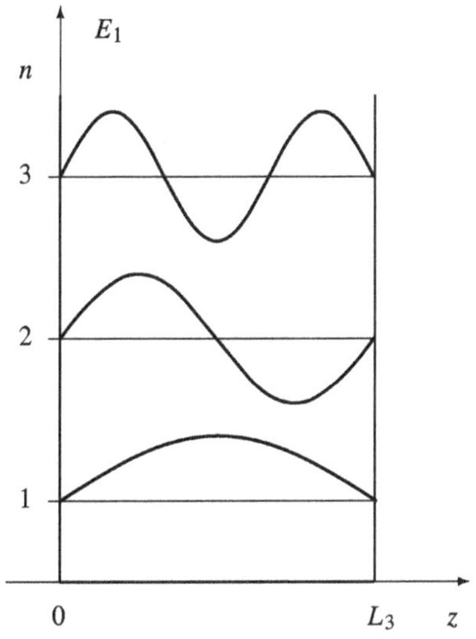

Figure 21.2 The eigenmodes of the cavity resonator are standing waves. As an example, the the component $E_1(z)$ of the electric field (21.21)–(21.23) is shown for the oscillation numbers $n = 1$, 2 and 3. For the eigenvibrations of a clamped string or the wave function of a particle in the box, the same picture is obtained.

In real systems, the vibrations are damped because the field penetrates somewhat into the metal wall inducing currents and thus ohmic losses. This damping can be overcome by suitable external excitations.

The general solution of Maxwell's equation for the cuboidal cavity results as a linear combination of the solutions found in (21.21)–(21.23). This general solution contains an infinite number of constants, which may be determined by the initial conditions.

Schumann resonances

The surface of the Earth (radius $R \approx 6400\,\text{km}$) and the Ionosphere (partially ionized layer in a height of about 100 km) have finite conductivity. The conductivity σ is defined in Chapter 31; one finds $\sigma \approx 10^9\,\text{s}^{-1}$ for sea water and $\sigma \approx 10^3 \ldots 10^6\,\text{s}^{-1}$ for the ionosphere. The Earth surface and the ionosphere can roughly be approximated by concentric, conductive spheres. Thereby a cavity with the shape of a spherical shell is formed. The electromagnetic eigenmodes of this shell are called *Schumann resonances*. The lowest frequency is about

$$\nu = \frac{\omega}{2\pi} \approx \frac{1}{2\pi}\,\frac{c}{R} \approx 8\ \text{Hz} \tag{21.30}$$

These and the next higher resonances clearly stand out from the background of the electromagnetic noise in the atmosphere. As they are damped, they have finite widths (of a few Hertz).

Cavity radiation

At finite temperatures, the eigenmodes of a cavity are statistically excited. The frequency distribution of these cavity waves is described by Planck's radiation law.

For temperatures that are not too low, most excited modes have wavenumbers (l, m and n) which are much larger than 1. For the components of the wave vectors, this means $k_i \gg \varDelta k_i = \pi/L_i$. The discrete wave vectors are then so close together that sums over wavenumbers may be replaced by integrals:

$$\sum_{l,m,n} \cdots = \int_0^\infty \frac{dk_1}{\varDelta k_1} \int_0^\infty \frac{dk_2}{\varDelta k_2} \int_0^\infty \frac{dk_3}{\varDelta k_3} \cdots = \frac{V}{(2\pi)^3} \int d^3k \ \ldots \tag{21.31}$$

Here, $V = L_1 L_2 L_3$ is the volume of the cavity. The natural frequencies of a particular cavity resonator depend on its size and shape. For modes with $k \gg \Delta k$, the walls of the cavity are of secondary importance only. For counting these modes, a cavity of any shape can be used; we may use for example a cuboid. The summation (21.31) only depends on the volume V of the cavity.

For a mode with specific k_i values, two amplitudes can be chosen arbitrarily. This corresponds to the two polarization directions of a free wave (Chapter 20) or the two spin settings of a photon. In the sum over all possible states or modes, this is taken into account by a factor 2.

Using (21.31) and the factor 2, the energy of the electromagnetic waves in the cavity is calculated as follows:

$$E(T, V) = \frac{2V}{(2\pi)^3} \int d^3k \; \hbar\omega \; \overline{n(\omega)} = V \int_0^{\infty} d\omega \; w(\omega) \qquad (21.32)$$

Here, $\overline{n(\omega)} = \left[\exp(\hbar\omega/k_B T) - 1\right]^{-1}$ is the mean number of excitation quanta $\hbar\omega$ in a certain mode; k_B is the Boltzmann constant and T the temperature. The spectral energy density $w(\omega)$ introduced in (21.32) constitutes *Planck's radiation law*:

$$w(\omega) = \frac{\hbar}{\pi^2 c^3} \frac{\omega^3}{\exp(\hbar\omega/k_B T) - 1} \qquad (21.33)$$

A detailed derivation of this frequency distribution is presented in Chapter 34 of my book [4] on statistical physics.

Radiation with the frequency distribution (21.33) may pass through a small hole in the wall of cavity resonator. Since such a hole appears black (from the outside), the result is also known as the *black body radiation*. The application range of (21.33) goes far beyond cavity radiation. The model of the cavity resonator is primarily used for simple counting (21.31) of the modes.

In the plasma of the sun's surface, hot matter and electromagnetic radiation are in thermal equilibrium. Also this leads to the frequency distribution (21.33). Comparing the frequency distribution of sun light with (21.33), we obtain the temperature of the sun's surface, $T \approx 6000\,$Kelvin.

Waveguide

As a cavity, we now consider the waveguide that is unrestricted in the z-direction (Figure 21.1, right). The inside volume is defined by

$$0 \le x \le L_1, \quad 0 \le y \le L_2 \tag{21.34}$$

We start again with the separation approach (21.6). Since there is no boundary in the z-direction, the values of $k = k_3$ in $Z(z) = \exp(\pm \mathrm{i} k z)$ are continuous. The solution for $X(x)$, $Y(y)$ and $T(t)$ in the separation approach proceeds analogously to the cavity resonator. Instead of (21.21)–(21.23) we obtain the solution

$$E_1 = C_1 \cos\left(\frac{l\pi x}{L_1}\right) \sin\left(\frac{m\pi y}{L_2}\right) \exp\left[\mathrm{i}(kz - \omega t)\right] \tag{21.35}$$

$$E_2 = C_2 \sin\left(\frac{l\pi x}{L_1}\right) \cos\left(\frac{m\pi y}{L_2}\right) \exp\left[\mathrm{i}(kz - \omega t)\right] \tag{21.36}$$

$$E_3 = C_3 \sin\left(\frac{l\pi x}{L_1}\right) \sin\left(\frac{m\pi y}{L_2}\right) \exp\left[\mathrm{i}(kz - \omega t)\right] \tag{21.37}$$

We have restricted ourselves to the positive sign in the exponent of $Z(z) = \exp(+\mathrm{i}kz)$, i.e., to waves that propagate into the $+z$ direction. The condition (21.9) becomes

$$\omega^2 = c^2 \pi^2 \left(\frac{l^2}{L_1^2} + \frac{m^2}{L_2^2} + \frac{k^2}{\pi^2}\right) \tag{21.38}$$

From $\operatorname{div} \boldsymbol{E} = 0$ follows

$$C_1 \frac{l\pi}{L_1} + C_2 \frac{m\pi}{L_2} - \mathrm{i} C_3 k = 0 \tag{21.39}$$

The B_3 component can be determined as in (21.26),

$$\begin{aligned} B_3 &= -\frac{\mathrm{i}c}{\omega}\left(\frac{\partial E_2}{\partial x} - \frac{\partial E_1}{\partial y}\right) \\ &= -\frac{\mathrm{i}c}{\omega}\left(\frac{C_2\, l\pi}{L_1} - \frac{C_1\, m\pi}{L_2}\right) \cos\left(\frac{l\pi x}{L_1}\right) \cos\left(\frac{m\pi y}{L_2}\right) \\ &\quad \times \exp\left[\mathrm{i}(kz - \omega t)\right] \end{aligned} \tag{21.40}$$

The B_1 and B_2 components follow from (21.27). On the right-hand sides of (21.35)–(21.37) and (21.40), the real parts are to be taken. It is easy to check that the $\boldsymbol{E}$ and $\boldsymbol{B}$ fields fulfill all Maxwell's equations and the boundary conditions.

Discussion of the solutions

For free waves (Chapter 20), the electric and magnetic fields are perpendicular to the direction of propagation, i.e., $(E_3, B_3) = (0, 0)$ for $\boldsymbol{k} = k\,\boldsymbol{e}_z$. Waves with this property are called TEM waves, where TEM stands for transverse electromagnetic mode. First, we show that there are no TEM waves in the waveguide because from $(E_3, B_3) = (0, 0)$ it follows that $\boldsymbol{E} = \boldsymbol{B} = 0$.

For $E_3 = 0$, we get either $C_3 = 0$, or one of the wavenumbers (l or m) must be zero. Let us first consider the case $l = 0$ (the discussion for $m = 0$ runs completely parallel). From $l = 0$ and $B_3 = 0$ follows $C_1 m = 0$ and therefore $E_1 = 0$. Then all components of $\boldsymbol{E}$ are zero (E_2 disappears due to $l = 0$). For wave solutions, then $\boldsymbol{B} = 0$ follows from (21.3). Therefore, we can exclude $l = 0$ and $m = 0$.

We consider the alternative $C_3 = 0$. From (21.39) and $B_3 = 0$ follows

$$\frac{l\,\pi}{L_1}\,C_1 + \frac{m\,\pi}{L_2}\,C_2 = 0, \qquad \frac{m\,\pi}{L_2}\,C_1 - \frac{l\,\pi}{L_1}\,C_2 = 0 \tag{21.41}$$

From this we obtain the trivial solution $C_1 = C_2 = 0$. For a non-trivial solution, the determinant of the system of equations must vanish. For arbitrary L_1 and L_2 the determinant only vanishes for $l = m = 0$. This again results in $\boldsymbol{E} = \boldsymbol{B} = 0$.

For $(E_3, B_3) = (0, 0)$, there is therefore no wave solution; in the waveguide, no TEM wave can propagate. For a non-vanishing solution, we must require $(E_3, B_3) \neq (0, 0)$. The possible solutions can be divided into the two cases $E_3 = 0$, $B_3 \neq 0$ and $E_3 \neq 0$, $B_3 = 0$:

$$\text{Transverse electric wave (TE):} \quad E_3 = 0, \quad \boldsymbol{E} \perp k\,\boldsymbol{e}_z \tag{21.42}$$

$$\text{Transverse magnetic wave (TM):} \quad B_3 = 0, \quad \boldsymbol{B} \perp k\,\boldsymbol{e}_z \tag{21.43}$$

The designations indicate which field is perpendicular to the propagation direction $\boldsymbol{e}_z$ of the cavity waves. The general solution can be a linear combination of the two fields. In the following, we discuss a simple solution for each case.

Transverse electric wave

Here $E_3 = 0$ applies. Since $B_3 \neq 0$, we exclude $l = m = 0$. One of the simplest solutions is

$$l = 1, \quad m = 0 \tag{21.44}$$

For this we get

$$\omega = c\sqrt{\frac{\pi^2}{L_1^2} + k^2} \tag{21.45}$$

Since $m = 0$, $E_1 = 0$ applies. From $E_1 = E_3 = 0$ follows $B_2 = 0$, see (21.27). Overall, the following applies to this particular wave:

$$E_1 = E_3 = B_2 = 0, \quad E_2 \neq 0, \quad B_1 \neq 0, \quad B_3 \neq 0, \tag{21.46}$$

The frequency (21.45) fulfills the condition

$$\omega > \omega_{\text{cr}} = \frac{c\,\pi}{L_1} \tag{21.47}$$

In the waveguide, a wave can only propagate above the critical frequency ω_{kr}. The waveguide can therefore serve as a high-pass filter.

Transverse magnetic wave

Here, $B_3 = 0$ applies. Due to $E_3 \neq 0$, neither l nor m may vanish. The lowest node numbers are therefore

$$l = 1, \quad m = 1 \tag{21.48}$$

The associated frequency is

$$\omega = c\sqrt{\frac{\pi^2}{L_1^2} + \frac{\pi^2}{L_2^2} + k^2} \tag{21.49}$$

Due to the condition $E_3 \neq 0$, we get $C_3 \neq 0$. From (21.40) and $B_3 = 0$, it follows that $C_1 \propto C_2$. Excluding $C_1 = C_2 = 0$ yields $C_1 \neq 0$ and $C_2 \neq 0$. This means that all components except B_3 are unequal zero. Similar to (21.45), there is a lower limit ω_{cr} for the frequency of the wave.

Coaxial cable

A well-known example of a waveguide is the coaxial cable, which is used to transmit UHF signals from the antenna to the television set. The core and the shielding of the cable form concentric circular cylinders between which the wave propagates. Due to the different geometry, the solutions given above are not applicable. The dominant modes in this case are transverse electromagnetic waves (TEM). These waves have transverse components (i.e., $E_3 = B_3 = 0$) only. Then, one obtains $\boldsymbol{B} = \pm \boldsymbol{e}_3 \times \boldsymbol{E}$ like for a free wave. In contrast to TE and TM, there is no lower limit for the frequency.

Chapter 22

Transformation of the Fields

We deal with some consequences of the covariance of Maxwell's equations. The field strength tensor (18.27)

$$F = \left(F^{\alpha\beta}\right) = \begin{pmatrix} 0 & -E_x & -E_y & -E_z \\ E_x & 0 & -B_z & B_y \\ E_y & B_z & 0 & -B_x \\ E_z & -B_y & B_x & 0 \end{pmatrix} \tag{22.1}$$

is a Lorentz tensor. This determines the transformations of the ***E*** *and* ***B*** *fields for a transition into another inertial frame. As an application, we calculate the field of an uniformly moving charge. For a monochromatic plane wave,* $(k^\alpha) = (\omega/c, \boldsymbol{k})$ *is a Lorentz vector. From this follows the Doppler effect and the aberration of starlight.*

The following discussion relies on Chapter 18 on the covariance. Since $F^{\alpha\beta}$ is a Lorentz tensor, it transforms like

$$F'^{\alpha\beta} = \Lambda^\alpha_{\ \gamma}\, \Lambda^\beta_{\ \delta}\, F^{\gamma\delta} \quad \text{or} \quad F' = \Lambda\, F\, \Lambda^{\mathrm{T}} \tag{22.2}$$

for a transition into another IS$'$. The system IS$'$ moves relative to IS with the velocity $\boldsymbol{v} = v\, \boldsymbol{e}_x$, the coordinate axes are parallel, and the origins of IS and IS$'$ coincide at the time $t = t' = 0$ (Figure 22.1). Then IS and IS$'$ are connected by the special Lorentz transformation with

$$\Lambda = \left(\Lambda^\alpha_{\ \beta}\right) = \begin{pmatrix} \gamma & -\gamma\, v/c & 0 & 0 \\ -\gamma\, v/c & \gamma & 0 & 0 \\ 0 & 0 & 1 & 0 \\ 0 & 0 & 0 & 1 \end{pmatrix} \tag{22.3}$$

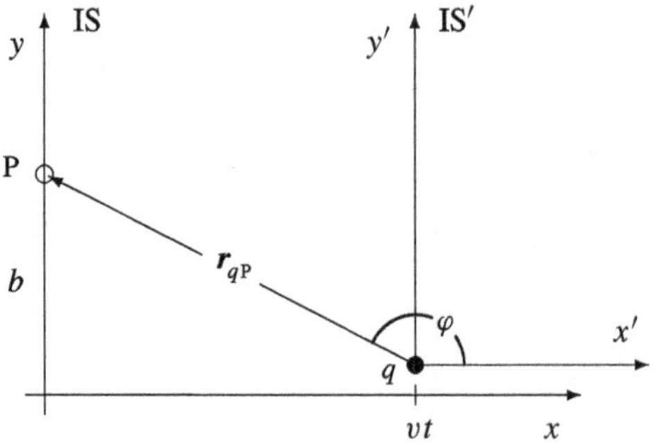

Figure 22.1 The electromagnetic field of a uniformly moving charge q shall be determined. In the inertial frame IS′, at the origin of which the charge is at rest, the field is known. Then a Lorentz transformation results in the wanted field in IS. Specifically, the field measured by an observer resting at point P in IS is calculated.

where $\gamma = 1/\sqrt{1 - v^2/c^2}$. By matrix multiplication we get

$$F' = \Lambda F \Lambda^{\mathrm{T}} = \begin{pmatrix} 0 & -E_x & -E_y\gamma + B_z\gamma\, v/c & -E_z\gamma - B_y\gamma\, v/c \\ & 0 & -B_z\gamma + E_y\gamma\, v/c & B_y\gamma + E_z\gamma\, v/c \\ & & 0 & -B_x \\ & & & 0 \end{pmatrix} \tag{22.4}$$

The elements to the left of the diagonal result from $F'^{\alpha\beta} = -F'^{\beta\alpha}$; they are not written on. Analogous to (22.1), the following applies in IS′:

$$F' = \left(F'^{\alpha\beta}\right) = \begin{pmatrix} 0 & -E'_x & -E'_y & -E'_z \\ & 0 & -B'_z & B'_y \\ & & 0 & -B'_x \\ & & & 0 \end{pmatrix} \tag{22.5}$$

By comparing (22.4) with (22.5), we obtain the transformation between the components of the fields $\boldsymbol{E}$ and $\boldsymbol{B}$ in IS′ and IS. The parts of these physical field, which are parallel or perpendicular to $\boldsymbol{v}$, transform differently:

$$\begin{aligned} \boldsymbol{E}'_{\parallel} &= \boldsymbol{E}_{\parallel}, \quad \boldsymbol{E}'_{\perp} = \gamma\left(\boldsymbol{E}_{\perp} + \frac{\boldsymbol{v}}{c} \times \boldsymbol{B}\right) \\ \boldsymbol{B}'_{\parallel} &= \boldsymbol{B}_{\parallel}, \quad \boldsymbol{B}'_{\perp} = \gamma\left(\boldsymbol{B}_{\perp} - \frac{\boldsymbol{v}}{c} \times \boldsymbol{E}\right) \end{aligned} \tag{22.6}$$

With $\boldsymbol{v} \to -\boldsymbol{v}$, one obtains the inverse transformation from this:

$$\boldsymbol{E}_{\parallel} = \boldsymbol{E}'_{\parallel}, \quad \boldsymbol{E}_{\perp} = \gamma \left(\boldsymbol{E}'_{\perp} - \frac{\boldsymbol{v}}{c} \times \boldsymbol{B}' \right)$$
$$\boldsymbol{B}_{\parallel} = \boldsymbol{B}'_{\parallel}, \quad \boldsymbol{B}_{\perp} = \gamma \left(\boldsymbol{B}'_{\perp} + \frac{\boldsymbol{v}}{c} \times \boldsymbol{E}' \right) \tag{22.7}$$

Uniformly Moving Charge

We determine the electromagnetic field of a uniformly moving charge. The constant velocity of the charge q is $\boldsymbol{v}$. We set up the inertial system in such a way that the charge moves along the x-axis (Figure 22.1). The observed field only depends on time and on the distance b from the rectilinear path of the charge. It is therefore sufficient to calculate the field at an observation point P on the y-axis.

The point charge rest at the origin of a comoving inertial system IS′ (Figure 22.1). IS and IS′ are connected by the transformation (22.3). For the vector from the charge to the observation point P we get

$$\boldsymbol{r}_{q\mathrm{P}} = (-v\,t,\, b, 0) \qquad \text{in IS}$$
$$\boldsymbol{r}'_{q\mathrm{P}} = \boldsymbol{r}' = (-v\,t',\, b, 0) \quad \text{in IS}' \tag{22.8}$$

The relation between t and t' is given by (22.3):

$$t' = \gamma \left(t - v\,x/c^2 \right) = \gamma\, t \tag{22.9}$$

Here, $x = 0$ was used for the point P. In the rest system IS′ of the charge, fields at position P are given by

$$\boldsymbol{E}' = q\, \frac{\boldsymbol{r}'}{r'^3} \quad \text{and} \quad \boldsymbol{B}' = 0 \tag{22.10}$$

The Cartesian components of the electric field are

$$E'_x = -q\, \frac{v\,t'}{r'^3}, \quad E'_y = q\, \frac{b}{r'^3}, \quad E'_z = 0 \tag{22.11}$$

where

$$r'^3 = \left| \boldsymbol{r}'_{q\mathrm{P}} \right|^3 = \left(b^2 + v^2 t'^2 \right)^{3/2} = \left(b^2 + \gamma^2 v^2 t^2 \right)^{3/2} \tag{22.12}$$

From (22.7) and $\boldsymbol{B}' = 0$ follows

$$E_x = E'_x, \quad E_y = \gamma\, E'_y, \qquad E_z = \gamma\, E'_z$$
$$B_x = 0, \qquad B_y = -\gamma\, \frac{v}{c}\, E'_z, \quad B_z = \gamma\, \frac{v}{c}\, E'_y \tag{22.13}$$

Here we insert (22.11) and (22.12):

$$\boxed{\begin{array}{ll} E_x = \dfrac{-q\,\gamma\,v\,t}{(b^2+\gamma^2 v^2 t^2)^{3/2}}, & E_y = \dfrac{\gamma\,q\,b}{(b^2+\gamma^2 v^2 t^2)^{3/2}} \\[2ex] E_z = B_x = B_y = 0, & B_z = \dfrac{\gamma\,q\,b\,v/c}{(b^2+\gamma^2 v^2 t^2)^{3/2}} \end{array}} \qquad (22.14)$$

These are the electric and magnetic fields of a uniformly moving charge. The magnetic field $\boldsymbol{B}'$ at the point P is perpendicular to the direction of motion (x-axis) and perpendicular to the connection vector $\boldsymbol{r}_{q\mathrm{P}}$; it is perpendicular to the image plane of Figure 22.1. The magnetic field lines are circles whose centers lie on the axis of motion.

Slowly moving charge

To discuss the result, we first consider the field of the moving charge in first order in v/c. From (22.14) follows

$$\boldsymbol{E} = q\,\frac{\boldsymbol{r}}{r^3}\,\big(1+\mathcal{O}(v^2/c^2)\big), \quad \boldsymbol{B} = \frac{q}{c}\,\frac{\boldsymbol{v}\times\boldsymbol{r}}{r^3}\,\big(1+\mathcal{O}(v^2/c^2)\big) \qquad (22.15)$$

where $\boldsymbol{r} = (-v\,t,\, b,\ 0)$. In this order, the electric field is the same as for the charge at rest. The magnetic field

$$\boldsymbol{B} = \frac{\boldsymbol{v}}{c}\times\boldsymbol{E} \quad (v \ll c) \qquad (22.16)$$

can be understood as a first-order relativistic effect in v/c,

$$|\boldsymbol{B}| = |\boldsymbol{E}|\cdot\mathcal{O}\Big(\frac{v}{c}\Big) \qquad (22.17)$$

Magnetic fields of slowly moving charges (or charge distributions) are small compared to the associated electric fields. For the forces between two charges, the following then applies

$$\frac{|\boldsymbol{F}_{\text{magn}}|}{|\boldsymbol{F}_{\text{Coulomb}}|} = \mathcal{O}\Big(\frac{v_1 v_2}{c^2}\Big) \qquad (22.18)$$

In this sense, magnetic effects are small, relativistic effects.

As an application, we consider a hydrogen atom in an semi-classical picture. The electron (mass m_{e}) moves on a circular orbit (radius r) around the proton. From the force equilibrium $e^2/r^2 = m_{\mathrm{e}}\,v_{\mathrm{e}}^2/r$ and an assumed angular momentum $\hbar = m_e v_e r$ we obtain the velocity $v_e/c = \alpha = e^2/(\hbar c) \approx 1/137$ for the electron. This is only a rough estimate, $v_e \sim \alpha c$. In the considered classical picture, the electron and the proton move around the common center of mass. In the center of mass system we obtain

$$v_e \sim \alpha c \quad \text{and} \quad v_p \sim \frac{m_e}{m_p} \alpha c \tag{22.19}$$

Thus, the relative strength of the magnetic interaction is

$$\left| \frac{\boldsymbol{F}_{\text{magn}}}{\boldsymbol{F}_{\text{Coulomb}}} \right| \sim \frac{v_e v_p}{c^2} \sim 10^{-7} \tag{22.20}$$

The magnetic forces cannot be eliminated by going into the rest system of the proton because this is not an inertial inertial system. These magnetic forces are so small that they can be neglected in most calculations. In the hydrogen atom, there is also an interaction between the magnetic dipole moments of the proton and electron. It is of the same size as (22.20), but it leads to an observable splitting of the ground state (hyperfine structure).

In the positronium (the bound system consisting of a positron and an electron) there is no factor m_e/m_p as in (22.19); the ratio $F_{\text{magn}}/F_{\text{Coulomb}}$ is then of the order $\alpha^2 = 10^{-4}$. In this case, the magnetic forces belong to the leading relativistic corrections. They have to be taken into account in addition to the corrections known from hydrogen problem (spin-orbit coupling, Darwin term (Zitterbewegung) and relativistic energy-momentum relation, Chapter 41 in [3]).

Relativistic effects

For $v \to c$, the $\boldsymbol{E}$ field in (22.14) deviates strongly from the spherical symmetry of the $\boldsymbol{E}'$ field. We use the vector $\boldsymbol{r} = (-v\,t, b, 0)$ from the charge to the observation point ($\boldsymbol{r}_{q\text{P}}$ in Figure 22.1). The field of the moving charge depends on the angle between $\boldsymbol{r}$ and the velocity $\boldsymbol{v}$ where

$$\cos\varphi = \frac{\boldsymbol{v}\cdot\boldsymbol{r}}{v\,r} = -\frac{v\,t}{r} \tag{22.21}$$

Using

$$r'^2 = b^2 + (\gamma\, v\, t)^2 = r^2 + (\gamma^2 - 1)(v\,t)^2 \tag{22.22}$$

$$= r^2\gamma^2\left(\frac{1}{\gamma^2} + \left(1 - \frac{1}{\gamma^2}\right)\cos^2\varphi\right) = r^2\gamma^2\left(1 - \frac{v^2}{c^2}\,\sin^2\varphi\right)$$

the $\boldsymbol{E}$ field from the first line of (22.14) becomes

$$\boldsymbol{E} = \frac{\gamma\, q\, \boldsymbol{r}}{r'^3} = \frac{q\, \boldsymbol{r}}{r^3}\, \frac{1-\dfrac{v^2}{c^2}}{\left(1-\dfrac{v^2}{c^2}\,\sin^2\varphi\right)^{3/2}} \tag{22.23}$$

This means that the field –compared to that of a charge at rest charge– is weaker parallel to $\boldsymbol{v}$ and stronger perpendicular to it is stronger (Figure 22.2). In particular, the following applies

$$\frac{|\boldsymbol{E}|}{q/r^2} = \begin{cases} 1-\dfrac{v^2}{c^2} & (\varphi = 0,\, \pi) \\ \dfrac{1}{\sqrt{1-v^2/c^2}} & (\varphi = \pm\pi/2, \pm 3\,\pi/2) \end{cases} \tag{22.24}$$

Doppler Effect

The vector potential of the monochromatic plane wave (20.16) is given by

$$\boldsymbol{A}(\boldsymbol{r}, t) = \mathrm{Re}\,\boldsymbol{A}_0\, \exp\left[\mathrm{i}(\boldsymbol{k}\cdot\boldsymbol{r} - \omega t)\right] = \mathrm{Re}\,\boldsymbol{A}_0\, \exp\left[-\,\mathrm{i}k^\alpha x_\alpha\right] \tag{22.25}$$

In the last expression, the phase

$$\omega t - \boldsymbol{k}\cdot\boldsymbol{r} = k^\alpha x_\alpha \tag{22.26}$$

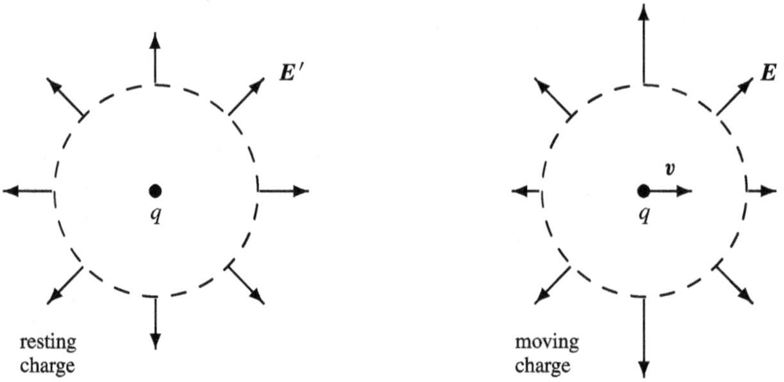

Figure 22.2 The electric field of a charge at rest is spherical (left). The field (22.23) of the moving charge, on the other hand, is reduced in the direction of motion and increased perpendicular to it (right). The field $\boldsymbol{E}'$ (left side) is that in the rest system IS′ of the charge, the field shown on the right is that in an inertial frame in which the charge is moving.

was written as the scalar product of $(x_\alpha) = (ct, -\boldsymbol{r})$ with the indexed quantity

$$(k^\alpha) = \left(\frac{\omega}{c},\, \boldsymbol{k}\right) \tag{22.27}$$

We substantiate that k^α is a Lorentz vector: A source emits a wave with a certain number of crests and troughs (or number of cycles). A Lorentz transformation cannot change this discrete number because each node of the wave (with $\boldsymbol{E} = \boldsymbol{B} = 0$) is preserved by a Lorentz transformation. Therefore, the phase $k^\alpha x_\alpha$ must be a Lorentz scalar. Since x_α is a Lorentz vector this must also apply for k^α, i.e.,

$$k'^\alpha = \Lambda^\alpha_{\ \beta}\, k^\beta \tag{22.28}$$

According to (20.17), we get

$$k^\alpha\, k_\alpha = \omega^2/c^2 - \boldsymbol{k}^2 = 0 \tag{22.29}$$

In another inertial frame IS′, the electromagnetic wave (22.25) is described by corresponding the LT transformed quantity. Thereby, the phase $k^\alpha x_\alpha = k'^\alpha x'_\alpha$ is invariant, and also (22.29) remains valid, $\omega^2 = c^2 k^2$ becomes $\omega'^2 = c^2 k'^2$.

As an application of (22.28), we consider the Doppler effect. A source emits a wave which, in the rest system RS = IS′ of the source has the frequency $\omega' = \omega_{\mathrm{RS}}$. The source and thus IS′, move relatively to IS with the velocity $\boldsymbol{v} = v\,\boldsymbol{e}_x = v\,\boldsymbol{e}_1$, Figure 22.3. An observer at rest in IS measures the frequency ω, which generally deviates from ω'. This frequency change is called *Doppler effect*. The Doppler effect is caused by the relative motion between the source and the observer.

For the special Lorentz transformation (22.3), we determine the 0-component of (22.28):

$$k'^0 = \frac{k^0 - v\,k^1/c}{\sqrt{1 - v^2/c^2}} \quad \text{or} \quad \omega_{\mathrm{RS}} = \gamma\,(\omega - v\,k^1) \tag{22.30}$$

Here $k'^0 = \omega'/c = \omega_{\mathrm{RS}}/c$ is the frequency in IS′, i.e., in the rest system of the source. On the other hand, $c\,k^0 = \omega$ is the frequency in IS, i.e., the frequency that an observer at rest in IS measures.

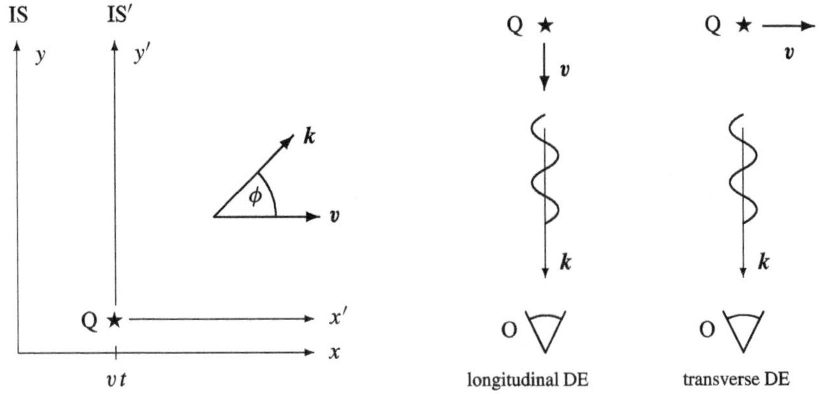

Figure 22.3 The Doppler effect (DE) is the frequency shift that an observer O measures for a moving source Q. In order to calculate the effect, one introduces an IS′ in which the source is momentarily at rest (left). A LT transforms the wave vector k'^{α} into the wave vector $(k^{\beta}) = (\omega/c, \boldsymbol{k})$ in the observer's IS. The frequency shift depends on the angle ϕ between $\boldsymbol{k}$ and the relative velocity $\boldsymbol{v}$. One considers the special cases of the longitudinal and transverse Doppler effect (right).

Using $\boldsymbol{v} = v\,\boldsymbol{e}_x$ we get

$$v\,k^1 = \boldsymbol{v}\cdot\boldsymbol{k} = v\,k\,\cos\phi = v\,\frac{\omega}{c}\,\cos\phi \tag{22.31}$$

Here ϕ is the angle that the observer in IS sees between the propagation direction $\boldsymbol{k}$ of the wave and the velocity $\boldsymbol{v}$ of the source. We insert (22.31) into (22.30) and solve for ω:

$$\boxed{\omega = \omega_{\text{RS}}\,\frac{\sqrt{1 - v^2/c^2}}{1 - (v/c)\,\cos\phi} \qquad \text{Doppler effect}} \tag{22.32}$$

The Doppler effect is the change $\omega_{\text{RS}} \to \omega$ due to the motion of the source. The factor $1 - (v/c)\,\cos\phi$ is a kinematic effect that already follows from the Galilean transformation. The factor $(1 - v^2/c^2)^{1/2}$, on the other hand, is a relativistic effect that corresponds to the time dilation. This factor becomes important in the relativistic limit.

In first order of v/c we obtain from (22.32) the linear or *longitudinal Doppler effect*,

$$\omega \approx \omega_{\text{RS}}\left(1 + \frac{v}{c}\,\cos\phi\right) \quad (v \ll c) \tag{22.33}$$

For $\phi = \pi/2$ the linear contribution vanishes and we get the quadratic or *transverse Doppler effect*,

$$\omega \approx \omega_{RS}\left(1 - \frac{v^2}{2c^2}\right) \qquad (v \ll c,\ \phi = \pi/2) \tag{22.34}$$

In the non-relativistic limit, the linear effect dominates.

We consider the relativistic limit $v \to c$ for $\phi = 0$ and $\phi = \pi$:

$$\omega = \omega_{RS} \cdot \begin{cases} \sqrt{\dfrac{1+v/c}{1-v/c}} \overset{v\to c}{\longrightarrow} \infty & (\phi = 0) \\[2ex] \sqrt{\dfrac{1-v/c}{1+v/c}} \overset{v\to c}{\longrightarrow} 0 & (\phi = \pi) \end{cases} \tag{22.35}$$

If a source moves towards the observer ($\phi = 0$), the observed frequency ω for $v \to c$ approaches infinity. If the light from the source moves away from the observer ($\phi = \pi$), the observed frequency approaches zero for $v \to c$; the light becomes less and less energetic. In the limit $v \to c$ the observer can no longer see the source.

The frequency shift may be expressed by the dimensionless *redshift* z expressed:

$$z = \frac{\lambda}{\lambda_{RS}} - 1 = \frac{\omega_{RS}}{\omega} - 1 \tag{22.36}$$

In the visible range (20.47) of the electromagnetic spectrum $z > 0$ means a shift towards the red end. Irrespective of this, terminus "redshift" is used in the entire spectrum for the statement $z > 0$. Occasionally the redshift parameter z itself is called "redshift".

Applications

One application of the Doppler shift is the police radar trap (or speed trap). A radar beam is directed towards a moving car. If the car is moving away at a speed of $v = |\boldsymbol{v}|$, it receives this beam with the redshift $z = v/c$. The metal of the car reflects a wave with the shifted frequency. Due to the motion of the car, the reflected wave arrives with a an additional shift at the radar trap, i.e., with $z = 2v/c$. For $v = 50\,\text{km/h}$ we get $z = 10^{-7}$. Using radar with $\nu_R = 10^{10}$ Hz the frequency shift is then $\Delta\nu = \nu_R\, z = 10^3$ Hz. A superposition of the emitted and the reflected wave leads to a signal with the frequency $\Delta\nu$. Such a frequency $\Delta\nu$ can be measured easily.

Doppler shifts also occur with sound waves. They can be observed for a car honking its horn as it drives past quickly. While the car is approaching, the horn sound frequency is higher, while it moves away, the sound frequency is lower. First the wave crests arrive in shorter time intervals (i.e., with a higher frequency), then in larger intervals (with a lower frequency). This is an everyday observation of the Doppler effect.

The analogy between sound and light waves only applies to some extent. In particular, light propagates in empty space, whereas sound requires a wave carrying medium (such as air). In the case of sound, the IS in which the medium rests is distinguished. The results then depend on the velocities of the source *and* the observer in this IS, and not only on the relative velocity as in (22.23). In the limiting case of small velocities, however, (22.33) also applies for the sound (where c is the speed of sound). For $v \to c$ there are significant differences.

Atoms (or molecules) emit and absorb light at specific frequencies in the electromagnetic spectrum; this leads to a line structure which is characteristic for the atom (or molecule). On earth, light reaches us light from stars and galaxies, which contains such line structures. Compared to the known line structure measured on Earth, one usually find a shift of this structure. One reason for this shift is the Doppler effect of a moving source. Light from a double star system, for example, shows periodic Doppler shifts. From this, information about the stars' orbits can be read off. If one of the stars is a pulsar, the observable Doppler shifts refer to the pulses of this neutron star.

For light from stars and galaxies, there are further causes for frequency shifts. On one hand, light from the surface of a star suffers a gravitational redshift when it passes through the star's gravitational field. On the other hand, due to the expansion of the universe causes a cosmological redshift in the light of distant galaxies. We summarize these three effects in the lowest approximation:

$$z = \begin{cases} v/c & \text{Doppler redshift} \\ -\Phi_{\text{grav}}/c^2 & \text{gravitational redshift} \\ HD/c & \text{cosmological redshift} \end{cases} \tag{22.37}$$

The first case applies to a source (star) that is moving away with the speed v. In the second case, $\Phi_{\text{grav}} < 0$ is the Newtonian gravitational potential at the surface of the star, from which the light emanates. In the third case, D is the distance of the source (galaxy) and H is the Hubble constant, which determines the current expansion of the universe ($1/H \approx$ 14 billion years).

Aberration

In IS a wave propagates along the z-axis (Figure 22.4). Its wave vector is

$$\left(k^{\alpha}\right) = (\omega/c, \boldsymbol{k}) = (k, 0, 0, -k) \tag{22.38}$$

Using the transformation (22.3), we determine this 4-vector in IS′:

$$\left(k'^{\alpha}\right) = \left(\omega'/c, \boldsymbol{k}'\right) = \left(k'^{0}, -\gamma\,\frac{v}{c}\,k, 0, -k\right) \tag{22.39}$$

In IS′, the vector $\boldsymbol{k}'$ then encloses an angle $\varphi_{\rm A}$ with the z' axis:

$$\tan\varphi_{\rm A} = \frac{k'^{1}}{k'^{3}} = \frac{v/c}{\sqrt{1 - v^2/c^2}} \tag{22.40}$$

This angle $\varphi_{\rm A}$ is the change in the direction of the wave due to the motion of an IS′-observer.

In the non-relativistic limit we get

$$\varphi_{\rm A} \approx \frac{v}{c} \qquad (v \ll c) \tag{22.41}$$

Just like the linear Doppler effect, this is a kinematic effect which also results from a Galilean transformation. In contrast to the Doppler effect, relativistic effects (i.e., the factor γ in (22.40)) usually plays no role.

A well-known application is the apparent change of star positions due to the motion of the earth around the sun. Let the system IS be the system

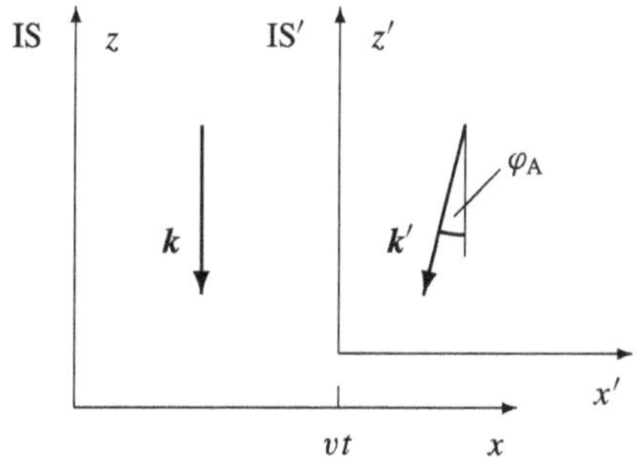

Figure 22.4 A plane wave with the wave vector $\boldsymbol{k}'$ in IS′ has a somewhat differently directed wave vector $\boldsymbol{k}'$ in IS′. Therefore, orbital motion of the earth causes a specific star to move apparently on a small circle. This effect is called *aberration*.

of the Sun, and IS′ is a system that is momentarily comoving with the Earth. For the Earth's orbit, $v/c \approx 10^{-4}$ applies. We consider a remote star which stands perpendicular to the Earth orbit plane. Then, its apparent position moves approximately on a circular cone with an opening angle $\varphi_{\mathrm{A}} \approx 10^{-4}\,(180 \cdot 3600''/\pi) \approx 20''$. The *aberration* (22.41) was detected by the astronomer Bradley in 1725. Smaller aberrations are caused by the Earth's rotation, and by the motion of our solar system within the Milky Way.

For nearby stars, there is also a change of the apparent star position due the stellar parallax. The stellar parallax means that the apparent star position depends on the positions on the Earth orbit (independent of the Earth's velocity). The relevant angle is given by the diameter of the Earth orbit divided by the star distance. For nearby stars this can be used to determine of their distances; for faraway stars this effect is negligible.

Exercises

22.1 *Invariants of the electromagnetic field*

How do $\boldsymbol{E}^2 - \boldsymbol{B}^2$ and $\boldsymbol{E} \cdot \boldsymbol{B}$ change under Lorentz transformations? Express these two quantities by the field strength tensor and its dual partner. Alternatively, use the transformation (22.6) of the electromagnetic fields.

22.2 *Fields of a passing charge*

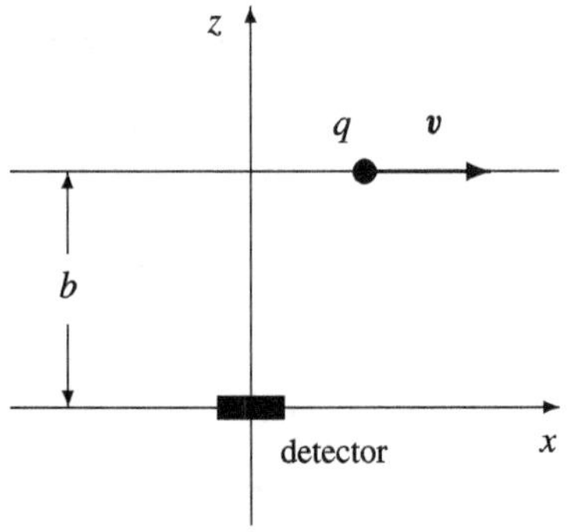

A relativistically moving charge q has the trajectory $\boldsymbol{r}(t) = v\,t\,\boldsymbol{e}_x + b\,\boldsymbol{e}_z$. Calculate the integrals

$$\int_{-\infty}^{\infty} dt\,\boldsymbol{E} \quad \text{and} \quad \int_{-\infty}^{\infty} dt\,\boldsymbol{B}$$

for the fields which a detector measures at $\boldsymbol{r} = 0$. To do this, split the fields into the parts that are parallel or perpendicular to $\boldsymbol{v}$.

22.3 *Energy flow of a uniformly moving charge*

A point charge q moves with a constant non-relativistic velocity $\boldsymbol{v}$. Calculate the energy density w_{em} and the Poynting vector $\boldsymbol{S}$ of the fields of this charge. How large is the energy current $\oint d\boldsymbol{a} \cdot \boldsymbol{S}$ through a spherical surface a, in the center of which the charge is momentarily located?

22.4 *Charge is independent of velocity*

A point charge q moves with a constant relativistic velocity $\boldsymbol{v}$. Calculate the integral $\oint_a d\boldsymbol{a} \cdot \boldsymbol{E}$ over the electric field $\boldsymbol{E}$ using a stationary, closed surface a, which encloses the charge. First of all, consider a spherical surface with the particle in the center. Then show that the result does not depend on the shape of the surface. The result $\oint_a d\boldsymbol{a} \cdot \boldsymbol{E} = 4\pi q$ implies that the charge is independent of the velocity.

22.5 *Electromagnetic mass of the moving charged sphere*

A homogeneously charged sphere (radius R, charge q) is considered as a model of a massive elementary particle. Specify the energy W_{em} of the electromagnetic fields of the sphere at rest. By $W_{\text{em}} = m_0\,c^2$ one may define the rest mass m_0.

The sphere now moves with a constant, non-relativistic velocity $\boldsymbol{v}$. Calculate the momentum $\boldsymbol{P}_{\text{em}}$ of the electromagnetic field of the moving sphere (neglect the terms of the order v^2/c^2). By $\boldsymbol{P}_{\text{em}} = \overline{m}\,\boldsymbol{v}$ another mass $\overline{m}$ is defined. Compare $\overline{m}$ and m_0.

22.6 *Aberration*

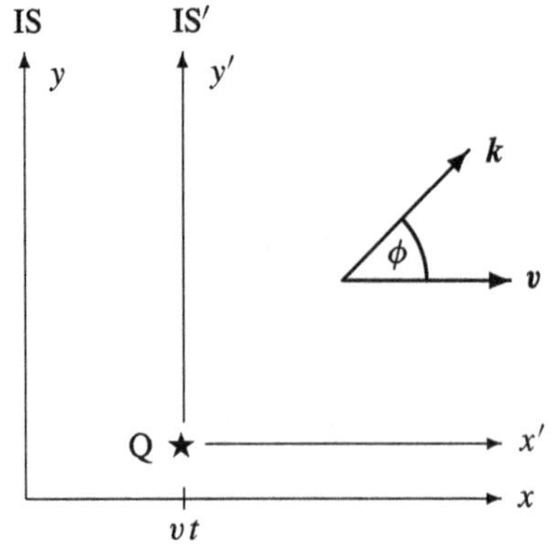

A source at rest in the inertial frame IS′ emits a plane wave with the frequency $\omega' = \omega_{\text{RS}}$ and with the wave vector

$$\boldsymbol{k}' = k' \cos\phi' \, \boldsymbol{e}_x + k' \sin\phi' \, \boldsymbol{e}_y$$

The source and IS′ move relative to an observer in IS with the velocity $\boldsymbol{v} = v\, \boldsymbol{e}_x$. At what angle ϕ to the x-axis does the observer see the wave? Establish the relation between ϕ and ϕ'.

Chapter 23

Accelerated Charge

We calculate the field of an arbitrarily moving point charge. In contrast to the uniformly moving charge considered in the previous chapter, an accelerated charge emits electromagnetic radiation. The angle distribution of this radiation is investigated. The radiation losses of linear and circular accelerators are compared.[1]

A point charge with the trajectory $\boldsymbol{r}_0(t)$ has the charge and current density

$$\varrho(\boldsymbol{r}, t) = q\ \delta(\boldsymbol{r} - \boldsymbol{r}_0(t)), \quad \boldsymbol{j}(\boldsymbol{r}, t) = q\ \dot{\boldsymbol{r}}_0(t)\ \delta(\boldsymbol{r} - \boldsymbol{r}_0(t)) \qquad (23.1)$$

A general way to calculate its radiation is as follows: The source terms (23.1) are inserted into the retarded potentials (17.32)–(17.33). These integrals are then evaluated. From the resulting potentials, the electromagnetic fields and eventually the Poynting vector are calculated. The derivation can be simplified by first determining the potentials in the momentary rest system of the charge.

In the retarded potentials $A^{\alpha}_{\text{ret}}(\boldsymbol{r}, t)$, the charge distribution enters at the retarded time:

$$t_{\text{ret}} = t - \frac{|\boldsymbol{r} - \boldsymbol{r}_0(t_{\text{ret}})|}{c} \qquad (23.2)$$

We denote the vector from the position Q of the charge to the observation point P (Figure 23.1) by

$$\boldsymbol{R}(t_{\text{ret}}) = \boldsymbol{r} - \boldsymbol{r}_0(t_{\text{ret}}) \qquad (23.3)$$

[1]This chapter may be skipped in a first reading. Following Chapter 24, the last section "Radiation loss" of this chapter should then be read and the results shown in Figures 23.3 and 23.4 should be noted.

For the vector

$$(R^{\alpha}) = \big(c\,(t - t_{\rm ret}),\ \boldsymbol{r} - \boldsymbol{r}_0(t_{\rm ret})\big) = (R, \boldsymbol{R}) \qquad (23.4)$$

and (23.2) we obtain:

$$R^{\alpha} R_{\alpha} = c^2 (t - t_{\rm ret})^2 - \big|\boldsymbol{r} - \boldsymbol{r}_0(t_{\rm ret})\big|^2 = 0 \qquad (23.5)$$

This applies in any IS. Therefore, $R_{\alpha} R^{\alpha}$ is is a Lorentz scalar and R_{α} is a Lorentz vector. For the zero component we get

$$R^0 = R_0 = c\,(t - t_{\rm ret}) = |\boldsymbol{R}| = R \qquad (23.6)$$

Figure 23.1 shows the position $\boldsymbol{r}_0(t_{\rm ret})$ of the charge. We now consider the inertial system IS′, which moves with $\boldsymbol{v} = \dot{\boldsymbol{r}}_0(t_{\rm ret})$ relative to IS; by $t_{\rm ret}$ from (23.2), IS′ depends on t and $\boldsymbol{r}$. In IS′ the charge is momentarily at rest. Therefore, the potentials in IS′ can easily be written on:

$$\Phi'_{\rm ret}(\boldsymbol{r}', t') = \frac{q}{|\boldsymbol{r}' - \boldsymbol{r}'_0(t'_{\rm ret})|} = \frac{q}{R'(t'_{\rm ret})}, \qquad \boldsymbol{A}'(\boldsymbol{r}', t') = 0 \qquad (23.7)$$

Formally, this can also be done by inserting $\varrho'(\boldsymbol{r}', t') = q\ \delta(\boldsymbol{r}' - \boldsymbol{r}'_0(t'))$ and $\boldsymbol{j}' = 0$ into (17.32) and (17.33). From (23.7) we obtain the potentials in IS by the Lorentz transformation. i.e., by $A^{\alpha} = \Lambda^{\alpha}_{\ \beta}(-\boldsymbol{v})\, A'^{\beta}$. We avoid the explicit execution of this transformation by writing (23.7) in covariant

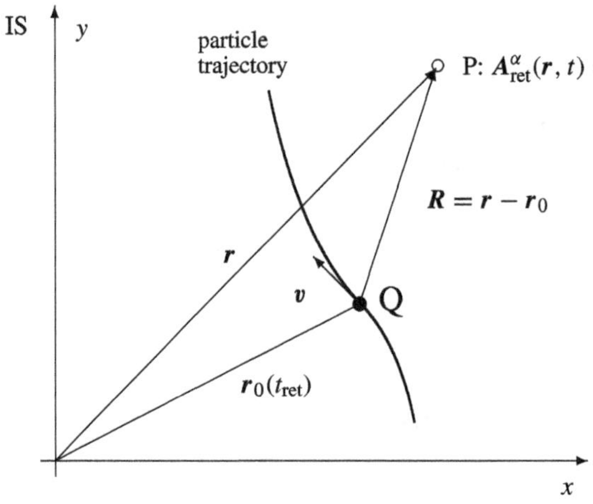

Figure 23.1 At position P, the potential $A^{\alpha}(\boldsymbol{r}, t)$ of an accelerated point charge shall be calculated. At the time $t_{\rm ret}$ the charge is at the position Q and has the velocity $\boldsymbol{v} = \boldsymbol{v}(t_{\rm ret})$.

form. For this covariant form, the 4-vectors R^{α} and u^{α} come into question. In the momentary rest system IS′ we obtain

$$(u'^{\alpha}) = (c, 0), \quad (R^{\beta} u_{\beta})' = c\, R' \quad \text{(in IS}'\text{)} \tag{23.8}$$

Using this we can write (23.7) as

$$A^{\alpha}(\boldsymbol{r}, t) = \frac{q\, u^{\alpha}}{u^{\beta} R_{\beta}}\bigg|_{\text{ret}} \quad \text{(Liénard–Wiechert potentials)} \tag{23.9}$$

In the first step, there are still primes because (23.7) refers to IS′. But these primes can be omitted because (23.9) is covariant, it is of the same form also in IS. The index "ret" indicates again that the quantity is to be taken at the retarded time t_{ret}.

The validity of the Liénard–Wiechert potentials is as follows:

1. Equation (23.9) reduces to (23.7) in the instantaneous rest system IS′. This means that (23.9) is correct in IS′.
2. Equation (23.9) is covariant. If (23.9) is valid in a specific inertial system, then it is valid in all inertial systems.

With $(A^{\alpha}) = (\Phi, \boldsymbol{A})$, $(u^{\alpha}) = \gamma\,(c, \boldsymbol{v})$ and $(R^{\alpha}) = (R, \boldsymbol{R})$, (23.9) becomes

$$\Phi(\boldsymbol{r}, t) = \frac{q}{R - \boldsymbol{R}\cdot\boldsymbol{v}/c}\bigg|_{\text{ret}}, \quad \boldsymbol{A}(\boldsymbol{r}, t) = \frac{q}{c}\,\frac{\boldsymbol{v}}{R - \boldsymbol{R}\cdot\boldsymbol{v}/c}\bigg|_{\text{ret}} \tag{23.10}$$

From this the electromagnetic fields

$$\boldsymbol{E} = -\operatorname{grad}\Phi(\boldsymbol{r}, t) - \frac{1}{c}\,\frac{\partial \boldsymbol{A}(\boldsymbol{r}, t)}{\partial t}, \quad \boldsymbol{B} = \operatorname{curl}\boldsymbol{A}(\boldsymbol{r}, t) \tag{23.11}$$

can be calculated. This requires some intermediate calculations because the potentials (23.10) are functions of

$$\boldsymbol{R} = \boldsymbol{R}(\boldsymbol{r}, t_{\text{ret}}) \quad \text{and} \quad \boldsymbol{v} = \boldsymbol{v}(t_{\text{ret}}) \tag{23.12}$$

For the fields (23.11), the potentials must be differentiated with respect to $\boldsymbol{r}$ and t; thereby $t_{\text{ret}} = t_{\text{ret}}(\boldsymbol{r}, t)$ must be taken into account.

The quantities $\boldsymbol{R}(\boldsymbol{r}, t_{\text{ret}})$ and $t_{\text{ret}}(\boldsymbol{r}, t)$ are defined in (23.2)–(23.4). We differentiate $\boldsymbol{R}^2 = R^2$ partially with respect to t_{ret},

$$2R\,\frac{\partial R}{\partial t_{\text{ret}}} = 2\boldsymbol{R}\cdot\frac{\partial \boldsymbol{R}}{\partial t_{\text{ret}}} = -2\boldsymbol{R}\cdot\dot{\boldsymbol{r}}_0(t_{\text{ret}}) = -2\boldsymbol{R}\cdot\boldsymbol{v} \tag{23.13}$$

This gives us

$$\frac{\partial R}{\partial t} = \frac{\partial R}{\partial t_{\rm ret}}\,\frac{\partial t_{\rm ret}}{\partial t} = -\,\frac{\boldsymbol{R}\cdot\boldsymbol{v}}{R}\,\frac{\partial t_{\rm ret}}{\partial t} \tag{23.14}$$

Since $R = R^0 = c\,(t - t_{\rm ret})$, the following holds:

$$\frac{\partial R}{\partial t} = c\left(1 - \frac{\partial t_{\rm ret}}{\partial t}\right) \tag{23.15}$$

From the last two equations, we obtain

$$\frac{\partial t_{\rm ret}}{\partial t} = \frac{1}{1 - \boldsymbol{R}\cdot\boldsymbol{v}/Rc} = \frac{1}{1 - \boldsymbol{\beta}\cdot\boldsymbol{e}_R} \tag{23.16}$$

where

$$\boldsymbol{\beta} = \frac{\boldsymbol{v}}{c} \quad \text{and} \quad \boldsymbol{e}_R = \frac{\boldsymbol{R}}{R} \tag{23.17}$$

We now differentiate $R = R^0 = c\,(t - t_{\rm ret})$ with respect to the coordinates,

$$\nabla R = \begin{cases} \nabla R^0 = \nabla c\,(t - t_{\rm ret}) = -c\,\nabla t_{\rm ret} \\[2ex] \nabla\left|\boldsymbol{r} - \boldsymbol{r}_0(t_{\rm ret})\right| = \dfrac{\boldsymbol{R}}{R} + \dfrac{\partial R}{\partial t_{\rm ret}}\,\nabla t_{\rm ret} \end{cases} \tag{23.18}$$

We equate the right-hand sides and use $\partial R/\partial t_{\rm ret} = -\boldsymbol{R}\cdot\boldsymbol{v}/R$ from (23.13). This gives us

$$\nabla t_{\rm ret} = -\frac{1}{c}\,\frac{\boldsymbol{R}}{R - \boldsymbol{R}\cdot\boldsymbol{v}/c} = -\frac{1}{c}\,\frac{\boldsymbol{e}_R}{1 - \boldsymbol{\beta}\cdot\boldsymbol{e}_R} \tag{23.19}$$

With (23.16) and (23.19), the prerequisites are given for differentiating (23.11) with (23.10). After a few intermediate calculations, one obtains

$$\boxed{\boldsymbol{E}(\boldsymbol{r},t) = \left.\frac{q\,(\boldsymbol{e}_R - \boldsymbol{\beta})}{R^2\gamma^2\,(1 - \boldsymbol{\beta}\cdot\boldsymbol{e}_R)^3}\right|_{\rm ret} + \left.\frac{q\,\boldsymbol{e}_R \times\,(\boldsymbol{e}_R - \boldsymbol{\beta}) \times \dot{\boldsymbol{\beta}}}{c\,R\,(1 - \boldsymbol{\beta}\cdot\boldsymbol{e}_R)^3}\right|_{\rm ret}} \tag{23.20}$$

and

$$\boldsymbol{B} = \boldsymbol{e}_R \times \boldsymbol{E} \tag{23.21}$$

The first term in (23.20) is equal to the field (22.14) of the uniformly moving charge; it is proportional to $1/R^2$. The second term is proportional to $1/R$ and vanishes for $\boldsymbol{\beta}$ = const. It leads to an energy current density $S \propto 1/R^2$ and to a finite value for the radiated power.

The coincidence of the first term in (23.20) with (22.14) is not obvious: In (23.20) the position Q of the charge q is taken at the time t_{ret} is included, whereas in (22.14) the position Q′refers to the observation time t. The geometric relations and the relevant quantities are sketched in Figure 23.2. This sketch shows the following:

$$\begin{aligned} \overline{\text{QQ}'} &= v\,(t - t_{\text{ret}}) = \beta R \\ \overline{\text{NQ}} &= \overline{\text{QQ}'}\,\cos\theta = (\boldsymbol{\beta}\cdot\boldsymbol{e}_R)\,R \\ \overline{\text{NP}} &= \; R - \overline{\text{NQ}} \; = (1 - \boldsymbol{\beta}\cdot\boldsymbol{e}_R)\,R \end{aligned} \tag{23.22}$$

This allows us to make the following connection:

$$\begin{aligned} \big((1-\boldsymbol{\beta}\cdot\boldsymbol{e}_R)\,R\big)^2 &= \overline{\text{NP}}^{\,2} = \overline{\text{Q}'\text{P}}^{\,2} - \overline{\text{NQ}'}^{\,2} = b^2 + v^2t^2 - (\beta R)^2\sin^2\theta \\ &= b^2 + v^2t^2 - \beta^2 b^2 \;=\; \frac{b^2 + \gamma^2 v^2 t^2}{\gamma^2} \end{aligned} \tag{23.23}$$

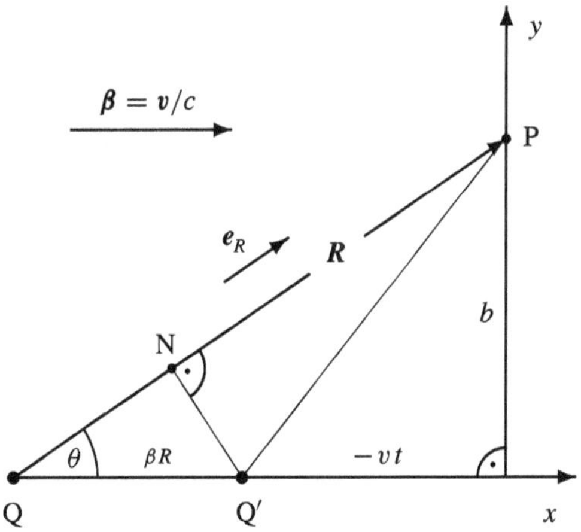

Figure 23.2 The position of a uniformly, with $\boldsymbol{v} = v\,\boldsymbol{e}_x$, moving charge at time t_{ret} is Q (as in Figure 23.1). At the time t at which the fields are to be calculated, the charge is then at a different point Q′. From the figure, the relations between the occurring lengths can be read off. At the time $t = 0$, the charge has the shortest distance to the observation point P; for the drawn position t is negative.

With $\boldsymbol{\beta} \cdot \boldsymbol{e}_y = 0$ and $R\,\boldsymbol{e}_R \cdot \boldsymbol{e}_y = b$ we get

$$E_y \overset{(\dot\beta=0)}{=} \frac{q\,(\boldsymbol{e}_R - \boldsymbol{\beta})_y}{R^2\gamma^2\,(1-\boldsymbol{\beta}\cdot\boldsymbol{e}_R)^3} = \frac{q\,\gamma\,b}{(b^2+\gamma^2 v^2 t^2)^{3/2}} \tag{23.24}$$

This shows the coincidence of (22.14) with (23.20) for E_y. The same follows for E_x follows using $R\,(\boldsymbol{e}_R - \boldsymbol{\beta})_x = \beta R - v\,t - \beta R = -v\,t$.

Radiation Field

The radiation field of an accelerated charge is described by the second term in (23.20):

$$\boldsymbol{E}_{\text{rad}}(\boldsymbol{r},t) = \left.\frac{q\,\boldsymbol{e}_R \times (\boldsymbol{e}_R - \boldsymbol{\beta}) \times \dot{\boldsymbol{\beta}}}{c\,R\,(1-\boldsymbol{\beta}\cdot\boldsymbol{e}_R)^3}\right|_{\text{ret}}, \qquad \boldsymbol{B}_{\text{rad}} = \boldsymbol{e}_R \times \boldsymbol{E}_{\text{rad}} \tag{23.25}$$

This yields the radial component of the energy flux density,

$$\boldsymbol{S}\cdot\boldsymbol{e}_R = \frac{c}{4\pi}\,\boldsymbol{e}_R \cdot \left(\boldsymbol{E}_{\text{rad}} \times \boldsymbol{B}_{\text{rad}}\right) = \frac{q^2}{4\pi c\,R^2}\left.\frac{|\boldsymbol{e}_R \times (\boldsymbol{e}_R - \boldsymbol{\beta}) \times \dot{\boldsymbol{\beta}}\,|^2}{(1-\boldsymbol{\beta}\cdot\boldsymbol{e}_R)^6}\right|_{\text{ret}} \tag{23.26}$$

The power $dP = \boldsymbol{S}\cdot\boldsymbol{e}_R\,R^2 d\Omega$ passes the surface element $R^2 d\Omega$, therefore

$$\frac{dP}{d\Omega} = R^2\boldsymbol{S}\cdot\boldsymbol{e}_R = \frac{q^2}{4\pi c}\left.\frac{|\boldsymbol{e}_R \times (\boldsymbol{e}_R - \boldsymbol{\beta}) \times \dot{\boldsymbol{\beta}}\,|^2}{(1-\boldsymbol{\beta}\cdot\boldsymbol{e}_R)^6}\right|_{\text{ret}} \tag{23.27}$$

Here, dP is the energy per time interval dt and per solid angle element $d\Omega$. From the point of view of the radiating particle, it is natural to refer to the time interval dt_{ret}. This energy flow dP' is

$$\frac{dP'}{d\Omega} = \frac{dP}{d\Omega}\,\frac{\partial t}{\partial t_{\text{ret}}} \overset{(23.16)}{=} \frac{q^2}{4\pi c}\,\frac{|\boldsymbol{e}_R \times (\boldsymbol{e}_R - \boldsymbol{\beta}) \times \dot{\boldsymbol{\beta}}\,|^2}{(1-\boldsymbol{\beta}\cdot\boldsymbol{e}_R)^5} \tag{23.28}$$

We evaluate this expression for the two special cases $\beta \ll 1$ and $\boldsymbol{\beta} \parallel \dot{\boldsymbol{\beta}}$. For this we use the angles θ and θ':

$$\cos\theta = \frac{\boldsymbol{e}_R\cdot\boldsymbol{\beta}}{\beta}, \qquad \cos\theta' = \frac{\boldsymbol{e}_R\cdot\dot{\boldsymbol{\beta}}}{\dot\beta} \tag{23.29}$$

In the non-relativistic limit case, Equations (23.27) and (23.28) reduce to

$$\boxed{\frac{dP}{d\Omega} = \frac{dP'}{d\Omega} = \frac{q^2}{4\pi c^3}\,\dot{\boldsymbol{v}}^2\,\sin^2\theta' \quad (v \ll c)} \tag{23.30}$$

For $\boldsymbol{\beta} \parallel \dot{\boldsymbol{\beta}}$ we get $\theta = \theta'$. Then (23.28) becomes

$$\boxed{\frac{dP'}{d\Omega} = \frac{q^2}{4\pi c^3}\,\dot{v}^2\,\frac{\sin^2\theta}{(1-\beta\,\cos\theta)^5} \qquad (\boldsymbol{v} \parallel \dot{\boldsymbol{v}})} \tag{23.31}$$

These angular distributions are shown in Figures 23.3 and 23.4. With increasing (relativistic) velocity the radiation becomes increasingly forward directed. At the same time the radiated power increases drastically.

Accelerated electrons are used to generate X-rays: In a vacuum tube, a high voltage (about 5 keV to 10 MeV) is applied between the cathode and anode. The electrons emerge from the heated cathode, pass through the potential difference V and hit the anode with high energy. There they are deflected and rapidly slowed down. Approximately 1% of their energy is emitted as *Bremsstrahlung* or X-rays; the remaining energy is converted into heat. The German word Bremsstrahlung (verbatim breaking radiation) is quite common also in English texts. The abrupt deceleration of electrons is the standard way to generate X-rays or Röntgen rays.

The Bremsstrahlung has a continuous frequency distribution, which breaks off at $\hbar\,\omega_{\text{max}} = |eV|$ where V is the applied voltage. Note that slanted fonts are used for physical quantities (like the charge e and the

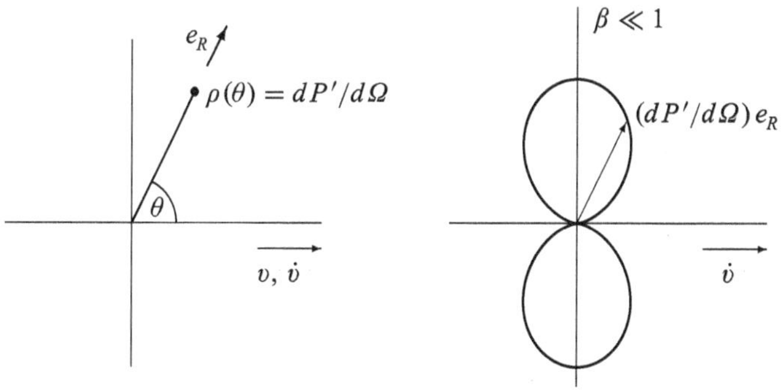

Figure 23.3 The radiated power (23.31) is depicted as the polar coordinate $\rho = dP'/d\Omega$ in the ρ-θ diagram (left). Here θ is the angle between $\boldsymbol{v} \parallel \dot{\boldsymbol{v}}$ and the direction of radiation $\boldsymbol{e}_R$. On the right is the radiation power $dP'/d\Omega \propto \sin^2\theta$ is depicted for $\beta \ll 1$. In the direction of $\boldsymbol{e}_r$, a vector with a length proportional to $dP'/d\Omega \propto \dot{\boldsymbol{v}}^2$ is shown. For $\beta \ll 1$, the radiation is independent of the direction of $\boldsymbol{v}$. The radiated power is proportional to $\dot{\boldsymbol{v}}^2$.

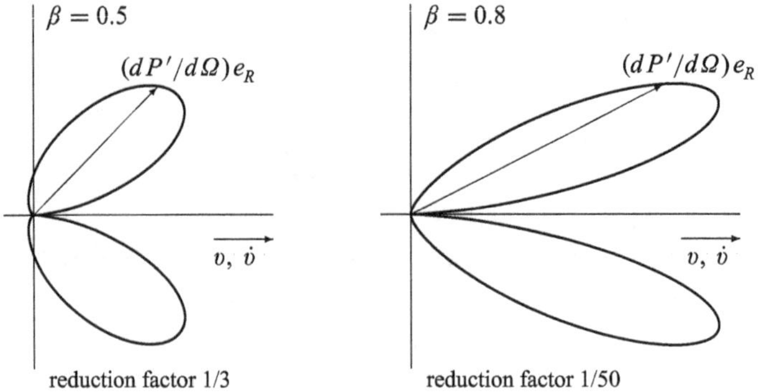

Figure 23.4 Angular dependence of the radiation (23.31) of an accelerated charge for $\beta = 0.5$ and $\beta = 0.8$, in each case for $\boldsymbol{v} \parallel \dot{\boldsymbol{v}}$. With increasing acceleration the radiation grows, and it is more and more forwardly directed. Figures 23.3 and 23.4 scale with the value of $\dot{\boldsymbol{v}}^2$. The curves in Figure 23.4 had to be enlarged in all directions by a factor of 3 or 50, respectively, if they are drawn oh the right in Figure 23.3 (assuming the same value for $\dot{\boldsymbol{v}}^2$).

voltage V). In eV for electron volt, the roman letter e stands for "electron", and V for the unit volt.

Radiation Loss

The angular integration of (23.30) results in

$$P = \frac{2q^2}{3c^3}\,\dot{\boldsymbol{v}}^2 \quad (v \ll c) \tag{23.32}$$

For relativistic velocities, the radiation power P is obtained either by integrating (23.27) or by the covariant generalization of (23.32). We take the second, easier approach.

The radiated power P is electromagnetic field energy per time, i.e., $dE_{\text{rad}} = P\,dt = P\,dx^0/c$. The energy E_{rad} of the electromagnetic (radiation) field is the 0-component of a 4-vector, (18.60). Since dx^0 is also such a 0-component, we conclude

$$P \text{ is a Lorentz scalar} \tag{23.33}$$

We are now looking for a Lorentz scalar that reduces to (23.32) for $v/c \to 0$. The relativistic generalization of $d\boldsymbol{v}/dt$ is $du^\alpha/d\tau$; where u^α is the 4-velocity and $d\tau$ the proper time interval. The obvious generalization of (23.32) is therefore

$$P = -\frac{2q^2}{3c^3}\,\frac{du^\alpha}{d\tau}\,\frac{du_\alpha}{d\tau} \tag{23.34}$$

This expression is correct because it is covariant and because for $v/c \to 0$ it reduces to the Equation (23.32), which is known to be correct.

We calculate the derivative of the 4-velocity:

$$\begin{aligned}\frac{d(u^\alpha)}{d\tau} = \gamma\,\frac{d(u^\alpha)}{dt} &= \gamma\,\frac{d}{dt}\,(\gamma\,c,\gamma\,\boldsymbol{v}) = \gamma^4\,\frac{\boldsymbol{v}\cdot\dot{\boldsymbol{v}}}{c^2}\,(c,\boldsymbol{v}) + \gamma^2\,(0,\dot{\boldsymbol{v}})\\ &= \gamma^4\left(\frac{\boldsymbol{v}\cdot\dot{\boldsymbol{v}}}{c},\,\frac{\boldsymbol{v}\cdot\dot{\boldsymbol{v}}}{c^2}\,\boldsymbol{v} + \left(1-\frac{v^2}{c^2}\right)\dot{\boldsymbol{v}}\right)\end{aligned} \tag{23.35}$$

Thus, (23.34) becomes

$$P = \frac{2q^2}{3c^3}\,\gamma^6\left(\dot{\boldsymbol{v}}^2 - \frac{(\boldsymbol{v}\times\dot{\boldsymbol{v}})^2}{c^2}\right) \tag{23.36}$$

From this follow the two special cases:

$$\boxed{P = \frac{2q^2}{3c^3}\,\gamma^6\left(\frac{d\boldsymbol{v}}{dt}\right)^2 = \frac{2q^2}{3m^2c^3}\left(\frac{d\boldsymbol{p}}{dt}\right)^2 \qquad (\boldsymbol{v}\parallel\dot{\boldsymbol{v}})} \tag{23.37}$$

$$\boxed{P = \frac{2q^2}{3c^3}\,\gamma^4\left(\frac{d\boldsymbol{v}}{dt}\right)^2 = \frac{2q^2}{3m^2c^3}\,\gamma^2\left(\frac{d\boldsymbol{p}}{dt}\right)^2 \qquad (\boldsymbol{v}\perp\dot{\boldsymbol{v}})} \tag{23.38}$$

The first expressions in both equations result directly from (23.36). The last expressions with the relativistic momentum $\boldsymbol{p} = m\gamma\,\boldsymbol{v}$ are obtained if $d\gamma/dt = \gamma^3\boldsymbol{v}\cdot\dot{\boldsymbol{v}}/c^2$ is taken into account. These expressions show: For the same magnitude of the accelerating force (same $|d\boldsymbol{p}/dt|$), the emitted radiation is greater for $\boldsymbol{v}\perp\dot{\boldsymbol{v}}$ than for $\boldsymbol{v}\parallel\dot{\boldsymbol{v}}$. It is greater by a factor of γ^2 which might become quite large.

In accelerators, particles are brought to high energies. The radiation power P reduces the energy of the particles; therefore one speaks of radiation losses. We discuss the relativistic radiation losses for a linear and a circular accelerator.

Linear accelerator

We estimate the radiation loss of a relativistic electron ($v \approx c$, $q = -e$, mass m_e) in a linear accelerator. At the linear accelerator SLAC in Stanford/USA electrons are accelerated over a distance of 3 km to an energy of 50 GeV. The length ℓ over which the electron energy increases by $m_e c^2 = 0.5\,\text{MeV}$ is

$$\ell = \frac{m_e c^2}{dE/d\ell} = \frac{0.5\,\text{MeV}}{50\,\text{GeV}/3\,\text{km}} = 3\,\text{cm} \tag{23.39}$$

For the highly relativistic case ($E \gg m_e c^2$) we obtain $E = (m_e^2 c^4 + c^2 p^2)^{1/2} \approx c\,p$ and $v \approx c$. From this follows

$$\frac{dp}{dt} \approx \frac{d(E/c)}{d\ell/c} = \frac{dE}{d\ell} = \frac{m_e c^2}{\ell} \tag{23.40}$$

We insert this into (23.37) and calculate the radiated energy $\varDelta E_{\text{rad}}$ during a travel over $\ell \approx 3\,\text{cm}$,

$$\begin{aligned}\varDelta E_{\text{rad}} &\approx P\,\frac{\ell}{c} \approx \frac{2e^2}{3m_e^2 c^3}\left(\frac{m_e c^2}{\ell}\right)^2 \frac{\ell}{c} = \frac{2}{3}\frac{e^2}{\ell} \approx \frac{2}{3}\frac{14.4\,\text{eV}}{3\,\text{cm}/\text{Å}} \\ &\approx 3 \cdot 10^{-8}\,\text{eV}\end{aligned} \tag{23.41}$$

Here $e^2/\text{Å} = 14.4\,\text{eV}$ was used. The resulting energy loss is negligibly small (compared to the energy gain $\varDelta E = m_e c^2 \approx 5 \cdot 10^5\,\text{eV}$). This also applies to other particles and energies in a linear accelerator.

Storage ring

The motion of charged particles (velocity $|\boldsymbol{v}|$) on a circle is an accelerated motion and leads to the emission of electromagnetic radiation. This radiation is called *synchrotron radiation* after the accelerator type with this name. The radiation losses due to the circular acceleration are a central problem of circular accelerators and storage rings in high-energy physics. On the other hand synchrotron radiation is an important means of investigation for other branches of physics, in particular for solid-state physics.

The acceleration of a relativistic particle on a circular orbit (velocity $v = |\boldsymbol{v}| = \text{const.} \approx c$, radius R) is given by

$$\left|\frac{d\boldsymbol{v}}{dt}\right| = \frac{v^2}{R_o} \approx \frac{c^2}{R_o} \tag{23.42}$$

With this, Equation (23.38) becomes

$$P \approx \frac{2c\,q^2}{3}\,\frac{\gamma^{\,4}}{R_o^2} \tag{23.43}$$

The radiation losses increase with the fourth power of γ or the energy. As an example, we consider electrons ($m_e c^2 = 0.5\,\text{MeV}$) in the Hadron Electron Ring Facility (HERA) in Hamburg (in operation until 2007). HERA had a radius of $R_o \approx 1\,\text{km}$. For electrons with the energy $E = \gamma\, m_e c^2 = 30\,\text{GeV}$, we get $\gamma^{\,4} \approx 1.3 \cdot 10^{19}$. This means that the energy loss $\varDelta E_{\text{rad}}$ after one cycle reads

$$\varDelta E_{\text{rad}} = P\,\frac{2\pi R_o}{v} \approx \frac{4\pi}{3}\,\gamma^{\,4}\,\frac{e^2}{R_o} \approx 5.4 \cdot 10^{19}\,\frac{14.4\,\text{eV}}{\text{km}/\text{Å}} \approx 80\,\text{MeV} \tag{23.44}$$

This energy has to be transferred during a cycle time $2\pi R_o/c \approx 2 \cdot 10^{-5}\,\text{s}$. Only then the energy of the electron can be kept constant.

Consider a storage ring or the circular accelerator and particles with a given energy $E = \gamma\, m c^2$. Then, according to (23.43), the radius R_o is the only adjustable parameter. It may be increased in order to reduce the energy losses. The Large Electron Positron Storage ring (LEP) at CERN in Geneva, which operated until 2000, had a radius of about $R_o \approx 4.3\,\text{km}$ and reached particle energies of 100 GeV.

In 2010, the Large Hadron Collider (LHC) went into operation in the existing tunnel facility of the LEP. Depending on the operating mode, protons or lead nuclei are accelerated and brought to collision in the LHC (two beams travel in opposite directions, the circular motion is enforced by strong magnetic field maintained by superconducting electromagnets). Due to the greater mass of the hadrons, they lose less energy by synchrotron radiation and can therefore reach higher energies (like 13 TeV for protons). A major success of the LHC was the detection of the Higgs boson in 2012.

Due to the large radiation losses on circular orbits, linear accelerators are planned for still higher energies. Electrons and positrons with energies between 200 and 500 GeV are to collide in the planned International Linear Collider (ILC).

Exercises

23.1 *Retarded potentials of the uniformly moving charge*

Determine the retarded potentials Φ_{ret} and $\boldsymbol{A}_{\mathrm{ret}}$ of a uniformly moving point charge. Establish also the electromagnetic fields.

Chapter 24

Dipole Radiation

We calculate the radiation emitted by an oscillating charge distribution. The oscillating charge distribution can be, for example, an FM transmitting antenna or a classical electron on its circular orbit in an atom (Figure 24.1). In this chapter, we restrict ourselves to the non-relativistic limit.

The source distribution shall be periodic:

$$j^{\alpha}(\boldsymbol{r}', t) = \mathrm{Re}\left(j^{\alpha}(\boldsymbol{r}') \exp(-\mathrm{i}\omega t)\right) \tag{24.1}$$

For the spatial part $j^{\alpha}(\boldsymbol{r}')$, we do not introduce a new letter; moreover, we do not include the Re-sign in the following. We use this notation for the electromagnetic fields and their potentials, too. The charge distribution shall be spatially limited:

$$j^{\alpha}(\boldsymbol{r}') = \begin{cases} \text{arbitrary} & (r' < R_0) \\ 0 & (r' > R_0) \end{cases} \tag{24.2}$$

We want to determine the field at large distances r. To do this, we expand the retarded potentials into powers of R_0/r. The lowest contribution comes then from the dipole moment of the oscillating charge distribution.

We insert (24.1) in the retarded vector potential (17.33),

$$\begin{aligned} \boldsymbol{A}_{\mathrm{ret}}(\boldsymbol{r}, t) &= \frac{1}{c} \int d^3r' \, \frac{\boldsymbol{j}(\boldsymbol{r}', t - |\boldsymbol{r} - \boldsymbol{r}'|/c)}{|\boldsymbol{r} - \boldsymbol{r}'|} \\ &= \frac{1}{c} \exp(-\mathrm{i}\omega t) \int d^3r' \, \boldsymbol{j}(\boldsymbol{r}') \, \frac{\exp(\mathrm{i}k|\boldsymbol{r} - \boldsymbol{r}'|)}{|\boldsymbol{r} - \boldsymbol{r}'|} = \boldsymbol{A}(\boldsymbol{r}) \exp(-\mathrm{i}\omega t) \end{aligned} \tag{24.3}$$

with the wavenumber

$$k = \frac{\omega}{c} \tag{24.4}$$

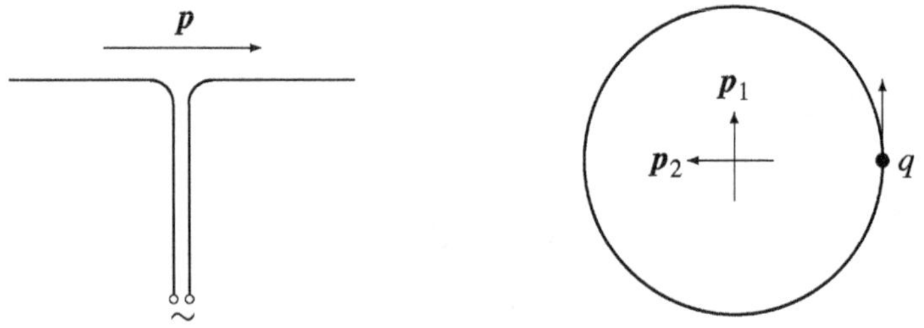

dipole antenna charge on circle

Figure 24.1 Two examples of an oscillating charge distribution (24.1): Due to a high frequency alternating current, the two wires shown on the left are charged in antiphase. The shape of the two wires is typical for an VHF antenna. The circular motion of a charge shown on the right can be understood as overlay of two oscillating dipoles, $\boldsymbol{p}_1$ and $\boldsymbol{p}_2$. We use this picture for the semi-classical calculation of the radiation by an atom.

The spatial part of the vector potential is

$$\boldsymbol{A}(\boldsymbol{r}) = \frac{1}{c}\int d^3r'\,\boldsymbol{j}(\boldsymbol{r}')\;\frac{\exp(\mathrm{i}k\,|\boldsymbol{r}-\boldsymbol{r}'|)}{|\boldsymbol{r}-\boldsymbol{r}'|} \tag{24.5}$$

Here and in the following we omit the index "ret". The spatial part of the magnetic field is $\boldsymbol{B}(\boldsymbol{r}) = \operatorname{curl}\boldsymbol{A}(\boldsymbol{r})$. Outside the charge distribution, the Maxwell equation $\partial_t \boldsymbol{E}(\boldsymbol{r},t) = c\,\operatorname{curl}\boldsymbol{B}(\boldsymbol{r},t)$ holds yielding $-\mathrm{i}\,\omega\,\boldsymbol{E}(\boldsymbol{r}) = c\,\operatorname{curl}\boldsymbol{B}(\boldsymbol{r})$. In the range $r > R_0$ all fields can therefore be expressed by (24.5):

$$\boldsymbol{B}(\boldsymbol{r}) = \nabla\times\boldsymbol{A}(\boldsymbol{r}),\quad \boldsymbol{E}(\boldsymbol{r}) = \frac{\mathrm{i}}{k}\;\nabla\times\boldsymbol{B}(\boldsymbol{r})\quad (r > R_0) \tag{24.6}$$

The time-dependent fields $\boldsymbol{E}(\boldsymbol{r},t)$ and $\boldsymbol{B}(\boldsymbol{r},t)$ result from this as in (24.1). The retarded potential Φ_{ret} is not used; due to the Lorenz gauge (17.10), it follows from $\boldsymbol{A}_{\text{ret}}$. From (24.6), the Poynting vector and thus the radiated power can be calculated.

For the following calculations we assume

$$\boxed{R_0 \ll \lambda \ll r \quad \text{(requirements)}} \tag{24.7}$$

where $\lambda = 2\pi/k = 2\pi c/\omega$. The condition $r \gg \lambda$ is fulfilled if the observation (measurement of the fields with a detector, reception of the radio waves with an antenna) is made at a sufficiently large distance r.

The condition $R_0 \ll \lambda$ restricts the charge distribution (24.1) by

$$R_0\,\omega \ll c \qquad \text{or} \qquad v_{\max} \ll c \tag{24.8}$$

The distribution (24.1) may contain charges that oscillate back and forth with frequency ω and amplitude R_0. Their maximum velocity is then $v_{\max} = R_0\,\omega$. The assumption (24.7) implies that we restrict ourselves to the non-relativistic case, $v \ll c$.

In the classical hydrogen atom, the electron moves with the speed $v_e \approx 10^{-2}\,c$, so that (24.8) is fulfilled. In the case of an FM or television antenna, the frequency ω is fixed. Then $R_0 \ll \lambda$ restricts the size of the dipole antenna. An antenna like the one on the left in Figure 24.1 is, however, is most effective for $R_0 \approx \lambda/4$. A smaller antenna radiates less power or absorbs less power as a receive antenna.

We now evaluate (24.5) under the restriction (24.7). In (24.5) the vectors $\boldsymbol{r}'$ are constrained by $r' \le R_0$. Due to $R_0 \ll r$, we can expand the distance $|\boldsymbol{r}-\boldsymbol{r}'|$ into powers of r':

$$|\boldsymbol{r}-\boldsymbol{r}'| = r + \sum_{i=1}^{3} \frac{\partial\,|\boldsymbol{r}-\boldsymbol{r}'|}{\partial x_i'}\bigg|_{r'=0} \cdot x_i' + \cdots = r - \boldsymbol{e}_r\cdot\boldsymbol{r}' + \cdots \tag{24.9}$$

The unit vector

$$\boldsymbol{e}_r = \frac{\boldsymbol{r}}{r} \tag{24.10}$$

points from the origin of the charge distribution to the location at which the fields are to be calculated. We expand the relevant parts of the integrand in (24.5):

$$\frac{1}{|\boldsymbol{r}-\boldsymbol{r}'|} = \frac{1}{r}\left(1 + \frac{\boldsymbol{e}_r\cdot\boldsymbol{r}'}{r} + \cdots\right) = \frac{1}{r}\Big(1 + \mathcal{O}(r'/r)\Big) \tag{24.11}$$

$$\exp(\mathrm{i}k\,|\boldsymbol{r}-\boldsymbol{r}'|) = \exp(\mathrm{i}kr)\,\exp\left(-\,\mathrm{i}k\,\boldsymbol{e}_r\cdot\boldsymbol{r}'\right)\Big(1 + \mathcal{O}(r'/r)\Big) \tag{24.12}$$

Neglecting the terms of the order r'/r, (24.5) becomes

$$\boldsymbol{A}(\boldsymbol{r}) = \frac{1}{c}\,\frac{\exp(\mathrm{i}kr)}{r}\int d^3r'\,\boldsymbol{j}(\boldsymbol{r}')\,\exp(-\mathrm{i}k\,\boldsymbol{e}_r\cdot\boldsymbol{r}') \tag{24.13}$$

So far we have utilized $r \gg R_0$. The other condition in (24.7), namely $\lambda \gg R_0$ enables the following *long wavelength approximation*:

$$\exp(-\mathrm{i}k\,\boldsymbol{e}_r\cdot\boldsymbol{r}') = 1 - \mathrm{i}k\,\boldsymbol{e}_r\cdot\boldsymbol{r}' \pm \cdots = 1 + \mathcal{O}(R_0/\lambda) \approx 1 \tag{24.14}$$

This gives us

$$A(r) = \frac{1}{c}\,\frac{\exp(\mathrm{i}kr)}{r}\int d^3r'\,j(r') \tag{24.15}$$

From the continuity equation follows

$$-\frac{\partial\varrho(r,t)}{\partial t} = \mathrm{div}\,j(r,t) \quad \stackrel{(24.1)}{\longrightarrow} \quad \mathrm{i}\omega\varrho(r) = \nabla\cdot j(r) \tag{24.16}$$

We reshape the integral in (24.15),

$$\int d^3r\,j(r) = \int d^3r\,(j\cdot\nabla)\,r \stackrel{\text{p.i.}}{=} -\int d^3r\,r\,(\nabla\cdot j) = -\mathrm{i}\omega\int d^3r\,r\,\varrho(r) \tag{24.17}$$

and introduce the dipole moment of the charge distribution:

$$p = \int d^3r\,r\,\varrho(r), \quad p(t) = \int d^3r\,r\,\varrho(r,t) = \mathrm{Re}\left(p\,\exp(-\mathrm{i}\omega t)\right) \tag{24.18}$$

The notation for p and $p(t)$ is used in the same way as for $\varrho(r)$ and $\varrho(r,t)$. We now insert (24.17) with (24.18) into (24.15):

$$A(r) = -\mathrm{i}kp\,\frac{\exp(\mathrm{i}kr)}{r} \tag{24.19}$$

This describes a *spherical wave*. The dipole moment p determines the amplitude of the wave, the frequency of the oscillating charge distribution determines the wavenumber $k = \omega/c$. If we had started in (24.3) with the advanced potential A_{av}, we would have obtained an incoming spherical wave with $\exp(-\mathrm{i}kr)/r$. In this case, the distribution (24.1) would absorb the radiation; it could be the current distribution in a receive antenna.

Due to

$$\frac{d}{dr}\,\frac{\exp(\mathrm{i}kr)}{r} = \frac{\exp(\mathrm{i}kr)}{r}\left(\mathrm{i}k - \frac{1}{r}\right) \approx \mathrm{i}k\,\frac{\exp(\mathrm{i}kr)}{r} \qquad (\lambda \ll r) \tag{24.20}$$

the nabla operator applied to (24.19) can be replaced by $\mathrm{i}ke_r$. With (24.19) and $\nabla \to \mathrm{i}ke_r$, we evaluate (24.6):

$$\boxed{\begin{aligned} B(r) &= k^2\,(e_r\times p)\,\frac{\exp(\mathrm{i}kr)}{r} \\ E(r) &= k^2\,(e_r\times p)\times e_r\,\frac{\exp(\mathrm{i}kr)}{r} \end{aligned} \qquad (R_0 \ll \lambda \ll r)} \tag{24.21}$$

The physical fields result if the factor $\exp(-\mathrm{i}\omega t)$ is added, and the real part is taken, for example

$$\boldsymbol{B}(\boldsymbol{r}, t) = k^2 \,\mathrm{Re}\left[(\boldsymbol{e}_r \times \boldsymbol{p})\, \frac{\exp(\mathrm{i}(kr - \omega t))}{r}\right] \tag{24.22}$$

In (24.1), $j^\alpha(\boldsymbol{r})$ is in general complex; this applies then also for the dipole moment $\boldsymbol{p}$.

The fields $\boldsymbol{E}$ and $\boldsymbol{B}$ from (24.21) are, as for the plane wave, are perpendicular to the direction of propagation $\boldsymbol{e}_r$ and perpendicular to each other. This is sketched in Figure 24.2 for $\boldsymbol{e}_r \perp \boldsymbol{p}$. At a large distance from the source, the wave can be locally approximated by a plane wave.

The energy flux density (energy per time and area) $\boldsymbol{S} = c\boldsymbol{E} \times \boldsymbol{B}/4\pi$ points in the direction of $\boldsymbol{e}_r = \boldsymbol{r}/r$, i.e., outwards from the oscillating charge distribution. The radiated power (energy per time) through the solid angle $d\Omega = d\cos\theta\, d\phi$ is then $dP = \langle \boldsymbol{S}\rangle\, \boldsymbol{e}_r\, r^2 d\Omega$. Using (20.43) we calculate the time averaged radiation power:

$$\begin{aligned}
\frac{dP}{d\Omega} &= \frac{c r^2 \boldsymbol{e}_r}{4\pi} \cdot \big\langle \boldsymbol{E}(\boldsymbol{r}, t) \times \boldsymbol{B}(\boldsymbol{r}, t)\big\rangle = \frac{c r^2 \boldsymbol{e}_r}{8\pi} \cdot \Big[\mathrm{Re}\, \boldsymbol{E}(\boldsymbol{r}) \times \boldsymbol{B}^*(\boldsymbol{r})\Big] \\
&= \frac{c k^4}{8\pi}\, \boldsymbol{e}_r \cdot \Big[\big((\boldsymbol{e}_r \times \boldsymbol{p}) \times \boldsymbol{e}_r\big) \times \big(\boldsymbol{e}_r \times \boldsymbol{p}^*\big)\Big] \\
&= \frac{\omega^4}{8\pi c^3} \big(\boldsymbol{e}_r \times \boldsymbol{p}\big) \cdot \big(\boldsymbol{e}_r \times \boldsymbol{p}^*\big) = \frac{\omega^4}{8\pi c^3} \big|\boldsymbol{e}_r \times \boldsymbol{p}\big|^2
\end{aligned} \tag{24.23}$$

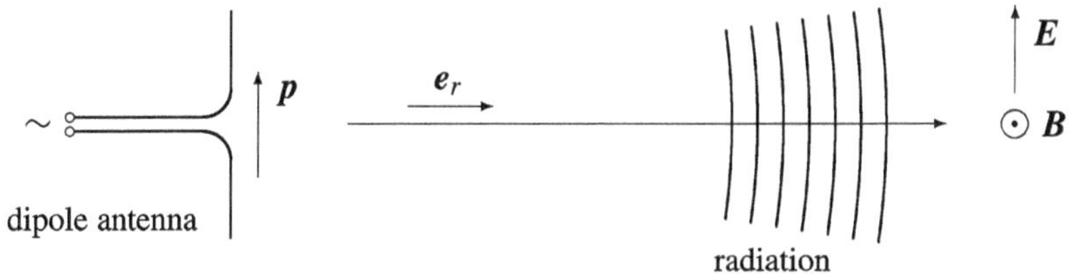

Figure 24.2 For the radiation field we have $\boldsymbol{E} \perp \boldsymbol{e}_r$, $\boldsymbol{B} \perp \boldsymbol{e}_r$ and $\boldsymbol{B} \perp \boldsymbol{E}$. The energy current density $\boldsymbol{S} = c\boldsymbol{E} \times \boldsymbol{B}/4\pi$ points in the direction of $\boldsymbol{e}_r$. For the picture, $\boldsymbol{e}_r$ was chosen in the direction of the maximum radiation power, $\boldsymbol{e}_r \perp \boldsymbol{p}$. Then $\boldsymbol{E} \parallel \boldsymbol{p}$ holds and the magnetic field $\boldsymbol{B}$ is perpendicular to the image plane ($\odot$). For optimum reception, a receiving antenna must be aligned parallel to $\boldsymbol{E}$.

The factors $\boldsymbol{e}_r$, $(\boldsymbol{e}_r \times \boldsymbol{p}) \times \boldsymbol{e}_r$ and $\boldsymbol{e}_r \times \boldsymbol{p}^*$ of the scalar triple product (2nd line) can be interchanged cyclically. This yields the scalar product of $(\boldsymbol{e}_r \times \boldsymbol{p}) \times \boldsymbol{e}_r$ with $(\boldsymbol{e}_r \times \boldsymbol{p}^*) \times \boldsymbol{e}_r$.

The asymptotic behavior $\boldsymbol{E} \propto 1/r$ and $\boldsymbol{B} \propto 1/r$ is characteristic for radiation fields. This behavior implies that $\boldsymbol{S} \cdot \boldsymbol{e}_r\, r^2$ becomes independent of r for $r \to \infty$. This means that the energy flow in a certain solid angle is independent of r. In contrast, the fields of an uniformly moving charge decrease with $1/r^2$; such a charge does not radiate.

The discussion of the angular dependence of (24.23) is simplified if we assume that all components of $\boldsymbol{p}$ have the same phase:

$$\begin{pmatrix} p_1 \\ p_2 \\ p_3 \end{pmatrix} = \begin{pmatrix} |p_1| \\ |p_2| \\ |p_3| \end{pmatrix} \exp(\mathrm{i}\delta) \quad \text{or} \quad \boldsymbol{p} = \boldsymbol{p}_\mathrm{r} \exp(\mathrm{i}\delta) \tag{24.24}$$

In general, a complex vector defines two different directions, namely $\mathrm{Re}\,\boldsymbol{p}$ and $\mathrm{Im}\,\boldsymbol{p}$. For (24.24) these two directions coincide. We introduce the angle θ between the (real) vectors $\boldsymbol{p}_\mathrm{r} = (|p_1|, |p_2|, |p_3|)$ and $\boldsymbol{e}_r$ by

$$\cos\theta = \frac{\boldsymbol{p}_\mathrm{r} \cdot \boldsymbol{e}_r}{|\boldsymbol{p}_\mathrm{r}|} \tag{24.25}$$

We evaluate (24.23) with (24.24) and (24.25):

$$\boxed{\frac{dP}{d\Omega} = \frac{\omega^4}{8\pi c^3}\, |\boldsymbol{p}|^2 \sin^2\theta \qquad \text{radiation of an oscillating dipole}} \tag{24.26}$$

Here, $|\boldsymbol{p}|^2 = \boldsymbol{p} \cdot \boldsymbol{p}^* = \boldsymbol{p}_\mathrm{r}^2$. The angular dependence of the dipole radiation is sketched in Figure 24.3.

The total radiated power is obtained by integrating over the full solid angle:

$$\boxed{P = \frac{\omega^4}{3c^3}\, |\boldsymbol{p}|^2 \qquad \text{dipole formula}} \tag{24.27}$$

The dipole radiation is characterized by the proportionality to ω^4 and to $|\boldsymbol{p}|^2$, and by the angular dependence with $\sin^2\theta$. The polarization of the radiation follows from (24.21), namely $\boldsymbol{E} \perp \boldsymbol{e}_r$, $\boldsymbol{B} \perp \boldsymbol{e}_r$ and $\boldsymbol{E} \perp \boldsymbol{B}$.

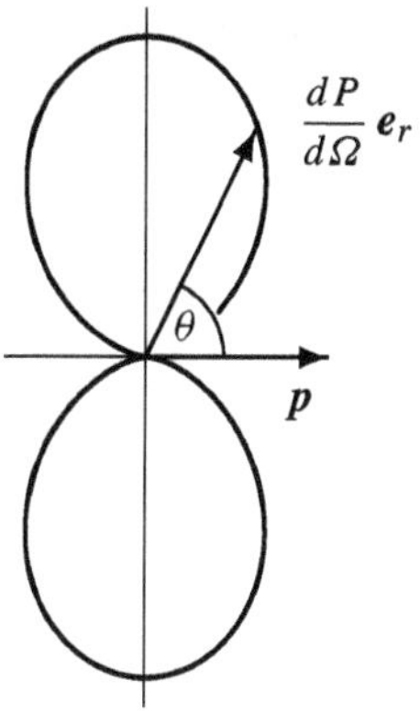

Figure 24.3 The angle between the oscillating dipole $\boldsymbol{p}$ and the direction of radiation $\boldsymbol{e}_r$ is denoted by θ. In the direction of $\boldsymbol{e}_r$, a vector with the magnitude $dP/d\Omega$ is plotted. The end points of this vector result in the shown curve.

Taking into account further terms in the expansion (24.14) leads to the magnetic dipole and the electric quadrupole (Exercise 24.4). For the magnetic dipole, we obtain analogously to (24.27),

$$P = \frac{\omega^4}{3c^3}\,\big|\boldsymbol{m}\big|^2 \quad \text{(magnetic dipole)} \tag{24.28}$$

The next term in the expansion (20.14) is proportional to kr'. This additional factor in the integral (24.18) leads to the quadrupole moment Q instead of the dipole moment. The accompanying factor $k = \omega/c$ is squared. This results in $P = \mathcal{O}(1)\,(\omega^6/c^5)\,Q^2$ for the radiation power of an oscillating quadrupole. This result can also be obtained by a superposition of two oscillating dipoles (Exercise 24.5).

The different radiation fields are also designated by E1 (electric dipole field), E2 (electric quadrupole field), M1 (magnetic dipole field) and so on.

Oscillating Point Charge

We consider a harmonic oscillation of a point charge. The trajectory $\boldsymbol{r}(t)$ and acceleration $\dot{\boldsymbol{v}}(t)$ are

$$\boldsymbol{r}(t) = \boldsymbol{r}_0\,\cos(\omega t), \qquad \dot{\boldsymbol{v}}(t) = -\boldsymbol{r}_0\,\omega^2\cos(\omega t) \tag{24.29}$$

This charge generates the time-dependent dipole moment

$$\boldsymbol{p}(t) = \int d^3r\; \boldsymbol{r}\,\varrho(\boldsymbol{r},t) = q\,\boldsymbol{r}_0\cos(\omega t) = \mathrm{Re}\Big(q\,\boldsymbol{r}_0\exp(-\mathrm{i}\omega t)\Big) \tag{24.30}$$

This is of the form assumed in (24.18). We insert the time-independent amplitude $\boldsymbol{p} = q\,\boldsymbol{r}_0$ in (24.27),

$$P = \frac{\omega^4 q^2 r_0^{\,2}}{3\,c^3} = \frac{2\,q^2}{3\,c^3}\left\langle \dot{\boldsymbol{v}}^2\right\rangle \tag{24.31}$$

According to (23.32), $P = (2q^2/3c^3)\,\dot{\boldsymbol{v}}^2$ applies for an arbitrarily accelerated point charge with $v \ll c$. In this chapter, the result was found under additional assumptions (harmonic motion and time averaging).

For a charge moving with the angular velocity ω on a circle (radius R), we get $|\dot{\boldsymbol{v}}| = \omega^2 R$. This turns (24.31) into

$$P = \frac{2\,\omega^4}{3\,c^3}\,q^2 R^2 \quad \text{(charge on a circular path)} \tag{24.32}$$

In Exercise 24.2 it is shown that the circular motion can be understood as a superposition of two oscillating dipoles (Figure 24.1, right).

Lifetime of Atomic States

We start from the semi-classical Bohr model of the hydrogen atom: An electron (mass m_{e}, charge $-e$) orbits a proton (mass m_{p}, charge e), whereby the angular momentum is required to be a multiple of $\hbar$. For a circular orbit, the force equilibrium and the angular momentum quantization read

$$\frac{m_{\mathrm{e}}\,v^2}{r} = \frac{e^2}{r^2}, \qquad m_{\mathrm{e}}\,v\,r = \hbar\,n \tag{24.33}$$

The reduced mass was approximated by m_{e} (because of $m_{\mathrm{p}} \gg m_{\mathrm{e}}$). The quantum number n may adopt the values 1, 2, ... For $n = 1$ we obtain from (24.33)

$$r = a_{\mathrm{B}} = \frac{\hbar^2}{m_{\mathrm{e}}\,e^2}, \qquad v = v_{\mathrm{at}} = \frac{e^2}{\hbar} = \frac{e^2}{\hbar c}\,c = \alpha\,c \tag{24.34}$$

Here we have introduced *Bohr's radius* a_{B} and the *fine structure constant* α:

$$a_{\mathrm{B}} \approx 0.53\,\text{Å}, \qquad \alpha = \frac{e^2}{\hbar c} \approx \frac{1}{137} \tag{24.35}$$

Due to $v/c = \alpha \approx 10^{-2}$, the condition $R/\lambda \ll 1$ is fulfilled. For transitions in the visible range, we obtain $R/\lambda \approx a_B/\lambda_{\text{visible}} \approx 10^{-4}$. From (24.34) follows the angular frequency

$$\omega_{\text{at}} = \frac{v_{\text{at}}}{a_B} = \frac{m_e e^4}{\hbar^3} \approx 4 \cdot 10^{16}\ \text{s}^{-1} \tag{24.36}$$

For the circular motion, we insert $q = -e$, $\omega = \omega_{\text{at}}$ and $R = a_B$ in (24.32):

$$P = \frac{2}{3}\frac{e^2}{a_B}\,\alpha^3\,\omega_{\text{at}} \tag{24.37}$$

Here we have used $\omega_{\text{at}} = v_{\text{at}}/a_B = \alpha\, c/a_B$.

Without quantization of the orbits, the electron would continuously emit this power and would eventually submerge onto the proton. Within the framework of Bohr's model, the orbits are restricted by the quantization condition $m\,v\,r = \hbar\,n$. Therefore, only orbits with the discrete energies

$$E_n = \frac{e^2}{a_B}\frac{1}{2n^2} = \frac{E_{\text{at}}}{2n^2} = \frac{\hbar\,\omega_{\text{at}}}{2n^2} \qquad (n = 1, 2, \ldots) \tag{24.38}$$

are allowed. The energy scale is characterized by the atomic energy unit

$$E_{\text{at}} = \hbar\,\omega_{\text{at}} = \frac{e^2}{a_B} = \alpha^2\,m_e c^2 \approx 27.2\,\text{eV} \tag{24.39}$$

An excited atomic state (with an electron orbit with $n \geq 2$) may jump into a lower state by radiating a photon. Therefore, the excited state has a finite lifetime τ. The lifetime τ can be estimated by the time it takes to emit the required energy:

$$\tau \sim \frac{E_{\text{at}}}{P} \approx \frac{1}{\alpha^3}\frac{1}{\omega_{\text{at}}} \tag{24.40}$$

Roughly speaking, it takes about $\alpha^{-3} \approx 10^6$ orbit cycles to radiate this energy. With $\omega_{\text{at}} \approx 4 \cdot 10^{16}\,\text{s}^{-1}$ Equation (24.40) yields the value $\tau \sim 10^{-10}\,\text{s}$.

The quantum mechanical calculation also gives a result of the form (24.40). The frequency of the emitted photon is given by the energy difference between the two states:

$$\omega_{\text{at}} \;\rightarrow\; \omega_{mn} = \frac{E_m - E_n}{\hbar} \tag{24.41}$$

The frequencies ω_{mn} can be significantly smaller than ω_{at}. For visible light, we have $\hbar\,\omega_{mn} \approx 2...3\,\text{eV}$ as compared to $\hbar\,\omega_{\text{at}} \approx 27\,\text{eV}$. This means that the quantum mechanical (as well as the actual) lifetimes of excited atomic states are thus significantly longer than the estimate (24.40) suggests. They are of the size

$$\tau \sim 10^{-8}\,\mathrm{s} \tag{24.42}$$

The photon emitted during this time corresponds to a wave packet with the extension

$$l_{\mathrm{c}} \sim c\,\tau \approx 3\,\mathrm{m} \tag{24.43}$$

This length is also called coherence length. Waves are *coherent* if their phases are in a fixed relation to each other. This is the case for different parts of a wave packet of length l_{c}, or for the two wave packets that result from the scattering of light at a double slit (Chapter 36). Natural light consists of a mixture of wave packets of size l_{c}. The centers, phases and polarizations of these wave packets are statistically distributed; the mixture is therefore incoherent.

Radiation Force

According to (18.47), a charged particle in a constant, homogeneous magnetic field moves on a circular path. Such a motion would imply that the energy of the particle is constant. However, the particle actually emits electromagnetic radiation with the power (24.32) because the circular motion is an accelerated one. In the non-relativistic limiting case we supplement the equation of motion by a *radiation force* $\boldsymbol{F}_{\mathrm{rad}}$,

$$m\,\dot{\boldsymbol{v}} = \boldsymbol{F}_{\mathrm{ext}} + \boldsymbol{F}_{\mathrm{rad}} \tag{24.44}$$

This radiation force shall describe the energy loss due to the radiation on the accelerated particle. The force $\boldsymbol{F}_{\mathrm{ext}}$ describes other, external forces. In the example considered, $\boldsymbol{F}_{\mathrm{ext}} = q\,(\boldsymbol{v}/c) \times \boldsymbol{B}$ is the force due to the homogeneous magnetic field $\boldsymbol{B}$.

We now determine the radiation force $\boldsymbol{F}_{\mathrm{rad}}$ in such a way that the energy balance is satisfied on average. A force $\boldsymbol{F}$ transfers the energy $\boldsymbol{F} \cdot d\boldsymbol{r}$ to the particle if it moves by $d\boldsymbol{r}$. This means that $\boldsymbol{F} \cdot \boldsymbol{v}$ is the power (energy per time) transferred to the particle. The energy emitted by the particle per time due to the radiation force is therefore $-\boldsymbol{F}_{\mathrm{rad}} \cdot \boldsymbol{v}$. We assume a harmonic motion and require that the radiation force equals in the (average) radiated power (24.31):

$$\frac{2q^2}{3c^3}\left\langle \dot{\boldsymbol{v}}^2 \right\rangle = -\langle \boldsymbol{F}_{\rm rad} \cdot \boldsymbol{v} \rangle \tag{24.45}$$

The averaging is performed by integration over one oscillation period $T = 2\pi/\omega$:

$$\left\langle \dot{\boldsymbol{v}}^2 \right\rangle = \frac{1}{T}\int_0^T dt\; \dot{\boldsymbol{v}} \cdot \dot{\boldsymbol{v}} = -\frac{1}{T}\int_0^T dt\; \ddot{\boldsymbol{v}} \cdot \boldsymbol{v} = -\left\langle \ddot{\boldsymbol{v}} \cdot \boldsymbol{v} \right\rangle \tag{24.46}$$

The boundary terms due to the partial integration disappear because the motion is periodic. It follows from the last two equations that

$$\boldsymbol{F}_{\rm rad} = \frac{2q^2}{3c^3}\,\ddot{\boldsymbol{v}} \tag{24.47}$$

guarantees the *average* energy balance. To ensure that the deviation from the periodic motion caused by $\boldsymbol{F}_{\rm rad}$ is small, the condition

$$\left|\boldsymbol{F}_{\rm rad}\right| \ll \left|\boldsymbol{F}_{\rm ext}\right| \tag{24.48}$$

must be fulfilled. The assumed periodic motion excludes, for example, $\dot{\boldsymbol{v}} = \text{const.}$ (which would result in $\boldsymbol{F}_{\rm rad} = 0$ although radiation is emitted). Also $\boldsymbol{F}_{\rm ext} = 0$ is not permitted; otherwise this would lead to nonsensical solutions of the equation of motion (24.44), see Exercise 24.7.

The force $\boldsymbol{F}_{\rm rad}$ has its origin in the radiation. It therefore results from the interaction of the time-dependent charge distribution with the generated electromagnetic radiation field. For the point charge considered here, the radiation field (i.e., $\boldsymbol{A}_{\rm ret}$ or also $\boldsymbol{A}_{\rm av}$) is, however, infinite at the point of the charge. The attempt to determine the retroactive effect of the radiation field on the accelerated charge therefore leads to problems. This is comparable to the problem of the infinite field energy of a point charge in electrostatics.

Exercises

24.1 *Periodic charge density*

Consider a periodic charge density $\varrho(\boldsymbol{r}, t+T) = \varrho(\boldsymbol{r}, t)$. Write the charge density in the form $\varrho(\boldsymbol{r}, t) = \text{Re}\sum_n \varrho_n(\boldsymbol{r}) \exp(-\mathrm{i}\,\omega_n t)$, and determine the quantities $\varrho_n(\boldsymbol{r})$ and ω_n.

24.2 *Charged particle on a circular orbit*

A particle with the charge q moves with the angular velocity ω on a circle (radius $R \ll c/\omega$). The charge density is then

$$\varrho(\boldsymbol{r}, t) = q\,\delta\big(x - R\,\cos(\omega t + \alpha)\big)\,\delta\big(y - R\,\sin(\omega t + \alpha)\big)\,\delta(z)$$

Calculate the dipole moment $\boldsymbol{p}(t)$ of this charge density and determine the complex amplitude $\boldsymbol{p}$ of $\boldsymbol{p}(t) = \mathrm{Re}\big[\boldsymbol{p}\,\exp(-\mathrm{i}\,\omega t)\big]$. Calculate the radiation power $dP/d\Omega$ and P.

24.3 *Charged particles moving on a circle*

On a circle (radius R) there are N equidistantly distributed charges q. They move around the circle with the angular velocity ω (where $R \ll c/\omega$). The charge density is then

$$\varrho(\boldsymbol{r}, t) = q \sum_{\nu=0}^{N-1} \big[\,\delta\big(x - R\,\cos(\omega t + \alpha_\nu)\big)\,\delta\big(y - R\,\sin(\omega t + \alpha_\nu)\big)\big]\,\delta(z)$$

where $\alpha_\nu = 2\pi\nu/N$ (with $\nu = 0, 1, \ldots N-1$ for $N \geq 2$). Show that this configuration does not emit dipole radiation. To do this, use the results from Exercise 24.2 and the superposition principle.

24.4 *Magnetic dipole and electric quadrupole radiation*

The radiation of an oscillating charge distribution $\varrho(\boldsymbol{r}, t) = \varrho(\boldsymbol{r})\,\exp(-\mathrm{i}\,\omega t)$ shall be investigated. The charge distribution shall have no electric dipole moment. Calculate the contributions to the vector potential

$$\boldsymbol{A}(\boldsymbol{r}) = \frac{1}{c}\,\frac{\exp(\mathrm{i}\,kr)}{r} \int d^3r'\,\boldsymbol{j}(\boldsymbol{r}')\,\exp(-\mathrm{i}\,k\,\boldsymbol{e}_r\cdot\boldsymbol{r}')$$

in the far zone. Using

$$\frac{1}{c}\,(\boldsymbol{e}_r\cdot\boldsymbol{r}')\boldsymbol{j}(\boldsymbol{r}') = \frac{1}{2c}\Big[(\boldsymbol{e}_r\cdot\boldsymbol{r}')\boldsymbol{j}(\boldsymbol{r}') + \big(\boldsymbol{e}_r\cdot\boldsymbol{j}(\boldsymbol{r}')\big)\,\boldsymbol{r}'\Big] + \frac{1}{2c}\,\big(\boldsymbol{r}'\times\boldsymbol{j}(\boldsymbol{r}')\big)\times\boldsymbol{e}_r$$

one obtains two contributions, $\boldsymbol{A}(\boldsymbol{r}) = \boldsymbol{A}_{\mathrm{el}}(\boldsymbol{r}) + \boldsymbol{A}_{\mathrm{mag}}(\boldsymbol{r})$. Verify the relation

$$\frac{1}{2c}\int d^3r'\,\big[(\boldsymbol{e}_r\cdot\boldsymbol{r}')\boldsymbol{j}(\boldsymbol{r}') + \big(\boldsymbol{e}_r\cdot\boldsymbol{j}(\boldsymbol{r}')\big)\,\boldsymbol{r}'\big] = -\frac{\mathrm{i}\,k}{2}\int d^3r'\,\varrho(\boldsymbol{r}')\,(\boldsymbol{e}_r\cdot\boldsymbol{r}')\,\boldsymbol{r}'$$

and simplify the electrical part. Calculate the electromagnetic radiation fields and the radiated power.

24.5 *Antenna with alternating voltage*

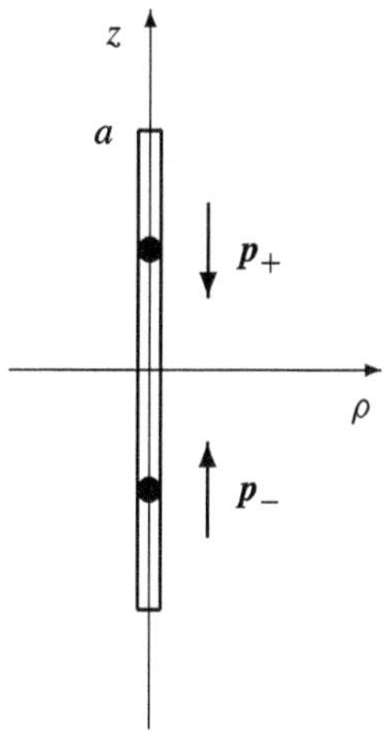

In a wire of length $2a$, an alternating voltage generates the oscillating charge distribution

$$\varrho(\boldsymbol{r}, t) = \varrho(\boldsymbol{r}) \exp(-\mathrm{i}\,\omega t)$$

$$\varrho(\boldsymbol{r}) = \frac{q}{2a}\ \delta(x)\ \delta(y)\ \cos(\pi z/a)\ \Theta\big(a - |z|\big)$$

Here $a \ll c/\omega$ is assumed.

How large is the dipole moment of the charge distribution? Replace the charge distribution by two dipoles and superimpose the two radiation fields for $r \gg \lambda$. Determine $\boldsymbol{E}$, $\boldsymbol{B}$ and the radiated power $dP/d\Omega$ and P.

24.6 *Antenna grid*

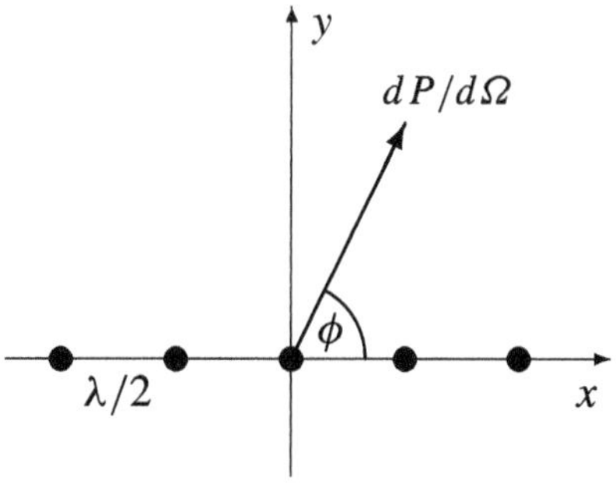

Along the x-axis, $(2N + 1)$ antennas are arranged at distances of $a = \lambda/2$. Each antenna radiates as a dipole with the frequency $\omega = 2\pi/\lambda$. All dipole vectors are aligned in the z-direction. Calculate the radiated power $dP/d\Omega$ for large distances $|\boldsymbol{r}| \gg \lambda$ as a function of the angles θ and ϕ.

Now divide the power $dP/d\Omega$ of the $2N + 1$ in-phase (i.e., coherent) radiating dipoles by $(2N+1)(dP/d\Omega)_1$, where $(dP/d\Omega)_1$ is the radiation of a dipole at the origin. In the x-y plane (i.e., for $\theta = \pi/2$), this ratio depends only on the angle ϕ. Sketch and discuss this angle dependence dependency for $N = 0$, 1 and 2.

24.7 *Equation of motion with radiation force*

Show that the equation of motion $m\,\dot{\boldsymbol{v}} = \boldsymbol{F}_{\text{ext}} + \boldsymbol{F}_{\text{rad}}$ with the radiation force $\boldsymbol{F}_{\text{rad}} = 2\,q^2\,\ddot{\boldsymbol{v}}/(3\,c^3)$ and with vanishing external force $\boldsymbol{F}_{\text{ext}}$ leads to nonsensical solutions.

Chapter 25

Scattering of Light

In the field of an incoming electromagnetic wave, a charged particle oscillates back and forth. This oscillating particle itself then emits electromagnetic radiation. This process means that the incoming electromagnetic wave is scattered. We calculate the cross-section for the scattering of light by atoms (Figure 25.1). We deal with the Thomson scattering, the Rayleigh scattering, the (recoilless) resonance fluorescence and the transition between coherent and incoherent scattering.

An electron (charge $q = -e$, mass m_e) moves in a harmonic oscillator potential (frequency ω_0, damping Γ). In addition, the fields

$$\boldsymbol{E}(\boldsymbol{r}, t) = \mathrm{Re}\, \boldsymbol{E}_0 \exp\big[\mathrm{i}(\boldsymbol{k} \cdot \boldsymbol{r} - \omega t)\big] \tag{25.1}$$

and $\boldsymbol{B}(\boldsymbol{r}, t) = (\boldsymbol{k}/k) \times \boldsymbol{E}$ of an electromagnetic wave (20.21) act on the electron. The equation of motion for the trajectory $\boldsymbol{r}_0(t)$ of the electron reads

$$m_e \ddot{\boldsymbol{r}}_0 + m_e \Gamma \dot{\boldsymbol{r}}_0 + m_e \omega_0^2 \boldsymbol{r}_0 = -e\, \boldsymbol{E}_0 \exp\big[\mathrm{i}(\boldsymbol{k} \cdot \boldsymbol{r}_0 - \omega t)\big] + \mathcal{O}(v/c) \tag{25.2}$$

The velocity $\boldsymbol{v} = \dot{\boldsymbol{r}}_0(t)$ shall be non-relativistic, $v \ll c$. Then the force of the magnetic field is of relative size $v/c \ll 1$; it will be neglected. From equation (25.2), the real part is to be taken; the real part formation is not written on explicitly.

The oscillator model (25.2) can be applied to an electron bound in the atom where ω_0 equals frequency $\omega_{at} = m_e e^4/\hbar^3$, (24.36). This results in a model for the scattering of light by atoms, Figure 25.1.

The frictional force $-m_e \Gamma \dot{\boldsymbol{r}}_0$ in (25.2) describes all processes by which the electron loses energy. Such processes are collisions of an atom with

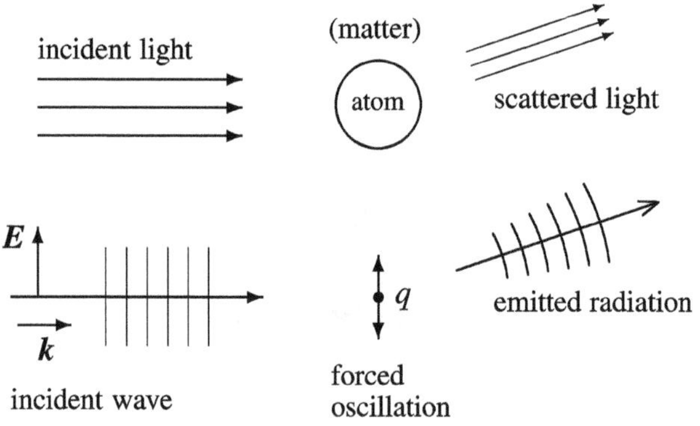

Figure 25.1 Light can be scattered by matter (top). We investigate scattering processes that are caused by individual atoms. In the theoretical treatment (below), the incident light is described by a plane electromagnetic wave. A harmonically bound particle simulates the electron bound in the atom. The electromagnetic forces of the wave excite oscillations of the charged particle. The oscillating particle represents an oscillating dipole. This dipole radiates in different directions according to the dipole formula.

other atoms or the radiation emitted by the electron. For the second case, we can use $\varGamma$ from the radiation force (24.47). Compared with the other forces, the radiation force is small; its effect is proportional to $\alpha^3 \sim 10^{-6}$, see (24.37). In such a small term, we can use the undisturbed solution $\boldsymbol{r}_0(t) = \boldsymbol{a}\,\exp(-\mathrm{i}\omega_0 t)$, i.e.,

$$\boldsymbol{F}_{\text{rad}} = \frac{2e^2}{3c^3}\,\ddot{\boldsymbol{v}} \approx -\frac{2e^2}{3c^3}\,\omega_0^2\,\boldsymbol{v} = -m_{\text{e}}\,\varGamma_{\text{rad}}\,\boldsymbol{v} \tag{25.3}$$

In the last step we have set $\boldsymbol{F}_{\text{rad}}$ equal to the damping force in (25.2) with $\varGamma = \varGamma_{\text{rad}}$. From this we obtain

$$\varGamma_{\text{rad}} = \frac{2e^2\omega_0^2}{3m_{\text{e}}c^3}, \qquad \varGamma_{\text{rad}} \overset{\omega_0 = \omega_{\text{at}}}{=} \frac{2}{3}\,\alpha^3\,\omega_{\text{at}} \sim 10^{-6}\,\omega_{\text{at}} \tag{25.4}$$

In the following, we use of $\varGamma$ and ω_0 as model parameters, and leave theirs values open for the time being.

The wavelength $\lambda = 2\pi c/\omega$ of the incident light shall be much larger than the amplitudes of the particle,

$$\lambda \gg r_0(t) \quad \text{or} \quad \lambda \gg a_{\text{B}} \tag{25.5}$$

For the electron in the atom, the wavelength must be large compared to Bohr's radius a_{B}, for the free electron ($\omega_0 = 0$), the deflection r_0 caused by the wave must not be too large. With

$$\exp(\mathrm{i}\boldsymbol{k}\cdot\boldsymbol{r}_0) = 1 + \mathcal{O}(r_0/\lambda) \approx 1 \tag{25.6}$$

(25.2) becomes

$$m_\mathrm{e}\,\ddot{\boldsymbol{r}}_0 + m_\mathrm{e}\,\Gamma\,\dot{\boldsymbol{r}}_0 + m_\mathrm{e}\,\omega_0^2\,\boldsymbol{r}_0 = -e\,\boldsymbol{E}_0\,\exp(-\mathrm{i}\omega t) \tag{25.7}$$

This linear, inhomogeneous differential equation describes the well-known case of a damped harmonic oscillator with a periodic force. The general solution is the sum of the general homogeneous solution and a particular solution:

$$\boldsymbol{r}_0(t) = \boldsymbol{r}_{0,\,\mathrm{hom}}(t) + \boldsymbol{r}_{0,\,\mathrm{part}}(t) \tag{25.8}$$

The homogeneous solution decays due to the damping, $\boldsymbol{r}_{0,\mathrm{hom}} \approx 0$ for $t \gg 1/\Gamma$. We therefore only consider the particular solution, i.e., the forced oscillation. The approach

$$\boldsymbol{r}_0(t) = \boldsymbol{r}_{0,\mathrm{part}}(t) = \boldsymbol{a}\,\exp(-\mathrm{i}\omega t) \tag{25.9}$$

leads in (25.7) to

$$\left(-\omega^2 - \mathrm{i}\,\Gamma\omega + \omega_0^2\right)\boldsymbol{a} = -\frac{e}{m_\mathrm{e}}\,\boldsymbol{E}_0 \tag{25.10}$$

From this we obtain the time-dependent dipole moment $\boldsymbol{p}(t)$ of the of the electron:

$$\boldsymbol{p}(t) = -e\,\boldsymbol{r}_0(t) = -e\,\boldsymbol{a}\,\exp(-\mathrm{i}\omega t) = \frac{e^2\,\boldsymbol{E}_0/m_\mathrm{e}}{\omega_0^2 - \omega^2 - \mathrm{i}\,\Gamma\omega}\,\exp(-\mathrm{i}\omega t) \tag{25.11}$$

The dipole moment $\boldsymbol{p}(t)$ is proportional to the field external $\boldsymbol{E} = \boldsymbol{E}_0\,\exp(-\mathrm{i}\omega t)$; it is *induced* by the field. The proportionality factor

$$\boxed{\alpha_\mathrm{e}(\omega) = \frac{e^2/m_\mathrm{e}}{\omega_0^2 - \omega^2 - \mathrm{i}\,\Gamma\omega}} \tag{25.12}$$

in $\boldsymbol{p} = \alpha_\mathrm{e}\,\boldsymbol{E}$ ist the electric *polarizability*. This quantity plays an important role for electrodynamics in matter (Part VI).

We assume that all components of $\boldsymbol{E}_0$ have the same phase so that (24.24) is fulfilled. We substitute (25.11) into (24.26):

$$\frac{dP}{d\Omega} = \frac{\omega^4\,|\boldsymbol{p}|^2\,\sin^2\theta}{8\pi\,c^3} = \frac{c}{8\pi}\left(\frac{e^2}{m_e c^2}\right)^2 \frac{\omega^4\,\sin^2\theta}{(\omega_0^2-\omega^2)^2+\omega^2\,\Gamma^2}\,|\boldsymbol{E}_0|^2 \tag{25.13}$$

Here θ is the angle between $\boldsymbol{E}_0$ and the direction of the emitted radiation:

$$\theta = \sphericalangle(\boldsymbol{p},\boldsymbol{e}_r) = \sphericalangle(\boldsymbol{E}_0,\boldsymbol{e}_r) \tag{25.14}$$

The incident light induces a time-dependent dipole moment; this induced dipole moment radiates with the same frequency. Therefore, this process means an *elastic scattering* of the light. The term "elastic" means "without energy loss", in the case under consideration therefore "without change in frequency". In contrast to this, the Compton effect is the inelastic scattering of light by free electrons.

We define the differential *cross-section* by

$$\frac{d\sigma}{d\Omega} = \frac{\text{scattered particles/time}/d\Omega}{\text{incident particles/time/area}} = \frac{\text{radiated power per } d\Omega}{\text{incident power per area}} = \frac{dP/d\Omega}{\langle|\boldsymbol{S}|\rangle} \tag{25.15}$$

We referred to the usual definition by "particles per time". If we multiply the numerator and denominator with the energy $\hbar\omega$ of a photon, one obtains "energy per time" or "power" instead of "particle per time" According to (20.45), (20.44) the incident energy flux density (power per area) is

$$\langle|\boldsymbol{S}|\rangle = c\,\langle w_{\text{em}}\rangle = \frac{c}{8\pi}\,|\boldsymbol{E}_0|^2 \tag{25.16}$$

This gives us

$$\frac{d\sigma}{d\Omega} = \frac{dP/d\Omega}{\langle|\boldsymbol{S}|\rangle} = \left(\frac{e^2}{m_e c^2}\right)^2 \frac{\omega^4}{(\omega_0^2-\omega^2)^2+\omega^2\,\Gamma^2}\,\sin^2\theta \tag{25.17}$$

The angle θ is defined by (25.14). From (25.17) we obtain the total cross-section for the scattering of light by an electron:

$$\boxed{\sigma = \sigma(\omega) = \frac{8\pi}{3}\left(\frac{e^2}{m_e c^2}\right)^2 \frac{\omega^4}{(\omega_0^2-\omega^2)^2+\omega^2\,\Gamma^2}} \tag{25.18}$$

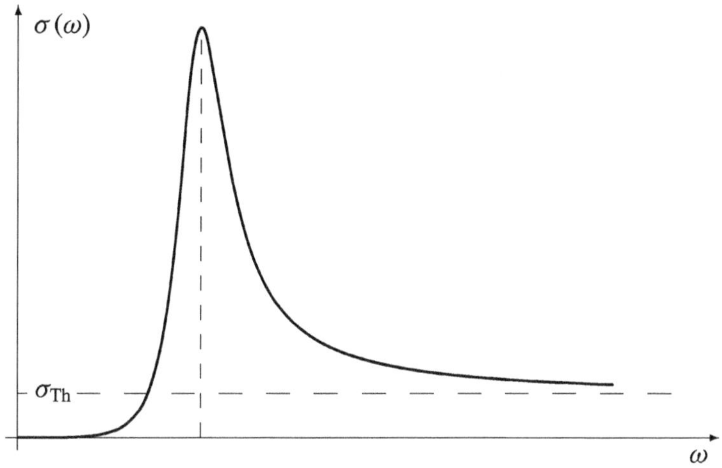

Figure 25.2 The cross-section $\sigma(\omega)$ for the scattering of light by atoms was calculated in the oscillator model (25.2). The frequency dependence is shown for $\Gamma/\omega_0 = 1/3$. Deviating from this sketch, the ratio Γ/ω_0 is usually much smaller; the resonance curve is then correspondingly narrower and higher. If Γ contains the radiation damping (25.4) only, then $\Gamma_{\text{rad}}/\omega_{\text{at}} \sim \alpha^3 \sim 10^{-6}$. The very sharp resonance curve then has its *natural line width.*

For specific parameter values, the function $\sigma(\omega)$ is shown in Figure 25.2. The cross-section σ is the area at which the incident current density (here energy current density) is effectively scattered. This area is determined by the factor $(e^2/m_\text{e}c^2)^2$, which is given numerically in the following.

Polarization of the Scattered Light

We discuss the polarization of the scattered light. For this we look again at Figure 25.1 The induced dipole moment $\boldsymbol{p}$ is parallel to the field vector $\boldsymbol{E}$ of the incident wave. According to (24.21), there is no radiation in the direction of $\boldsymbol{p}$, i.e., in the direction “up” in the figure. However, if the field vector $\boldsymbol{E}$ is perpendicular to the image plane, this also applies to $\boldsymbol{p}$, and the radiation in the “up” direction does not disappear. This means for incoming non-polarized light: For a scattering angle of 90° the scattered light is polarized. The field vector $\boldsymbol{E}_{\text{rad}}$ of the scattered light is perpendicular on the plane formed by $\boldsymbol{k}$ (incident wave) and $\boldsymbol{e}_r$ (outgoing direction). For other scattering angles one obtains partially polarized light.

Thomson Scattering

We consider some limiting cases of the cross-section (25.18). The case of very high frequencies is called *Thomson scattering* is designated:

$$\sigma_{\mathrm{Th}} = \frac{8\pi}{3}\left(\frac{e^2}{m_{\mathrm{e}}c^2}\right)^2 = 0.665 \cdot 10^{-24}\,\mathrm{cm}^2 \quad \text{Thomson scattering } (\omega \gg \omega_0) \tag{25.19}$$

For high frequencies, the electron follows the alternating field of the wave less and less, $r_0 \propto 1/\omega^2$ and thus $p^2 \propto 1/\omega^4$. Since the radiated dipole power P increases with ω^4, the cross-section approaches the finite value σ_{Th}.

Quantum electrodynamics (QED) yields for free, non-relativistic electrons ($\omega_0 = 0$, $v/c \ll 1$):

$$\sigma\,(\mathrm{QED}) = \sigma_{\mathrm{Th}}\big(1 + \mathcal{O}(v/c)\big) \quad (\text{free electrons, } \hbar\omega \ll m_{\mathrm{e}}c^2) \tag{25.20}$$

For sufficiently high frequencies, the precondition (25.5) of our derivation is no longer fulfilled; (25.20) does not apply either. Experimentally, the elastic effective cross-section falls off for frequencies $\hbar\omega \sim m_{\mathrm{e}}c^2$ and approaches zero for $\omega \to \infty$. This behavior — as well as all other experiments with electrons and photons — are correctly described by the relativistic QED.

If one sets σ_{Th} equal to the geometric cross-section for the scattering on a sphere with the radius R_0, the result is

$$\sigma_{\mathrm{Th}} = \pi R_0^2 \quad \longrightarrow \quad R_0 \approx 4.6\,\mathrm{fm} = \mathcal{O}(R_{\mathrm{e}}) \tag{25.21}$$

Apart from numerical factors, R_0 is equal to the classical electron radius $R_{\mathrm{e}} = 3e^2/(5m_{\mathrm{e}}c^2) \approx 1.7\,\mathrm{fm}$ from (6.35).

Rayleigh Scattering

For small frequencies, (25.18) yields the limiting case of *Rayleigh scattering*

$$\sigma = \sigma_{\mathrm{Th}}\,\frac{\omega^4}{\omega_0^4} \quad \text{Rayleigh scattering } (\omega \ll \omega_0) \tag{25.22}$$

Due to

$$\frac{\hbar\,\omega_{\text{visible}}}{\hbar\,\omega_{\text{at}}} \approx 0.1 \tag{25.23}$$

we can apply (25.22) to the scattering of visible light by atoms. The ω^4-dependence implies that blue sunlight in the atmosphere is scattered more strongly than red light:

$$\frac{\sigma_{\text{blue}}}{\sigma_{\text{red}}} = \frac{\omega_{\text{blue}}^4}{\omega_{\text{red}}^4} \approx 10 \tag{25.24}$$

Here we have $\omega_{\text{blue}}/\omega_{\text{red}} = \lambda_{\text{red}}/\lambda_{\text{blue}} \approx 1.8$ is used. The scattered light contains about ten times more blue light than red light. This is why the sky appears blue. A sunset, on the other hand, can appear red because the blue component is reduced by the scattering.

Resonance Fluorescence

At $\omega = \omega_0$ the cross-section (25.18) has a maximum of the strength

$$\boxed{\sigma_{\text{res}} = \frac{8\pi}{3}\left(\frac{e^2}{m_e c^2}\right)^2 \frac{\omega_0^2}{\Gamma^2} \qquad \text{resonance scattering } (\omega = \omega_0)} \tag{25.25}$$

This limiting case arises when radiation from a specific atomic transition is absorbed again by an atom of the same type. This *resonance fluorescence* can also occur for transitions in the atomic nucleus (Figure 25.3).

An atom, a molecule or an atomic nucleus generally have many resonance frequencies, which can be determined by measuring the cross-section for electromagnetic waves. The resonance frequencies $\hbar\,\omega_j$ correspond to the energy differences of states of the system under consideration.

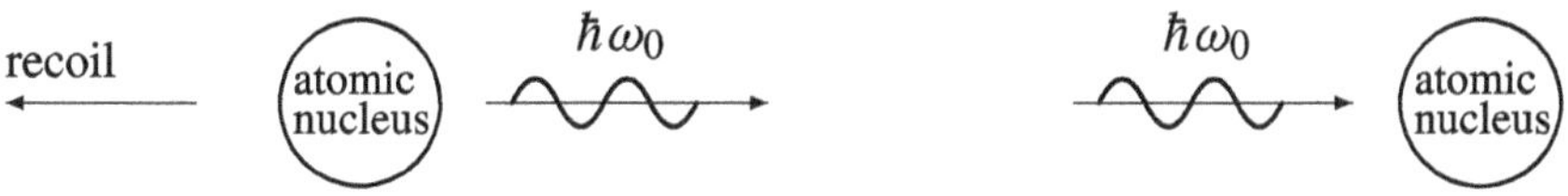

Figure 25.3 An atomic nucleus emits a γ-quantum with a specific frequency (for example $\hbar\omega_0 = 14.4\,\text{keV}$ for ^{57}Fe). An atomic nucleus of the same type can absorb this γ-quantum again. This resonance fluorescence is hindered by the fact that the recoil on the emitting nucleus leads to a frequency shift. Under suitable conditions there is a finite probability, that the recoil momentum is absorbed by the entire crystal lattice; the frequency shift is then practically zero. This recoilless resonance fluorescence is known as the Mößbauer effect.

The width of the resonance corresponds to the uncertainty $\Delta E \sim \hbar/\tau$ of these energy states due to their finite lifetime τ.

We insert $\omega_0/\Gamma = \omega_{\text{at}}/\Gamma_{\text{rad}} = 3/(2\alpha^3)$, (25.4), and $(e^2/a_{\text{B}})/m_{\text{e}}c^2 = \alpha^2$ into (25.25):

$$\sigma_{\text{res}} = \frac{8\pi}{3}\left(\frac{e^2/a_{\text{B}}}{m_{\text{e}}c^2}\right)^2 a_{\text{B}}^2 \left(\frac{3}{2\alpha^3}\right)^2 = \frac{6\pi}{\alpha^2}\, a_{\text{B}}^2 \tag{25.26}$$

This corresponds to the geometric cross-section $\sigma = \pi R_0^2$ on a sphere with a radius $R_0 \approx 3 \cdot 10^2\, a_{\text{B}}$. This is about 10^5 times larger than the cross-sectional area of the atom. This case of a very sharp, very high maximum only arises when the width Γ is equal to Γ_{rad}; in this case no other effects besides the radiation contribute to Γ. This width is also called *natural line width.*

The width of a resonance is increased by all processes that shorten the associated lifetime. In addition, there is a Doppler broadening due to the recoil (opposite reactional momentum) that an atom (or atomic nucleus) suffers during emission (Figure 25.3). In the Mößbauer effect this recoil is absorbed by the crystal in which the atomic nucleus is embedded. This results in *recoilless resonance fluorescence*, called Mößbauer effect. In this case, the emitted γ radiation has its natural line width. The extraordinary sharpness of such a γ line makes it possible to measure frequencies with extremely high precision.

Coherent and Incoherent Scattering

We now consider the scattering of light at N scattering centers of the same kind at the positions $\boldsymbol{r}_j$. The incident wave induces the dipole moments $\boldsymbol{p}_j(t) \propto \boldsymbol{E}(\boldsymbol{r}_j, t)$. This means that the $\boldsymbol{p}_j$ contain the phase factor $\exp(\mathrm{i}\boldsymbol{k} \cdot \boldsymbol{r}_j)$, which reflects path differences of the incident wave to the scattering centers. Another factor $\exp(-\mathrm{i}k\,\boldsymbol{e}_r \cdot \boldsymbol{r}_j)$ results from the path differences from the scattering centers to the observation point. We add the scattered waves of the individual oscillating dipoles:

$$\left(\frac{d\sigma}{d\Omega}\right)_N = \frac{d\sigma}{d\Omega}\left|\sum_{j=1}^{N}\exp(\mathrm{i}\boldsymbol{q}\cdot\boldsymbol{r}_j)\right|^2 = \frac{d\sigma}{d\Omega}\,\big|F(\boldsymbol{q})\big|^2 \quad \text{with } \boldsymbol{q} = \boldsymbol{k} - k\,\boldsymbol{e}_r \tag{25.27}$$

The cross-section $d\sigma/d\Omega$, (25.18), for a single scattering center is multiplied with the squared magnitude of the *form factor* $F(\boldsymbol{q})$. This form factor is the Fourier transform of the density $\sum_j \delta(\boldsymbol{r}-\boldsymbol{r}_j)$ of the scattering centers. From the experimental cross-section $(d\sigma/d\Omega)_N$ on can deduce $F(\boldsymbol{q})$ and thus the spatial distribution of the scattering centers (Exercise 25.2). In the following we consider a statistical distribution of the scattering centers (like the molecules of a gas). We evaluate $|F(\boldsymbol{q})|^2$:

$$|F(\boldsymbol{q})|^2 = \sum_{i=1}^{N}\sum_{j=1}^{N} \exp\left[\mathrm{i}\boldsymbol{q}\cdot(\boldsymbol{r}_i-\boldsymbol{r}_j)\right] = N + 2\sum_{i=2}^{N}\sum_{j=1}^{i-1} \cos\left[\boldsymbol{q}\cdot(\boldsymbol{r}_i-\boldsymbol{r}_j)\right] \tag{25.28}$$

The terms with $i = j$ result in the contribution N; for $i \neq j$ the (i, j) and the (j, i) terms were combined.

A large number $N \gg 1$ of scattering centers shall distributed over an region with the radius R, so that $|\boldsymbol{r}_j| \leq R$. From (25.28) then follow the two limiting cases:

$$|F(\boldsymbol{q})|^2 \approx \begin{cases} N^2 & (R \ll \lambda, \text{coherent sdattering}) \\ N & (R \gg \lambda, \text{incoherent scattering}) \end{cases} \tag{25.29}$$

For $R \ll \lambda$ we obtain $\boldsymbol{q}\cdot(\boldsymbol{r}_i-\boldsymbol{r}_j) \ll 1$ and $\cos \boldsymbol{q}\cdot(\boldsymbol{r}_i-\boldsymbol{r}_j) \approx 1$. For $R \gg \lambda$, on the other hand, positive and negative values of the cosine function occur with the same weight; then only the contributions form $i = j$ survive.

For coherent scattering, the individual dipoles oscillate in phase; then the fields superpose coherently and the effective cross-section gets a factor N^2. For incoherent scattering, on the other hand, the individual cross-section add up (factor N).

As an example, let us consider the scattering of visible light by water vapor in the atmosphere (i.e., by individual water molecules in air). Coherently, only those molecules can scatter that are in a volume that is small compared to λ^3. Otherwise the scattering is incoherent.

Under suitable conditions (such as a change in temperature) the water vapor contained in the air condenses into fog or clouds. Condensation begins with water droplets whose diameter d is small compared to the wavelength. As an example we consider a spherical droplet with a radius of $d/2 = 200\,\text{Å}$, which contains $N \approx 10^6$ molecules.[1] Due to $d \ll \lambda$, the N molecules of this droplet scatter coherently.

The formation of a mist of small droplets results in the transition from incoherent (water vapor) to coherent (within a droplet) scattering. This means an increase of the scattering cross-section by a factor of N, i.e., by several orders of magnitude in the above example. Although the amount of water does not change, the air becomes suddenly opaque when water vapor condenses into fog.

As the effect is proportional to N, it increases with the size of the water droplets. However, as soon as the diameter d is comparable to λ, the molecules in different places in the drop no longer oscillate coherently, and no further amplification occurs. A typical cloud consists of droplets with diameter d is in the range of $2 \ldots 50 \cdot 10^{-6}\,\text{m}$ (i.e., $d \gg \lambda$). Such a cloud is opaque due to the refraction and reflection of light (Chapter 37) at the surface of the many water droplets. Homogeneous water, on the other hand, is relatively transparent again (Figure 34.4).

Exercises

25.1 *Classical hydrogen atom*

An electron moves classically on a circular orbit with radius r around a proton. It experiences the Coulomb force $\boldsymbol{F} = -\boldsymbol{e}_r\, e^2/r^2$. Express the energy E and the angular momentum L as a function of the orbital radius r. Calculate the radiated power P.

The radiated power leads to a decrease of the radius $r(t)$; the electron moves along a spiral path until it hits the proton. Set up a differential equation for $r(t)$ and integrate it with the initial condition $r(0) = a_{\text{B}}$ (Bohr radius). Estimate the spiral time τ, after which the electron falls onto the proton. Discuss the time-dependence of the energy $E(t)$ and the angular momentum $L(t)$.

---

[1]The volume of the droplet is $V = \pi d^3/6 \approx 3 \cdot 10^7\,\text{Å}^3$. For water vapor, one finds one molecule per volume $v = 18\,\text{cm}^3/6 \cdot 10^{23} \approx 30\,\text{Å}^3$. This results in $N = V/v \approx 10^6$. At 100% air humidity and a temperature of 20 °C, the air volume V contains only about 30 water molecules.

Chapter 26

Resonant Circuit

An active resonant circuit (Figure 26.1) represents an oscillating charge distribution and therefore emits electromagnetic waves. In many practical cases, one can approximately neglect the feedback effect of this radiation on the circuit. This is done in the quasi-static approximation, in which certain time derivatives in the Maxwell equations are classified as small terms and are omitted. We investigate this quasi-static approximation for the oscillating circuit. Subsequently, the radiation losses of a resonant circuit are estimated.

Figure 26.1 shows a resonant circuit consisting of a coil (inductor) and a capacitor. This arrangement is also called LC circuit, or oscillating circuit or resonant circuit. The capacitance C of the capacitor was defined in (8.22), the self inductance L of the coil (with N_S windings) in (14.28):

$$Q = C\,U \tag{26.1}$$

$$I = \frac{N_\mathrm{S}}{c\,L}\,\Phi_\mathrm{m} \tag{26.2}$$

These equations were established in the frame of electrostatics and magnetostatics. The common treatment of the resonant circuit is that these equations are also used for the time-dependent case:

$$Q(t) \approx C\,U(t) \qquad \text{(quasi-static)} \tag{26.3}$$

$$I(t) \approx \frac{N_\mathrm{S}}{c\,L}\,\Phi_\mathrm{m}(t) \quad \text{(quasi-static)} \tag{26.4}$$

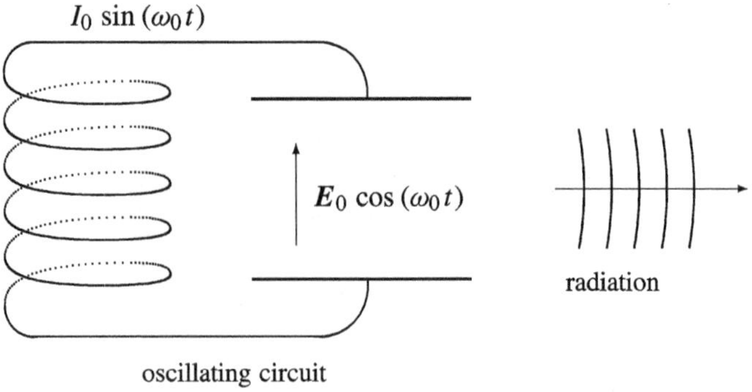

Figure 26.1 An inductor (coil) and a capacitor, which are connected in parallel, form a resonant circuit. In this circuit, the energy oscillate back and forth between the coil and the capacitor, i.e., between magnetic and electric field energy. As an oscillating charge distribution, the circuit emits electromagnetic radiation. The oscillating circuit is usually treated in the quasi-static approximation, in which the feedback of the radiation on the circuit itself is neglected.

According to Faraday's law of induction (16.12), the time change of the magnetic flux Φ_m induces a voltage U at the ends of the coil:

$$U(t) = -\frac{N_\mathrm{S}}{c}\,\frac{d\Phi_\mathrm{m}}{dt} \tag{26.5}$$

In the parallel resonant circuit, this is equal to the voltage of the capacitor. The change in charge $\dot{Q} = dQ/dt$ on the capacitor equals the current I through the coil. From $I = \dot{Q}$ and (26.3)–(26.5) we obtain

$$U(t) = -\frac{N_\mathrm{S}}{c}\,\frac{d\Phi_\mathrm{m}}{dt} = -L\,\frac{dI}{dt} = -L\,\ddot{Q} = -L\,C\,\ddot{U}(t) \tag{26.6}$$

The solution of this differential equation is $U = U_0\cos(\omega_0 t + \varphi)$. The resonant circuit therefore represents an oscillator with the characteristic frequency

$$\boxed{\omega_0 = \frac{1}{\sqrt{LC}} \quad \text{eigenfrequency of resonant circuit}} \tag{26.7}$$

In the resonant circuit, the energy oscillates between electrical (capacitor) and magnetic (coil) field energy. This can be compared to a mechanical pendulum, in which the energy alternates between kinetic and potential energy.

We now examine how the equations for the resonant circuit can be justified by the Maxwell equations

$$\operatorname{div} \boldsymbol{E}(\boldsymbol{r}, t) = 4\pi\, \varrho(\boldsymbol{r}, t), \quad \operatorname{curl} \boldsymbol{E}(\boldsymbol{r}, t) + \underbrace{\frac{1}{c}\frac{\partial \boldsymbol{B}(\boldsymbol{r}, t)}{\partial t}}_{\text{induction}} = 0 \tag{26.8}$$

$$\operatorname{curl} \boldsymbol{B}(\boldsymbol{r}, t) - \underbrace{\frac{1}{c}\frac{\partial \boldsymbol{E}(\boldsymbol{r}, t)}{\partial t}}_{\text{displacement current}} = \frac{4\pi}{c}\, \boldsymbol{j}(\boldsymbol{r}, t), \quad \operatorname{div} \boldsymbol{B}(\boldsymbol{r}, t) = 0 \tag{26.9}$$

In the *static* case, (26.1) can be derived from (26.8) and (26.2) from (26.9). This derivation can be carried over to (26.3) and (26.4) when we allow that these terms depend on time t as a parameter. Specifically, this means:

- For the capacitor: The term "induction" is neglected. Then (26.8) has the structure of the electrostatic field equations; from this follows (26.3).

  In fact, the displacement current causes a magnetic field in the capacitor. However, this magnetic field is small compared to the electric field.
- For the coil: The term "displacement current" is neglected. Then (26.9) has the structure of the magnetostatic field equations; from this follows (26.4).

  In fact, the induction causes an electric field in the coil. However, this electric field is small compared to the magnetic field.

Each of these approximations assumes that the occurring changes are sufficiently slow. The *quasi-static approximation* consists in the neglect of one of the two terms, the displacement current or the inductance. Which term is to be neglected depends on the situation under consideration. It is not possible to omit both terms; in particular, we need the term "induction" for (26.5).

In the literature, the neglect of the displacement current is often referred to as *the* quasi-static approximation. This is due to the outstanding practical importance of the Faraday's law of induction; for generators, electric motors and transformers, the term "induction" is important, the "displacement current" mostly not. From the point of view of electrodynamics, however, both approximations (neglecting the displacement current or the induction) are to be seen in parallel to each other. Both terms together lead to radiation for time-dependent processes.

Capacitor

With the approximation

$$\operatorname{curl} \boldsymbol{E}(\boldsymbol{r}, t) = -\frac{1}{c}\,\frac{\partial \boldsymbol{B}(\boldsymbol{r}, t)}{\partial t} \approx 0 \tag{26.10}$$

Equation (26.8) becomes the field equations of electrostatics, in which the time t only appears as a parameter. In this approximation, one obtains (26.3), i.e., the relation known from electrostatics between the charge $Q(t)$ and the voltage $U(t)$.

The approximation (26.10) is equivalent to

$$\boldsymbol{E}(\boldsymbol{r}, t) = -\operatorname{grad} \Phi(\boldsymbol{r}, t) - \frac{1}{c}\,\frac{\partial \boldsymbol{A}(r, t)}{\partial t} \approx -\operatorname{grad} \Phi(\boldsymbol{r}, t) \tag{26.11}$$

This assumption is justified if

$$\frac{1}{c}\left|\frac{\partial \boldsymbol{A}}{\partial t}\right| \ll |\boldsymbol{E}| \quad \begin{array}{l}\text{(quasi-static approximation}\\ \text{for the capacitor)}\end{array} \tag{26.12}$$

To examine this condition, we assume a localized, periodic field and charge configuration. The spatial extension boundary is characterized by the length ℓ, the frequency is ω. Specifically, imagine a plate capacitor with the area $A_{\mathrm{C}} = \ell^2$ and the plate spacing $d = \ell$, to which a periodic voltage is applied. We now assume

$$\frac{\omega \ell}{c} \ll 1 \tag{26.13}$$

and show that then (26.12) applies.

The periodicity of the sources and fields is described by the factor $\exp(-\mathrm{i}\omega t)$. Then the following applies

$$\frac{\partial \boldsymbol{E}}{\partial t} = -\mathrm{i}\omega \boldsymbol{E}, \qquad \frac{\partial \boldsymbol{B}}{\partial t} = -\mathrm{i}\omega \boldsymbol{B} \tag{26.14}$$

The fields associated with the charge distribution change significantly on the length scale ℓ. Using this, we estimate the size of the spatial derivatives:

$$\left|\frac{\partial E_i}{\partial x_j}\right| \sim \frac{E}{\ell}, \quad \left|\frac{\partial B_i}{\partial x_j}\right| \sim \frac{B}{\ell}, \quad \left|\frac{\partial A_i}{\partial x_j}\right| \sim \frac{A}{\ell} \tag{26.15}$$

On the right-hand side are the mean magnitudes of the fields E, B and A. This estimate shows the order of magnitude of the respective fields derivatives, as long as they do not vanish (like $\operatorname{div} \boldsymbol{B} = 0$). Electromagnetic fields vary independent of the charge distribution also with the wavelength λ. However, because of (26.13), $\lambda \gg \ell$ such contributions (as $|\partial E_i/\partial x_j| \sim E/\lambda \ll E/\ell$) can be neglected in (26.15).

In the capacitor, due to $\dot{\boldsymbol{E}} \neq 0$ and the displacement current, a magnetic field is created. For $\omega \to 0$, this magnetic field disappears, for slow oscillations it is weak. With the help of (26.14) and (26.15) we estimate the magnitude of the magnetic field in the capacitor:

$$\operatorname{curl} \boldsymbol{B} \overset{j=0}{=} \frac{1}{c}\frac{\partial \boldsymbol{E}}{\partial t} \quad \longrightarrow \quad \frac{B}{\ell} \sim \frac{\omega}{c}\, E \tag{26.16}$$

From $\boldsymbol{B} = \operatorname{curl} \boldsymbol{A}$ follows $B \sim A/\ell$. With this we obtain:

$$\frac{1}{c}\left|\frac{\partial \boldsymbol{A}}{\partial t}\right| = \frac{\omega}{c}\, A \sim \frac{\omega \ell}{c}\, B \sim \frac{\omega^2 \ell^2}{c^2}\, |\boldsymbol{E}| \tag{26.17}$$

This shows that (26.12) is valid under the condition (26.13).

Coil

With the approximation

$$\operatorname{curl} \boldsymbol{B}(\boldsymbol{r}, t) = \frac{1}{c}\frac{\partial \boldsymbol{E}(\boldsymbol{r}, t)}{\partial t} + \frac{4\pi}{c}\,\boldsymbol{j}(\boldsymbol{r}, t) \approx \frac{4\pi}{c}\,\boldsymbol{j}(\boldsymbol{r}, t) \tag{26.18}$$

Equation (26.9) becomes the field equations of magnetostatics, in which the time t only appears as a parameter. In this approximation, we obtain (26.4) is obtained, i.e., the relation known from magnetostatics between the current $I(t)$ and the magnetic flux $\Phi_{\mathrm{m}}(t)$.

The prerequisite for the approximation (26.18) is $|\partial \boldsymbol{E}/\partial t| \ll |\boldsymbol{j}|$ or

$$\frac{1}{c}\left|\frac{\partial \boldsymbol{E}}{\partial t}\right| \ll \left|\operatorname{curl} \boldsymbol{B}\right| \qquad \begin{array}{l}\text{(quasi–static approximation}\\ \text{for the coil)}\end{array} \tag{26.19}$$

To examine this condition, we again assume a localized field and charge configuration. The spatial extension is characterized by the length ℓ, the frequency is ω. Specifically, imagine a coil with the diameter diameter ℓ and length ℓ through which a periodic current flows. We assume (26.13).

In the coil, because of $\dot{\boldsymbol{B}} \neq 0$ and the term "induction", an electric field is created. For $\omega \to 0$ the electric field disappears, so it is weak for slow oscillations. With the help of (26.14) and (26.15), we obtain for the size of the electric field in the coil

$$\operatorname{curl} \boldsymbol{E} = -\frac{1}{c}\frac{\partial \boldsymbol{B}}{\partial t} \quad \longrightarrow \quad \frac{E}{\ell} \sim \frac{\omega}{c}\, B \tag{26.20}$$

With this we estimate the left-hand side of (26.19):

$$\frac{1}{c}\left|\frac{\partial \boldsymbol{E}}{\partial t}\right| = \frac{\omega}{c}\, E \sim \frac{\omega^2 \ell}{c^2}\, B \sim \frac{\omega^2 \ell^2}{c^2}\, \big|\operatorname{red} \boldsymbol{B}\big| \tag{26.21}$$

This shows that (26.18) is valid under the condition (26.13).

Quasi-Static Approximation

In both cases, the capacitor and the inductor, the respective neglected terms are of the relative size $\omega^2\ell^2/c^2$. In both cases, the condition for the quasi-static approximation reads

$$\boxed{\frac{\omega \ell}{c} \ll 1 \quad \text{condition for quasi-static approximation}} \tag{26.22}$$

The frequency ω and the length ℓ correspond to the velocity $v = \omega\ell$. This is the maximum effective speed of charge carriers that oscillate back and forth in the circuit. (The drift velocities of individual electrons in the wires are much smaller). With $v = \omega\ell$ (26.22) becomes $v/c \ll 1$. The quasi-static approximation can therefore be regarded as a non-relativistic approximation.

The meaning of the quasi-static approximation in Maxwell's equations depends — as discussed above — on the considered part of the configuration. Depending on this part (coil or capacitor), different terms must be neglected because

$$\text{magnetic field in the capacitor:} \quad B \sim \frac{\omega \ell}{c}\, E \ll E \tag{26.23}$$

$$\text{electric field in the coil:} \quad E \sim \frac{\omega \ell}{c}\, B \ll B \tag{26.24}$$

Both approximations are complementary to each other.

In the transition from (26.1)–(26.2) to (26.3)–(26.4), the time t is only considered as a parameter but not as a variable that influences on the dynamics of the processes. This corresponds to the neglect of the retardation in

$$A^{\alpha}_{\text{ret}}(\boldsymbol{r},t) = \frac{1}{c}\int d^3r'\ \frac{j^{\alpha}(\boldsymbol{r}',t-|\boldsymbol{r}-\boldsymbol{r}'|/c)}{|\boldsymbol{r}-\boldsymbol{r}'|} \approx \frac{1}{c}\int d^3r'\ \frac{j^{\alpha}(\boldsymbol{r}',t)}{|\boldsymbol{r}-\boldsymbol{r}'|} \tag{26.25}$$

With this approximation, the connections $\varrho \leftrightarrow \Phi$ (i.e., $Q \leftrightarrow U$) and $\boldsymbol{j} \leftrightarrow \boldsymbol{A}$ (i.e., $I \leftrightarrow \Phi_{\text{m}}$) can be calculated as in the static case.

The approximation (26.25) can be justified by (26.22). For a localized field and charge configuration, $r \lesssim \ell$, $r' \lesssim \ell$ and thus $|\boldsymbol{r}-\boldsymbol{r}'| \lesssim \ell$. For a periodic charge distribution, the retardation results in a phase shift $\Delta\varphi = \omega\,\Delta t_{\text{ret}} = \omega|\boldsymbol{r}-\boldsymbol{r}'|/c \sim \omega\ell/c$. Due to $\Delta\varphi \ll 1$, then the approximation $j^{\alpha}(\boldsymbol{r}',t-|\boldsymbol{r}-\boldsymbol{r}'|/c) \approx j^{\alpha}(\boldsymbol{r}',t)$ is allowed.

Physically, the retardation is the basis of radiation phenomena (Chapter 24). Their neglect in (26.25) therefore means that we do not take into account the feedback effect of the radiation on the resonant circuit.

Radiation of the Circuit

The resonant circuit represents an oscillating charge distribution. Therefore, it emits electromagnetic radiation. As long as the energy losses due to this radiation are small, the radiation can be calculated using (26.3) and (26.4).

From (26.6) and (26.7) follows $\ddot{Q} = -\omega_0^2\, Q$. For the later discussion we include a damping term in this differential equation:

$$\ddot{Q}(t) + \Gamma\,\dot{Q}(t) + \omega_0^2\, Q(t) = 0 \tag{26.26}$$

For $\Gamma < \omega_0$ the general solution of this differential equation (Chapter 24 in [1]) reads

$$Q(t) = Q_0\, \exp(-\Gamma t/2)\, \cos(w_0 t + \varphi) \tag{26.27}$$

where $w_0^2 = \omega_0^2 - \Gamma^2/4$. For $\Gamma \ll \omega_0$, Equation (26.27) describes a weakly damped oscillation with $w_0 \approx \omega_0$. In a first step, we consider the undamped oscillation $Q(t) = Q_0\, \cos(\omega_0 t + \varphi)$ and apply the dipole formula to it.

The capacitance of the plate capacitor is known from (8.41), the inductance of the cylindrical coil from (14.29):

$$C = \frac{A_{\rm C}}{4\pi d}, \quad L = \frac{4\pi}{c^2}\, N_{\rm S}^2\, \frac{A_{\rm S}}{\ell} \tag{26.28}$$

For the sake of simplicity, we assume the same length scale for all quantities,

$$A_{\rm C} = \ell_0^2, \quad d = \ell_0, \quad A_{\rm S} = \ell_0^2, \quad \ell = \ell_0 \tag{26.29}$$

Then the eigenfrequency of the resonant circuit is

$$\omega_0 = \frac{1}{\sqrt{LC}} = \frac{c}{\ell_0}\,\frac{1}{N_{\rm S}} \tag{26.30}$$

The condition (26.22) for the quasi-static approximation thus becomes

$$\frac{\omega_0\,\ell_0}{c} = \frac{1}{N_{\rm S}} \ll 1 \tag{26.31}$$

This condition is fulfilled for $N_{\rm S} \gg 1$. As a special example we consider

$$\ell_0 = 1\,\text{cm}, \quad N_{\rm S} = 100 \quad \longrightarrow \quad \nu = \frac{\omega_0}{2\pi} \approx 50\ \text{MHz} \tag{26.32}$$

A radio wave of this frequency is in the FM range. The resonant circuit of an FM receiver is dimensioned differently, for example $A_{\rm C} = 10\,\text{cm}^2$, $d = 0.1$ cm, $N_{\rm S} = 10$, $A_{\rm S}$ and $\ell_{\rm coil}$ as above.

The condition (26.31) implies $\lambda \gg \ell_0$. This fulfills the condition for the long-wave approximation (24.14). Therefore, the radiation losses can indeed be calculated using the dipole formula. The capacitor represents an oscillating dipole of strength $p \sim Q_0\,\ell_0$. The coil is an oscillating magnetic dipole (15.8) of the strength

$$m = \frac{1}{2c}\left|\int d^3r\ \boldsymbol{r}\times\boldsymbol{j}\right| \sim \frac{1}{c}\,\ell_0^2\, I\, N_{\rm S} \sim \ell_0\,\frac{I}{\omega_0} \sim \ell_0\, Q_0 \sim p \tag{26.33}$$

According to (24.27) and (24.28), an electric and magnetic dipole radiate equally strongly. The following therefore applies to the total radiated power

$$P \sim \frac{\omega_0^4\, Q_0^2\, \ell_0^2}{c^3} \tag{26.34}$$

The energy of the resonant circuit is equal to the maximum energy (8.28) of the capacitor:

$$E_0 = \frac{Q_0^2}{2C} \sim \frac{Q_0^2}{\ell_0} \tag{26.35}$$

During a period $T = 2\pi/\omega_0$ the resonant circuit radiates the energy PT. We calculate the number $n_{\rm osz}$ of the oscillations after which an energy of the size E_0 is radiated:

$$n_{\rm osc} = \frac{E_0}{PT} = \frac{\omega_0 E_0}{2\pi P} \sim \frac{c^3}{\omega_0^3 l_0^3} \sim N_{\rm S}^3 = 10^6 \tag{26.36}$$

This radiation can be described by the damping term in (26.26) and (26.27) with

$$\Gamma = \Gamma_{\rm rad} \sim \frac{\omega_0}{N_{\rm S}^3} \tag{26.37}$$

In addition to radiation damping, all processes contribute that extract energy from the resonant circuit. Thereby, the ohmic losses are often more important than the radiation losses.

By designing the resonant circuit appropriately, the radiation can be reduced. For example, for the resonant circuit of an FM receiver on may chose $A_{\rm C} = 10\,{\rm cm}^2$, $d = 0.1\,{\rm cm}$, $N_{\rm S} = 10$, $A_{\rm S} = 1\,{\rm cm}^2$ and $\ell = 1\,{\rm cm}$. Compared to the case considered above, the capacitance is greater by a factor of 100 and the inductance is smaller by a factor of 100. This results in the same eigenfrequency of the circuit. The electric and magnetic dipole moments are both reduced by a factor of 10, so that the radiation is reduced by a factor 100. In the resonant circuit of a radio receiver the radiation is undesirable because it worsens the quality of the resonant circuit by causing disturbing feedback within the device.

If the capacitor shown in Figure 26.1 is connected by a wire half-circle ($N_{\rm S} \sim 1$), then $\omega_0 \ell/c \sim 1$. Then, the condition for the quasi-static approximation is no longer fulfilled. From (26.37) we obtain $\Gamma_{\rm rad} \sim \omega_0$, i.e., a very strong radiation (however, the derivation of (26.37) is also no longer valid).

For $\omega\ell/c \gg 1$, the capacitor plates are only spacial boundaries for waves, which propagate in between; the formula (26.3) has then completely lost its meaning.

Exercises

26.1 *Magnetic field in the capacitor*

Two parallel circular disks (distance d, radius R, $R \gg d$) form a capacitor. The disks are charged with $Q(t) = Q_0 \cos(\omega t)$ and $-Q(t)$.

Determine the electric field in quasi-static approximation and under neglect of boundary effects. Determine the associated magnetic field from the Maxwell equations.

Calculate the Poynting vector in the area between the plates. Use this to determine the energy flow P (energy per time) through the cylindrical surface $2\pi R d$ defined by the borders of the two disks. Compare this with the power supplied to the capacitor from outside.

26.2 *Resonant circuit*

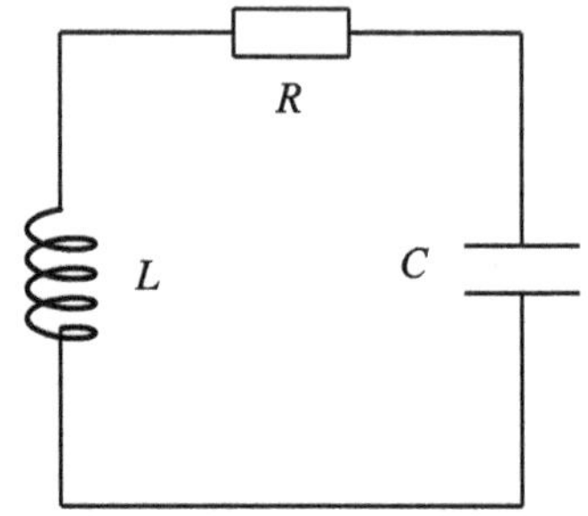

A resonant circuit consists of a coil (self inductance L), a capacitor (capacitor C) and a resistor. The resistor causes a voltage drop $U_R(t) = R\, I(t)$. Set up the differential equation for the charge $Q(t)$ on the capacitor in quasi-static approximation. What is the general solution in the case $R < 2\sqrt{L/C}$?

PART VI

Electrodynamics in Matter

Chapter 27

Microscopic Maxwell Equations

The Maxwell equations are fundamental laws of nature with a wide range of validity; they also apply in matter. Often, one is interested in the reaction of matter due to external electromagnetic fields and not in the fields in the undisturbed matter itself. From this point of view, we derive the Maxwell equations in matter.

For this purpose, the fields are divided into (i) the fields of the matter, (ii) additional external fields (perturbation of the matter) and (iii) induced fields (reaction of matter). For all these fields, microscopic Maxwell equations are set up at first.

There are various manifestations of matter like plasmas, gases, liquids or solids. We often refer to a solid body but also to other aggregate states. Only elementary knowledge about the solid state structure is assumed.

We write the previous Maxwell equations (16.5) in the form

$$\begin{aligned} \operatorname{div} \boldsymbol{E}_{\text{tot}} &= 4\pi\, \varrho_{\text{tot}}, \quad \operatorname{curl} \boldsymbol{E}_{\text{tot}} + \frac{1}{c}\frac{\partial \boldsymbol{B}_{\text{tot}}}{\partial t} = 0 \\ \operatorname{div} \boldsymbol{B}_{\text{tot}} &= 0, \quad \operatorname{curl} \boldsymbol{B}_{\text{tot}} - \frac{1}{c}\frac{\partial \boldsymbol{E}_{\text{tot}}}{\partial t} = \frac{4\pi}{c}\, \boldsymbol{j}_{\text{tot}} \end{aligned} \tag{27.1}$$

The fields and sources depend on the arguments $\boldsymbol{r}$ and t. The index "tot" (for total) refers to the inclusion of all sources and fields in the considered the system (the matter under consideration and all additional fields or sources). The fields $\boldsymbol{E}_{\text{tot}}$, $\boldsymbol{B}_{\text{tot}}$ and sources $\varrho_{\text{tot}}, \boldsymbol{j}_{\text{tot}}$ are therefore the actual fields and sources. The designations $\boldsymbol{E}$, $\boldsymbol{B}$, ϱ and $\boldsymbol{j}$ used so far are assigned to *others* quantities in the following. For a distinction, we refer to (27.1) as the Maxwell equations in matter, and to (16.5) as the Maxwell equations in vacuum.

We consider an initially unspecified division $\varrho = \varrho_A + \varrho_B$ and $\boldsymbol{j} = \boldsymbol{j}_A + \boldsymbol{j}_B$ of the sources. The associated fields are labeled accordingly. Since the Maxwell equations *linear* in the fields and the source terms, we can write them for the individual components, for example

$$\operatorname{div}\boldsymbol{E} = 4\pi\,\varrho \;\xleftrightarrow{\varrho=\varrho_A+\varrho_B}\; \begin{array}{l} \operatorname{div}\boldsymbol{E}_A = 4\pi\,\varrho_A \\ \operatorname{div}\boldsymbol{E}_B = 4\pi\,\varrho_B \end{array} \tag{27.2}$$

Due to their linearity, this works for all Maxwell equations. In the following, we choose a division of the sources and fields that is adapted to the experimental situation.

In the context of *electrodynamics in matter*, an experiment consists of exposing the matter to additional fields, for example:

- The matter is placed in the electric field of a capacitor or the magnetic field of a coil.
- Additional charges or currents can be brought onto or into the matter; they generate an electromagnetic field in matter. This example shows that the additional sources are not necessarily external ones.
- A piece of matter is placed in the field of an electromagnetic wave. In this case, the sources that generate the wave are not explicitly considered.

The experimental situation suggests the following division of sources and fields: First of all, there are the sources and fields of undisturbed matter (index matter (index "0"); the matter is in its ground state or in a thermodynamic equilibrium state. The *additional* (index "ext" for extra or external) fields *induce* (index "ind") or provoke a reaction of the matter. This gives us the following categorization (see also Table 27.1):

$$\begin{aligned} \varrho_{\text{tot}} &= \varrho_0 + \varrho_{\text{ext}} + \varrho_{\text{ind}} = \varrho_0 + \varrho \\ \boldsymbol{j}_{\text{tot}} &= \boldsymbol{j}_0 + \boldsymbol{j}_{\text{ext}} + \boldsymbol{j}_{\text{ind}} = \boldsymbol{j}_0 + \boldsymbol{j} \\ \boldsymbol{E}_{\text{tot}} &= \boldsymbol{E}_0 + \boldsymbol{E}_{\text{ext}} + \boldsymbol{E}_{\text{ind}} = \boldsymbol{E}_0 + \boldsymbol{E} \\ \boldsymbol{B}_{\text{tot}} &= \boldsymbol{B}_0 + \boldsymbol{B}_{\text{ext}} + \boldsymbol{B}_{\text{ind}} = \boldsymbol{B}_0 + \boldsymbol{B} \end{aligned} \tag{27.3}$$

Table 27.1 Designation of the sources and fields in matter, for the example of the charge density. An illustration of this categorization is sketched in Figure 28.1. The division of the sources and fields is carried over to the Maxwell equations.

ϱ_{tot}	total charge density, previously denoted by ϱ
ϱ_0	charge density of the *undisturbed matter*
ϱ_{ext}	additional charge density. The index "ext" stands for extra, but can often also be interpreted as external. This part constitutes the *disturbance* of matter.
ϱ_{ind}	induced charge density. This contribution describes the *reaction* of the matter to the disturbance.
ϱ	$= \varrho_{\text{ext}} + \varrho_{\text{ind}}$, designation change!

A charged particle (such as an electron) in matter is exposed to the Lorentz force $\boldsymbol{F}_{\text{L}} = \boldsymbol{F}_{\text{tot}} = \boldsymbol{F}_0 + \boldsymbol{F}$. Effective, however, is only the force

$$\boldsymbol{F} = q\left(\boldsymbol{E}(\boldsymbol{r},t) + \frac{\boldsymbol{v}}{c} \times \boldsymbol{B}(\boldsymbol{r},t)\right) \tag{27.4}$$

In the classical equilibrium state, all forces are balanced out, i.e., $\boldsymbol{F}_0 = 0$. The quantum mechanical ground state is stationary; the forces $\boldsymbol{F}_0$ are therefore effectively balanced. In experiments, one usually deals with a thermodynamic equilibrium state (finite temperature); imagine, for example, a classical gas with atom-atom collisions. Then the forces $\boldsymbol{F}_0$ cancel out in the statistical average. In this respect, the effective forces in matter are given by (27.4). This is also the reason for the new designations for the fields $\boldsymbol{E}$ and $\boldsymbol{B}$ that do not include the undisturbed fields. As in the Parts 2–4, the fields are defined by their force effects.

We use the division scheme (27.2) for $\varrho_{\text{tot}} = \varrho_0 + \varrho$. From this we obtain the Maxwell equations for the fields $\boldsymbol{E}$ and $\boldsymbol{B}$,

$$\begin{aligned} &\operatorname{div}\boldsymbol{E}(\boldsymbol{r},t) = 4\pi\,\varrho(\boldsymbol{r},t), \quad \operatorname{curl}\boldsymbol{E}(\boldsymbol{r},t) + \frac{1}{c}\,\frac{\partial \boldsymbol{B}(\boldsymbol{r},t)}{\partial t} = 0 \\ &\operatorname{div}\boldsymbol{B}(\boldsymbol{r},t) = 0, \quad \operatorname{curl}\boldsymbol{B}(\boldsymbol{r},t) - \frac{1}{c}\,\frac{\partial \boldsymbol{E}(\boldsymbol{r},t)}{\partial t} = \frac{4\pi}{c}\,\boldsymbol{j}(\boldsymbol{r},t) \end{aligned} \tag{27.5}$$

and corresponding equations for $\boldsymbol{E}_0$ and $\boldsymbol{B}_0$, which are not of interest in the following. According to $\boldsymbol{E} = \boldsymbol{E}_{\text{ext}} + \boldsymbol{E}_{\text{ind}}$ and $\boldsymbol{B} = \boldsymbol{B}_{\text{ext}} + \boldsymbol{B}_{\text{ind}}$ we divide the

Maxwell equations (27.5) into the corresponding parts:

$$\begin{aligned} \operatorname{div} \boldsymbol{E}_{\text{ext}} &= 4\pi\,\varrho_{\text{ext}}, & \operatorname{curl} \boldsymbol{E}_{\text{ext}} + \frac{1}{c}\frac{\partial \boldsymbol{B}_{\text{ext}}}{\partial t} &= 0 \\ \operatorname{div} \boldsymbol{B}_{\text{ext}} &= 0, & \operatorname{curl} \boldsymbol{B}_{\text{ext}} - \frac{1}{c}\frac{\partial \boldsymbol{E}_{\text{ext}}}{\partial t} &= \frac{4\pi}{c}\,\boldsymbol{j}_{\text{ext}} \end{aligned} \tag{27.6}$$

$$\begin{aligned} \operatorname{div} \boldsymbol{E}_{\text{ind}} &= 4\pi\,\varrho_{\text{ind}}, & \operatorname{curl} \boldsymbol{E}_{\text{ind}} + \frac{1}{c}\frac{\partial \boldsymbol{B}_{\text{ind}}}{\partial t} &= 0 \\ \operatorname{div} \boldsymbol{B}_{\text{ind}} &= 0, & \operatorname{curl} \boldsymbol{B}_{\text{ind}} - \frac{1}{c}\frac{\partial \boldsymbol{E}_{\text{ind}}}{\partial t} &= \frac{4\pi}{c}\,\boldsymbol{j}_{\text{ind}} \end{aligned} \tag{27.7}$$

So far, the derivation did neither use a spatial averaging nor another approximation. The Maxwell equations (27.5)–(27.7) are therefore still microscopic and just as exact as the Maxwell's equations in vacuum (27.1).

The solution procedure for the equations (27.6) is the same as was used in the previous chapters. For example, the charge density ϱ_{ext} on the plates of a capacitor generates generate a homogeneous field. The actual problem consists in determining the induced sources. The structure of matter and its reaction to additional electromagnetic fields are generally complex. In particular, the reaction of matter at a certain position depends on the effective fields $\boldsymbol{E}$ and $\boldsymbol{B}$ (and not just $\boldsymbol{E}_{\text{ext}}$ and $\boldsymbol{B}_{\text{ext}}$); however, the effective fields themselves depend on the induced sources and fields. By the induced sources are the equations (27.6) and (27.7) are coupled with each other (because to $\varrho_{\text{ind}} = \varrho_{\text{ind}}[\boldsymbol{E}, \boldsymbol{B}]$ and $\boldsymbol{j}_{\text{ind}} = \boldsymbol{j}_{\text{ind}}[\boldsymbol{E}, \boldsymbol{B}]$ are functional of the fields $\boldsymbol{E}$ and $\boldsymbol{B}$).

The theoretical treatment of the induced fields is usually carried out in the context of *linear response* (Chapter 28). Using further approximations, we finally find a practicable form of Maxwell equations for our purposes (Chapter 29).

Macroscopic and Microscopic Fields

We consider matter that consists of atoms or molecules. Such particles have a size of a few Ångström (1 Å $= 10^{-10}$ m). The structure and the corresponding fields vary considerably on this microscopic scale.

We refer to fields as *macroscopic*, if they are practically constant on the microscopic scale. However, if a field varies significantly in this range, we refer to it as *microscopic*. For the fields occurring in (27.5)–(27.7) we note:

$$\begin{aligned} &\text{macroscopic:} \quad \varrho_{\text{ext}},\, \boldsymbol{j}_{\text{ext}},\, \boldsymbol{E}_{\text{ext}},\, \boldsymbol{B}_{\text{ext}} \\ &\text{microscopic:} \quad \varrho_{\text{ind}},\, \boldsymbol{j}_{\text{ind}},\, \boldsymbol{E}_{\text{ind}},\, \boldsymbol{B}_{\text{ind}},\, \boldsymbol{E},\, \boldsymbol{B} \end{aligned} \tag{27.8}$$

The microscopic structure of matter varies strongly in the range of an atom or molecule. Therefore, ϱ_{ind} and $\boldsymbol{j}_{\text{ind}}$ generally change significantly on the microscopic scale, even if the fields $\boldsymbol{E}_{\text{ext}}$ and $\boldsymbol{B}_{\text{ext}}$ are constant (Figure 28.1). This means that $\boldsymbol{E}_{\text{ind}}$, $\boldsymbol{B}_{\text{ind}}$, $\boldsymbol{E} = \boldsymbol{E}_{\text{ext}} + \boldsymbol{E}_{\text{ind}}$ and $\boldsymbol{B} = \boldsymbol{B}_{\text{ext}} + \boldsymbol{B}_{\text{ind}}$ display microscopic structures. The Maxwell equations (27.5) and (27.7) are *microscopic* equations.

Often one is not interested in the microscopic details. Then these details can be eliminated by a spatial average a over many atomic distances or lattice constants. The averaging over a field $A = A(\boldsymbol{r}, t)$ is defined by a convolution (integral) with a suitable function $f(\boldsymbol{r})$,

$$\langle A\rangle(\boldsymbol{r}, t) = \int d^3r'\; A(\boldsymbol{r}', t)\; f(\boldsymbol{r} - \boldsymbol{r}') \quad \text{with} \quad \int d^3r'\; f(\boldsymbol{r} - \boldsymbol{r}') = 1 \tag{27.9}$$

Occasionally one finds the notation $\langle A(\boldsymbol{r}, t)\rangle$ instead of $\langle A\rangle(\boldsymbol{r}, t)$. However, when averaging over the argument, the result does not depend on *this* argument. Rather, there is a different spatial dependence of the averaged quantity $\langle A\rangle$. From a logical point of view, the notation chosen here is preferable.

Let the function $f(\boldsymbol{r})$ be localized at $r = 0$ and not negative. A possible choice is the Gaussian function

$$f(\boldsymbol{r} - \boldsymbol{r}') = \frac{1}{\pi^{3/2}\, b^3}\, \exp\left(-\frac{|\boldsymbol{r} - \boldsymbol{r}'|^2}{b^2}\right) \quad \text{with} \quad a \ll b \ll \lambda \tag{27.10}$$

Let the microscopic scale be determined by the length a; for a solid body crystal, a would be the lattice constant. The averaging length b should extend over many (at least several) elementary cells. On the other hand, b should be small compared to the scale on which the phenomena to be studied vary. For a wave, this scale is determined by the wavelength λ.

Other Presentations

When introducing Maxwell's equations into matter, many textbooks proceed differently in two respects:

1. The equations (27.1) are averaged spatially. Due to

$$\langle E_0 \rangle = 0, \quad \langle B_0 \rangle = 0 \tag{27.11}$$

one obtains Maxwell equations of the form (27.5) for the *spatially averaged* fields $\boldsymbol{E}$ and $\boldsymbol{B}$. From Chapter 29 on, we consider these macroscopic fields and the corresponding Maxwell equations.

The reason for the starting point chosen in this book is: If the fields are averaged from the outset, the microscopic structure of the effective fields $\boldsymbol{E}$ and $\boldsymbol{B}$ is eliminated. Then, it is also impossible to determine the exact reaction of the matter to an external field (i.e., the exact ϱ_{ind}); but this is exactly what solid-state physicists want to do today. Within the scope of this book, a solution of the microscopic Maxwell equations is out of the question. However, these equations are of fundamental importance, and they are accompanied by a physically adequate splitting of the fields.

2. The charges are often divided up slightly differently:

$$\varrho = \varrho_{\text{ext}} + \varrho_{\text{ind}} = \underbrace{\varrho_{\text{ext}} + \varrho_{\text{ind}}^{\text{free}}}_{\varrho_{\text{free}}} + \underbrace{\varrho_{\text{ind}}^{\text{geb}}}_{\varrho_{\text{pol}}} \tag{27.12}$$

The term "free" refers to the charges that contribute to the current when a external electric field is applied. All other charges are "bound". In a metal, for example, the electrons in the partially filled energy band (conduction band) are free charges, in the filled energy bands on finds the bound charges. (These designations must not be taken too literally: The electrons of the conduction band are not really free particles, and the wave functions wave functions of the electrons in the filled bands are not localized as, for example, bound electrons in the atom.)

The reason for our division $\varrho = \varrho_{\text{ext}} + \varrho_{\text{ind}}$ is: It corresponds to the physical division into the disturbance of matter by additional fields or charges, and into the reaction (response) of the matter. This division is adequate both for the experimental situation as well as the theoretical treatment. The alternative division (27.12) of the induced charge into "free" and "bound" is usually not unambiguous (for example for weakly bound charges in a semiconductor).

Chapter 28

Linear Response

The induced fields are often proportional to the applied external fields. This means a linear response of the matter to the external disturbance. The proportionality coefficients are introduced as the permittivity ε and the permeability μ.

We discuss the functional dependencies of the microscopic permittivity ε; the resulting function is called dielectric function. After that, we focus on spatially averaged fields and introduce the macroscopic dielectric function.

The division (27.3) of the fields, $\boldsymbol{E}_{\text{tot}} = \boldsymbol{E}_0 + \boldsymbol{E}_{\text{ext}} + \boldsymbol{E}_{\text{ind}}$ and $\boldsymbol{B}_{\text{tot}} = \boldsymbol{B}_0 + \boldsymbol{B}_{\text{ext}} + \boldsymbol{B}_{\text{ind},}$ reflects the experimental situation. At the same time, this division is particularly suitable for the theoretical treatment because in many cases the extra fields (the perturbation) can be regarded as *small*. Then a *linear* relation between the reaction and the disturbance of the system can be assumed.

The undisturbed electric field $\boldsymbol{E}_0$ in the unit cell of a solid or in an atom depends is of the size $E_0 \sim e/(1\,\text{Å})^2 \sim 10^9\,\text{V/cm}$, where e is the elementary charge. Compared to this, the additional (extra) fields are generally very small. Imagine, for example, a plate capacitor that generates a field of strength $E_{\text{ext}} = 10^3\,\text{V/cm}$ (Figure 28.1). The external disturbance $\boldsymbol{E}_{\text{ext}}$ is the cause for a reaction of the system. Due to $E_{\text{ext}} \ll E_0$ one can expect that the reaction of the system (e.g., the deviation of the charge distribution from the equilibrium distribution) depends *linearly* on the disturbance, i.e., $\varrho_{\text{ind}} \propto \boldsymbol{E}_{\text{ext}}$. This implies $\boldsymbol{E}_{\text{ind}} \propto \boldsymbol{E}_{\text{ext}}$ and $\boldsymbol{E} = \boldsymbol{E}_{\text{ext}} + \boldsymbol{E}_{\text{ind}} \propto \boldsymbol{E}_{\text{ext}}$. The assumption of a *linear response* to a small disturbance is used in many areas of physics.

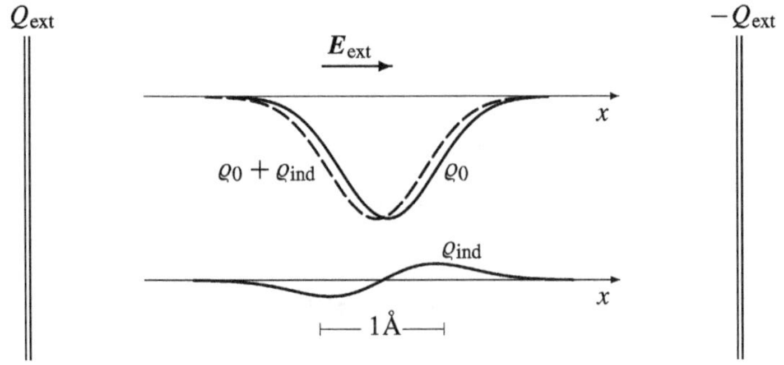

Figure 28.1 The charge density ϱ_{ext} on the plates of a capacitor generates the external field $\boldsymbol{E}_{\text{ext}}$; the capacitor plates shall be much larger and much further apart from each other than in the sketch. Matter is placed in the homogeneous field $\boldsymbol{E}_{\text{ext}}$. We are looking specifically at the charge distribution $\varrho_0 = -e\,|\psi(\boldsymbol{r})|^2$ of an electron in an atom (let $\psi(\boldsymbol{r})$ be the bound wave function of the electron). The external field now leads to a change in this charge distribution (solid line) to $\varrho_0 + \varrho_{\text{ind}}$ (dashed line). For the sketch, it was assumed that the change consists of a shift of the charge distribution. The following points are important: (i) The induced charge density varies on a microscopic scale; therefore $\varrho = \varrho_{\text{ext}} + \varrho_{\text{ind}}$ and $\boldsymbol{E} = \boldsymbol{E}_{\text{ext}} + \boldsymbol{E}_{\text{ind}}$ contain microscopic structures. (ii) The external field is generally very small, about $E_{\text{ext}}/E_0 \sim 10^{-6}$ for $E_{\text{ext}} = 10^3$ V/cm. The relative displacement and $|\varrho_{\text{ind}}/\varrho_0|$ are then also of the size 10^{-6} (and not 10^{-1} as in the sketch).

The linear relationship $\boldsymbol{E} \propto \boldsymbol{E}_{\text{ext}}$ and its magnetic analogue can be written in the form

$$\boxed{\boldsymbol{E} = \varepsilon^{-1}\boldsymbol{E}_{\text{ext}}, \quad \boldsymbol{B} = \mu\,\boldsymbol{B}_{\text{ext}}} \tag{28.1}$$

If the disturbance is made stronger (lets say by factor of 2) then reaction of the matter is correspondingly stronger (also by a factor of 2). Otherwise, the notation (28.1) is rather symbolic. In particular, the fields $\boldsymbol{E}$ and $\boldsymbol{E}_{\text{ext}}$ generally have different position and time dependencies and are not parallel to each other. This is discussed in detail in the next section.

In (28.1), the cause (disturbance) is on the right and the response (reaction, effect) on the left. The quantities ε^{-1} and μ are therefore called *response functions*. For historical reasons, ε (and not the response function ε^{-1}) was given the name *permittivity*. The quantity μ is called *permeability* (or magnetic permeability). The quantities are defined in such a way that $\varepsilon^{-1} = 1$ and $\mu = 1$ apply if there is no reaction of the matter or if there is no matter at all; this implies $\boldsymbol{E}_{\text{ind}} = 0$ and $\boldsymbol{B}_{\text{ind}} = 0$.

Microscopic Response Function

As a proxy for the occurring response relationships we examine the general form of $\boldsymbol{E} = \varepsilon^{-1} \boldsymbol{E}_{\text{ext}}$. We consider one after the other the tensor structure, the linearity, the dependence of state variables (e.g., temperature) and on position and time. We start with the first two points:

(i) The induced field $\boldsymbol{E}_{\text{ind}}$ is generally not parallel to the additional field $\boldsymbol{E}_{\text{ext}}$. Then also $\boldsymbol{E} = \boldsymbol{E}_{\text{ext}} + \boldsymbol{E}_{\text{ind}}$ is not parallel to $\boldsymbol{E}_{\text{ext}}$. Therefore, ε^{-1} represents a tensor of second rank with the components $\varepsilon^{-1}_{ij} = (\varepsilon^{-1})_{ij}$ (we refer to the 3-dimensional space).

(ii) For strong fields (such as in laser light), nonlinear terms can be important. Such terms lead to specific effects such as the scattering of waves from each other or a frequency doubling.

These two effects can be expressed in Cartesian components by

$$E_i = \sum_{j=1}^{3} \varepsilon^{-1}_{ij}\, E_{\text{ext},\,j} + \sum_{j,k=1}^{3} \gamma_{ijk}\, E_{\text{ext},\,j}\, E_{\text{ext},\,k} + \cdots \tag{28.2}$$

Within the framework of the linear response model, only the linear term is taken into account. This is sufficient for many applications.

We add the possible functional dependencies:

$$E_i(\boldsymbol{r}, t) = \sum_{j=1}^{3} \int d^3r' \int dt'\, \varepsilon^{-1}_{ij}(\boldsymbol{r}, \boldsymbol{r}', t - t'; T, P)\, E_{\text{ext},\,j}(\boldsymbol{r}', t') \tag{28.3}$$

This describes the following effects:

(iii) The reaction of matter to disturbances depends on the thermodynamic state of matter. In the simplest case, the thermodynamic state of homogeneous matter is determined by the temperature T and the pressure P. The permittivity therefore depends (like other material parameters) on on the temperature and pressure. In the following, these dependencies are no longer explicitly displayed.

(iv) The disturbance of matter at the point $\boldsymbol{r}'$ at time t' can cause reactions at other positions $\boldsymbol{r}$ and at other (later) times t. As an analogy, consider a piano: The disturbance at a certain point (pushing down on a piano key) causes a reaction at other places and at later times (a damped vibration of the piano string and the sound resonator).

Due to the homogeneity of time, the relationship between excitation and reaction can only depend on the time difference $t - t'$. (The state of the undisturbed matter is assumed to be time-independent). Matter with its microscopic structure, on the other hand, is not spatially homogeneous; therefore ε^{-1}_{ij} may depend on $\boldsymbol{r}$ and $\boldsymbol{r}'$. Both points can be illustrated by the piano example.

The considered piece of matter should be so large that we can neglect surfaces effects. Then the spatial integration extends in (28.3) over the entire space. The time integration runs from $-\infty$ to $+\infty$.

The reaction ($\boldsymbol{E}_{\text{ind}}$) of the system can occur only *after* the disturbance ($\boldsymbol{E}_{\text{ext}}$); this condition is also called *causality*. If one subtracts in (28.3) on both sides $E_{\text{ext},\,i}(\boldsymbol{r}, t)$, the left-hand side becomes $E_{\text{ind},\,i}(\boldsymbol{r}, t)$, and on the right-hand side $\varepsilon^{-1}_{ij} - \delta_{ij}\,\delta(\boldsymbol{r} - \boldsymbol{r}')\,\delta(t - t')$ takes the place of ε^{-1}_{ij}. Due to the causality, $E_{\text{ind},\,i}(\boldsymbol{r}, t)$ must disappear before the perturbation, i.e., for $t < t'$. This means

$$\varepsilon^{-1}_{ij}(\boldsymbol{r}, \boldsymbol{r}', t - t') = 0 \quad \text{for } t < t' \quad \text{(causality)} \tag{28.4}$$

In the crystalline solid, the spatial dependence of the (microscopic) dielectric function may be complicated. In particular, a constant external $\boldsymbol{E}_{\text{ext}}$-field also leads to internal $\boldsymbol{E}$-fields displaying variations within a unit cell. In the following, we limit ourselves to a simple macroscopic approximation.

Macroscopic Response Function

We perform the spatial averaging (27.9) in (28.3):

$$\langle E_i \rangle(\boldsymbol{r}, t) = \sum_{j=1}^{3} \int d^3r' \int dt' \; \langle \varepsilon^{-1}_{ij} \rangle(\boldsymbol{r} - \boldsymbol{r}', t - t') \; E_{\text{ext},\,j}(\boldsymbol{r}', t') \tag{28.5}$$

The averaging is denoted by the brackets at ε^{-1}_{ij} and $E - i$. The r'-integration describes the potential non-locality; a disturbance $E_{\text{ext},\,j}(\boldsymbol{r}', t')$ at $\boldsymbol{r}'$ may lead to reactions in E_i at $\boldsymbol{r}$.

After the spatial averaging, the matter represents a *homogeneous* medium[1] We consider only one type of substance and no boundaries

[1]The *medium* is what continuously fills the space in which the cause and effect (like waves) take place. Air, for example, is a medium for sound waves.

of the matter (which would violate homogeneity). Due to this spatial homogeneity, the response function can then only depend on the difference $\boldsymbol{r} - \boldsymbol{r}'$.

In the spatial and temporal Fourier transform

$$f(\boldsymbol{k}, \omega) = \int d^3r \int dt \; f(\boldsymbol{r}, t) \exp(\mathrm{i}\boldsymbol{k} \cdot \boldsymbol{r}) \exp(\mathrm{i}\omega t) \tag{28.6}$$

$$f(\boldsymbol{r}, t) = \frac{1}{(2\pi)^4} \int d^3k \int d\omega \; f(\boldsymbol{k}, \omega) \exp(-\mathrm{i}\boldsymbol{k} \cdot \boldsymbol{r}) \exp(-\mathrm{i}\omega t) \tag{28.7}$$

we use the same symbol f for the function and its Fourier transform; the distinction between both quantities is given by the arguments. We perform this transformation on both sides in (28.5). The convolution on the right-hand side becomes a product:

$$\langle E_i \rangle(\boldsymbol{k}, \omega) = \sum_{j=1}^{3} \langle \varepsilon^{-1}_{ij} \rangle(\boldsymbol{k}, \omega) \; E_{\text{ext}, j}(\boldsymbol{k}, \omega) \tag{28.8}$$

We now neglect the $\boldsymbol{k}$-dependence of the response function,

$$\langle \varepsilon^{-1}_{ij} \rangle(\boldsymbol{k}, \omega) \approx \langle \varepsilon^{-1}_{ij} \rangle(0, \omega) \tag{28.9}$$

Thus, Equation (28.8) becomes

$$\langle E_i \rangle(\boldsymbol{r}, \omega) = \sum_{j=1}^{3} \langle \varepsilon^{-1}_{ij} \rangle(0, \omega) \; E_{\text{ext}, j}(\boldsymbol{r}, \omega) \tag{28.10}$$

Due to (28.9), the fields $\boldsymbol{E}_{\text{ext}}$ and $\boldsymbol{E}$ have the same spatial dependency. With this approximation, we neglect non-localities in (28.5); in concrete terms, this means, for example, the spatial propagation of a localized disturbance.

Previously, the perturbation ($\boldsymbol{E}_{\text{ext}}$) was always on the right and the reaction ($\boldsymbol{E}_{\text{ind}}$ on the left (contained in $\boldsymbol{E}$)). However, it is generally more common to express $\boldsymbol{E}_{\text{ext}}$ as a function of $\boldsymbol{E}$. To do this, we introduce the matrix that is inverse to $\langle \varepsilon^{-1} \rangle = (\langle \varepsilon^{-1}_{ij} \rangle)$:

$$\varepsilon^{\text{macro}}_{ij}(\omega) = \left(\frac{1}{\langle \varepsilon^{-1} \rangle(0, \omega)} \right)_{ij} \tag{28.11}$$

The index "macro" indicates that this is now a macroscopic quantity. With (28.11), (28.10) becomes

$$\begin{aligned} E_{\text{ext},\,i}(\boldsymbol{r},\omega) &= \sum_{j=1}^{3} \varepsilon_{ij}^{\text{macro}}(\omega)\,\langle E_j\rangle(\boldsymbol{r},\omega) \\ E_{\text{ext},\,i}(\boldsymbol{r},\omega) &= \sum_{j=1}^{3} \varepsilon_{ij}(\omega)\,E_j(\boldsymbol{r},\omega) \quad \text{(simplified notation!)} \end{aligned} \tag{28.12}$$

The first line shows the result in the previous notation. To simplify the notation, we omit from now on the averaging bracket and the index "macro" (second line).

As a final simplification, we restrict ourselves to isotropic media. If no direction is extinguished, $\boldsymbol{E}$ must be parallel to $\boldsymbol{E}_{\text{ext}}$, i.e.,

$$\varepsilon_{ij}(\omega) = \varepsilon(\omega)\,\delta_{ij} \quad \text{(isotropic medium)} \tag{28.13}$$

Isotropic media are in particular liquids and gases. Solid crystalline bodies on the other hand, generally have polarizabilities that are different for the different crystal axes. Specifically for a cubic crystal, however, (28.13) applies, too. With (28.13), (28.12) becomes

$$\boxed{\boldsymbol{E}_{\text{ext}}(\boldsymbol{r},\omega) = \varepsilon(\omega)\,\boldsymbol{E}(\boldsymbol{r},\omega) \quad \text{(homogeneous, isotropic case)}} \tag{28.14}$$

We denote $\varepsilon(\omega)$ as the *dielectric function.* This dielectric function contains much less information than in the microscopic permittivity ε in (28.3). The dielectric function $\varepsilon(\omega)$ is dimensionless. Due to the use of the factor $\exp(-\mathrm{i}\omega t)$, $\varepsilon(\omega)$ is generally complex. The limiting case $\omega \to 0$ is called *dielectric constant.*

$$\varepsilon = \varepsilon(0) \quad \text{(dielectric constant)} \tag{28.15}$$

The dielectric function $\varepsilon(\omega)$ is a material property. For water, the experimental dielectric function is sketched in Figure 34.3 and 34.4.

In the magnetic case we have $\mu_{ij}^{\rm macro} = \langle \mu_{ij} \rangle$ and $\boldsymbol{B} = \mu \boldsymbol{B}_{\rm ext}$; here $\boldsymbol{B}$ and $\boldsymbol{B}_{\rm ext}$ have their logical position from the beginning (i.e., cause at right, effect at left, in contrast to (28.11)). Analogous to (28.14) we obtain

$$\boldsymbol{B}(\boldsymbol{r}, \omega) = \mu(\omega)\, \boldsymbol{B}_{\rm ext}(\boldsymbol{r}, \omega) \quad \text{(homogeneous, isotropic case)} \qquad (28.16)$$

The assumptions under which we have derived the macroscopic response relations (28.14) and (28.16) are summarized by "homogeneous, isotropic case". Thereby, it is assumed that the matter fills the entire space because the boundaries would violate homogeneity and isotropy.

Chapter 29

Macroscopic Maxwell Equations

A spatial averaging leads from the microscopic to the macroscopic Maxwell equations. For the occurring quantities, we introduce simple approximations: The induced charge distribution is represented by the dipole moments of the individual atoms (or molecules, or unit cells). The dielectric function $\varepsilon = 1 + 4\pi n_0 \alpha_e$ *is given by the density* n_0 *and the electric polarizability* α_e *of the atoms. Eventually, the energy balance of a system of electromagnetic fields and charged particles is investigated.*

The spatial averaging (27.9) interchanges with the partial differentiations:

$$\frac{\partial \langle A \rangle}{\partial x} = \int d^3r'\; A(\boldsymbol{r}', t)\; \frac{\partial f(\boldsymbol{r} - \boldsymbol{r}')}{\partial x} \;=\; -\int d^3r'\; A(\boldsymbol{r}', t)\; \frac{\partial f(\boldsymbol{r} - \boldsymbol{r}')}{\partial x'}$$

$$\stackrel{\text{p.i.}}{=} \int d^3r'\; \frac{\partial A(\boldsymbol{r}', t)}{\partial x'}\; f(\boldsymbol{r} - \boldsymbol{r}') \;=\; \left\langle \frac{\partial A}{\partial x} \right\rangle \qquad (29.1)$$

Therefore, we obtain $\langle \operatorname{div} \boldsymbol{A} \rangle = \operatorname{div} \langle \boldsymbol{A} \rangle$ and $\langle \operatorname{curl} \boldsymbol{A} \rangle = \operatorname{curl} \langle \boldsymbol{A} \rangle$. Of course, the spatial averaging also interchanges with a time derivative. A spatial averaging over the Maxwell equations (27.5) or (27.6), (27.7) results in Maxwell's equations of the same form but for the macroscopic quantities.

We average over the Maxwell equations (27.5) using $\varrho = \varrho_{\text{ext}} + \varrho_{\text{ind}}$ and $\boldsymbol{j} = \boldsymbol{j}_{\text{ext}} + \boldsymbol{j}_{\text{ind}}$:

$$\operatorname{div} \langle \boldsymbol{E} \rangle = 4\pi \big(\langle \varrho_{\text{ext}} \rangle + \langle \varrho_{\text{ind}} \rangle \big), \quad \operatorname{curl} \langle \boldsymbol{E} \rangle + \frac{1}{c}\, \frac{\partial \langle \boldsymbol{B} \rangle}{\partial t} = 0$$

$$\operatorname{div} \langle \boldsymbol{B} \rangle = 0, \quad \operatorname{curl} \langle \boldsymbol{B} \rangle - \frac{1}{c}\, \frac{\partial \langle \boldsymbol{E} \rangle}{\partial t} = \frac{4\pi}{c} \big(\langle \boldsymbol{j}_{\text{ext}} \rangle + \langle \boldsymbol{j}_{\text{ind}} \rangle \big) \qquad (29.2)$$

Polarization

We derive a approximation for the averaged charge density $\langle \varrho_{\rm ind} \rangle$. To do this, we divide the matter into microscopic units. In a gas or a liquid, we consider the individual atoms or molecules as units, in a crystal the elementary cells. The units shall be neutral, i.e., their total charge shall disappear. The change in the charge distribution in the νth unit (due to the additional fields) is denoted by $\varDelta\varrho_\nu(\boldsymbol{r} - \boldsymbol{r}_\nu, t)$. Here $\boldsymbol{r}_\nu$ is the vector to the center of the νth unit (the center of mass for an atom). This vector may be time-dependent, $\boldsymbol{r}_\nu = \boldsymbol{r}_\nu(t)$.

The induced charge is obtained by summing over all units:

$$\varrho_{\rm ind}(\boldsymbol{r}, t) = \sum_\nu \varDelta\varrho_\nu(\boldsymbol{r} - \boldsymbol{r}_\nu, t) \tag{29.3}$$

The index ν of $\varDelta\varrho_\nu$ allows for different kinds of units (atoms). If all units are identical, then this index is canceled.

The vector $\boldsymbol{r}' = \boldsymbol{r} - \boldsymbol{r}_\nu$ in the argument of $\varrho(\boldsymbol{r}', t)$ points from the center of the unit to the position under consideration. Due to the finite size of the microscopic units, the following applies:

$$\varDelta\varrho_\nu(\boldsymbol{r}', t) = \begin{cases} 0 & |\boldsymbol{r}'| \gg a \\ \text{arbitrary} & |\boldsymbol{r}'| \lesssim a \end{cases} \tag{29.4}$$

Here a denotes the lengths extension of the microscopic units (atom, molecule, elementary cell).

An additional field can change the shape of the charge distribution, but not the total charge of a unit. Therefore we get

$$\int d^3r\ \varDelta\varrho_\nu(\boldsymbol{r}, t) = 0 \tag{29.5}$$

Taking into account the properties (29.4) and (29.5), we perform the spatial averaging over the induced charge distribution (29.3):

$$\begin{aligned} \langle \varrho_{\rm ind} \rangle(\boldsymbol{r}, t) &= \Big\langle \sum_\nu \varDelta\varrho_\nu \Big\rangle = \sum_\nu \int d^3\tilde{r}\ \varDelta\varrho_\nu(\tilde{\boldsymbol{r}} - \boldsymbol{r}_\nu, t)\ f(\boldsymbol{r} - \tilde{\boldsymbol{r}}) \\ &= \sum_\nu \int d^3r'\ \varDelta\varrho_\nu(\boldsymbol{r}', t)\ f(\boldsymbol{r} - \boldsymbol{r}_\nu - \boldsymbol{r}') \end{aligned}$$

$$= \sum_{\nu} f(\boldsymbol{r} - \boldsymbol{r}_\nu) \underbrace{\int d^3r' \; \Delta\varrho_\nu(\boldsymbol{r}', t)}_{=0}$$

$$- \sum_{\nu} \nabla f(\boldsymbol{r} - \boldsymbol{r}_\nu) \cdot \underbrace{\int d^3r' \, \boldsymbol{r}' \; \Delta\varrho_\nu(\boldsymbol{r}', t)}_{=\boldsymbol{p}_\nu(t)} + \cdots \tag{29.6}$$

The function $f(\boldsymbol{r} - \boldsymbol{r}_\nu - \boldsymbol{r}')$ was expanded into powers of $\boldsymbol{r}'$. Due to $\partial^n f/\partial x^n \sim f/b^n$ and $r' \lesssim a$, this is a expansion into powers of $a/b \ll 1$, where b is the range of the averaging function.

The first non-vanishing contribution of the expansion (29.6) can be expressed by the dipole moments $\boldsymbol{p}_\nu(t)$ of the microscopic units (atom, elementary cell). Neglecting the next terms (higher multipole moments) yields

$$\langle \varrho_{\text{ind}} \rangle(\boldsymbol{r}, t) = -\sum_{\nu} \boldsymbol{p}_\nu(t) \cdot \nabla f(\boldsymbol{r} - \boldsymbol{r}_\nu) = -\nabla \cdot \Big(\sum_{\nu} \boldsymbol{p}_\nu(t)\, f(\boldsymbol{r} - \boldsymbol{r}_\nu) \Big)$$

$$= -\nabla \cdot \int d^3r' \sum_{\nu} \boldsymbol{p}_\nu \, \delta(\boldsymbol{r}' - \boldsymbol{r}_\nu)\, f(\boldsymbol{r} - \boldsymbol{r}')$$

$$= -\nabla \cdot \Big\langle \sum_{\nu} \boldsymbol{p}_\nu \, \delta\big(\boldsymbol{r}' - \boldsymbol{r}_\nu\big) \Big\rangle (\boldsymbol{r}, t) \tag{29.7}$$

The quantity to be averaged depends on $\boldsymbol{r}'$, the averaged quantity depends on $\boldsymbol{r}$. The result may also depend on the time, namely via $\boldsymbol{p}_\nu = \boldsymbol{p}_\nu(t)$ and via $\boldsymbol{r}_\nu = \boldsymbol{r}_\nu(t)$. By

$$\boldsymbol{P}(\boldsymbol{r}, t) = \Big\langle \sum_{\nu} \boldsymbol{p}_\nu \, \delta\big(\boldsymbol{r}' - \boldsymbol{r}_\nu\big) \Big\rangle (\boldsymbol{r}, t) = \frac{\text{electric dipole moment}}{\text{volume}} \tag{29.8}$$

we introduce the *polarization* $\boldsymbol{P}$. Sometimes, the dipole moment may be zero (for certain symmetries, for example for the cells of a silicon crystal). Then the expansion (29.6) might be continued to higher multipole moments.

From (29.7) and (29.8) follows

$$\operatorname{div} \boldsymbol{P} = -\langle \varrho_{\text{ind}} \rangle \tag{29.9}$$

Magnetization

The analogous treatment of $\langle \boldsymbol{j}_{\text{ind}} \rangle = \langle \sum \boldsymbol{j}_\nu \rangle$ results in

$$\operatorname{curl} \boldsymbol{M} = \frac{1}{c}\, \langle \boldsymbol{j}_{\text{ind}} \rangle - \frac{1}{c}\frac{\partial \boldsymbol{P}}{\partial t} \tag{29.10}$$

with the *magnetization*

$$\boldsymbol{M}(\boldsymbol{r}, t) = \Big\langle \sum_\nu \boldsymbol{\mu}_\nu\, \delta(\boldsymbol{r}' - \boldsymbol{r}_\nu) \Big\rangle (\boldsymbol{r}, t) = \frac{\text{magnetic dipole moment}}{\text{volume}} \tag{29.11}$$

The main contribution to $\langle \boldsymbol{j}_{\text{ind}} \rangle$ comes from the magnetic moments $\boldsymbol{\mu}_\nu = \int d^3r\, \boldsymbol{r} \times \boldsymbol{j}_\nu / 2c$ of the microscopic units.

The Fields *D* and *H*

In (29.2) we omit the averaging brackets for the fields,

$$\langle \boldsymbol{E} \rangle \to \boldsymbol{E}, \quad \langle \boldsymbol{B} \rangle \to \boldsymbol{B} \quad \text{(simplified notation!)} \tag{29.12}$$

The additional sources are macroscopic from the outset, i.e., $\langle \varrho_{\text{ext}} \rangle = \varrho_{\text{ext}}$ and $\langle \boldsymbol{j}_{\text{ext}} \rangle = \boldsymbol{j}_{\text{ext}}$. By

$$\boldsymbol{E} = \boldsymbol{D} - 4\pi \boldsymbol{P}, \quad \boldsymbol{B} = \boldsymbol{H} + 4\pi \boldsymbol{M} \tag{29.13}$$

we define the *electric displacement* (also electric displacement field or electric induction) $\boldsymbol{D}$ and the *magnetic field strength* $\boldsymbol{H}$. The designations are historically conditioned and partly misleading. It should therefore be emphasized once again that $\boldsymbol{E}$ and $\boldsymbol{B}$ are the basic fields (because of (27.4), or (16.7) in vacuum).

We insert $\langle \varrho_{\text{ind}} \rangle = -\operatorname{div} \boldsymbol{P}$ and $\langle \boldsymbol{j}_{\text{ind}} \rangle = c \operatorname{curl} \boldsymbol{M} + \dot{\boldsymbol{P}}$ into (29.2). With the notation (29.12) and the designations (29.13), we obtain the *macroscopic Maxwell equations*

$$\boxed{\begin{aligned} &\operatorname{div} \boldsymbol{D} = 4\pi\, \varrho_{\text{ext}}, \quad && \operatorname{curl} \boldsymbol{E} + \frac{1}{c}\frac{\partial \boldsymbol{B}}{\partial t} = 0 \\ &\operatorname{div} \boldsymbol{B} = 0, && \operatorname{curl} \boldsymbol{H} - \frac{1}{c}\frac{\partial \boldsymbol{D}}{\partial t} = \frac{4\pi}{c}\, \boldsymbol{j}_{\text{ext}} \end{aligned}} \tag{29.14}$$

As already discussed, the position of the fields do not reflect their logical relation. Rather, the basic fields $\boldsymbol{E}$ and $\boldsymbol{B}$ correspond to each other, and on the other hand, the fields $\boldsymbol{D} = \boldsymbol{E}_{\text{ext}}$ and $\boldsymbol{H} = \boldsymbol{B}_{\text{ext}}$.

Macroscopic Maxwell Equations with ε and μ

In the following we use the quantities ε and μ as in (29.22), (29.23), even if the homogeneous, isotropic case is not realized. This usage corresponds to the textbooks common presentation of macroscopic electrodynamics. In the homogeneous, isotropic case (29.20), this coincides with the use in Chapter 28.

From $\boldsymbol{D} = \varepsilon \boldsymbol{E}$, $\boldsymbol{D} = \boldsymbol{E} + 4\pi \boldsymbol{P}$, and $\boldsymbol{P} = \chi_{\text{e}} \boldsymbol{E}$ follows

$$\varepsilon = 1 + 4\pi\, \chi_{\text{e}} \tag{29.24}$$

Using the electric susceptibility of (29.17) we get

$$\varepsilon(\boldsymbol{r}, \omega) = 1 + 4\pi\, n_0(\boldsymbol{r})\, \alpha_{\text{e}}(\omega) \tag{29.25}$$

The spatial dependence of $n_0(\boldsymbol{r})$ takes into account inhomogeneities or spatial boundaries of the matter. We can also admit a position dependence of $\alpha_{\text{e}}(\boldsymbol{r}, \omega)$ to allow for different polarizabilities of various materials. In the following, we use the form

$$\boldsymbol{D}(\boldsymbol{r}, \omega) = \varepsilon(\boldsymbol{r}, \omega)\, \boldsymbol{E}(\boldsymbol{r}, \omega) \tag{29.26}$$

$$\boldsymbol{B}(\boldsymbol{r}, \omega) = \mu(\boldsymbol{r}, \omega)\, \boldsymbol{H}(\boldsymbol{r}, \omega) \tag{29.27}$$

of the response relations. The time-dependent fields then result from

$$\boldsymbol{D}(\boldsymbol{r}, t) = \frac{1}{2\pi} \int_0^{\infty} d\omega\; \varepsilon(\boldsymbol{r}, \omega)\, \boldsymbol{E}(\boldsymbol{r}, \omega) \exp(-\mathrm{i}\omega t) \tag{29.28}$$

and $\boldsymbol{E}(\boldsymbol{r}, t) = \int_0^{\infty} d\omega\, \boldsymbol{E}(\boldsymbol{r}, \omega) \exp(-\mathrm{i}\omega t)/2\pi$.

For (29.28) we write $\boldsymbol{D} = \varepsilon \boldsymbol{E}$ for short. With this and $\boldsymbol{H} = \boldsymbol{B}/\mu$, Equation (29.14) becomes

$$\boxed{\begin{aligned} &\operatorname{div}(\varepsilon \boldsymbol{E}) = 4\pi\, \varrho_{\text{ext}}, \quad && \operatorname{curl} \boldsymbol{E} + \frac{1}{c}\frac{\partial \boldsymbol{B}}{\partial t} = 0 \\ &\operatorname{div} \boldsymbol{B} = 0, && \operatorname{curl} \frac{\boldsymbol{B}}{\mu} - \frac{1}{c}\frac{\partial(\varepsilon \boldsymbol{E})}{\partial t} = \frac{4\pi}{c}\, \boldsymbol{j}_{\text{ext}} \end{aligned}} \tag{29.29}$$

We consider the sources ϱ_{ext}, $\boldsymbol{j}_{\text{ext}}$ and the material parameters $\varepsilon(\boldsymbol{r}, \omega)$ and $\mu(\boldsymbol{r}, \omega)$ as given. Then, Equations (29.29) together with (29.26) and (29.27) represent is a closed system of equations for the fields $\boldsymbol{E}$ and $\boldsymbol{B}$. In all following applications we apply the Maxwell equations in this form.

Energy Balance

We consider the Lorentz force on a point charge q; this charge is part of the charge density ϱ_{ext}. A spatial averaging over the in effective force in matter (27.4) yields $\langle \boldsymbol{F} \rangle = q\,(\langle \boldsymbol{E} \rangle + \boldsymbol{v} \times \langle \boldsymbol{B} \rangle / c)$. As for the macroscopic Maxwell equations we omit the averaging brackets:

$$\boldsymbol{F} = q\left(\boldsymbol{E}(\boldsymbol{r}, t) + \frac{\boldsymbol{v}}{c} \times \boldsymbol{B}(\boldsymbol{r}, t)\right) \tag{29.30}$$

Since this Lorentz force has the usual form, the considerations of the section "Energy balance" in Chapter 16 can be easily transferred. Analogous to (16.18), $\boldsymbol{j}_{\text{ext}} \cdot \boldsymbol{E}$ is expressed by the fields with the help of Maxwell's equations (29.14):

$$\begin{aligned} \boldsymbol{j}_{\text{ext}} \cdot \boldsymbol{E} &= \frac{c}{4\pi}\, \boldsymbol{E} \cdot \operatorname{curl} \boldsymbol{H} - \frac{1}{4\pi}\, \boldsymbol{E} \cdot \frac{\partial \boldsymbol{D}}{\partial t} \\ &= -\frac{c}{4\pi}\, \operatorname{div}(\boldsymbol{E} \times \boldsymbol{H}) + \frac{c}{4\pi}\, \boldsymbol{H} \cdot \operatorname{curl} \boldsymbol{E} - \frac{1}{4\pi}\, \boldsymbol{E} \cdot \frac{\partial \boldsymbol{D}}{\partial t} \\ &= -\operatorname{div} \boldsymbol{S} - \frac{\partial w_{\text{em}}}{\partial t} \end{aligned} \tag{29.31}$$

In the last step, we have introduced the Poynting vector

$$\boldsymbol{S}(\boldsymbol{r}, t) = \frac{c}{4\pi}\, \boldsymbol{E} \times \boldsymbol{H} \quad \text{(energy flux density)} \tag{29.32}$$

and $\partial w_{\text{em}}/\partial t = (\boldsymbol{E} \cdot \dot{\boldsymbol{D}} + \boldsymbol{H} \cdot \dot{\boldsymbol{B}})/4\pi$. For $\boldsymbol{D} = \varepsilon \boldsymbol{E}$ and $\boldsymbol{H} = \boldsymbol{B}/\mu$ with time-independent quantities ε and μ we get

$$w_{\text{em}} = \frac{1}{8\pi}\left(\boldsymbol{E} \cdot \boldsymbol{D} + \boldsymbol{H} \cdot \boldsymbol{B}\right) \quad \text{(energy density)} \tag{29.33}$$

The result of (29.31)

$$\frac{\partial w_{\text{em}}}{\partial t} + \operatorname{div} \boldsymbol{S} = -\boldsymbol{j}_{\text{ext}} \cdot \boldsymbol{E} \quad \text{(Poynting theorem)} \tag{29.34}$$

is the Poynting theorem in matter.

Analogous to (16.16) and (16.17), we obtain $dE_{\text{mat}}/dt = \int d^3r\, \boldsymbol{j}_{\text{ext}} \cdot \boldsymbol{E}$ for the kinetic energy change of the material charge carriers. We integrate this over a coherent volume V with the surface $a(V)$:

$$\frac{dE_{\text{em}}}{dt} + \frac{dE_{\text{mat}}}{dt} = -\oint_{a(V)} d\boldsymbol{a} \cdot \boldsymbol{S}(\boldsymbol{r}, t) \tag{29.35}$$

The prerequisite for this is that only an electromagnetic energy flow through the surface $a(V)$ (no matter flow).

We consider a closed system and choose the volume V such that the system lies completely within V. Then the integral over the surface $a(V)$ disappears, and (29.35) becomes

$$E = E_{\text{mat}} + E_{\text{em}} = \text{const.} \quad \text{(closed system)} \tag{29.36}$$

The volume V could also be the entire room. For the vanishing of the surface integral it is then sufficient that the fields fall off like $1/r^2$ for $r \to \infty$.

The quantity (29.36) has the dimension of an energy. The energy quantity, which is constant for a closed system, is the total energy of the system. The system consists of fields and particles, and it is known

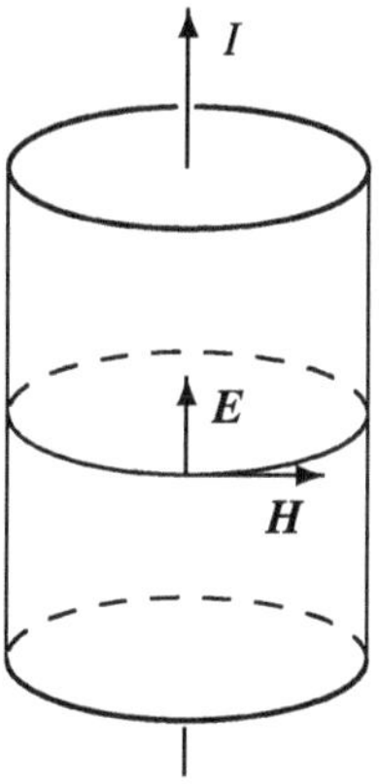

Figure 29.1 A constant current I flows homogeneously through a wire. At the cylindrical surface of the wire the energy current density is $\boldsymbol{S} = c\boldsymbol{E} \times \boldsymbol{H}/4\pi$ is directed inwards. The incoming field energy is transferred to the mobile charges in the form of kinetic energy. Subsequently, the charges transfer their energy to the metal by collisions.

that E_{mat} is the energy of the particles. From this follows that E_{em} is the energy and w_{em} is the *energy density* of the electromagnetic field. The corresponding *energy current density* is then $\boldsymbol{S}$.

As an application of (29.35), we consider a wire in Figure 29.1 with the constant current density j_{ext}. The time-independent magnetic field is known from Chapter 14; there is also an electric field necessary to keep the current flowing. In this case, $\boldsymbol{S}$ points inwards everywhere on the surface of the wire; field energy therefore flows into the wire. Since the fields are constant, $\partial w_{\text{em}}/\partial t = 0$. According to (29.35), the field energy flowing into the wire is converted to the energy E_{mat} of the electrons that carry the current. The electrons themselves pass on the energy to the lattice (or phonons) by scattering. This second step is not included in the balance equation (29.35). The energy flowing into the wire is finally converted into heat.

In the stationary case, the field energy flowing into the wire (and being eventually transformed into heat) is provided, for example, by a battery connected to the ends of a real wire.

Chapter 30

First Applications

We present first applications of the macroscopic Maxwell equations. For different materials (for example, air and glass), we derive boundary conditions for the fields at the interface between two media. Then we solve two instructive problems of the macroscopic electrostatics.

Boundary Conditions

Two different media shall fill the space and have a common interface (boundary surface) R, see Figure 30.1. We want to establish the boundary conditions for the fields at the common boundary of the two media. On this boundary, we allow for an external surface charge σ_{ext} and an external surface current $\boldsymbol{J}_{\text{ext}}$. The surface current equals $\boldsymbol{J}_{\text{ext}} = \sigma_{\text{ext}}\,\boldsymbol{v}$ if all charge carriers in σ_{ext} have the same velocity $\boldsymbol{v}$ (parallel to the surface).

From two small surface elements a that are parallel to the boundary area, we may form a volume element ΔV (Figure 30.1). The distance between the two surface elements shall be arbitrarily small. We apply the Gaussian theorem to $\operatorname{div}\boldsymbol{D} = 4\pi\,\varrho_{\text{ext}}$ and ΔV:

$$\oint_{A(\Delta V)} d\boldsymbol{A}\cdot\boldsymbol{D} = a\boldsymbol{n}\cdot(\boldsymbol{D}_2 - \boldsymbol{D}_1) = 4\pi\int_{\Delta V} d^3r\;\varrho_{\text{ext}} = 4\pi\,q_{\text{ext}} \qquad (30.1)$$

Here, $\boldsymbol{n}$ is the normal vector shown in Figure 30.1, and q_{ext} is the charge contained in ΔV. With the surfaces charge $\sigma_{\text{ext}} = q_{\text{ext}}/a$, we obtain from this

$$(\boldsymbol{D}_2 - \boldsymbol{D}_1)\cdot\boldsymbol{n} = 4\pi\,\sigma_{\text{ext}} \qquad (30.2)$$

An analogous application of Gaussian theorem to $\operatorname{div}\boldsymbol{B} = 0$ results in

$$(\boldsymbol{B}_2 - \boldsymbol{B}_1)\cdot\boldsymbol{n} = 0 \qquad (30.3)$$

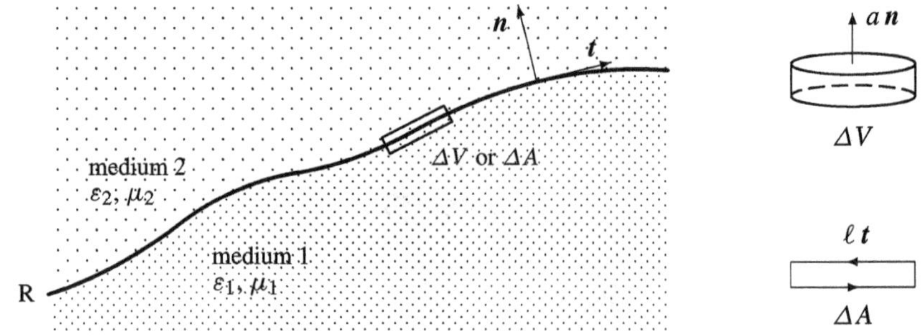

Figure 30.1 At the interface R between two homogeneous media, there are boundary conditions for the fields. For their derivation, a volume element ΔV and a surface element ΔA are used, the shapes of which are sketched on the right.

From two small line elements of length ℓ, which are parallel to the boundary surface, we form the rectangular area ΔA (Figure 30.1). The distance between the two line elements shall be arbitrarily small. For this area, we use Stokes' theorem for $\operatorname{curl} \boldsymbol{H} = (4\pi/c)\boldsymbol{j}_{\text{ext}} + \dot{\boldsymbol{D}}/c$:

$$\oint_C d\boldsymbol{s} \cdot \boldsymbol{H} = \ell\, \boldsymbol{t} \cdot (\boldsymbol{H}_2 - \boldsymbol{H}_1) = \frac{1}{c} \int_{\Delta A} d\boldsymbol{A} \cdot \left(4\pi\, \boldsymbol{j}_{\text{ext}} + \frac{\partial \boldsymbol{D}}{\partial t} \right) = \frac{4\pi}{c}\, i_{\text{ext}} \tag{30.4}$$

Let the quantity $\partial \boldsymbol{D}/\partial t$ be finite at the boundary; then this contribution goes to zero together with ΔA. On the right-hand side, i_{ext} is the current through ΔA. After division by ℓ, the surfaces current on the right-hand side is $J_{\text{ext}} = i_{\text{ext}}/\ell$ (current/length = (charge/surface) × velocity). There are two independent tangent vectors $\boldsymbol{t}$ and thus two conditions of the form (30.4), that can be summarized by

$$\boldsymbol{n} \times (\boldsymbol{H}_2 - \boldsymbol{H}_1) = \frac{4\pi}{c}\, \boldsymbol{J}_{\text{ext}} \tag{30.5}$$

Analogously, the application of Stokes' theorem to $\operatorname{curl} \boldsymbol{E} = -\dot{\boldsymbol{B}}/c$ leads to

$$\boldsymbol{n} \times (\boldsymbol{E}_2 - \boldsymbol{E}_1) = 0 \tag{30.6}$$

We summarize the conditions for the case that there are there are no extra charges on the boundary surface:

1. The tangential components of $\boldsymbol{E}$ are continuous.
2. The normal component of $\boldsymbol{D}$ is continuous ($\sigma_{\text{ext}} = 0$).

3. The normal component of $\boldsymbol{B}$ is continuous.
4. The tangential components of $\boldsymbol{H}$ are continuous ($\boldsymbol{J}_{\text{ext}} = 0$).

Electrostatic

From (29.29) we obtain the electrostatic case:

$$\text{div}\left(\varepsilon(\boldsymbol{r})\,\boldsymbol{E}(\boldsymbol{r})\right) = 4\pi\,\varrho_{\text{ext}} \tag{30.7}$$

$$\text{curl}\,\boldsymbol{E}(\boldsymbol{r}) = 0 \tag{30.8}$$

These are the basic equations of electrostatics in matter. The first equation may also be written in the form $\text{div}\,\boldsymbol{D} = 4\pi\,\varrho_{\text{ext}}$.

In the following, we specifically consider a vacuum region ($\varepsilon = 1$) and a region filled by matter with $\varepsilon > 1$. In this context, we refer to the matter as *dielectric*. Table 31.1 lists the dielectric constants ε of some substances. The dielectric is assumed to be an insulator (a non-conducting medium). At the interface between vacuum and dielectric, the boundary conditions (30.2) and (30.6) apply.

Dielectric in the capacitor

In a plate capacitor (plate area A_C, distance d) there is a cuboidal dielectric, Figure 30.2. The arrangement shall extend perpendicular to the z-direction far enough so that we can neglect any boundary effects. Then we can

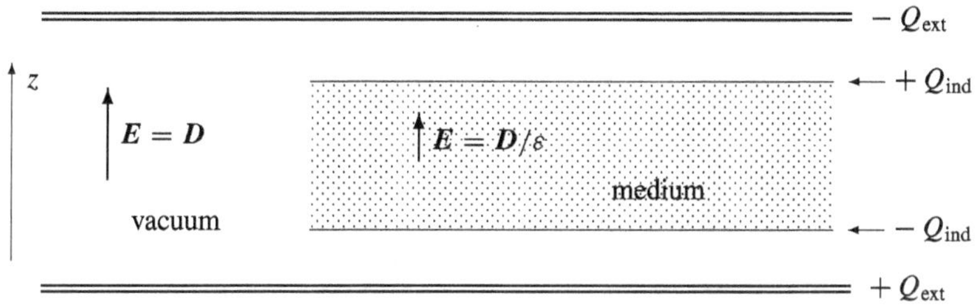

Figure 30.2 The charges $\pm Q_{\text{ext}}$ on the capacitor plates generate the field $\boldsymbol{D} = \boldsymbol{E}_{\text{ext}}$. A cuboidal piece of matter is placed in this field. Due to the boundary condition (30.2) the field $\boldsymbol{D}$ is the same in the vacuum and in the medium (no external surface charge at the dielectric). The polarization of this medium leads to the induced charges $\pm Q_{\text{ind}}$ at the surfaces; where is $Q_{\text{ind}} = Q_{\text{ext}}(\varepsilon - 1)/\varepsilon$. The induced charges partially shield the $\boldsymbol{E}$ field for $\varepsilon > 1$; the $\boldsymbol{E}$ field is weakened in the medium.

set $\boldsymbol{E} = E(z)\,\boldsymbol{e}_z$ for the electric field and $\boldsymbol{D} = D(z)\,\boldsymbol{e}_z$ for the electric displacement field.

We consider the (thin) capacitor plates as surface, to which we can apply (30.2). With $\boldsymbol{D}_1 = 0$ (outside the capacitor) and $\boldsymbol{D}_2 = \boldsymbol{D}$ (inside), and from $\sigma_{\text{ext}} = Q_{\text{ext}}/A_{\text{C}}$ (lower plate) then follows

$$\boldsymbol{D} = 4\pi\,\frac{Q_{\text{ext}}}{A_{\text{C}}}\,\boldsymbol{e}_z \quad \text{(with or without medium)} \tag{30.9}$$

In a first step, we consider the a vacuum only. According to (30.7), $\operatorname{div}\boldsymbol{D} = 0$ holds within the capacitor. For $\boldsymbol{D} = D(z)\,\boldsymbol{e}_z$ this means $dD(z)/dz = 0$, i.e., $D(z) = \text{const}$. This yields (30.9) applies everywhere between the plates. Now we admit the medium: The electric displacement field $\boldsymbol{D}$ continuous at the interface between the vacuum and the dielectric (no external surface charge density σ_{ext}). Therefore, (30.9) is valid in both, the vacuum and the dielectric.

With $\varepsilon_{\text{vac}} = 1$ and $\varepsilon_{\text{med}} = \varepsilon$ the field are $\boldsymbol{P} = (\boldsymbol{D} - \boldsymbol{E})/4\pi$ und $\boldsymbol{E} = \boldsymbol{D}/\varepsilon$:

$$\boldsymbol{E} = \begin{cases} \boldsymbol{D} \\ \dfrac{\boldsymbol{D}}{\varepsilon} \end{cases} \quad \text{and} \quad \boldsymbol{P} = \begin{cases} 0 & \text{(vacuum)} \\ \dfrac{\varepsilon - 1}{4\pi\,\varepsilon}\,\boldsymbol{D} & \text{(medium)} \end{cases} \tag{30.10}$$

At the dielectric-vacuum interface, we evaluate $\operatorname{div}\boldsymbol{P} = -\varrho_{\text{ind}}$ analogous to (30.1) and (30.2). This results in $(\boldsymbol{P}_2 - \boldsymbol{P}_1)\cdot\boldsymbol{n} = -\sigma_{\text{ind}}$, i.e., the induced surfaces charges

$$\sigma_{\text{ind}} = \pm\,\frac{\varepsilon - 1}{\varepsilon}\,\frac{Q_{\text{ext}}}{A_{\text{C}}} \tag{30.11}$$

The plus sign applies to the upper surface of the medium in Figure 30.2, the minus sign for the lower one.

The polarization $\boldsymbol{P}$ of the dielectric and the induced surface charges σ_{ind} are related as follows: Due to the external field, the positive and negative charges in each microscopic unit (atom, elementary cell) are slightly shifted against each other. In the macroscopic approximation, the positive and negative charges each form a homogeneously charged cuboid. The polarization is due to the small displacements of these two oppositely charged cuboids. As a result, the positive and negative charges cancel out in the interior; at the boundary, however, they lead to an induced surfaces charge σ_{ind}. These induced charges are the visible expression of the homogeneous volume polarization of the dielectric.

The induced charge Q_{ind} on the dielectric is proportional to the charge Q_{ext} on the capacitor (in both cases we refer to some finite area):

$$Q_{\text{ind}} = \frac{\varepsilon - 1}{\varepsilon}\, Q_{\text{ext}} \stackrel{\varepsilon\to\infty}{\longrightarrow} Q_{\text{ext}} \tag{30.12}$$

For $\varepsilon = 2$ is $Q_{\text{ind}} = Q_{\text{ext}}/2$; the field of the capacitor charges is just shielded to the half. The dielectric constant of a metal is infinite, (31.30). In this case, $Q_{\text{ind}} = Q_{\text{ext}}$. The field of the capacitor charges is completely shielded, so that $\boldsymbol{E} = 0$ in the metal. According to (30.11), there is a continuous transition from $\sigma_{\text{ind}} = 0$ (vacuum, $\varepsilon = 1$) to $\sigma_{\text{ind}} = \pm\sigma_{\text{ext}}$ for $\varepsilon \to \infty$ (metal, complete shielding).

We consider the voltage U of the capacitor charged with $\pm Q_{\text{ext}}$ (plate area A_{C}, distance d). Without medium, the voltage reads $U_{\text{vac}} = E_{\text{vac}}\, d$. The medium shall now fills the whole space between the plates. Then the voltage drops to $U_{\text{med}} = E_{\text{med}}\, d = U_{\text{vac}}/\varepsilon$. From the ratio of the voltages or capacitance values, it is easy to read off the dielectric constant ε:

$$\varepsilon = \frac{U_{\text{vac}}}{U_{\text{med}}} = \frac{C_{\text{med}}}{C_{\text{vac}}} \tag{30.13}$$

One may also apply an alternating voltage $U = U_0 \cos(\omega t)$ and measure the maximum charge Q_0 with and without medium. If the quasi-static approximation (Chapter 26) is permissible, then the measurement yields the capacitance $C = Q_0/U_0$ and the dielectric function $\varepsilon(\omega) = C_{\text{med}}/C_{\text{vac}}$.

Point charge and dielectric

Let the half-space $x \geq 0$ be filled by a dielectric with $\varepsilon > 1$ (Figure 30.3). In the vacuum range there is a point charge of strength q at $-a\,\boldsymbol{e}_x$. The electrostatic field in the entire space shall be determined. With $\varrho_{\text{ext}} = q\,\delta(\boldsymbol{r} + a\,\boldsymbol{e}_z)$, the Maxwell equations (30.7), (30.8) read

$$\operatorname{div}\big(\varepsilon(\boldsymbol{r})\,\boldsymbol{E}(\boldsymbol{r})\big) = 4\pi\, q\,\delta(\boldsymbol{r} + a\,\boldsymbol{e}_x), \quad \operatorname{curl}\boldsymbol{E}(\boldsymbol{r}) = 0 \tag{30.14}$$

where

$$\varepsilon(\boldsymbol{r}) = \begin{cases} 1 & (x < 0) \\ \varepsilon & (x > 0) \end{cases} \tag{30.15}$$

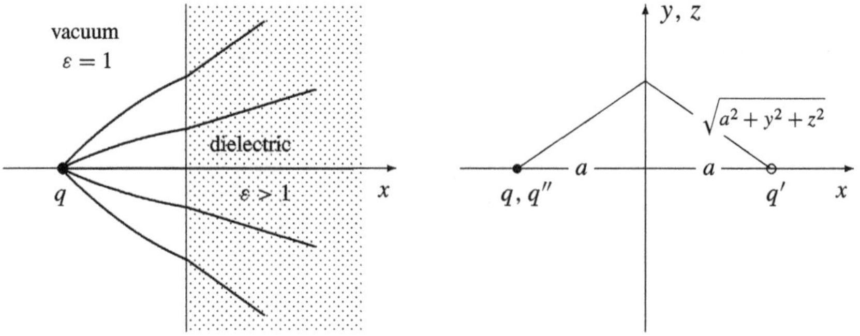

Figure 30.3 The range $x \geq 0$ is filled by a dielectric (dotted). A point charge polarizes the dielectric. The field can be determined with the help of image charges, whose position are shown on the right. Some field lines are sketched on the left.

The field of the point charge q is equal to $\boldsymbol{E}_{\text{ext}} = q\,(\boldsymbol{r} + a\,\boldsymbol{e}_x)/|\boldsymbol{r} + a\,\boldsymbol{e}_x|^3$. In the homogeneous isotropic case (29.20), we can identify $\boldsymbol{D}$ with $\boldsymbol{E}_{\text{ext}}$. However, the present case is inhomogeneous (30.15), implying $\boldsymbol{D} \neq \boldsymbol{E}_{\text{ext}}$. The jump in $\varepsilon(\boldsymbol{r})$ means that the second equation in (30.14) leads to $\operatorname{curl}\boldsymbol{D} \neq 0$.

The equations (30.14) can be replaced by the corresponding equations in the ranges $x > 0$ and $x < 0$ *and* the boundary conditions at $x = 0$. We write (30.14) separately in the two ranges:

$$\begin{aligned} \operatorname{div}\boldsymbol{E}(\boldsymbol{r}) &= 4\pi q\,\delta(\boldsymbol{r} + a\,\boldsymbol{e}_x), & \operatorname{curl}\boldsymbol{E} &= 0 \quad (x < 0) \\ \varepsilon\,\operatorname{div}\boldsymbol{E}(\boldsymbol{r}) &= 0, & \operatorname{curl}\boldsymbol{E} &= 0 \quad (x > 0) \end{aligned} \tag{30.16}$$

The problem can be solved with the help of image charges. One first convinces oneself that the approach

$$\boldsymbol{E}(\boldsymbol{r}) = \begin{cases} q\,\dfrac{\boldsymbol{r} + a\,\boldsymbol{e}_x}{|\boldsymbol{r} + a\,\boldsymbol{e}_x|^3} + q'\,\dfrac{\boldsymbol{r} - a\,\boldsymbol{e}_x}{|\boldsymbol{r} - a\,\boldsymbol{e}_x|^3} & (x < 0) \\[2ex] q''\,\dfrac{\boldsymbol{r} + a\,\boldsymbol{e}_x}{|\boldsymbol{r} + a\,\boldsymbol{e}_x|^3} & (x > 0) \end{cases} \tag{30.17}$$

solves the equations (30.16) for any values of q' and q''. As we see in a moment, the boundary conditions can then be fulfilled by a suitable choice of q' and q''.

At the boundary $x = 0$, there is no external surface charge, $\sigma_{\text{ext}} = 0$. Therefore, the normal component of $\boldsymbol{D}$, i.e., D_x, is continuous here.

With (30.15) and (30.17), we obtain from this

$$q\,\frac{a}{|a^2+y^2+z^2|^{3/2}}+q'\,\frac{-a}{|a^2+y^2+z^2|^{3/2}}=\varepsilon\,q''\,\frac{a}{|a^2+y^2+z^2|^{3/2}} \tag{30.18}$$

In addition, the tangential components of $\boldsymbol{E}$, i.e., E_y and E_z, must be continuous at $x=0$. For E_y this means

$$q\,\frac{y}{|a^2+y^2+z^2|^{3/2}}+q'\,\frac{y}{|a^2+y^2+z^2|^{3/2}}=q''\,\frac{y}{|a^2+y^2+z^2|^{3/2}} \tag{30.19}$$

The continuity of E_z leads to an equivalent condition. Thus all boundary conditions are fulfilled if $q-q'=\varepsilon\,q''$ and $q+q'=q''$. From this follows

$$q'=-q\,\frac{\varepsilon-1}{\varepsilon+1}\quad\text{and}\quad q''=\frac{2q}{\varepsilon+1} \tag{30.20}$$

One may check following limit cases:

- $\varepsilon=1$ (vacuum everywhere). Then $q'=0$ and $q''=q$. The solution $\boldsymbol{E}=\boldsymbol{E}_{\text{ext}}=q\,(\boldsymbol{r}+a\,\boldsymbol{e}_x)/|\boldsymbol{r}+a\,\boldsymbol{e}_x|^3$ is valid in the whole space.
- $\varepsilon=\infty$ (dielectric is a metal). Then $q'=-q$ and $q''=0$. Since $q''=0$, the field in the dielectric disappears. This problem has already been solved in Chapter 8.

Exercises

30.1 *Point charge and dielectric*

Determine the induced charge surface density ϱ_{ind} for the arrangement shown in Figure 30.3.

30.2 *Potential from external charge density and polarization*

In a dielectric, the charge density $\varrho_{\text{ext}}(\boldsymbol{r})$ and the polarization $\boldsymbol{P}(\boldsymbol{r})$ are given. Show that the electrostatic potential

$$\Phi(\boldsymbol{r})=\Phi_{\text{ext}}(\boldsymbol{r})+\Phi_{\text{ind}}(\boldsymbol{r})=\int d^3r'\,\frac{\varrho_{\text{ext}}(\boldsymbol{r}')}{|\boldsymbol{r}-\boldsymbol{r}'|}+\int d^3r'\,\frac{\boldsymbol{P}(\boldsymbol{r}')\cdot(\boldsymbol{r}-\boldsymbol{r}')}{|\boldsymbol{r}-\boldsymbol{r}'|^3} \tag{30.21}$$

solves the macroscopic Maxwell equation $\operatorname{div}\boldsymbol{D} = \operatorname{div}(\boldsymbol{E} + 4\pi\boldsymbol{P}) = 4\pi\,\varrho_{\text{ext}}$.

30.3 *Homogeneously polarized sphere*

Determine the electric field $\boldsymbol{E}$ of a homogeneously polarized sphere (radius R, $\varrho_{\text{ext}} = 0$). Sketch the course of the field and calculate the induced charge density. Use (30.21).

Chapter 31

Dielectric Function

We discuss the Lorentz model for the macroscopic dielectric function $\varepsilon(\omega)$. The Kramers–Kronig relations relate the real and the imaginary part of $\varepsilon(\omega)$ to one another. The conductivity σ is introduced within the Lorentz model. The values of the dielectric constant $\varepsilon(0)$ of materials with and without a permanent electric dipole moment are estimated.

Lorentz Model

In Chapter 25, we have considered the oscillator model (natural frequency ω_0, damping Γ) for a bound electron (charge $-e$, mass m_e). An oscillating field $\boldsymbol{E}$ with the frequency ω leads to forced excitations of this electron; it induces a dipole moment $\boldsymbol{p} = \alpha_e \boldsymbol{E}$. According to (25.12), the electrical polarizability α_e is equal to

$$\alpha_e(\omega) = \frac{e^2/m_e}{\omega_0^2 - \omega^2 - i\omega\Gamma} \tag{31.1}$$

The polarizability α_e is complex because the time dependency is described by the factor $\exp(-i\omega t)$. The magnitude of α_e determines the amplitude of the forced oscillation, the imaginary part the dissipation. With (29.25) and a constant density n_0 of the atoms, we obtain

$$\varepsilon(\omega) = 1 + 4\pi n_0\, \alpha_e(\omega) = 1 + \frac{4\pi\, n_0\, e^2/m_e}{\omega_0^2 - \omega^2 - i\omega\Gamma} \tag{31.2}$$

We generalize this to the case that in each atom there are $Z = \sum f_j$ bound electrons in various states j. The f_j electrons in the state j have the same natural frequency ω_j and damping Γ_j coefficient:

$$\boxed{\varepsilon(\omega) = 1 + \frac{4\pi n_0 e^2}{m_e} \sum_j \frac{f_j}{\omega_j^2 - \omega^2 - \mathrm{i}\omega\Gamma_j}} \tag{31.3}$$

This model for the dielectric function $\varepsilon(\omega)$ is called *Lorentz model*. For inhomogeneous density $n_0(\boldsymbol{r})$ or for atoms with different polarizability $\alpha_e(\boldsymbol{r}, \omega)$ this can be generalized to a model for $\varepsilon(\boldsymbol{r}, \omega)$.

An expression of the form (31.3) can also be found in the quantum mechanical perturbation theory. In this case, the $\hbar\omega_j$ are then the energy differences between two electron states,

$$\hbar\omega_j = E_{n'} - E_n \tag{31.4}$$

The binding energies of the electrons in the Coulomb field of the Z-fold charged atomic nucleus are $E_n = -(Z^2/2n^2)\,\hbar\omega_{\mathrm{at}}$. In the quantum mechanical case, the f_j are the relative transition strengths, which are normalized according to $\sum_j f_j = 1$; the electron number Z is then added as a factor in (31.3).

The form (31.3) can also be applied to other systems in which excitations with the natural frequencies ω_j and with the relative strengths f_j. Therefore, the Lorentz model (31.3) can be considered as a general form of the dielectric function $\varepsilon(\omega)$ with initially unspecified parameters ω_j, f_j and Γ_j. In the crystal lattice of a solid, the ω_j may describe the distance between two energy bands. Another example are molecule vibrations: The atoms of a molecule can vibrate with characteristic frequencies ω_j; these frequencies lie in the infrared. When the atoms of a molecule have an effective charge (for example in the ionic bond), the vibrations can be excited by an electric field. The strengths f_j of the coupling to the electric field depend on the effective charge of the atoms and the type of oscillation.

After a more formal section (Kramers–Kronig relations), we investigate some practical consequences of the Lorentz model (conductivity, limiting case $\varepsilon(0)$). The frequency dependence of the dielectric function in real materials is investigated in Chapter 34.

Kramers–Kronig Relations

We derive dispersion laws which relate the real part $\mathrm{Re}\,\varepsilon(\omega)$ and the imaginary part $\mathrm{Im}\,\varepsilon(\omega)$ of the dielectric function to each other. This section is somewhat more formal and uses relations of the complex analysis (theory of complex functions). Since the results are not required in the following, this section may be skipped.

We consider $\varepsilon(\omega)$ as a complex-valued function of the complex variable ω. Observables are real quantities; only the real and imaginary parts of $\varepsilon(\omega)$ with a real value of ω may be connected to such observables. From the (relatively general) form (31.3) of the dielectric function follows:

$$\varepsilon(\omega) \quad \text{is analytic for } \mathrm{Im}(\omega) \geq 0 \tag{31.5}$$

$$\varepsilon^*(\omega) = \varepsilon(-\omega) \tag{31.6}$$

Analytic means that a power expansion converges locally. In implies that the complex function $\varepsilon(\omega)$ is differentiable. It is easy to check that the poles of (31.2) are in the lower ω-plane (i.e., at $\mathrm{Im}(\omega) < 0$); only at these points the function $\varepsilon(\omega)$ is not differentiable.

These statements (31.5) and (31.6) can alternatively be derived from relatively weak assumptions: In $\boldsymbol{P} = \chi_{\mathrm{e}}\,\boldsymbol{E}$ we include all time arguments: $\boldsymbol{P}(t) = \int dt'\,\chi_{\mathrm{e}}(t-t')\,\boldsymbol{E}(t')$; here all quantities are real. The effective field $\boldsymbol{E}$ can be regarded as the cause of the polarization $\boldsymbol{P}$. Due to the causality, $\chi_{\mathrm{e}}(t-t') = 0$ for $t - t' < 0$ must apply. This determines the lower integral limit in $\varepsilon(\omega) = 1 + 4\pi \int_0^\infty dt \;\; \chi_{\mathrm{e}}(t)\,\exp(\mathrm{i}\omega t)$. From this expression for $\varepsilon(\omega)$ (with real χ_{e}) follow (31.5) and (31.6), too.

We now calculate an integral along the contour C shown in Figure 31.1 shown in Figure 31.1, which is composed of the three paths C_1, C_2 and C_3:

$$\begin{aligned}\oint_C d\omega' \; \frac{\varepsilon(\omega') - 1}{\omega' - \omega} &= \mathrm{P}\int_{-\infty}^{\infty} \cdots + \int_{C_2} \cdots + \int_{C_3} \cdots \\ &= \mathrm{P}\int_{-\infty}^{\infty} \cdots - \frac{1}{2}\int_{C_4} \cdots = 0\end{aligned} \tag{31.7}$$

Here, P denotes the Cauchy principal value integral. The integral over C_2 has been replaced by the small circle C_4 (times minus 1/2). The integral over C_3 disappears, as explained in the following paragraph. The closed

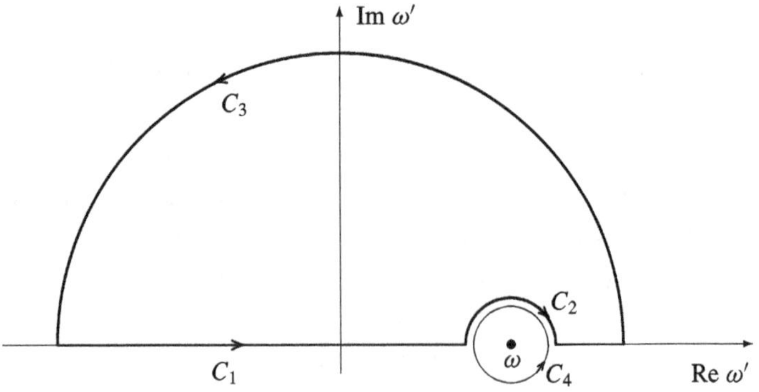

Figure 31.1 The contour C for the complex integration (31.7) consists of the partial paths C_1, C_2 and C_3. The radius of the semicircle C_3 shall approach infinity and that of C_2 can be made arbitrarily small. The contour C_2 is replaced by $-(1/2) \times C_4$.

integral over C disappears because the integrand in this region is analytic; in particular, there is no pole within the contour C.

For the radius $R_0 = |\omega|$ of the semicircle C_3, we use $R_0 \to \infty$. According to (31.3), $\varepsilon(\omega) - 1$ decreases for $\omega \to \infty$ like $1/\omega^2$. On the path C_3, the integrand in (31.7) therefore behaves like $1/R_0^3$ for $R_0 \to \infty$. The integration path has the length πR_0. This means that the integral over C_3 vanishes for $R_0 \to \infty$.

The integral over C_4 yields according to Cauchy's residue theorem[1] the value $2\pi\mathrm{i}\,\varepsilon(\omega) - 1$. Thus (31.7) yields

$$\varepsilon(\omega) - 1 = \frac{1}{\mathrm{i}\,\pi}\,\mathrm{P}\int_{-\infty}^{\infty} d\omega'\,\frac{\varepsilon(\omega') - 1}{\omega' - \omega} \tag{31.8}$$

From this equation we take the real and imaginary part:

$$\mathrm{Re}\;\varepsilon(\omega) = 1 + \frac{1}{\pi}\,\mathrm{P}\int_{-\infty}^{\infty} d\omega'\,\frac{\mathrm{Im}\;\varepsilon(\omega')}{\omega' - \omega} \tag{31.9}$$

[1]The relation $2\pi\mathrm{i}\,f(z_0) = \oint dz\, f(z)/(z - z_0)$ is valid, if the function $f(z)$ is differentiable in the region of the contour, and if z_0 lies within the contour. The initially arbitrary contour can be deformed to a small circle (with the radius ϵ) around z_0 because closed integrals over differentiable functions result in zero. For the small circle, we can write $z = z_0 + \epsilon\exp(\mathrm{i}\varphi)$ and $dz = \mathrm{i}\epsilon\exp(\mathrm{i}\varphi)\,d\varphi$. The function $f(z)$ can be drawn as $f(z_0 + \mathcal{O}(\epsilon))$ in front of the integral. The remaining integral is $\oint dz/(z - z_0) = \mathrm{i}\int d\varphi = 2\pi\mathrm{i}$. And in the limit $\epsilon \to 0$, the factor $f(z_0 + \mathcal{O}(\epsilon))$ becomes $f(z_0)$.

$$\mathrm{Im}\ \varepsilon(\omega) = -\frac{1}{\pi}\ \mathrm{P}\int_{-\infty}^{\infty} d\omega'\ \frac{\mathrm{Re}\ \varepsilon(\omega') - 1}{\omega' - \omega} \tag{31.10}$$

From (31.6) follows

$$\mathrm{Re}\ \varepsilon(-\omega) = \mathrm{Re}\ \varepsilon(\omega), \quad \mathrm{Im}\ \varepsilon(-\omega) = -\mathrm{Im}\ \varepsilon(\omega) \tag{31.11}$$

Herewith and with $\int_{-\infty}^{\infty} d\omega\ f(\omega) = \int_0^{\infty} d\omega\,\big(f(\omega) + f(-\omega)\big)$ we get the final form of the *dispersion laws*:

$$\mathrm{Re}\ \varepsilon(\omega) = 1 + \frac{2}{\pi}\ \mathrm{P}\int_0^{\infty} d\omega'\ \frac{\omega'\ \mathrm{Im}\ \varepsilon(\omega')}{\omega'^2 - \omega^2} \tag{31.12}$$

$$\mathrm{Im}\ \varepsilon(\omega) = -\frac{2\,\omega}{\pi}\ \mathrm{P}\int_0^{\infty} d\omega'\ \frac{\mathrm{Re}\ \varepsilon(\omega') - 1}{\omega'^2 - \omega^2} \tag{31.13}$$

These *Kramers–Kronig relations* provide a connection between the real and imaginary parts of ε. This can be used to calculate one quantity from the other, or to check the consistency of measured values for $\mathrm{Re}\,\varepsilon(\omega)$ and $\mathrm{Im}\,\varepsilon(\omega)$.

Conductivity

We show that the conductivity of a material can be expressed by its dielectric function. In doing so, we refer first of all to a metal. In the crystal lattice of a metal there are electrons that can (largely) move freely; usually there is one such electron per atom. For a free electron the eigenfrequency in the oscillator model (25.2) is zero. For a free and $Z - 1$ bound electrons per atom, we therefore insert the frequencies

$$\omega_1 = 0, \quad \omega_j \neq 0 \quad \text{for } j = 2, 3, \ldots \tag{31.14}$$

and $\Gamma_1 = \Gamma$ into the Lorentz model (31.3):

$$\varepsilon(\omega) = \varepsilon_{\mathrm{o}}(\omega) - \frac{4\pi n_0\, e^2}{m_{\mathrm{e}}}\ \frac{1}{\omega(\omega + \mathrm{i}\,\Gamma)} = \varepsilon_{\mathrm{o}}(\omega) + \mathrm{i}\ \frac{4\pi\,\sigma}{\omega} \tag{31.15}$$

This special Lorentz model is also called *Drude model*. The first term $\varepsilon_{\mathrm{o}}(\omega)$ includes the 1 and the contribution of the bound electrons. In the second term, the $1/\omega$ factor (which dominates for $\omega \to 0$) was displayed explicitly, and the coefficient was abbreviated by $4\pi\,\mathrm{i}\,\sigma$. A comparison of the second terms in (31.15) shows

$$\sigma(\omega) = \frac{n_0\, e^2}{m_e \varGamma}\,\frac{1}{1 - \mathrm{i}\omega/\varGamma} \qquad \text{(conductivity)} \tag{31.16}$$

This quantity is called *conductivity*; the reason for this will immediately become clear. We consider fields of the form $\boldsymbol{E}(\boldsymbol{r}, t) = \mathrm{Re}\,\boldsymbol{E}(\boldsymbol{r}) \exp(-\mathrm{i}\omega t)$. For this a time derivative yields the factor $-\mathrm{i}\omega$. Then the last Maxwell equation in (29.29) becomes

$$\mathrm{curl}\,\frac{\boldsymbol{B}}{\mu} + \frac{\mathrm{i}\omega}{c}\underbrace{\left(\varepsilon_\mathrm{o} + \mathrm{i}\,\frac{4\pi\sigma}{\omega}\right)}_{=\,\varepsilon}\boldsymbol{E} = \frac{4\pi}{c}\,\boldsymbol{j}_\mathrm{ext} \tag{31.17}$$

The term with σ is now written on the right-hand side:

$$\mathrm{curl}\,\frac{\boldsymbol{B}}{\mu} + \frac{\mathrm{i}\omega}{c}\left(\varepsilon_\mathrm{o}\,\boldsymbol{E}\right) = \frac{4\pi}{c}\left(\boldsymbol{j}_\mathrm{ext} + \sigma\,\boldsymbol{E}\right) \tag{31.18}$$

This result means that the free electrons contribute to the current density with

$$\boxed{\boldsymbol{j}_\mathrm{ind}^\mathrm{free} = \sigma\,\boldsymbol{E} \quad \text{Ohm's law}} \tag{31.19}$$

A material is *conductive* if an applied electric field induces a current density. The designation *Ohm's law* is used for a *linear* dependence of the current density on the electric field. In real materials, one will often find deviations from this linearity.

The current density (31.19) is part of the induced sources: For $\mu = 1$ and $\varepsilon_\mathrm{o} = 1$, the left-hand side in (31.18) becomes $\mathrm{curl}\,\boldsymbol{B} - \dot{\boldsymbol{E}}/c$. According to (29.2), this is equal to $4\pi\,(\boldsymbol{j}_\mathrm{ext} + \boldsymbol{j}_\mathrm{ind})/c$; therefore $\sigma\,\boldsymbol{E} = \boldsymbol{j}_\mathrm{ind}$. In general (i.e., for $\varepsilon_\mathrm{o} \neq 1$, $\mu \neq 1$), the term $\sigma\,\boldsymbol{E}$ is only a part of $\boldsymbol{j}_\mathrm{ind}$; this has been expressed in (31.19) by the additional index "free". In the Lorentz model the "free" part comes from the electrons with $\omega_1 = 0$, the other parts from $\omega_j \neq 0$. This also shows that the transition insulator–conductor is flowing; the decisive factor is the binding energy of the weakest bound electron. Materials in the intermediate range are semiconductors.

For a static electric field ($\omega = 0$) one obtains the is the static conductivity,

$$\sigma(0) = \frac{n_0\, e^2}{m_e\,\varGamma} \qquad \text{(static conductivity)} \tag{31.20}$$

We mention the experimental values for copper:

$$\sigma(0) \approx 6 \cdot 10^{17}\,\mathrm{s}^{-1}, \quad \Gamma \approx 4 \cdot 10^{13}\,\mathrm{s}^{-1} \quad (\text{copper}, T = 0^\circ\mathrm{C}) \qquad (31.21)$$

The value for Γ was calculated from (31.20) with $n_0 \approx 1/(12\,\text{Å}^3)$. The static conductivity is a real quantity.

The material parameter σ is subject to all dependencies discussed in Chapter 28 for the permittivity. In the macroscopic approximation, as in (28.14), we only take into account the dependence on the frequency. The frequency dependence (31.16) can be neglected in the UHF range (approximately up to $\omega \sim 10^{10}\,\mathrm{s}^{-1}$) because here $\omega \ll \Gamma$ and $\sigma(\omega) \approx \sigma(0)$. In this range, $\sigma(\omega)$ has an imaginary part of the relative size $\omega/\Gamma \ll 1$; it leads to a phase shift between the oscillating electric field and the induced current. For $\omega > \Gamma$, the dielectric function (31.15) of the metal is discussed in more detail in Chapter 34.

Mean collision time

The Lorentz model is based on the equation of motion (25.2). This equation has solutions of the form $\boldsymbol{v}(t) = \boldsymbol{v}(0)\,\exp(-\Gamma t)$ for the velocity of the free electrons (with $\omega_0 = 0$). This means that

$$\tau = \frac{1}{\Gamma} \quad (\text{mean collision time}) \qquad (31.22)$$

the time during which the electron loses a significant part of its initial velocity. This quantity can be considered as *mean collision time* for the gas of free electrons. In a simple kinetic gas model, the conductivity can be derived directly from the collision time τ (Chapter 43 in [4]). The crucial point here is that each particle (mass m, charge q) is accelerated between two collisions by the electric field. As a result, it reaches an average drift velocity $\overline{\boldsymbol{v}} \approx q\boldsymbol{E}\tau/m$. This results in the current density $\boldsymbol{j} = n_0\,q\,\overline{\boldsymbol{v}} = \sigma\boldsymbol{E}$ with $\sigma = n_0\,q^2\,\tau/m$. This model is also applicable to the conductivity in a common, partially ionized gas.

In $\boldsymbol{j} = \sigma\boldsymbol{E}$, the field $\boldsymbol{E}$ is the effective electric field $\boldsymbol{E} = \boldsymbol{E}_{\text{ext}} + \boldsymbol{E}_{\text{ind}}$. If $\boldsymbol{E}_{\text{ind}}$ does not contribute to the drift velocity, $\boldsymbol{j} = \sigma\boldsymbol{E}_{\text{ext}}$ is obtained. This usually applies to a static external field.

Joule heating

For a current-carrying wire (length ℓ, cross-sectional area a) we assume the voltage $U = E\,\ell$ and the current $I = j\,a$. For $j = \sigma\,E$ this yields

$$I = \frac{U}{R} \quad \text{with} \quad R = \frac{\ell}{\sigma\,a} \tag{31.23}$$

The relationship $I = U/R$ is also called Ohm's law. The ratio U/I defines the electric *resistance*. In general, a resistor is a two-terminal electrical component that implements resistance as a circuit element. R. In the SI or MKSA system, the unit of resistance $R = \mathrm{V/A} = \Omega$ is denoted by Ohm. For real resistances the linear dependence of the current on the voltage can be a useful approximation.

The applied field accelerates the electrons, so it transfers energy to them. After the means collision time τ, the electrons collide and lose their additional energy. As a result, their kinetic energy is converted into other energy forms (in particular lattice vibrations). Eventually, this energy distributes itself statistically over all the available degrees of freedom and thus becomes heat. If the charge ΔQ crosses the resistor, it takes on the energy $\Delta E = U\,\Delta Q$. The power transferred to the charge is then $P = \Delta E/\Delta t = U\,\Delta Q/\Delta t$. With $I = \Delta Q/\Delta t$ this becomes

$$P = UI = R\,I^2 = \frac{U^2}{R} \tag{31.24}$$

The heat generated in the resistor is called *Joule heating*.

Superconductor

An important, completely different type of conductor is the superconductor. In it currents can flow without friction. The equation of motion $m\,\dot{v} = q\boldsymbol{E}$ then leads to

$$\frac{\partial \boldsymbol{j}}{\partial t} = n_0\,q\,\dot{v} = \frac{n_0\,q^2}{m}\,\boldsymbol{E} \tag{31.25}$$

In the superconductor, an applied force $q\boldsymbol{E}$ leads to constant acceleration. Therefore, in the stationary case, there is no electric field in the superconductor. In an ohmic conductor, on the other hand, there is friction; the collisions discussed above correspond to the frictional force $-m\,\boldsymbol{v}/\tau$ in the equation of motion $m\,(\dot{v} + \boldsymbol{v}/\tau) = q\boldsymbol{E}$. Here, the applied force $q\boldsymbol{E}$ leads to the stationary drift velocity $\overline{\boldsymbol{v}} = q\boldsymbol{E}\,\tau/m$.

Dielectric Constant

In this section, we deal with the *dielectric constant*, i.e., the static limiting case $\varepsilon(0)$ of the dielectric function $\varepsilon(\omega)$. We first consider an *insulator*, which is described by the Lorentz model (31.3) with $\omega_j \sim \omega_{\text{at}}$. Then, $\varepsilon(\omega) \approx \varepsilon(0)$ applies for $\omega \ll \omega_{\text{at}}$.

For $\omega_0 = \omega_{\text{at}} = m_{\text{e}}\, e^4/\hbar^3$, (24.36), we obtain the static polarizability from (31.1)

$$\alpha_{\text{e}}(0) = \left(\frac{\hbar^2}{m_{\text{e}}\, e^2}\right)^3 = a_{\text{B}}^3 \tag{31.26}$$

where a_{B} is the Bohr radius. We consider two alternative models: For a conducting metal sphere, we got $\alpha_{\text{e}}(0) = R^3$, (10.44). In quantum mechanics, one calculates polarizability within the framework of perturbation theory (quadratic Stark effect in part VII of [3]). For the hydrogen atom, one obtains $\alpha_{\text{e}}(0) = 4.5\, a_{\text{B}}^3$.

Surprisingly, these very different models provide comparable results. This is often the case in physics and can be made plausible by dimension considerations. Since $\varepsilon = 1 + 4\pi\, n_0\, \alpha_{\text{e}}$ dimensionless, one may infer $\alpha_{\text{e}} = 1/n_0 = (\text{length})^3$. If the considered model for the atom (such as a classical oscillator, a conducting sphere, or bound quantum states in the Coulomb potential) contains only one length ℓ, then the result is $\alpha_{\text{e}} \propto \ell^3$ is mandatory. As a rule, occurring numerical factors are of the order 1, i.e., $\alpha_{\text{e}} \sim \ell^3$. For a meaningful result, it is sufficient that the model parameter ℓ correctly reflects the size of the atom.

We estimate the dielectric constant for hydrogen gas (at room temperature and normal pressure). The density of the H_2 molecules in the gas is $6 \cdot 10^{23}/(22 \cdot 10^3\ \text{cm}^3)$; the density n_0 of the atoms is then twice as large. From (31.26) and $a_{\text{B}} \approx 5 \cdot 10^{-11}$ m we obtain

$$\varepsilon(0) = 1 + 4\pi\, n_0\, \alpha_{\text{e}}(0) \approx 1.0001 \tag{31.27}$$

With the quantum mechanical value $\alpha_{\text{e}}(0) = 4.5\, a_{\text{B}}^3$ for the hydrogen atom one obtains $\varepsilon(0) = 1 \approx 1.00045$. The experimental value (Table 31.1) for hydrogen gas lies between these two values. The quoted quantum mechanical calculation refers to the polarizability of hydrogen atoms; this can only be an approximation for the polarizability of the H_2 molecules.

Table 31.1 Dielectric constants for some solid, liquid and gaseous substances. The values apply to a temperature of 20 °C and a pressure of 1 bar.

Solid	$\varepsilon(0)$	Liquid	$\varepsilon(0)$	Gas	$\varepsilon(0)$
diamond	5.5	water	80	helium	1.00007
glass	4...10	acetone	22	air	1.00059
silicon	11.7	ethanol	25	hydrogen	1.00026

For atoms with many electrons we use the Lorentz model (31.3). For $\omega = 0$, $\omega_j \approx \omega_{\rm at}$ and $\sum f_j = Z$ we obtain

$$\varepsilon(0) = 1 + \frac{4\pi n_0 e^2}{m_{\rm e}} \sum_j \frac{f_j}{\omega_j^2} \sim 1 + \omega_{\rm P}^2 \frac{Z}{\omega_{\rm at}^2} \approx 1 + 0.2\, Z \tag{31.28}$$

The prefactor was replaced by $\omega_{\rm P}^2$. With $n_0 \sim 0.1\,\text{Å}^{-3}$ (condensed matter density), $a_{\rm B} = 0.53\,\text{Å}$, $c/a_{\rm B} = \omega_{\rm at}/\alpha$ and $e^2/a_{\rm B} = \alpha^2 m_{\rm e} c^2$ we get

$$\omega_{\rm P}^2 = \frac{4\pi n_0 e^2}{m_{\rm e}} = \frac{4\pi}{10} \frac{e^2}{\text{Å}} \frac{1}{m_{\rm e} c^2} \frac{c^2}{\text{Å}^2} \approx 0.2\, \frac{e^2/a_{\rm B}}{m_{\rm e} c^2} \left(\frac{c}{a_{\rm B}}\right)^2 = 0.2\, \omega_{\rm at}^2 \tag{31.29}$$

This was used in the last expression in (31.28). In a realistic model, the ω_j are the excitation frequencies of the system. This means that $\varepsilon(0)$ is dominated by the behavior of the weakest bound electrons; the electrons in the lower shells or energy bands contribute only to a minor extent. The coarse estimate $\varepsilon(0) \sim 1 + 0.2\, Z$ makes the size of the experimental values of Table 31.1 plausible (for condensed matter). For semiconductors such as silicon some ω_j are smaller and one obtains higher values $\varepsilon \sim 10$. For gases (right column), $\varepsilon(0) - 1$ is much smaller due to the lower density. In liquids, the permanent electric dipole moments of the molecules give the main contribution (paraelectrics, see the following section).

The dielectric constant can be determined by measuring the capacitance of a capacitor with and without dielectric, (30.13).

Metal

For free electrons in a static field ($\omega = 0$), the displacement in the direction of the field is arbitrarily large, therefore

$$\varepsilon(0) = +\infty \tag{31.30}$$

In practice, the shift is limited by the boundary of the metal body.

Approach the static case ($\omega \to 0$), the collisions of electrons play the decisive role. From (31.15) follows $\varepsilon = \varepsilon_o + 4\pi\,\mathrm{i}\,\sigma$ with the conductivity σ. In Chapter 34 the Drude model (31.15) for metals is investigated further.

Paraelectric

In a *paraelectric* there are rotatable molecules that possess a permanent electric dipole moment. The standard example of a paraelectric is water. In the H_2O molecule the electrons of the two hydrogen atoms tend to stay near the oxygen atom. This results in an electron configuration similar to that of the noble gas neon. The energy of such configuration is lower; as a result the molecule is bound. The directions from the oxygen atom to the two hydrogen atoms form an angle of about 105°. As the hydrogen atoms are effectively positive, but the oxygen atom is effectively negative, the molecule has a dipole moment of the size $p \sim e\,(1\,\text{Å})$.

We assume that the molecules under consideration are rotatable; this is usually only the case in the liquid or gaseous state. Then only the magnitude $p = |\boldsymbol{p}|$, but not the direction of the individual dipole moments is fixed. The potential energy of the dipole is given by (12.30):

$$W = -\boldsymbol{p}\cdot\boldsymbol{E} = -\,p\,E\,\cos(\theta) = W(\theta) \tag{31.30}$$

where $\boldsymbol{E}$ is the field acting in the matter. We consider first of all a static field.

The electric field tries to align the dipole so that the energy W is minimal; this is the case for $\boldsymbol{p} \parallel \boldsymbol{E}$. A finite temperature counteracts this tendency to alignment. In statistical equilibrium, the probability $w(\theta)\,d\theta$ that a dipole direction lies between the angles θ and $\theta + d\theta$ is given by the Boltzmann factor:

$$w(\theta)\,d\theta = \frac{1}{z(T)}\,\exp\Big(-\frac{W(\theta)}{k_\mathrm{B}T}\Big)\,d(\cos\theta) \tag{31.31}$$

Here T is the temperature and k_B is the Boltzmann constant. The normalization $\int d\theta\; w(\theta) = 1$ determines the partition function $z(T)$,

$$z(T) = \int_{-1}^{1} d(\cos\theta)\,\exp\Big(-\frac{W(\theta)}{k_\mathrm{B}T}\Big) \tag{31.32}$$

We place the z-axis of the coordinate system in the direction of $\boldsymbol{E}$ and calculate the mean value $\overline{p_z}$ of the dipole component $p_z = p\,\cos\theta$:

$$\overline{p_z} = \frac{p}{z(T)} \int_{-1}^{1} d(\cos\theta)\ \cos\theta\ \exp\Big(-\frac{W(\theta)}{k_B T}\Big) \tag{31.33}$$

The integration can be carried out elementarily. We restrict ourselves to high temperatures

$$pE \ll k_B T \tag{31.34}$$

We set $t = \cos\theta$ and expand the exponential function in (31.34):

$$\overline{p_z} = \frac{p \int_{-1}^{1} dt\ t\,(1 + t\,(pE/k_B T) + \cdots)}{\int_{-1}^{1} dt\ \ (1 + t\,(pE/k_B T) + \cdots)} = \frac{p^2 E}{3 k_B T}\left(1 + \mathcal{O}\Big(\frac{pE}{k_B T}\Big)\right) \tag{31.35}$$

The corrections $\mathcal{O}(pE/k_B T)$ are neglected in the following. For the components perpendicular to the field a $\cos\theta$ on (31.33) is replaced by $\sin\theta$ in (31.34) yielding

$$\overline{p_x} = \overline{p_y} = 0 \tag{31.36}$$

This means that the mean dipole moment is parallel to the field, i.e., $\overline{\boldsymbol{p}} = \alpha_e \boldsymbol{E}$. From (31.36) we can read off the static polarizability:

$$\alpha_e(0) = \frac{\overline{p_z}}{E} = \frac{p^2}{3 k_B T} \tag{31.37}$$

We use the molecular density $n_0 \approx 0.03\,\text{Å}^{-3}$ of water, the dipole moment $p \sim e\,(1\,\text{Å})$ of the molecule, $k_B T = \text{eV}/40$ and $e^2/(1\,\text{Å}) = 14.4\,\text{eV}$:

$$\varepsilon = 1 + 4\pi\, n_0\, \alpha_e(0) = 1 + 4\pi\ \frac{0.03\, e^2}{3\,\text{Å}}\ \frac{40}{\text{eV}} \approx 73 \tag{31.38}$$

The experimental value is $\varepsilon_{\text{exp}} \approx 80$. There are also other liquids with values that are large compared to 1 (Table 31.1, middle column).

In the model considered here, it was assumed that the molecules can rotate freely and adjust independently of each other. In real materials, these conditions are fulfilled only approximately. In Table 31.2 we compare

Table 31.2 The temperature dependence of the dielectric constant of water (at normal pressure). The experimental values $\varepsilon_{\rm exp}$ are consistent with the $1/T$-dependence shown in (31.38) for $\varepsilon - 1$. For the second line, the absolute value was adjusted to the experimental one at 20 °C (i.e., at $T = 293.15$ K).

Temperature	0 °C	10 °C	20 °C	30 °C	40 °C	50 °C
$\varepsilon_{\rm exp}$ for water	87.8	83.9	80.1	76.5	73.0	69.7
$\varepsilon - 1 = {\rm const.}/T$	86.0	82.9	80.1	77.5	75.0	72.8

the model prediction for the temperature dependence of $\varepsilon - 1$ with the experimental results for water. The experimental temperature dependence is slightly stronger than the theoretical one. Due to the hydrogen bridges, the water molecules are not completely free and independent of each other.

Our derivation referred to a static electric field. For an oscillating field (such as an electromagnetic wave), the alignment of the dipoles is hindered by the friction and the inertia of the rotational motion of the molecules (in addition to the temperature). The inertia implies that for high frequencies, the alignment of the dipoles can no longer follow the field; the transition is in the microwave range. The friction implies that a electromagnetic wave in water is damped. These effects are indeed observed in water (Figures 34.3 and 34.4).

Ferroelectricity

There are molecular crystals in which the molecules with permanent electric dipole moments can change their orientation more or less easily. For high temperatures these solids show paraelectric behavior, as discussed in the last section for water. At lower temperatures, the interaction between neighboring dipoles can lead to a permanent alignment. This results in a spontaneous polarization, i.e., a polarization without external field. This phenomenon is called ferroelectricity. Chapter 32 investigates in more detail the comparable magnetic case, the ferromagnetism.

Bodies with a permanent electric dipole field are called electrets. Certain resins (with polar molecules) can form an electret when they solidify under the influence of a strong electric field. Analogous to the permanent magnet, an electret can also be formed by a global spontaneous alignment of the molecular dipoles.

Exercises

31.1 *Dipole alignment in thermal equilibrium*

A permanent electric dipole $\boldsymbol{p}$ has the potential energy $W(\theta) = -\boldsymbol{p} \cdot \boldsymbol{E} = -pE \cos\theta$ in the electric field $\boldsymbol{E} = E\,\boldsymbol{e}_z$. The mean value of the dipole component $p_z = p \cos\theta$ im thermal equilibrium follows from

$$\overline{p_z} = \frac{p}{z(T)} \int_{-1}^{1} d\cos\theta \;\cos\theta\; \exp\left(-\frac{W(\theta)}{k_\mathrm{B}T}\right)$$

$$z(T) = \int_{-1}^{1} d\cos\theta\; \exp\left(-\frac{W(\theta)}{k_\mathrm{B}T}\right)$$

Determine $\overline{p_z}$ as a function of the temperature T. Discuss the result graphically.

31.2 *Conductivity in SI units*

The expression for the static conductivity

$$\sigma = \frac{n_0\, e^2}{m_\mathrm{e}\, \Gamma}$$

applies in both the Gauss and SI systems. Start from $n_0 \approx 1/(12\,\text{Å}^3)$ and $\Gamma \approx 4 \cdot 10^{13}\,\mathrm{s}^{-1}$ (for copper), and determine the conductivity in SI units. A copper wire of 10 m length and a cross-sectional area of 1 mm^2 represents an ohmic resistance R. What is the value of R in ohms?

Chapter 32

Permeability Constant

We present simple atomistic models of paramagnetism (permanent magnetic dipole moments) and diamagnetism (induced magnetic dipole moments). In each case, the size of the magnetic susceptibility is estimated. We restrict ourselves to the static macroscopic permeability, i.e., to the permeability constant. The ferromagnetism is discussed briefly.

In simple cases, the induced magnetic dipole moment is $\boldsymbol{\mu}$ of an atom or molecule is proportional to the effective field, i.e., $\boldsymbol{\mu} = \alpha_{\mathrm{m}} \boldsymbol{B}$; the coefficient α_{m} is the magnetic polarizability. According to (29.11), the magnetization $\boldsymbol{M}$ is equal to the dipole moment per volume. With the density n_0 of the atoms (molecules, unit cells), we therefore obtain

$$\boldsymbol{M} = n_0\, \boldsymbol{\mu} = n_0\, \alpha_{\mathrm{m}}\, \boldsymbol{B} \tag{32.1}$$

The *magnetic susceptibility* χ_{m} is given by

$$\boldsymbol{M} = \chi_{\mathrm{m}}\, \boldsymbol{H} \tag{32.2}$$

We insert $\boldsymbol{H} = \boldsymbol{B} - 4\pi \boldsymbol{M}$ and solve for $\boldsymbol{M}$:

$$\boldsymbol{M} = \frac{\chi_{\mathrm{m}}}{1 + 4\pi\, \chi_{\mathrm{m}}}\, \boldsymbol{B} \approx \chi_{\mathrm{m}}\, \boldsymbol{B} \quad (|\chi_{\mathrm{m}}| \ll 1) \tag{32.3}$$

For para- and diamagnetism considered here $|\chi_{\mathrm{m}}| \ll 1$ holds. With $\boldsymbol{M} = \chi_{\mathrm{m}}\, \boldsymbol{B}$, we have achieved a form analogous to $\boldsymbol{P} = \chi_{\mathrm{e}}\, \boldsymbol{E}$. The comparison of (32.1) and (32.3) results in

$$\chi_{\mathrm{m}} = n_0\, \alpha_{\mathrm{m}} \quad (|\chi_{\mathrm{m}}| \ll 1) \tag{32.4}$$

From $\boldsymbol{B} = \boldsymbol{H} + 4\pi \boldsymbol{M} = \mu\, \boldsymbol{H}$ and (32.2), we obtain

$$\mu = 1 + 4\pi\, \chi_{\mathrm{m}} \tag{32.5}$$

In (29.17) we have admitted a macroscopic permeability function of the form $\mu(\boldsymbol{r}, \omega)$. In the following, we restrict ourselves here to the homogeneous and static case, i.e., to the permeability constant.[1]

Depending on the sign of the susceptibility, we designate a substance as paramagnetic or diamagnetic:

$$\begin{aligned} \text{paramagnetic:}\quad \chi_{\mathrm{m}} &> 0 \\ \text{diamagnetic:}\quad \chi_{\mathrm{m}} &< 0 \end{aligned} \tag{32.6}$$

Paramagnetism is caused by the alignment of existing magnetic moments (due to spin and/or orbital angular momentum of electrons). Paramagnetism only occurs if the atoms have unpaired electrons. Diamagnetism is caused by atomic currents induced by the applied magnetic field. Diamagnetism occurs in all atoms.

In ferromagnetic materials, it comes to significantly stronger magnetizations than for dia- or paramagnetism. Provided that the linear relation $\boldsymbol{M} = \chi_{\mathrm{m}} \boldsymbol{H}$ is applicable, a susceptibility $\chi_{\mathrm{m}} \gg 1$ is possible.

Paramagnetism

Due to their spin and their orbital angular momentum, electrons in atoms or molecules may exhibit permanent magnetic moments $|\boldsymbol{\mu}| = \text{const}$. We assume that the individual dipoles are independent of each other. The effective magnetic field effective $\boldsymbol{B}$ tries to align the dipoles, the temperature works towards a distribution in all directions. Similar to paraelectrics (Chapter 31), this results in a temperature dependent magnetization in the direction of the applied field, i.e., in $\chi_{\text{para}} > 0$.

Essential for a paramagnetic behavior is that interaction between the magnetic dipoles can be neglected. In the ferromagnetic case, it is precisely this interaction that may lead to a spontaneous magnetization.

---

[1] We use the letter μ with indices that clarify whether the permeability (μ_{para}, μ_{dia}) or components of the magnetic dipole moment $\boldsymbol{\mu}$ (e.g., μ_z) are meant.

In (15.19) we established a relationship between the angular momentum $\boldsymbol{L}$ and the magnetic moment $\boldsymbol{\mu}$ of charged particles:

$$\boldsymbol{\mu} = \frac{g\,q}{2mc}\,\boldsymbol{L} \tag{32.7}$$

The angular momentum $\boldsymbol{L}$ stems from the spin and/or the orbital angular momentum of the considered particle. Since the magnetic moment is inversely proportional to the mass we disregard the magnetic moments of the nucleons and consider the electrons only. For the sake of simplicity, we imagine an atom with closed shells and one additional electron. Closed electron shells have the total angular momentum and spin zero. An electron has a spin $\hbar/2$ and the gyromagnetic factor $g = 2$. The contributing electron shall have the orbital angular momentum zero; otherwise, the coupling between spin and orbital angular momentum results in a more complex expression for the g-factor.

The energy of the dipole in the external field is given by (15.28), $W = -\boldsymbol{\mu}\cdot\boldsymbol{B}$. The mean value of the magnetic moment in direction of the magnetic field $\boldsymbol{B} = B\,\boldsymbol{e}_z$ is obtained as in (31.31)–(31.36):

$$\overline{\mu_z} = \overline{\boldsymbol{\mu}\cdot\boldsymbol{e}_z} = \frac{|\boldsymbol{\mu}|^2 B}{3k_{\rm B}T} \qquad (\mu B \ll k_{\rm B}T) \tag{32.8}$$

From this follows the static polarizability

$$\alpha_{\rm para}(0) = \frac{|\boldsymbol{\mu}|^2}{3k_{\rm B}T} \approx \frac{e^2\hbar^2}{12\,m_{\rm e}^2c^2}\,\frac{1}{k_{\rm B}T} \tag{32.9}$$

In the last expression, we used $|\boldsymbol{\mu}| \approx e\hbar/2m_{\rm e}c$; according to (32.7) this is the expected size for electrons. For a numerical estimation of the susceptibility, we use $n_0 \sim 0.1\,\text{Å}^{-3}$ (solid state density), $e^2/(1\,\text{Å}) = 14.4\,\text{eV}$, $\hbar c/e^2 = 137$, $m_{\rm e}c^2 = 5\cdot 10^5\,\text{eV}$ and $k_{\rm B}T = \text{eV}/40$ (room temperature):

$$\chi_{\rm para} = n_0\,\alpha_{\rm para}(0) \sim \frac{0.1}{12}\,\frac{e^2}{\text{Å}}\left(\frac{\hbar c}{\text{Å}}\right)^2\left(\frac{1}{m_{\rm e}c^2}\right)^2\frac{40}{\text{eV}} \approx 7\cdot 10^{-5} \tag{32.10}$$

From this follows the permeability constant

$$\mu_{\rm para} - 1 = 4\pi\,\chi_{\rm para} \sim 10^{-3} \tag{32.11}$$

Diamagnetism

If a magnetic field is switched on in the area of a conductor loop, a voltage is generated in the loop (Faraday's law of induction). This voltage is directed in such a way that the induced current weakens the applied magnetic field (Lenz's rule). This means that the magnetization $\boldsymbol{M}$ is opposite to the $\boldsymbol{B}$ field and the susceptibility is negative, $\chi_{\text{dia}} < 0$.

For a quantitative estimate of diamagnetism, we consider the classical equation of motion of a particle with mass m and charge q. An electrostatic field $\boldsymbol{E} = -\operatorname{grad} \Phi$ and a external magnetic field $\boldsymbol{B}$ act on the particle:

$$m\,\dot{v} = -q \operatorname{grad} \Phi + \frac{q}{c}\, \boldsymbol{v} \times \boldsymbol{B} \quad \text{(in IS)} \tag{32.12}$$

This equation of motion refers to an inertial system (IS). We now consider this motion in a coordinate system S′, which is rotates relative to IS with the frequency $\boldsymbol{\omega}$. In S′ the particle is subject to the Coriolis force and to the centrifugal force (of the size $\mathcal{O}(\omega^2)$), see Chapter 6 in [1]:

$$m\,\dot{v}' = -\, q \operatorname{grad} \Phi + \frac{q}{c}\, \boldsymbol{v}' \times \boldsymbol{B} + 2m\, \boldsymbol{v}' \times \boldsymbol{\omega} + \mathcal{O}(\omega^2) \tag{32.13}$$

We set $\boldsymbol{\omega}$ equal to the *Larmor frequency*

$$\boldsymbol{\omega}_{\text{L}} = -\frac{q}{2mc}\, \boldsymbol{B} \tag{32.14}$$

Then the second and third term on the r.h.s. of (32.13) cancel each other. Therefore, the Coriolis force compensates the magnetic force, and (32.13) becomes

$$m\,\dot{v}' = -\, q \operatorname{grad} \Phi + \mathcal{O}\left(\omega_{\text{L}}^2\right) \quad \text{(in S}'\text{)} \tag{32.15}$$

As a rule, $\boldsymbol{B}$ and therefore $\boldsymbol{\omega}_{\text{L}}$ are so small that the centrifugal force can be neglected. In S′ we then see the undisturbed motion, i.e., as if there were no magnetic field. Now S′ rotates relative to IS with $\boldsymbol{\omega}_{\text{L}}$. In IS we see a rotation with $\boldsymbol{\omega}_{\text{L}}$ as compared to the undisturbed motion. The magnetic field therefore leads to a precession with the rotational frequency $\boldsymbol{\omega}_{\text{L}} \parallel -q\boldsymbol{B}$. This induces a circular current that generates a magnetic field (see Figure 15.1), which is opposite to the original magnetic field. This statement is independent of the sign of the charge.

We now apply this model to an electron (mass $m = m_{\text{e}}$, charge $q = -e$) in an atom. We consider a classical electron on a circle with the Bohr radius

a_B; the orbital plane shall be perpendicular to the magnetic field. Then the Larmor precession gives the additional angular momentum $\boldsymbol{L} = m_\mathrm{e}\, a_\mathrm{B}^2\, \boldsymbol{\omega}_\mathrm{L}$ to the electron. We insert $\boldsymbol{L}_\mathrm{L} = m_\mathrm{e}\, a_\mathrm{B}^2\, \boldsymbol{\omega}_\mathrm{L}$, $q = -e$ and $g = 1$ into (32.7) and multiply the result by the number Z of electrons (per atom). From this we obtain the *induced* magnetic moment of an atom

$$\boldsymbol{\mu} = Z\,\frac{q}{2mc}\,\boldsymbol{L}_\mathrm{L} = \frac{-Ze}{2m_\mathrm{e}c}\,m_\mathrm{e}\,a_\mathrm{B}^2\,\boldsymbol{\omega}_\mathrm{L} = -\frac{Ze^2a_\mathrm{B}^2}{4m_\mathrm{e}c^2}\,\boldsymbol{B} = \alpha_\mathrm{dia}\,\boldsymbol{B} \qquad (32.16)$$

For a numerical estimate we use $n_0 \sim 0.1\,\text{Å}^{-3}$ (solid state density), $a_\mathrm{B} = 0.53\,\text{Å}$, $e^2/(1\,\text{Å}) = 14.4\,\text{eV}$, $m_\mathrm{e}c^2 = 5\cdot 10^5\,\text{eV}$ and $Z = 10$:

$$\chi_\mathrm{dia} = n_0\,\alpha_\mathrm{dia} = -\,n_0\,\frac{Ze^2a_\mathrm{B}^2}{4m_\mathrm{e}c^2} \sim -\frac{0.1\,Z}{4}\,\frac{e^2}{\text{Å}}\left(\frac{a_\mathrm{B}}{\text{Å}}\right)^2\frac{1}{m_\mathrm{e}c^2} \approx -2\cdot 10^{-6} \qquad (32.17)$$

The permeability constant is then

$$\mu_\mathrm{dia} - 1 = 4\pi\,\chi_\mathrm{dia} \sim -3\cdot 10^{-5} \qquad (32.18)$$

The values (32.11) and (32.18) give the typical order of magnitude for liquids and solids. For gases, the density n_0 and thus the susceptibility χ_m are about a factor 10^4 smaller.

If there are unpaired electrons, the paramagnetic effects prevail; from (32.10) and (32.17) it follows $|\chi_\mathrm{para}| \gg |\chi_\mathrm{dia}|$. In this case, the substance is effectively paramagnetic. If there are no unpaired electrons, the substance is diamagnetic.

Ferromagnetism

Ferromagnetism occurs in special materials, for example in iron, cobalt, nickel and a number of alloys. In a *ferromagnet* the spins of the unpaired electrons are aligned if the temperature is below a critical value T_c. This magnetization occurs without an external magnetic field, it constitutes a *spontaneous magnetization* $\boldsymbol{M}_\mathrm{s}$.

The reason for this behavior is the interaction between the spins of neighboring electrons. This interaction may be expressed in the form $W = -I\,\boldsymbol{s}_i\cdot\boldsymbol{s}_j$ (Heisenberg model). The cause of this is the so-called exchange interaction between the two electrons.

For temperatures $T \gg T_c$ the material is paramagnetic. When approaching T_c from above, the susceptibility behaves like

$$\chi_{\text{ferro}} = \frac{\text{const.}}{T - T_c} \quad (T > T_c) \tag{32.19}$$

This *Curie–Weiss law* is derived in Chapter 36 of [4].

For $T < T_c$, the global magnetization of a ferromagnet is initially zero because the spontaneous magnetization occurs in finite ranges (Weiss' domains), with statistical directions of the spontaneous magnetization $\boldsymbol{M}_s$. An applied magnetic field can cause an alignment the magnetization of these domains. This may yield a global magnetization $\boldsymbol{M} = \chi_{\text{ferro}} \boldsymbol{H}$ with

$$\chi_{\text{ferro}} \gg 1 \quad (T < T_c) \tag{32.20}$$

However, the proportionality between $\boldsymbol{M}$ and $\boldsymbol{H}$ is of only limited validity. For strong fields, the alignment of all domains results in a saturation magnetization. In addition, the alignment of the individual districts depends on the history of the sample (hysteresis).

Exercises

32.1 *Vector potential from external current density and magnetization*

In a magnetic medium, the current density $\boldsymbol{j}_{\text{ext}}(\boldsymbol{r})$ and the magnetization $\boldsymbol{M}(\boldsymbol{r})$ are given. Show that the vector potential

$$\boldsymbol{A}(\boldsymbol{r}) = \boldsymbol{A}_{\text{ext}}(\boldsymbol{r}) + \boldsymbol{A}_{\text{ind}}(\boldsymbol{r}) = \frac{1}{c}\int d^3r' \, \frac{\boldsymbol{j}_{\text{ext}}(\boldsymbol{r}')}{|\boldsymbol{r}-\boldsymbol{r}'|} + \int d^3r' \, \frac{\boldsymbol{M}(\boldsymbol{r}') \times (\boldsymbol{r}-\boldsymbol{r}')}{|\boldsymbol{r}-\boldsymbol{r}'|^3} \tag{32.21}$$

solves the macroscopic Maxwell equation $\operatorname{curl} \boldsymbol{H} = \operatorname{curl}(\boldsymbol{B} - 4\pi \boldsymbol{M}) = 4\pi \boldsymbol{j}_{\text{ext}}/c$.

32.2 *Homogeneously magnetized sphere*

Determine the magnetic field $\boldsymbol{B}$ of a homogeneously magnetized sphere (radius $R, \boldsymbol{j}_{\text{ext}} = 0$). Sketch the field lines and calculate the induced current density. Use (32.21).

32.3 *Magnetization due to external field*

A sphere (radius R, permeability μ) is located in an external homogeneous magnetic field $\boldsymbol{B}_0$. The field induces a homogeneous magnetization $\boldsymbol{M}_0$ of the sphere. Determine $\boldsymbol{M}_0$ from $\boldsymbol{B}_0$ and μ. To do this, start from

$$\boldsymbol{H} = \boldsymbol{B} - 4\pi \boldsymbol{M} = \boldsymbol{B}/\mu$$

Sketch the field lines for the case $\mu > 1$. Which strength has the $\boldsymbol{H}$-field inside the sphere for $\mu \gg 1$?

32.4 *Magnetization of a spherical shell in an external field*

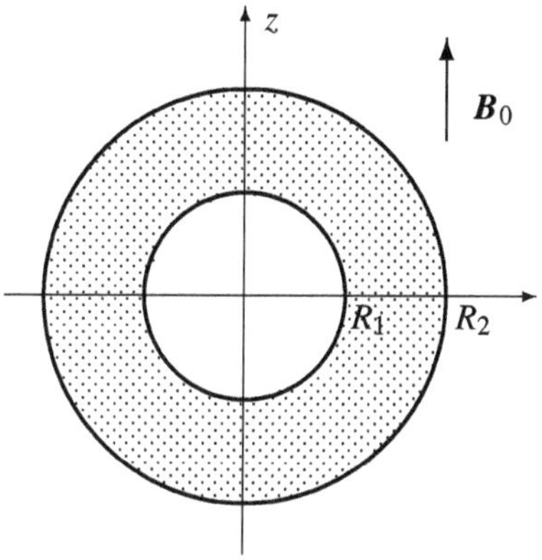

A spherical shell (radii R_1 and R_2) with the permeability μ is located in an external homogeneous magnetic field $\boldsymbol{B}_0 = B_0\, \boldsymbol{e}_z$. In the individual ranges one can use $\boldsymbol{H} = -\text{grad}\, \Psi$ with a magnetic potential

$$\Psi(r, \theta) = \sum_{l=0}^{\infty} \left(a_l\, r^l + \frac{b_l}{r^{l+1}} \right) P_l(\cos\theta)$$

Justify this approach and determine the coefficients a_l and b_l. Discuss the case $\mu \gg 1$.

Chapter 33

Wave Solutions

We investigate electromagnetic waves in matter. For the monochromatic plane wave, we discuss the differences (polarization, phase shift, damping) to the vacuum solution. We investigate the dispersion of wave packets caused by the frequency dependence of the response functions.

Monochromatic Plane Wave

We consider a homogeneous and isotropic medium with the macroscopic response functions

$$\varepsilon = \varepsilon(\omega), \qquad \mu = \mu(\omega) \tag{33.1}$$

We first examine solutions with a fixed frequency ω:

$$\begin{aligned} \boldsymbol{E}(\boldsymbol{r}, t) &= \operatorname{Re} \boldsymbol{E}(\boldsymbol{r}) \exp(-\mathrm{i}\omega t), \ \ \boldsymbol{D}(\boldsymbol{r}, t) = \operatorname{Re} \varepsilon(\omega)\, \boldsymbol{E}(\boldsymbol{r}) \exp(-\mathrm{i}\omega t) \\ \boldsymbol{B}(\boldsymbol{r}, t) &= \operatorname{Re} \boldsymbol{B}(\boldsymbol{r}) \exp(-\mathrm{i}\omega t), \ \ \boldsymbol{H}(\boldsymbol{r}, t) = \operatorname{Re} \frac{\boldsymbol{B}(\boldsymbol{r})}{\mu(\omega)} \exp(-\mathrm{i}\omega t) \end{aligned} \tag{33.2}$$

Here, $\omega =$ real. The quantities $\boldsymbol{E}(\boldsymbol{r})$, $\boldsymbol{B}(\boldsymbol{r})$, ε and μ are generally complex. The sign Re for the formation of the real part is mostly suppressed in the following.

We put the fields (33.2) into the source-free ($\varrho_{\text{ext}} = 0, \boldsymbol{j}_{\text{ext}} = 0$) macroscopic Maxwell equations (29.29):

$$\begin{aligned} &\operatorname{div} \boldsymbol{E}(\boldsymbol{r}) = 0, \ \ \operatorname{curl} \boldsymbol{E}(\boldsymbol{r}) - \frac{\mathrm{i}\omega}{c}\, \boldsymbol{B}(\boldsymbol{r}) = 0 \\ &\operatorname{div} \boldsymbol{B}(\boldsymbol{r}) = 0, \ \ \operatorname{curl} \boldsymbol{B}(\boldsymbol{r}) + \frac{\mathrm{i}\omega\, \varepsilon(\omega)\, \mu(\omega)}{c}\, \boldsymbol{E}(\boldsymbol{r}) = 0 \end{aligned} \tag{33.3}$$

With

$$\begin{aligned}\operatorname{curl} \boldsymbol{B}(\boldsymbol{r}) &= -\Delta \boldsymbol{B} + \operatorname{grad}\operatorname{div} \boldsymbol{B} = -\Delta \boldsymbol{B}(\boldsymbol{r}) \\ \operatorname{curl} \boldsymbol{E}(\boldsymbol{r}) &= -\Delta \boldsymbol{E} + \operatorname{grad}\operatorname{div} \boldsymbol{E} = -\Delta \boldsymbol{E}(\boldsymbol{r})\end{aligned} \tag{33.4}$$

we get

$$\left(\Delta + \frac{\omega^2 \varepsilon \mu}{c^2}\right)\boldsymbol{B}(\boldsymbol{r}) = 0 \quad \text{and} \quad \left(\Delta + \frac{\omega^2 \varepsilon \mu}{c^2}\right)\boldsymbol{E}(\boldsymbol{r}) = 0 \tag{33.5}$$

These equations can be solved by the approach

$$\boldsymbol{E}(\boldsymbol{r}) = \boldsymbol{E}_0 \exp(\mathrm{i}\boldsymbol{k}\cdot\boldsymbol{r}), \quad \boldsymbol{B}(\boldsymbol{r}) = \boldsymbol{B}_0 \exp(\mathrm{i}\boldsymbol{k}\cdot\boldsymbol{r}) \tag{33.6}$$

with arbitrary amplitudes $\boldsymbol{E}_0$ and $\boldsymbol{B}_0$, provided that the wave vector $\boldsymbol{k}$ fulfills the condition

$$\omega^2 = \frac{c^2 \boldsymbol{k}^2}{\varepsilon \mu} = \frac{c^2 k^2}{n^2} \tag{33.7}$$

The *refractive index n* introduced here is generally complex:

$$\boxed{n = \sqrt{\varepsilon\mu} = n_\mathrm{r} + \mathrm{i}\kappa = |n| \exp(\mathrm{i}\delta) \quad \text{refractive index}} \tag{33.8}$$

Occasionally, the real part n_r alone is referred to as the refractive index. The real quantities n_r and κ are also called *optical constants* (of the considered material). Actually, these quantities are not really constants. They depend in particular on the frequency of the wave.

With n, the wave vector $\boldsymbol{k}$ is also complex; whereas the other quantities (c and ω) in (33.7) are real. The vector $\boldsymbol{k} = \boldsymbol{k}_\mathrm{r} + \mathrm{i}\,\boldsymbol{k}_\mathrm{i}$ contains two real vectors $\boldsymbol{k}_\mathrm{r}$ and $\boldsymbol{k}_\mathrm{i}$, which generally have different directions. For the wave of the form

$$\exp(\mathrm{i}\boldsymbol{k}\cdot\boldsymbol{r}) = \exp(\mathrm{i}\boldsymbol{k}_\mathrm{r}\cdot\boldsymbol{r}) \exp(-\boldsymbol{k}_\mathrm{i}\cdot\boldsymbol{r}) \tag{33.9}$$

the surface of equal phase, $\boldsymbol{k}_\mathrm{r}\cdot\boldsymbol{r} = \text{const.}$, are different from the surface $\boldsymbol{k}_\mathrm{i}\cdot\boldsymbol{r} = \text{const.}$ In the following we restrict ourselves to the case $\boldsymbol{k}_\mathrm{r} \parallel \boldsymbol{k}_\mathrm{i}$. For this we set

$$\boldsymbol{k} = n\,\boldsymbol{k}_0 \tag{33.10}$$

where $\boldsymbol{k}_0$ is a real vector with an arbitrary direction. From (33.7) follows $k_0 = \omega/c$. On the planes $\boldsymbol{k}_0\cdot\boldsymbol{r} = \text{const.}$, the complex wave has the same value everywhere; hence the designation as a "plane wave". On such a plane, both the amplitude and the phase are constant.

The Maxwell equations also provide conditions for the amplitudes $\boldsymbol{E}_0$ and $\boldsymbol{B}_0$ in addition to (33.5). In order to obtain these conditions, we insert (33.6) into all four equations (33.3) using $\varepsilon\,\mu = n^2$ and $\boldsymbol{k} = n\,\boldsymbol{k}_0$:

$$\begin{aligned} \boldsymbol{k}_0 \cdot \boldsymbol{E}_0 = 0, \;\; \boldsymbol{k}_0 \times (n\,\boldsymbol{E}_0) = k_0\,\boldsymbol{B}_0 \\ \boldsymbol{k}_0 \cdot \boldsymbol{B}_0 = 0, \;\; \boldsymbol{k}_0 \times \boldsymbol{B}_0 = -k_0\,(n\,\boldsymbol{E}_0) \end{aligned} \tag{33.11}$$

We now calculate the real fields $\boldsymbol{E}$ and $\boldsymbol{B}$ from (33.2), taking into account (33.8), (33.10) and (33.11):

$$\boldsymbol{E} = \mathrm{Re}\left(\boldsymbol{E}_0 \exp \mathrm{i}(n_\mathrm{r}\boldsymbol{k}_0 \cdot \boldsymbol{r} - \omega t)\right) \exp(-\kappa\,\boldsymbol{k}_0 \cdot \boldsymbol{r}) \tag{33.12}$$

$$\boldsymbol{B} = |n|\,(\boldsymbol{k}_0/k_0) \times \mathrm{Re}\left(\boldsymbol{E}_0 \exp \mathrm{i}(n_\mathrm{r}\boldsymbol{k}_0 \cdot \boldsymbol{r} - \omega t + \delta)\right) \exp(-\kappa\,\boldsymbol{k}_0 \cdot \boldsymbol{r}) \tag{33.13}$$

For specifying this solution, a wave vector $\boldsymbol{k}_0$ (real, arbitrary direction, $k_0 = \omega/c$) and an amplitude vector $\boldsymbol{E}_0$ (generally complex, $\boldsymbol{k}_0 \cdot \boldsymbol{E}_0 = 0$) have to be chosen. The refractive index (33.8) and thereby the optical parameters n_r, κ and δ are assumed to be given.

Especially for a real amplitude $\boldsymbol{E}_0$, Equations (33.12), (33.13) describe a linearly polarized wave:

$$\boldsymbol{E} = \boldsymbol{E}_0 \cos(n_\mathrm{r}\boldsymbol{k}_0 \cdot \boldsymbol{r} - \omega t) \exp(-\kappa\,\boldsymbol{k}_0 \cdot \boldsymbol{r}) \tag{33.14}$$

$$\boldsymbol{B} = |n|\,(\boldsymbol{k}_0/k_0) \times \boldsymbol{E}_0 \cos(n_\mathrm{r}\boldsymbol{k}_0 \cdot \boldsymbol{r} - \omega t + \delta) \exp(-\kappa\,\boldsymbol{k}_0 \cdot \boldsymbol{r}) \tag{33.15}$$

Here, $\boldsymbol{k}_0$, $\boldsymbol{E}_0$ and $\boldsymbol{B}_0$ form three orthogonal vectors.

We discuss the properties of the matter wave (33.12), (33.13) in detail. In each case, we point out the similarities and the differences to the vacuum solution (with $n = 1$, i.e., $n_\mathrm{r} = 1$ and $\kappa = 0$):

1. *Transversality:* The following applies

$$\boldsymbol{E}(\boldsymbol{r}, t) \perp \boldsymbol{k}_0, \quad \boldsymbol{B}(\boldsymbol{r}, t) \perp \boldsymbol{k}_0 \tag{33.16}$$

   For the first statement, one multiplies (33.12) by $\boldsymbol{k}_0$, puts the real vector $\boldsymbol{k}_0$ in front of the real part sign and takes into account $\boldsymbol{E}_0\,\boldsymbol{k}_0 = 0$ from (33.11). The second statement follows immediately by multiplying (33.13) with $\boldsymbol{k}_0$.

The electromagnetic wave is therefore transverse, just like the wave in a vacuum.

2. *Polarization:* We compare the matter solution (33.12) with the vacuum solution (20.24). The differences (real factor n_r in the wave vector and the attenuation factor $\exp(-\kappa\, \boldsymbol{k}_0 \cdot \boldsymbol{r})$) have no influence on the directions of $\boldsymbol{E}$, $\boldsymbol{B}$ or $\boldsymbol{k}_0$. Therefore, the polarization can be treated as in Chapter 20. The general solution (33.12) therefore represents an elliptically polarized wave.

   Equations (33.14) and (33.15) describe a linearly polarized wave.
3. *Phase velocity:* For $\boldsymbol{k}_0 = k_0 \boldsymbol{e}_z$ and $n_r k_0 z - \omega t = 0$, (33.12) has a fixed phase. The maxima of (33.14) occur for specific values $z = z_{\rm max}$. The position of a maximum shifts with the *phase velocity* $v_{\rm P} = z_{\rm max}/t$, i.e., with

$$v_{\rm P} = \frac{c}{n_r} \tag{33.17}$$

   At this speed, the points with a specific phase (such as the maxima) move in $\boldsymbol{k}_0$-direction. As we in the following, this is not the speed of a light signal.
4. *Wavelength:* Two neighboring maxima are separated by the phase 2π from each other. The corresponding spatial distance is the wavelength λ. From $n_r k_0 \lambda = 2\pi$ follows

$$\lambda = \frac{\lambda_0}{n_r} \tag{33.18}$$

   Here $\lambda_0 = 2\pi/k_0$ is the vacuum wavelength.
5. *Phase shift and amplitude ratio:* From (33.12), (33.13) follows

$$\frac{|\boldsymbol{B}(\boldsymbol{r}, t + \delta/\omega)|}{|\boldsymbol{E}(\boldsymbol{r}, t)|} = |n| \tag{33.19}$$

   Compare to the $\boldsymbol{E}$-field, the phase of the $\boldsymbol{B}$-field is shifted by

$$\delta = \arctan \frac{\kappa}{n_r} \tag{33.20}$$

   Dur to this phase shift, the maxima of $\boldsymbol{E}$ and $\boldsymbol{B}$ are not at the same position.

   The ratio of the maximum amplitude of the magnetic and electric field is equal to $|n|$. For the vacuum wave, $\boldsymbol{B}$ and $\boldsymbol{E}$ are equal in phase and have the same strength, (20.27) and Figure 20.1.

6. *Damping:* The electromagnetic wave in matter is damped for $\kappa \neq 0$. The *absorption coefficient* α,

$$\alpha = 2k_0\kappa \tag{33.21}$$

determines the attenuation of the energy density (29.33),

$$w_{\text{em}} = \frac{1}{8\pi}\left(\varepsilon\,|\boldsymbol{E}|^2 + |\boldsymbol{B}|^2/\mu\right) \propto \exp(-\alpha\,\ell) \tag{33.22}$$

Here $\ell = \boldsymbol{k}_0 \cdot \boldsymbol{r}/k_0$. On the length $\ell = 1/\alpha$ the intensity of the wave drops to the eth part. Usually, $\text{Im}\,\varepsilon \ll \text{Re}\,\varepsilon$ and $\mu \approx 1$ hold. Then the absorption coefficient (33.21) is proportional to the imaginary part of ε,

$$\alpha(\omega) = \frac{2\omega}{c}\,\text{Im}\sqrt{\varepsilon\mu} \approx \frac{\omega}{c}\,\frac{\text{Im}\,\varepsilon(\omega)}{\sqrt{\text{Re}\,\varepsilon(\omega)}} \tag{33.23}$$

Damping

Damping or attenuation of the wave means that it loses energy. This energy is transferred to the matter. We study this relation in the oscillator model (25.2).

As a representative of matter, we consider a particle (mass m, charge q) that is harmonically bound (natural frequency ω_0, damping Γ). The displacement $\boldsymbol{r}_0(t)$ of a particle in the field of a wave is determined by (25.7):

$$m\left(\ddot{\boldsymbol{r}}_0 + \Gamma\,\dot{\boldsymbol{r}}_0 + \omega_0^2\,\boldsymbol{r}_0\right) = q\,\boldsymbol{E}_0\,\exp(-\mathrm{i}\omega t) \tag{33.24}$$

We assume $\lambda \gg r_0$ so that $\exp(-\mathrm{i}\boldsymbol{k}\boldsymbol{r}) \approx 1$. For the sake of simplicity we restrict the discussion to $\boldsymbol{E}_0 =$ real. In the steady state, the solution is

$$\begin{aligned}\boldsymbol{r}_0(t) &= \frac{q\boldsymbol{E}_0}{m}\,\text{Re}\left(\frac{\exp(-\mathrm{i}\omega t)}{\omega_0^2 - \omega^2 - \mathrm{i}\omega\Gamma}\right)\\ &= \frac{q\boldsymbol{E}_0}{m}\left(\frac{(\omega_0^2-\omega^2)\,\cos(\omega t)}{(\omega_0^2-\omega^2)^2+\omega^2\Gamma^2} + \frac{\omega\Gamma\,\sin(\omega t)}{(\omega_0^2-\omega^2)^2+\omega^2\Gamma^2}\right)\end{aligned} \tag{33.25}$$

Due to the force

$$\boldsymbol{F} = q\,\text{Re}\,\boldsymbol{E}_0\,\exp(-\mathrm{i}\omega t) = q\boldsymbol{E}_0\,\cos(\omega t) \tag{33.26}$$

the power $P = \boldsymbol{F} \cdot \dot{\boldsymbol{r}}_0$ is transferred to the particle. We average this power over a period $T = 2\pi/\omega$:

$$P = \big\langle \boldsymbol{F} \cdot \dot{\boldsymbol{r}}_0 \big\rangle = \frac{1}{T} \int_t^{t+T} dt\; \boldsymbol{F}(t) \cdot \dot{\boldsymbol{r}}_0(t) = \frac{q^2 \boldsymbol{E}_0^2}{2m}\, \frac{\omega^2 \Gamma}{(\omega_0^2 - \omega^2)^2 + \omega^2 \Gamma^2} \tag{33.27}$$

The oscillator model results in the dielectric function (31.2),

$$\varepsilon(\omega) = 1 + \frac{4\pi\, n_0\, q^2/m}{\omega_0^2 - \omega^2 - \mathrm{i}\omega\Gamma} \tag{33.28}$$

where n_0 is the number of oscillators (e.g., atoms) per volume. Then the absorbed power per volume is $n_0 P$. It can be expressed by the imaginary part of the dielectric function:

$$n_0\, P = \frac{1}{8\pi}\, \boldsymbol{E}_0^2\, \omega\, \mathrm{Im}\; \varepsilon(\omega) \tag{33.29}$$

This power is transferred from the wave to the particle and disappears via the friction term in (33.24) into not specified other degrees of freedom (radiation in other directions, thermal disorder). At the same time the wave loses energy; it is damped.

Telegraph equation

For a conductor (metal) the Lorentz model yields $\varepsilon = \varepsilon_\mathrm{o}(\omega) + 4\pi\mathrm{i}\sigma/\omega$, (31.15). If we insert this into the wave equation (33.5) for the electric field, we obtain

$$\left(\Delta + \frac{\omega^2 \varepsilon_\mathrm{o}\, \mu}{c^2}\right) \boldsymbol{E}(\boldsymbol{r}) = -4\pi\mathrm{i}\, \frac{\sigma\, \omega\, \mu}{c^2}\, \boldsymbol{E}(\boldsymbol{r}) \tag{33.30}$$

Using $-\mathrm{i}\omega \to \partial_t$ we may to go back to the time-dependent form. The right-hand side is then proportional to $\partial \boldsymbol{E}(\boldsymbol{r}, t)/\partial t$. An analogous form applies for the magnetic field. This form of the wave equation with a linear time derivative is called *telegraph equation.*

The solution of (33.30) can be carried out as above, except that the complex refractive index is now given by $n^2 = \mu\,(\varepsilon_\mathrm{o} + 4\pi\mathrm{i}\sigma/\omega)$. For high-frequency waves, the main contribution for the damping comes from the conductivity. In fact such waves penetrate the metal very little; the metallic wire of a telegraph line essentially serves as a guide for the waves. The penetration depth is discussed quantitatively in the next chapter (section about metal) as the skin effect.

Wave Packet

A superposition of the solutions (33.6) with different $\boldsymbol{k}$ vectors is again a solution; this follows from the linearity of the Maxwell equations (29.29). In this section we discuss some properties of such superpositions or wave packets. The discussion also applies to other waveforms such as sound waves.

In the considered frequency range, let the refractive index be real:

$$n = n(\omega) = \text{real} \quad \text{(transparent medium)} \tag{33.31}$$

A medium with real n is transparent because the damping coefficient disappears, $\alpha = 2k_0 \operatorname{Im} n = 0$. According to (33.7), the wave number $k = |\boldsymbol{k}|$ is then also real. We resolve (33.7), $k^2 = \omega^2\, n(\omega)^2/c^2$, for ω:

$$\omega = \omega(k) \quad (\text{using } n = n(\omega)) \tag{33.32}$$

The relationship between frequency and wave number is called *dispersion relation*. Dispersion relations can also be specified for other waves, for example:

$$\omega(k) = \begin{cases} ck & \text{light in a vacuum} \\ \sqrt{\omega_{\mathrm{P}}^2 + c^2k^2} & \text{plasma wave (34.25), cavity wave (21.45), (21.49)} \\ \hbar k^2/(2m) & \text{free Schrödinger wave} \\ c_{\mathrm{s}}k & \text{sound wave, acoustic phonons} \\ \approx \text{const.} & \text{optical phonons} \end{cases} \tag{33.33}$$

A deviation from the linear relationship $\omega \propto k$ leads to *dispersion effects*. For electromagnetic waves in matter $n(\omega) \neq$ const. implies such effects.

Dispersion effects occur if there are different k or ω values in a wave packet. Dispersion literally means to run apart or spread from one another. This means specifically:

- The different parts of the wave packet move with different phase velocity $v_{\mathrm{P}} = c/n(\omega)$. This causes the wave packet to diverge spatially over time. This is investigated in more detail in the following.
- The different parts of the wave packet are refracted by different degrees at a boundary between two media. Thereby a light ray may be separated into various parts (Figure 34.1).

For the sake of simplicity, we consider just one component of the fields $\boldsymbol{E}$ or $\boldsymbol{B}$, which we denote by ψ. In the following discussion, ψ can also be the wave field of one of the waves mentioned in (33.33). We restrict ourselves to $\boldsymbol{k} = k\,\boldsymbol{e}_z$. The general form of such a wave packet is then

$$\psi(z,t) = \frac{1}{\sqrt{2\pi}} \int_{-\infty}^{\infty} dk\; A(k)\, \exp\big[\mathrm{i}\big(kz - \omega(k)\,t\big)\big] \qquad (33.34)$$

This is a solution of the wave equation if $\omega(k)$ fulfills the associated dispersion relation.

We start by considering a single k component of the wave packet, i.e., $\psi_k = A(k)\exp\mathrm{i}(kz - \omega(k)\,t)$. The position z_{max} of a wave crest has a constant phase, $k\,z_{\text{max}} - \omega(k)\,t = \text{const}$. The crest of the wave propagates with the *phase velocity* $v_{\text{P}} = z_{\text{max}}/t$:

$$v_{\text{P}} = \frac{\omega(k)}{k} \quad \text{(phase velocity)} \qquad (33.35)$$

For plasma waves with $\omega^2 = \omega_{\text{P}}^2 + c^2k^2$ this means for example

$$v_{\text{P}} = c\,\sqrt{1 + \frac{\omega_{\text{P}}^2}{c^2k^2}} \qquad (33.36)$$

At time $t = 0$, the wave packet (33.34) has the form

$$\psi_0(z) = \psi(z,0) = \widetilde{A}(z) \qquad (33.37)$$

Here $\widetilde{A}(z)$ is the Fourier transform of $A(k)$,

$$\widetilde{A}(z) = \frac{1}{\sqrt{2\pi}} \int_{-\infty}^{\infty} dk\; A(k)\, \exp(\mathrm{i}kz) \qquad (33.38)$$

In the following, we consider a wave packet that is localized around $k = k_{\text{o}}$. As a concrete example, we use

$$A(k) = C\, \exp\big(-a\,(k-k_{\text{o}})^2\big) \qquad (33.39)$$

In this case, $|\psi_0(z)|$ is also a Gaussian function. The real part $\text{Re}\,\psi_0$ and the amplitude function $A(k)$ are sketched in Figure 33.1.

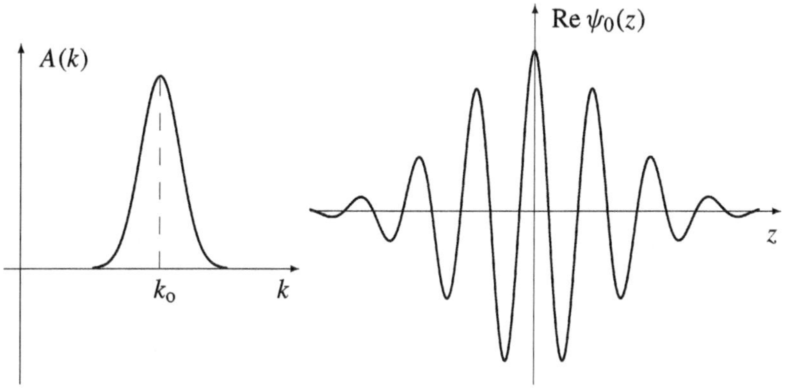

Figure 33.1 On th left, the amplitude function $A(k)$ with $a^{1/2}\,k_o = 6$ is shown. On the right the real part of the associated wave function (33.34) for $t = 0$ is displayed; the function $\psi_0(z) = \psi(z, 0)$ results from (33.47).

Due to the assumed localization of $A(k)$, it makes sense to expand $\omega(k)$ in (33.34) around k_o:

$$\omega(k) = \omega(k_o) + \left(\frac{d\omega}{dk}\right)_{k_o} (k - k_o) + \frac{1}{2}\left(\frac{d^2\omega}{dk^2}\right)_{k_o} (k - k_o)^2 + \cdots$$

$$= \omega_o + v_G\,(k - k_o) + \beta\,(k - k_o)^2 + \cdots \tag{33.40}$$

The expansion coefficients are the *group velocity*

$$v_G = \left(\frac{d\omega}{dk}\right)_{k_o} \qquad \text{(group velocity)} \tag{33.41}$$

and the *dispersion parameter*,

$$\beta = \frac{1}{2}\left(\frac{d^2\omega}{dk^2}\right)_{k_o} \qquad \text{(dispersion parameter)} \tag{33.42}$$

In the lowest approximation, we neglect the dispersion ($\beta = 0$) and insert $\omega \approx \omega_o + v_G\,(k - k_o)$ into (33.34):

$$\psi(z, t) \approx \exp\big[\mathrm{i}\,(k_o v_G - \omega_o)t\big]\,\frac{1}{\sqrt{2\pi}}\int_{-\infty}^{\infty} dk\; A(k)\,\exp\big[\mathrm{i}k\,(z - v_G t)\big]$$

$$= \exp\big[\mathrm{i}\,(k_o v_G - \omega_o)t\big]\;\psi_0(z - v_G t) \tag{33.43}$$

The intensity of the wave is

$$\big|\psi(z,t)\big|^2 \approx \big|\psi_0(z - v_G t)\big|^2 \quad \begin{array}{c}\text{(linear approximation of}\\ \text{the dispersion relation)}\end{array} \tag{33.44}$$

In the electromagnetic case, ψ is one of the components of $\boldsymbol{E}$ or $\boldsymbol{B}$, and $|\psi|^2$ is proportional to the energy density.

In this linear approximation (33.43), the wave packet propagates with the velocity v_G in the z-direction without changing its shape (i.e., without dispersion). This velocity v_G, with which the entire wave group (wave packet) moves, is called *group velocity.* In general (without the linear approximation), v_G the velocity of the center of energy the wave packet; this shows up in (33.48). The wave packet could be used as a signal; therefore v_G is also the signal velocity.

As an example, we use the dispersion relation (33.33) for plasma or cavity waves in (33.41):

$$v_G = \left(\frac{d\omega}{dk}\right)_{k_o} = \frac{c}{\sqrt{1 + \omega_P^2/c^2 k_o^2}} \tag{33.45}$$

For a linear dispersion relation we obtain

$$v_G = v_P = c, \qquad (\text{for } \omega = ck) \tag{33.46}$$

In this case, (33.44) applies exactly, i.e., the wave propagates without dispersion. This applies in particular to electromagnetic waves in a vacuum.

In the case of a non-linear dispersion relation, the wave packet broadens over time, it disperses. This means that the individual components (frequencies) of the wave packet move with different speed. The deviation from the linearity is characterized by the dispersion parameter (33.42). We calculate the dispersion of the wave packet (33.34) using (33.39) and (33.40):

$$\begin{aligned}\psi(z,t) &= \frac{C}{\sqrt{2\pi}} \int_{-\infty}^{\infty} dk\, \exp\big[-a\,(k-k_o)^2\big]\, \exp(\mathrm{i}kz)\\ &\quad\cdot \exp\big(-\mathrm{i}\big[\omega_o + v_G(k-k_o) + \beta(k-k_o)^2\big]t\big)\\ &= \frac{C}{\sqrt{2}}\, \frac{\exp\mathrm{i}(k_o z - \omega_o t)}{\sqrt{a + \mathrm{i}\beta t}}\, \exp\!\left(-\frac{(z - v_G t)^2}{4\,(a + \mathrm{i}\beta t)}\right)\end{aligned} \tag{33.47}$$

The wave has the intensity

$$\boxed{\left|\psi(z,t)\right|^2 = \frac{|C|^2}{2\sqrt{a^2+\beta^2 t^2}} \exp\left(-\frac{a\,(z-v_G\,t)^2}{2\,(a^2+\beta^2\,t^2)}\right)} \tag{33.48}$$

First of all, we can see that the center of the wave packet moves with the group velocity v_G. In addition, the wave packet becomes wider over time, it *disperses*. This behavior is sketched in Figure 33.2.

The width of a Gaussian packet $\exp(-z^2/2\,\Delta z^2)$ is characterized by Δz. For (33.48) we get

$$\Delta z(t) = \sqrt{\frac{a^2+\beta^2 t^2}{a}} \tag{33.49}$$

The parameter $\sqrt{a}$ defines the width $\Delta z(0)$ at time zero. Over time, the packet broadens according to

$$\frac{\Delta z(t)}{\Delta z(0)} = \sqrt{1+\frac{\beta^2 t^2}{a^2}} \tag{33.50}$$

This effect is called *dispersion of the wave packet*.

Since $n = n(\omega)$, electromagnetic waves in matter usually show dispersion. For the free Schrödinger, the broadening of a wave packet

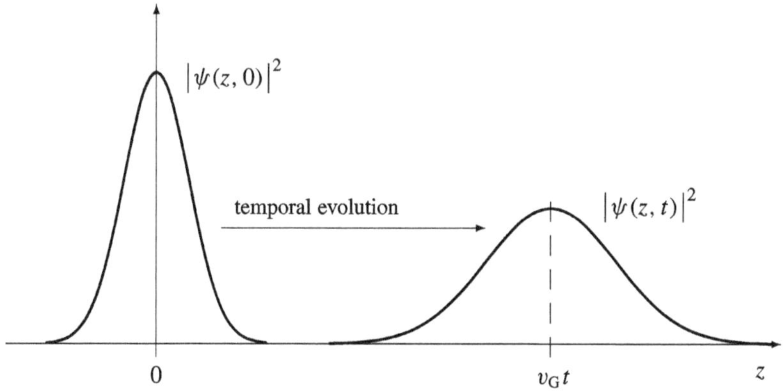

Figure 33.2 The intensity distribution (33.48) of the wave packet (33.34) with (33.39) is shown for two different times. The center of the wave packet moves with the group velocity v_G in the z-direction. At the same time the wave packet becomes wider, it disperses. Only for a linear dispersion relation, i.e., for $\omega \propto k$, a wave propagates without dispersion.

follows from the quantum mechanical dispersion relation $\beta = \hbar/2m$. From the point of view of the expansion (33.40), dispersion free propagation should be regarded as a special case. Light propagates in a vacuum exactly and in air approximately dispersion free. In the audible range, the dispersion of sound waves in air (water, iron) is small.

Maximum signal velocity

In the course of the discussion, we have introduced the phase velocity v_{P} and the group velocity v_{G},

$$v_{\mathrm{P}} = \frac{\omega(k)}{k}, \qquad v_{\mathrm{G}} = \left(\frac{d\omega}{dk}\right)_{k_0} \tag{33.51}$$

The phase velocity is the propagation velocity for the maxima or minima of the wave. The group velocity was identified as the velocity of the center of the wave packet.

In the dispersion free case, $v_{\mathrm{P}} = v_{\mathrm{G}} = c$. In matter, the phase velocity $v_{\mathrm{P}} = c/n_{\mathrm{r}}$, (33.17), of an electromagnetic wave in matter is mostly less than c. As the example of the plasma wave shows, $v_{\mathrm{P}} > c$ is also possible, (33.36). The group velocity (33.41) of a plasma wave on the other hand, is less than c.

According to the special theory of relativity, signals cannot travel faster than the speed of light. A signal necessarily has a finite spatial extension; it therefore always consists of a wave packet. For the wave packet examined above (33.47), the group velocity is also the signal velocity. Under the given conditions, we obtain $v_{\mathrm{G}} \leq c$. The prerequisites were a real dispersion relation and the expansion (33.40) (slowly changing $\omega(k)$).

Under certain circumstances (dispersion in the vicinity of resonances, "instantaneous" tunneling in a waveguide[1]) can result in group velocities that are larger than c; these are then not signal velocities. In these cases, the transport of energy must be investigated in more detail; the energy transport is the decisive criterion for a signal transmission.

---

[1]See G. Nimtz, Instantaneous tunneling, *Physikalische Blätter* 49 (1993) 1119, and subsequent discussion by P. Thoma, T. Weiland and G. Eilenberger under the title Wie real ist das instantaneous tunneling?, *Physikalische Blätter* 50 (1994) 313.

Chapter 34

Dispersion and Absorption

Dispersion is caused by the frequency dependence of the refractive index $n = n_{\mathrm{r}} + \mathrm{i}\kappa$. *Starting from the Lorentz model, we investigate this frequency dependence for different types of materials (insulator, metal, plasma) and discuss the resulting effects for electromagnetic waves in matter. The frequency dependence of the real part* $n_{\mathrm{r}}(\omega)$ *leads to the splitting of a light beam in the prism (Figure 34.1) and to the broadening of a wave packet (Figure 33.2); in both cases, there is a dispersion in the literal sense of the word, i.e., a diverging or a running apart. The imaginary part* κ *determines the absorption of an electromagnetic wave in matter.*

Insulator

For the permeability μ and the dielectric function ε of the medium, we assume

$$\mu = 1, \quad \varepsilon = \varepsilon(\omega), \quad \varepsilon(0) = \text{finite} \qquad (34.1)$$

The last condition means that the conductivity σ of the material disappears (Chapter 31). The medium under consideration is therefore an *insulator*. As an example, we consider water.

We start from the Lorentz model (31.3) for the dielectric function:

$$\varepsilon(\omega) = 1 + \frac{4\pi n_0 e^2}{m_{\mathrm{e}}} \sum_j \frac{f_j}{\omega_j^2 - \omega^2 - \mathrm{i}\omega\Gamma_j} \qquad (\omega_j \neq 0) \qquad (34.2)$$

The condition $\omega_j \neq 0$ guarantees that $\varepsilon(0)$ is finite.

As already discussed in Chapter 31, we can consider (34.2) as a general form of the dielectric function. Here, the ω_j are the frequencies of the

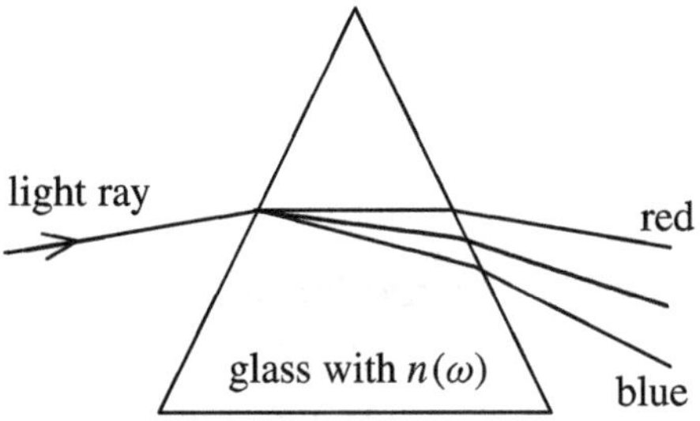

Figure 34.1 A beam of white light falls on a glass prism. As the refractive index is frequency dependent, each frequency or color of the light leads to slightly different refraction angle. This implies a broadening or *dispersion* of the light beam.

natural oscillations of the material. This can refer to the motion of the electrons (in the atom or molecule), or to molecular vibrations or to lattice vibrations in a solid body.

For the prefactor in (34.2) we introduce the *plasma frequency* ω_P, which was already numerically estimated in (31.29):

$$\omega_\mathrm{P}^2 = \frac{4\pi\, n_0\, e^2}{m_\mathrm{e}} \approx 0.2\, \omega_\mathrm{at}^2, \quad \omega_\mathrm{P} \approx 2 \cdot 10^{16}\,\mathrm{s}^{-1} \tag{34.3}$$

Here, the density $n_0 \sim 0.1\,\text{Å}^{-3}$ of a solid or a liquid was used.

We investigate the behavior of $\varepsilon(\omega)$ in the vicinity of a resonance at ω_j. The width Γ_j of the resonance shall be small compared to the interval $\Delta\omega \sim |\omega_i - \omega_j|$. Then, for $\omega \approx \omega_j$ the term with ω_j yields the main contribution to the sum in (34.2):

$$\mathrm{Re}\,\varepsilon(\omega) = \frac{\omega_\mathrm{P}^2\, f_j\, (\omega_j^2 - \omega^2)}{(\omega_j^2 - \omega^2)^2 + \omega^2\, \Gamma_j^2} + g(\omega) \qquad (\omega \approx \omega_j) \tag{34.4}$$

$$\mathrm{Im}\,\varepsilon(\omega) = \frac{\omega_\mathrm{P}^2\, f_j\, \omega\, \Gamma_j}{(\omega_j^2 - \omega^2)^2 + \omega^2\, \Gamma_j^2} + \mathcal{O}(\Gamma/\Delta\omega) \quad (\omega \approx \omega_j) \tag{34.5}$$

The function $g(\omega)$ consists of the 1 and the non-resonant contributions in the sum (34.2); under the given conditions $g(\omega)$ changes only weakly in the vicinity of the resonance. For weak damping ($\Gamma_i \ll \omega_i$), the imaginary part is limited to the environment of the resonance, Figure 34.2 sketches the frequency dependence of $\mathrm{Re}\,\varepsilon(\omega)$ and $\mathrm{Im}\,\varepsilon(\omega)$ for the case of two resonances.

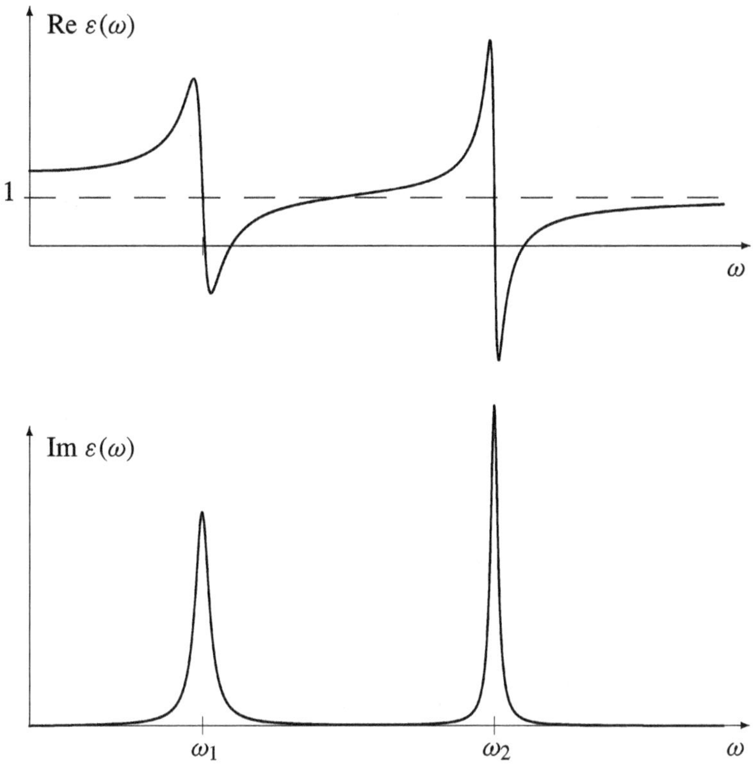

Figure 34.2 Possible course of the real and imaginary parts of the complex dielectric function $\varepsilon(\omega)$ for an insulator. Shown are the curves resulting from (34.2) for two resonances. The following model parameters were chosen: $\omega_2 = 8\,\omega_1/3$, $\Gamma_1 = \omega_1/10$, $\Gamma_2 = \Gamma_1/2$, $f_2/f_1 = 2$ and $\omega_p^2\, f_2 = \omega_2^2/4$.

The dielectric function determines the refractive index $n = \sqrt{\varepsilon}$. The significance of the refractive index for electromagnetic waves was discussed in Chapter 33. In particular, it determines

$$n_r(\omega) = \mathrm{Re}\, n \approx \sqrt{\mathrm{Re}\, \varepsilon(\omega)} \qquad \text{(refractive index)} \qquad (34.6)$$

$$\alpha(\omega) = 2k_0\, \mathrm{Im}\, n \approx \frac{\omega}{c}\, \frac{\mathrm{Im}\, \varepsilon(\omega)}{\sqrt{\mathrm{Re}\, \varepsilon(\omega)}} \qquad \text{(absorption coefficient)} \qquad (34.7)$$

The term "refractive index" is used here in the narrower sense for the real part of the (complex) refractive index n.

In Figure 34.2, $dn_r/d\omega > 0$ usually applies; this is referred to as *normal* dispersion. In the vicinity of a resonance, $dn_r/d\omega < 0$ occurs, i.e., *anomalous* dispersion. In concrete terms, normal dispersion means that blue light in the prism is refracted more strongly than red light (Figure 34.1).

Some insulators, such as water, have molecules with a permanent dipole moment. These dipole moments can align themselves in a external field. This leads to static polarizability (31.38) and thus to

$$\varepsilon(\omega) = 1 + 4\pi n_0 \, \frac{p^2}{3k_B T} \qquad (\omega \approx 0) \tag{34.8}$$

The estimate (31.39) resulted in $\varepsilon(0) \approx 73$ for water. For sufficiently large frequencies, the rotational motion of the molecules can no longer follow the alternating field. Then the contribution of the permanent dipoles goes to zero. The intermediate region is characterized by an increased energy transfer of the field to the rotational motion of the molecules; this absorption of energy is described by the imaginary part $\mathrm{Im}\,\varepsilon(\omega)$ of the dielectric function.

We compare discussed behavior of $\varepsilon(\omega)$, Figure 34.2, (34.2) and (34.8), with the actual behavior of water, Figures 34.3 and 34.4. H_2O molecules have a large permanent dipole moment. The frequency dependencies discussed above leads to

$$\varepsilon \approx \begin{cases} 80 & \text{for } \nu \lesssim 10^{10}\,\text{Hz} \qquad \text{(radio waves)} \\[1ex] 1.8 & \text{for } \nu \sim 5 \cdot 10^{14}\,\text{Hz} \quad \text{(visible light)} \end{cases} \tag{34.9}$$

While UHF waves with $n_r \approx 9$ are strongly refracted, the value for visible light is $n_r \approx 1.33$. The drop from 9 to values of quantity 1 is associated with strong absorption. In this range, the dispersion is anomalous, $dn_r/d\omega < 0$. Absorption occurs, as expected from (34.5), also in the resonance range.

The drop (34.9) in ε is associated with a sharp increase in absorption. This absorption is important for many technical applications:

- Bad weather impairs television reception via a satellite ($\lambda \approx 3\,$cm) usually more strongly than that from (not too distant) terrestrial transmitters ($\lambda \approx 30\,$cm). Therefore, the satellite dish should not be too small.
- The radar in aircraft navigation ($\lambda \approx 30\,$cm) also works in fog or rain. High resolution radar (approximately $\lambda \approx 1\,$cm), on the other hand, is more hindered.

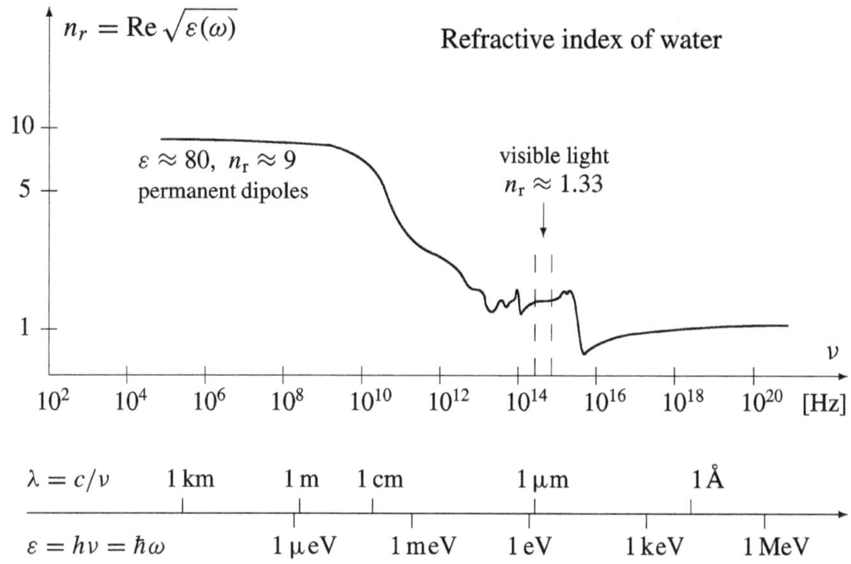

Figure 34.3 The experimental refractive index $n_r = \text{Re}\sqrt{\varepsilon}$ for water as a function of the frequency $\nu = \omega/2\pi$ (shown in a log-log plot). The scales at the bottom display are the energy $\varepsilon = h\nu$ of a photon and the wavelengths λ. In the visible range (wavelengths $\lambda \approx 4 \ldots 7 \cdot 10^{-7}$ m), the refractive index has the well-known value $n_\text{r} \approx 1.33$. At lower frequencies, especially in the radio wave range, the permanent dipoles of the water molecules align in the alternating electric field. This leads to a strong polarization and $n_\text{r} \approx 9$.

- In the microwave oven ($\lambda \approx 1$ cm), absorption is the wanted effect. The energy is first of all converted into the rotations of H_2O molecules in the food; other molecules with a permanent electric dipole moment may contribute, too (be careful with plastic dishes!). By means of the rotating back and forth molecules, the energy is transferred to other degrees of freedom. It finally distributes itself statistically to all available degrees of freedoms; the temperature of the food rises accordingly. As the energy is primarily pumped into the water-containing components the porcelain crockery heats up predominantly by normal heat conduction. Metal pots, on the other hand, shield the microwaves (skin effect (34.16)–(34.20)); they are not suitable for the microwave oven.

The light to which our eyes are sensitive is called *visible* light. Figure 34.4 shows the physical cause for the biological distinction of this

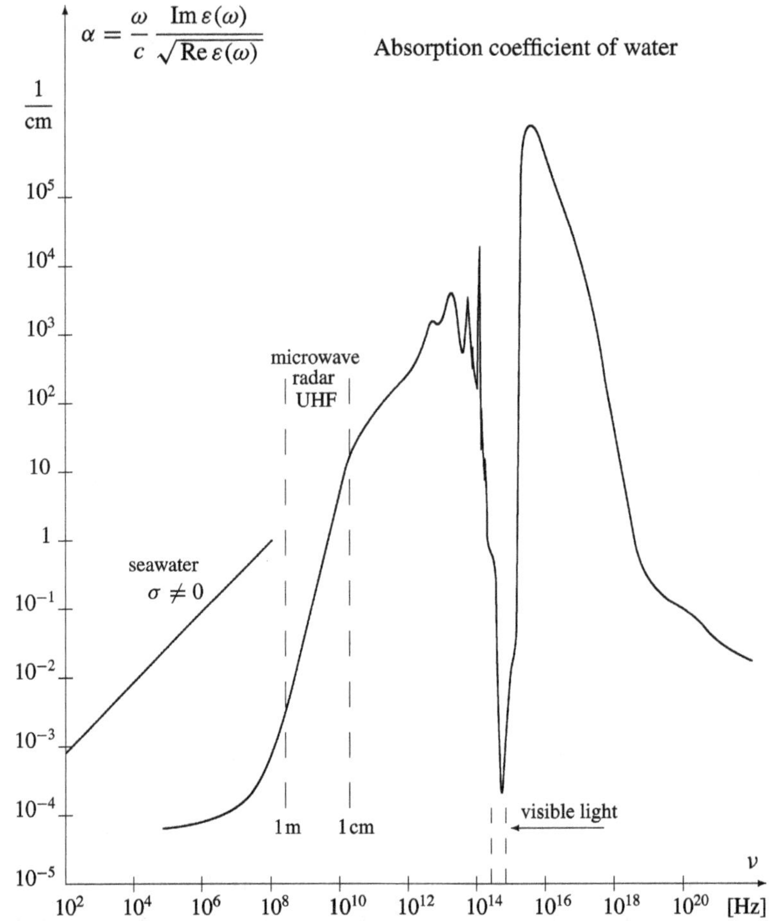

Figure 34.4 The experimental absorption coefficient $\alpha \propto \mathrm{Im}(\varepsilon)$ as a function of frequency $\nu = \omega/2\pi$ for water (shown in a log-log plot). Over the length $1/\alpha$, the intensity of an electromagnetic wave falls off to the eth fraction. In the range $\lambda \sim 100 \ldots 1$ cm the absorption increases drastically; this is due to the energy loss caused by the alternating alignment of the permanent dipoles. In this range there are many technical applications, such as microwaves ($\lambda \sim 30 \ldots 0.03$ cm), radar ($\lambda \sim 60 \ldots 0.8$ cm), television waves (UHF range: $\lambda \sim 100 \ldots 10$ cm) and the waves used by TV satellites ($\lambda \sim 6 \ldots 1$ cm). In the range of visible light ($\lambda \approx 4 \ldots 8 \cdot 10^{-7}$ m) the absorption coefficient decreases by many orders of magnitude.

frequency range: water is transparent here. The length $1/\alpha$, over which the intensity of a wave decreases by a factor of e is 10–100 m in the visible range; this refers to pure water. Outside the visible range $1/\alpha$ drops rapidly to fractions of a millimeter. Evolution has used this transparency window.

In contrast to condition (34.1), seawater has a certain conductivity due to the dissolved salt. For not too high frequencies, this contributes to the imaginary part of ε, (31.15). Due to this effect, radio waves are attenuated much more strongly in seawater than is the case with would be the case with fresh water; in Figure 34.4 the absorption for seawater is shown separately. This absorption hinders the communication between submarines by radio waves.

In the limiting case of very high frequencies $\omega \gg \omega_j$, (34.2) becomes

$$\varepsilon(\omega) \approx 1 - \frac{4\pi n_0 e^2}{m_\mathrm{e}} \sum_j \frac{f_j}{\omega^2} = 1 - \frac{Z\omega_\mathrm{P}^2}{\omega^2} \qquad (\omega \gg \omega_j) \tag{34.10}$$

The plasma frequency ω_P was given in (34.3). In the range $\omega^2 \gg Z\omega_\mathrm{P}^2 \sim \omega_\mathrm{at}^2$, $\varepsilon(\omega)$ of (34.10) asymptotically approaches 1. This behavior can be seen for $\mathrm{Re}\,\varepsilon$ in Figure 34.3. For $\omega \gg \omega_j$ the electrons behave practically like free particles. The imaginary part of ε then comes from the Thomson scattering (Chapter 25).

For still higher frequencies ($\hbar\omega \gtrsim 1\,\mathrm{MeV}$) the excitations of nucleons in atomic nuclei lead to further resonances. These phenomena are usually described by the cross-section for the scattering of γ-rays on atomic nuclei, but not by the dielectric function.

Metal

In a metal lattice, there is one mobile electron per unit cell. In the Lorentz model we insert $\omega_1 = 0$ and $\Gamma_1 = \Gamma$) for this electron. Using this the dielectric function (31.15) becomes

$$\varepsilon(\omega) = \varepsilon_\mathrm{o}(\omega) - \frac{4\pi n_0 e^2}{m_\mathrm{e}} \frac{1}{\omega(\omega + \mathrm{i}\Gamma)} = \varepsilon_\mathrm{o}(\omega) - \frac{\omega_\mathrm{P}^2}{\omega(\omega + \mathrm{i}\Gamma)} \tag{34.11}$$

The first term $\varepsilon_\mathrm{o}(\omega)$ includes the contributions of the bound electrons; for the associated effects we refer to the above section about insulators. In the following we exclude the ranges $\omega \sim \omega_j \pm \Gamma_j$ with resonances form $\varepsilon_\mathrm{o}(\omega)$.

Then we get approximately

$$\varepsilon_o(\omega) \approx \text{real}, \quad \varepsilon_o(\omega) \approx \text{const.}, \quad \varepsilon_o(\omega) = \mathcal{O}(1) \tag{34.12}$$

We assume $\varepsilon_o(\omega) = 1$ and split the dielectric function into its real and imaginary parts:

$$\text{Re}\,\varepsilon(\omega) = 1 - \frac{\omega_P^2}{\omega^2 + \Gamma^2}, \quad \text{Im}\,\varepsilon(\omega) = \frac{\omega_P^2\,\Gamma}{\omega\,(\omega^2 + \Gamma^2)} \tag{34.13}$$

These two functions are shown in Figure 34.5. According to (31.15)–(31.19), the amplitude of the $1/\omega$-term determines the conductivity. For $\omega \ll \Gamma$ we obtain from (31.20) the conductivity

$$\sigma = \sigma(0) = \frac{\omega_P^2}{4\pi\,\Gamma} = \frac{n_0\,e^2}{m_e\,\Gamma} \tag{34.14}$$

In the discussion we use the values (31.21) for copper:

$$\Gamma = \frac{1}{\tau} \approx 4 \cdot 10^{13}\,\text{s}^{-1}, \quad \sigma \approx 6 \cdot 10^{17}\,\text{s}^{-1}, \quad \omega_P \approx 2 \cdot 10^{16}\,\text{s}^{-1} \tag{34.15}$$

We discuss three frequency ranges and their corresponding characteristic effects:

1. $\omega \ll \Gamma$: this leads to the skin effect.

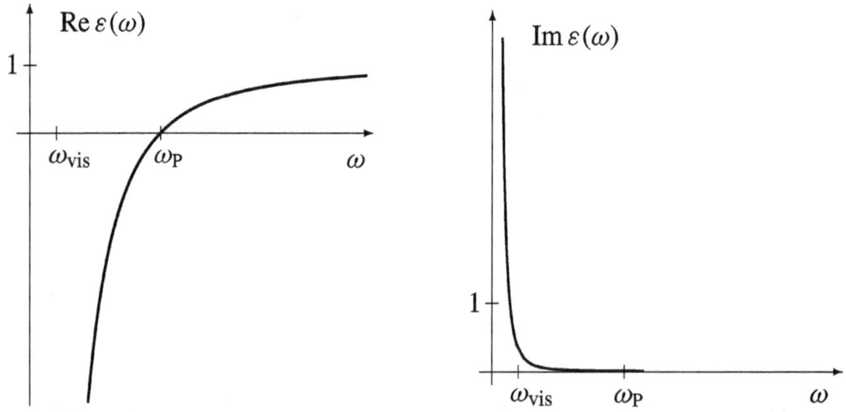

Figure 34.5 The real and imaginary part of the dielectric function $\varepsilon(\omega)$ of a metal, (34.13). For the ratio Γ/ω_P, the value for copper was chosen, i.e., approx. $3 \cdot 10^{-3}$. In the middle of the visible range, one obtains $\varepsilon(\omega_{\text{vis}}) \approx -30$. For $\omega \to 0$ we get $\text{Re}\,\varepsilon \to -\infty$ and $\text{Im}\,\varepsilon \to +\text{i}\,\infty$. This corresponds to large displacements of the mobile electrons.

2. $\Gamma \ll \omega < \omega_{\rm P}/\sqrt{\varepsilon_{\rm o}}$: visible light is almost completely reflected by metals.
3. $\omega > \omega_{\rm P}/\sqrt{\varepsilon_{\rm o}}$: metals become transparent in the ultraviolet.

Skin effect

For $\omega \ll \Gamma$ (34.11) becomes

$$\varepsilon(\omega) \approx \varepsilon_{\rm o}(\omega) + {\rm i}\,\frac{4\pi\sigma}{\omega} \qquad (\omega \ll \Gamma) \tag{34.16}$$

From $\omega \ll \Gamma$ and (34.15) follows $\omega \ll \sigma$. From $\varepsilon_{\rm o} = \mathcal{O}(1)$ we obtain $\sigma/\omega \gg \varepsilon_{\rm o}$. This gives us

$$n = \sqrt{\varepsilon} = \sqrt{\varepsilon_{\rm o} + {\rm i}\,\frac{4\pi\sigma}{\omega}} \approx \sqrt{{\rm i}\,\frac{4\pi\sigma}{\omega}} = (1+{\rm i})\sqrt{\frac{2\pi\sigma}{\omega}} \tag{34.17}$$

For a wave with $\boldsymbol{k} = n\,k_0\,\boldsymbol{e}_z$ this means

$$\exp({\rm i}\boldsymbol{k}\cdot\boldsymbol{r}) = \exp({\rm i}k_0\,n\,z) = \exp\left({\rm i}\,\frac{z}{d_{\rm skin}}\right)\,\exp\left(-\frac{z}{d_{\rm skin}}\right) \tag{34.18}$$

We evaluate the length $d_{\rm skin}$ using (34.17) with $k_0 = \omega/c$:

$$d_{\rm skin} = \frac{c}{\sqrt{2\pi\sigma\,\omega}} = \frac{c}{2\pi\,\sqrt{\sigma\,\nu}} \tag{34.19}$$

For copper with $\sigma = 5\cdot 10^{17}\ {\rm s}^{-1}$ we get in the UHF range

$$d_{\rm skin} = 2\cdot 10^{-4}\ {\rm cm} \quad \text{for } \nu = 10^9\ {\rm Hz} \tag{34.20}$$

The condition $\omega = 2\pi\nu \ll \Gamma$ is fulfilled. An electric field with this frequency causes the current density $\boldsymbol{j} = \sigma\boldsymbol{E}$. According to (34.18) and (34.20), the field can only penetrate the outer skin of the wire. This means that the current density $\boldsymbol{j}(\boldsymbol{r}) \propto \boldsymbol{E}(\boldsymbol{r})$ is limited to the external skin of the wire. This effect is called *skin effect.*

Due to the skin effect, a silver coated wire is well suitable for transporting high frequency currents; it is just as suitable as a solid silver wire. The conductivity of silver is higher than that of copper; however, silver is more expensive than copper. This is why silver coated copper wires are used in the receiver part of a television set.

Reflectivity in the visible–transparency in the ultraviolet

For $\omega \gg \Gamma$, (34.11) becomes

$$\varepsilon(\omega) = \varepsilon_o(\omega) - \frac{\omega_P^2}{\omega^2} = \begin{cases} \text{negative} & \left(\omega < \omega_P/\sqrt{\varepsilon_o}\right) \\ \text{positive} & \left(\omega > \omega_P/\sqrt{\varepsilon_o}\right) \end{cases} \tag{34.21}$$

Here, we assumed ε_o = real. The refractive index follows from (34.21):

$$n = \sqrt{\varepsilon(\omega)} = \begin{cases} \mathrm{i}\kappa = \mathrm{i}\sqrt{\omega_P^2/\omega^2 - \varepsilon_o} & (\omega < \omega_P/\sqrt{\varepsilon_o}, \text{ reflectivity}) \\ n_r = \sqrt{\varepsilon_o - \omega_P^2/\omega^2} & (\omega > \omega_P/\sqrt{\varepsilon_o}, \text{ transparency}) \end{cases} \tag{34.22}$$

These frequency dependencies are sketched in the left part of Figure 34.6.

When a wave encounters a boundary layer between two media (such as air and metal), part of the wave is generally reflected. The *reflection coefficient* R is the ratio between the reflected and the incident energy current density. An imaginary refractive index means that no wave can propagate in this medium. A wave with $\omega < \omega_P/\sqrt{\varepsilon_o}$, which hits a metal surface is therefore fully reflected, $R = 1$. For higher frequencies, $\omega > \omega_P/\sqrt{\varepsilon_o}$, the refractive index, on the other hand, is real and the metal

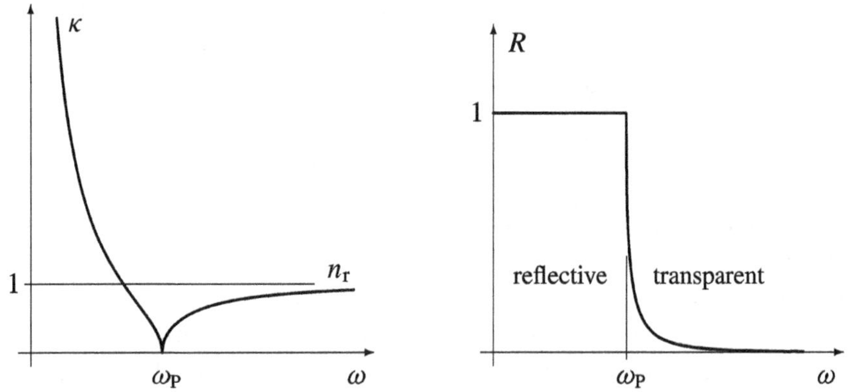

Figure 34.6 The imaginary part κ and the real part n_r of the refractive index n for the dielectric function (34.21) with $\varepsilon_o = 1$ (on the left). For frequencies below ω_P, we obtain $n = \mathrm{i}\kappa$, for frequencies above ω_P we get $n = n_r$. This determines the reflection coefficient (right). The change from reflective to transparent behavior is characteristic of a metal or a plasma.

is transparent. In this case, an incident wave partly propagates into the metal, and part of the wave is reflected. Figure 34.6 on the right sketches the frequency dependence of the reflection coefficient (for $\varepsilon_o = 1$). The quantitative expression for the reflection coefficient $R(\omega)$, which forms the basis of this figure, is given in Chapter 37.

In the Lorentz model as well as in real materials, the dielectric function $\varepsilon_o(\omega)$ has a small imaginary part for $\omega \sim \omega_P/\sqrt{\varepsilon_o}$. This implies that the dielectric function (34.21) nowhere disappears exactly. Therefore, the transition from reflective to transparent behavior is smeared out somewhat.

For the frequency of visible light, $\omega_{vis} \approx \omega_P/5 < \omega_P/\sqrt{\varepsilon_o}$ holds. Visible light is therefore almost completely reflected by metals ($R \approx 1$). The complete reflection also means that metals are opaque (non-transparent).

Due to $R \approx 1$, metals are suitable as mirrors. From (34.21) and $\varepsilon_o \approx 1$ it follows that $\varepsilon(\omega) \approx -15 \ldots -45$ in the visible range, i.e., $\kappa \sim 4 \ldots 7$. A wave only penetrates along the length $\mathcal{O}(\lambda/2\pi\kappa)$ into the metal. Therefore, a thin layer of silver on glass can be used as a mirror.

For $\omega > \omega_P/\sqrt{\varepsilon_o}$ we get $\kappa \approx 0$. The damping of a wave is therefore very small; the medium is transparent. The critical frequency for $\varepsilon_o = 1$ is about $\omega_P \approx 2 \cdot 10^{16}\,\mathrm{s}^{-1}$, i.e., in the ultraviolet. When the frequency ω crosses the value $\omega_P/\sqrt{\varepsilon_o}$, the previously opaque metal becomes transparent. This effect is referred to as *transparency in the ultraviolet.*

The actual value of the critical frequency $\omega_P/\sqrt{\varepsilon_o}$ depends on the density n_0 in (34.3) and on the value of $\varepsilon_o = \mathcal{O}(1)$. For the alkali metals lithium, sodium and potassium, the transition to transparency occurs at the frequencies $\omega \approx 1.2$, 0.9 and $0.6 \cdot 10^{16}\,\mathrm{s}^{-1}$ (for comparison $\omega_{violet} \approx 0.5 \cdot 10^{16}\,\mathrm{s}^{-1}$).

Plasma

Finally, we consider a plasma of free electrons and ions. Due to the free-moving charges, we can assume the form (34.11) with suitable parameters for the plasma.

In the dielectric function, the contribution of the free electrons dominates. Electrons are easier to displace than the ions (because of their smaller mass). We focus on the case $\omega \gg \Gamma = 1/\tau$; here τ is the mean collision time of the electrons. For this we use (34.21) with $\varepsilon_o = 1$:

$$\varepsilon(\omega) = 1 - \frac{\omega_P^2}{\omega^2} \quad (\omega \gg 1/\tau) \tag{34.23}$$

In

$$\omega_P^2 = \frac{4\pi n_0 e^2}{m_e} \tag{34.24}$$

the density n_0 is that of the free electrons. We insert $\varepsilon(\omega) = n^2 = c^2k^2/\omega^2$ in (34.23) and solve for ω^2:

$$\omega^2 = \omega_P^2 + c^2k^2 \quad \begin{pmatrix}\text{dispersion relation} \\ \text{for plasma waves}\end{pmatrix} \tag{34.25}$$

For high frequencies ($\omega \gg \omega_P$), the deviation from $n = 1$ and $\omega = ck$ is small. This applies, for example, to the dispersion of starlight in the interstellar space. In this case, n_0 and therefore ω_P are very small.

From (34.23) follows

$$n(\omega) = \sqrt{\varepsilon(\omega)} = \begin{cases} \text{imaginary} & (\omega < \omega_P, \ \text{reflective}) \\ \text{real} & (\omega > \omega_P, \ \text{transparent}) \end{cases} \tag{34.26}$$

For the transition from reflexive to transparent behavior we cite two examples:

- Due to the reflection at the ionosphere a worldwide broadcasting is possible in the shortwave range. In the VHF range, on the other hand, the ionosphere, is transparent (TV-signal transmission via satellite is therefore possible).
- Upon re-entry into the atmosphere, the radio connection between a spacecraft and the ground station is interrupted. During re-entry the air surrounding the spaceship heats up strongly. This increases the density n_0 of the ionized particles in this layer. The plasma frequency is then temporarily higher than the frequency of the radio waves, and the ionized air layer is intransparent for radio waves.

The interaction of plasma with electromagnetic fields is an extensive field of research, which is not even hinted at by the formulas given here. One area of application is the confinement and compression of hot plasma by electromagnetic fields with the aim of controlled fusion.

PART VII

Elements of Optics

Chapter 35

Huygens' Principle

Optics is the study of the propagation of light or, more generally, of electromagnetic waves. The starting point are therefore Maxwell's equations and their wave solutions in a vacuum (Chapter 20) or in matter (Chapter 33). Optics is a large and extensive field[1] *in which we take just a foray: We discuss the most basis features like diffraction and interference, reflection and refraction, and geometrical optics.*

An electromagnetic wave falls on an aperture, i.e., on an impermeable surface with openings. The wave field behind the aperture can be determined using Huygens' principle, a very useful approximation. According to this principle, a spherical wave emanates from each point of the aperture opening. Within the framework of Kirchhoff's diffraction theory, we make Huygens' principle plausible. The result can be simplified in the Fraunhofer approximation.

We consider a monochromatic vacuum wave with

$$\boldsymbol{E}(\boldsymbol{r},t) = \boldsymbol{E}(\boldsymbol{r})\,\exp(-\mathrm{i}\omega t), \qquad \boldsymbol{B}(\boldsymbol{r},t) = \boldsymbol{B}(\boldsymbol{r})\,\exp(-\mathrm{i}\omega t) \tag{35.1}$$

The real part formation is omitted in the notation. In the source free range, the Maxwell equations in vacuum give rise to the wave equations (21.4) and (21.5). For (35.1), they result in

$$\left(\Delta + k^2\right)\begin{pmatrix}\boldsymbol{E}(\boldsymbol{r})\\ \boldsymbol{B}(\boldsymbol{r})\end{pmatrix} = 0 \quad (k = \omega/c) \tag{35.2}$$

[1] A. Lipson, S. G. Lipson, H. Lipson, *Optical Physics*, 4th edition, Cambridge University Press, 2010.

For each component of the electromagnetic field (for example, for $\psi = E_x$), we obtain

$$\left(\Delta + k^2 \right) \psi(\boldsymbol{r}) = 0 \tag{35.3}$$

In the following, we restrict ourselves to this *scalar* (one-component) wave equation. We thus neglect all polarization effects. This is sufficient for many applications.

We consider the situation sketched in Figures 35.1 and 35.2. Before the aperture there is a point source at $\boldsymbol{r}''$, and we want to determine the wave field behind the aperture:

$$\psi(\boldsymbol{r}) = \begin{cases} C \, \dfrac{\exp(\mathrm{i}k\,|\boldsymbol{r}-\boldsymbol{r}''|)}{|\boldsymbol{r}-\boldsymbol{r}''|} & \text{before the aperture} \\ ? & \text{behind the aperture} \end{cases} \tag{35.4}$$

With the exception of the point $\boldsymbol{r}''$ of the source, $\psi(\boldsymbol{r})$ satisfies the wave equation (35.3). The position and shape of the aperture is described by a surface A. The surface A shall enclose a volume V; for this purpose, a plane aperture might be closed far away from the openings by a large hemisphere.

For calculating the field $\psi(\boldsymbol{r})$ behind the aperture, we start from the second Green's theorem (1.31)

$$\int_V d^3r \,\big(\psi(\boldsymbol{r})\,\Delta G - G\,\Delta\psi(\boldsymbol{r})\big) = \oint_A d\boldsymbol{A}\cdot\big(\psi(\boldsymbol{r})\,\boldsymbol{\nabla} G - G\,\boldsymbol{\nabla}\psi(\boldsymbol{r})\big) \tag{35.5}$$

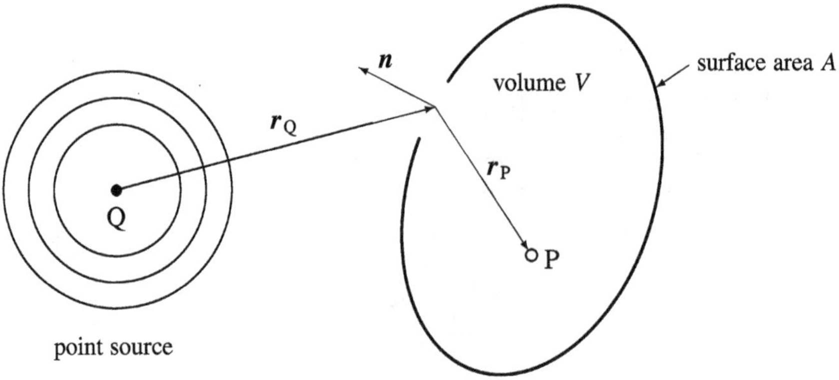

Figure 35.1 The spherical wave of a source Q falls onto an aperture (closed surface A with an opening). We are looking for the wave field at point P within the volume V.

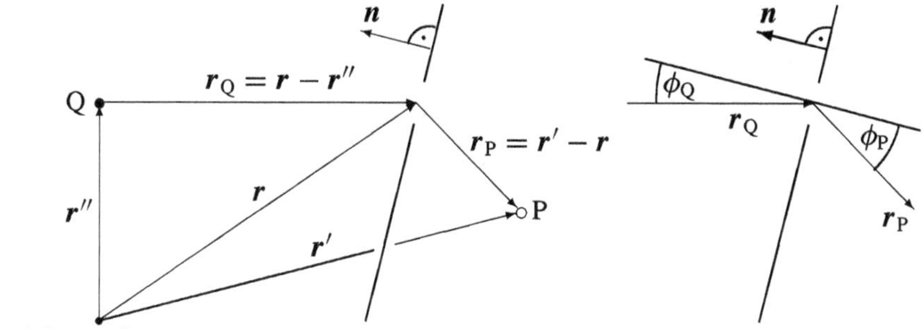

Figure 35.2 Supplementing to Figure 35.1, some vectors and angles are defined that occur in the calculation. The vector $\boldsymbol{r}$ points to a position of the aperture, $\boldsymbol{r}''$ is the position of the source, and $\boldsymbol{n}$ is a unit vector that is perpendicular to the surface. At the position $\boldsymbol{r}'$ the wave field shall be calculated. For the angles defined in the right part, we note $\boldsymbol{n}\cdot\boldsymbol{r}_Q = -r_Q \cos\phi_Q$ and $\boldsymbol{n}\cdot\boldsymbol{r}_P = -r_P \cos\phi_P$.

with

$$G(\boldsymbol{r},\boldsymbol{r}') = \frac{\exp(ik|\boldsymbol{r}-\boldsymbol{r}'|)}{|\boldsymbol{r}-\boldsymbol{r}'|} \tag{35.6}$$

According to (3.37), the following applies

$$\left(\Delta + k^2\right) G(\boldsymbol{r},\boldsymbol{r}') = -4\pi\,\delta(\boldsymbol{r}-\boldsymbol{r}') \tag{35.7}$$

On the left-hand side of (35.5), we can replace both Δ operators by $\Delta + k^2$ because the additional terms with k^2 cancel each other. Using (35.3) and (35.7), Equation (35.5) becomes thus

$$\psi(\boldsymbol{r}') = \frac{1}{4\pi}\oint_A dA\cdot\Big(G(\boldsymbol{r},\boldsymbol{r}')\,\nabla\psi(\boldsymbol{r}) - \psi(\boldsymbol{r})\,\nabla G(\boldsymbol{r},\boldsymbol{r}')\Big) \tag{35.8}$$

The vector $\boldsymbol{r}$ points to a point of the area A, over the integral runs. The vector $\boldsymbol{r}'$ points to a position inside of V where the wave field $\psi(\boldsymbol{r}')$ is to be calculated.

The aperture completely absorbs the light (outside the openings). This makes the following assumptions of *Kirchhoff's diffraction theory* for the field ψ on the surface A plausible:

1. On the aperture (without the openings), the field and its derivative disappear. The integration in (35.8) can therefore be restricted to the aperture.

2. In the openings (of the aperture), ψ is the undisturbed field of the source (first line in (35.4)).

Using the designations from Figure 35.2, we obtain $G = \exp(\mathrm{i}k r_{\mathrm{P}})/r_{\mathrm{P}}$ and $\psi = A\exp(\mathrm{i}k r_{\mathrm{Q}})/r_{\mathrm{Q}}$. With this and with the Kirchhoff's assumptions, (35.8) becomes

$$\psi(\boldsymbol{r}') \approx \frac{C}{4\pi}\int_{\text{openings}} dA\,\boldsymbol{n}$$
$$\cdot\left(\frac{\exp(\mathrm{i}k r_{\mathrm{P}})}{r_{\mathrm{P}}}\,\boldsymbol{\nabla}\,\frac{\exp(\mathrm{i}k r_{\mathrm{Q}})}{r_{\mathrm{Q}}} - \frac{\exp(\mathrm{i}k r_{\mathrm{Q}})}{r_{\mathrm{Q}}}\,\boldsymbol{\nabla}\,\frac{\exp(\mathrm{i}k r_{\mathrm{P}})}{r_{\mathrm{P}}}\right) \quad (35.9)$$

Mathematically, however, Kirchhoff's assumptions are inconsistent: If on a finite part of the surface $\psi = 0$ and $\boldsymbol{n}\cdot\nabla\psi = 0$ hold (where $\boldsymbol{n}$ is a normal vector), then $\psi \equiv 0$ follows. Therefore, we modify the Kirchhoff's first assumption: Outside the openings ψ and $\boldsymbol{n}\cdot\nabla\psi$ yield negligible contributions to the integral (35.8).

For the further evaluation we assume

$$r_{\mathrm{Q}} \gg \lambda, \quad r_{\mathrm{P}} \gg \lambda \quad (35.10)$$

We carry out the derivatives in the integrand of (35.9). With $\boldsymbol{r}_{\mathrm{Q}} = \boldsymbol{r} - \boldsymbol{r}''$ and the angle ϕ_{Q} from Figure 35.2, we obtain

$$\boldsymbol{n}\cdot\boldsymbol{\nabla}\,\frac{\exp(\mathrm{i}k r_{\mathrm{Q}})}{r_{\mathrm{Q}}} = \frac{\boldsymbol{n}\cdot\boldsymbol{r}_{\mathrm{Q}}}{r_{\mathrm{Q}}}\left(\mathrm{i}k - \frac{1}{r_{\mathrm{Q}}}\right)\frac{\exp(\mathrm{i}k r_{\mathrm{Q}})}{r_{\mathrm{Q}}}$$
$$\approx -\mathrm{i}k\cos\phi_{\mathrm{Q}}\,\frac{\exp(\mathrm{i}k r_{\mathrm{Q}})}{r_{\mathrm{Q}}} \quad (35.11)$$

The corresponding formula for $\boldsymbol{r}_{\mathrm{P}} = \boldsymbol{r}' - \boldsymbol{r}$ contains an additional minus sign from the differentiation ($\boldsymbol{\nabla}$ acts on $\boldsymbol{r}$). We inset the expressions for the derivatives into (35.9):

$$\psi(\boldsymbol{r}') \approx \frac{-\mathrm{i}kC}{4\pi}\int_{\text{openings}} dA\,\frac{\exp(\mathrm{i}k r_{\mathrm{P}})}{r_{\mathrm{P}}}\,\frac{\exp(\mathrm{i}k r_{\mathrm{Q}})}{r_{\mathrm{Q}}}\left(\cos\phi_{\mathrm{Q}} + \cos\phi_{\mathrm{P}}\right) \quad (35.12)$$

The aforementioned inconsistency of Kirchhoff's first assumption can be formally avoided by a modified Green's function $G(\boldsymbol{r},\boldsymbol{r}')$ (Section 9.8 in [6]). In this case, one obtains instead of $\cos\phi_{\mathrm{Q}} + \cos\phi_{\mathrm{P}}$ another angle factor, namely $2\cos\phi_{\mathrm{Q}}$ or $2\cos\phi_{\mathrm{P}}$ depending on the assumed approximation. In the following, we restrict ourselves to small angles and approximate the angle factor by $\cos\phi_Q + \cos\phi_P \approx 2$. In this approximation, (35.12) becomes

$$\boxed{\psi(\boldsymbol{r}') \approx \frac{-\mathrm{i}kC}{2\pi}\int_{\text{openings}} dA\ \frac{\exp(\mathrm{i}kr_Q)}{r_Q}\ \frac{\exp(\mathrm{i}kr_P)}{r_P}} \tag{35.13}$$

This is the *Huygens principle*: From each point of the opening, a spherical wave $\exp(\mathrm{i}kr_P)/r_P$ emanates. Its strength and phase is determined by the incident spherical wave $C\exp(\mathrm{i}kr_Q)/r_Q$. As a possible generalization, several sources can also be allowed; then $C\exp(\mathrm{i}kr_Q)/r_Q$ must be replaced by the superposition of the fields of all of these sources.

We further simplify (35.13) by assuming that the distance r_Q to the source is much greater than the diameter d of the aperture opening and that the wave is perpendicular to a flat aperture falls. Then the following applies in the area of the aperture

$$\frac{\exp(\mathrm{i}kr_Q)}{r_Q} \approx \text{const.} \quad (r_Q \gg d, \text{ vertical incidence}) \tag{35.14}$$

A constant phase in the area of the aperture means that the spherical wave is locally approximated here by a plane wave. This means that (35.13) to

$$\boxed{\psi(\boldsymbol{r}') \approx C'\int_{\text{openings}} dA\ \frac{\exp(\mathrm{i}kr_P)}{r_P} \qquad \begin{array}{l}\text{Huygens' principle}\\ \text{for vertical incidence}\\ \text{of a plane wave}\end{array}} \tag{35.15}$$

The prefactors were abbreviated by C'. In this form, Huygens' principle reads: From each point of the aperture a spherical wave emanates.

Huygens established his principle as early as 1679. The reasoning given here goes back to Kirchhoff (1882). Despite of the problems in the presented derivation, Huygens' principle is a rather good and useful approximation.

According to Huygens' principle, diffraction and scattering are related phenomena. In scattering, light falls on scattering centers (such as atoms or lines of a lattice). In Chapter 25 calculated (with polarization effects) how an atom is excited to dipole oscillations and thus in turn becomes the source of an outgoing spherical wave. This means that each atom (or scattering center), corresponds to a point of the opening in (35.13).

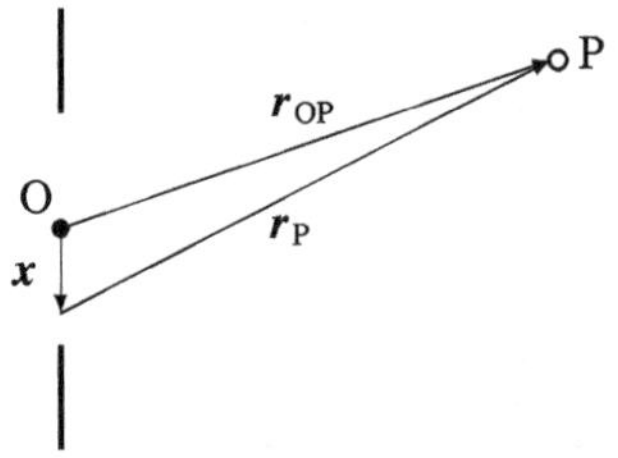

Figure 35.3 We are using the vector $\boldsymbol{r}_{OP}$ from a point O of the opening to the observation point P. Then $\boldsymbol{r}_P = \boldsymbol{r}_{OP} - \boldsymbol{x}$, where $\boldsymbol{x} = (\xi, \eta)$ is a (two-dimensional) vector in the plane of the aperture.

Fraunhofer's Diffraction

We expand $r_P^2 = (\boldsymbol{r}_{OP} - \boldsymbol{x})^2 = (x_{OP} - \xi)^2 + (y_{OP} - \eta)^2 + z_{OP}^2$ to powers of $\boldsymbol{x}$,

$$r_P^2 = r_{OP}^2 \left(1 - \frac{2\,\boldsymbol{k}_{OP} \cdot \boldsymbol{x}}{k\, r_{OP}} + \frac{x^2}{r_{OP}^2} \right) \tag{35.16}$$

where

$$\boldsymbol{k}_{OP} = k\, \frac{\boldsymbol{r}_{OP}}{r_{OP}} \tag{35.17}$$

We use the expansion (35.16) in the integrand of (35.15):

$$\begin{aligned} \frac{\exp(\mathrm{i}k r_P)}{r_P} &\approx \frac{\exp(\mathrm{i}k r_{OP})}{r_{OP}}\ \exp(-\mathrm{i}\boldsymbol{k}_{OP} \cdot \boldsymbol{x})\ \exp\left(\frac{\mathrm{i}k x^2}{2\, r_{OP}}\right) \\ &\approx \frac{\exp(\mathrm{i}k r_{OP})}{r_{OP}}\ \exp(-\mathrm{i}\boldsymbol{k}_{OP} \cdot \boldsymbol{x}) \quad \text{for } D \gg \frac{\pi a^2}{\lambda} \end{aligned} \tag{35.18}$$

For sufficiently large distanced D between aperture and screen the last exponential factor in the first line is approximately is equal to 1; where $D \approx r_P \approx r_{OP}$, and a is the size of the aperture opening. For $D \leq \pi a^2/\lambda$ this exponential factor must, however, be taken into account; this case is referred as *Fresnel diffraction.*

With (35.18), (35.15) becomes *Fraunhofer diffraction* (or Fraunhofer approximation):

$$\boxed{\psi(\boldsymbol{r}') \approx C'\ \frac{\exp(\mathrm{i}k r_{OP})}{r_{OP}} \int_{\text{opening}} d^2x\ \exp(-\mathrm{i}\boldsymbol{k}_{OP} \cdot \boldsymbol{x})} \tag{35.19}$$

If there is matter in the opening of the aperture, then the strength of the outgoing waves (Chapter 25) is modified. This can be described by replacing the prefactor C' by a function $B(\boldsymbol{x})$ in the integral:

$$\psi(\boldsymbol{r}') \approx \frac{\exp(\mathrm{i}k r_{\mathrm{OP}})}{r_{\mathrm{OP}}} \int_{\text{opening}} d^2x \; B(\boldsymbol{x}) \exp(-\mathrm{i}\boldsymbol{k}_{\mathrm{OP}} \cdot \boldsymbol{x}) \qquad (35.20)$$

The integral is the Fourier transform of $B(\boldsymbol{x})$, i.e., of the two-dimensional structure in the opening. From the diffraction pattern one obtains $|\psi|^2$ and thus the squared magnitude of this Fourier transform.

It is often the case that a scattering experiment yields the Fourier transform of the structure under investigation. An example is the Born approximation (Chapter 45 in [3]): Here, the scattering of particles at a target is described a potential (between the projectile and a target particle). In the Born approximation, the calculated cross-section is then proportional to the square of the Fourier transform of this potential. We have also obtained a comparable result in (25.27) for the scattering of light.

A standard problem of such experiments is that one can only determine the absolute value but not the phase of the Fourier transform. Moreover, an experiment only covers a finite range of $\boldsymbol{k}$ values. In order to obtain $B(\boldsymbol{x})$ (or the potential) one must therefore use additional assumptions. One could, for example make a plausible approach for $B(\boldsymbol{x})$ that depends on some parameters. Comparing the Fourier transform of this approach with the experiment then determines the parameters.

Chapter 36

Interference and Diffraction

The Huygens principle is applied to a few simple and informative cases. First of all, we deal with the interference effects that occur when light is scattered by a double slit or by two small openings of an aperture. We discuss the shadow formation and, in connection with this, the blurring of the shadow boundary due to diffraction effects. In the Fraunhofer approximation, we calculate the diffraction for a rectangular aperture opening.

Interference at the Double Slit

An aperture shall have two slit openings whose distance d is comparable to the wavelength λ. When light falls on this aperture, an interference pattern is observed on the screen (Figure 36.1). This experiment was carried out in 1801 by T. Young. The observed interference was in contradiction to Newton's generally accepted opinion that light consists of particle beams.

For a simplified calculation, we consider an aperture which has tow small circular openings (with the area a) instead of slits; Figure 36.1 also applies for this setup. A plan wave propagates vertically toward the aperture. For the openings if the aperture, we use Fraunhofer's approximation (35.19). The diameter of the apertures openings shall be so small that $\exp(-\mathrm{i}\boldsymbol{k}_{\mathrm{OP}} \cdot \boldsymbol{x}) \approx 1$ holds. Then, the integral in (35.19) gives the area a for each opening, and (35.19) becomes

$$\psi(\boldsymbol{r}) = C \left(\frac{\exp(\mathrm{i}kr_1)}{r_1} + \frac{\exp(\mathrm{i}kr_2)}{r_2} \right) \tag{36.1}$$

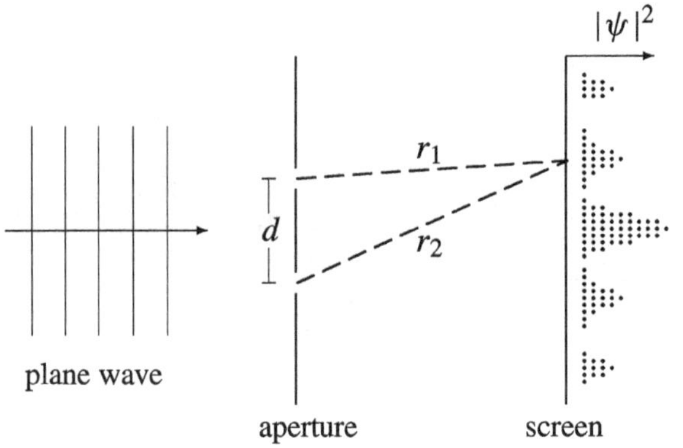

Figure 36.1 A wave falls on an aperture with two equal openings (hole or gap), whose distance d is comparable to the wavelength λ. According to Huygens' principle, a spherical wave emanates from each of the two openings. To the right of the screen, the measurable intensity $|\psi|^2$ is plotted schematically.

Here, r_1 and r_2 are the distances from the opening to a point on the screen. The prefactor was abbreviated as $C'a = C$. The intensity I on the screen (measured for example by the blackening of a photo plate) is then

$$I = |\psi|^2 = |C|^2 \left(\frac{1}{r_1^2} + \frac{1}{r_2^2} + \frac{2}{r_1 r_2} \cos\left[k\,(r_1 - r_2)\right] \right) \tag{36.2}$$

This results in the interference typical for waves:

$$I = \begin{cases} \dfrac{|C|^2\,(r_1 + r_2)^2}{r_2^2\, r_1^2} \approx \dfrac{4|C|^2}{r^2} & \text{for } \cos(...) = 1 \\[2ex] \dfrac{|C|^2\,(r_1 - r_2)^2}{r_2^2\, r_1^2} \approx 0 & \text{for } \cos(...) = -1 \end{cases} \tag{36.3}$$

Depending on the sign of the cosine, this results in an amplification or weakening. In the amplitude we used $r_1 \approx r_2 \approx r$. The approximation $r_1 \approx r_2 \approx r$ must not be used in the cosine function, however, because here small changes of the path difference $r_1 - r_2$ are decisive. Depending on the path difference $r_1 - r_2$, a wave crest meets another wave crest at the screen (constructive interference) or a wave crest meets a wave trough (destructive interference). For constructive interference, the intensity is twice as large as the sum of the intensities of the individual waves.

For arbitrary apertures, (35.13), (35.15) or (35.19) are the basis for calculating the interference effects: The waves $\exp(\mathrm{i}kr_{\mathrm{P}})/r_{\mathrm{P}}$ emanating from the openings are added. At the observation point, they generally have different phases. Their superposition therefore leads to an interference pattern.

The interference pattern is determined by the ratio of the wavelength λ to the distance d of the scattering centers. In order that interference phenomena occur, both quantities must be comparable (e.g., $\lambda \sim d$ or also $\lambda \sim 10\,d$). By scattering centers we mean here the openings of the aperture, from which, according to Huygens' principle, spherical waves emanate. Also the atoms of a crystal lattice or the lines of an optical grating might constitute suitable scattering centers. For this we refer to the scattering of Röntgen rays at the crystal lattice (Laue 1912) and to the Exercises 25.2 and 36.1.

Coherence length

We have assumed that a plane wave falls on the openings of the screen. However, natural light consists of individual wave packets that have statistically distributed phases and centers. The replacement of a wave packet by a plane wave (or spherical wave) is only permissible if the extension of the wave packet is greater than the relevant dimensions (such as the hole spacing d in Figure 36.1). According to (24.42), a photon in natural light can be considered as a wave packet with the size

$$\ell_{\mathrm{c}} \sim 3\,\mathrm{m} \tag{36.4}$$

It is said that light has the *coherence length* ℓ_{c}. If the relevant dimensions are small compared to ℓ_{c}, we can use infinitely extended waves (such as $\exp(\mathrm{i}\boldsymbol{kr}$ or $\exp(\mathrm{i}kr)/r$) for simplicity.

In the double-slit experiment, the individual wave packets lead to the interference shown in Figure 36.1 (due to $d \sim \lambda \ll \ell_{\mathrm{c}}$). Each individual wave packet can be interpreted as the quantum mechanical wave function of a single photon. In this sense, individual photons interfere with themselves. Indeed, the interference also occurs when the incident wave in Figure 36.1 has such a low intensity that individual photons arrive separated from each other in time.

The quantum mechanical wave function for photons becomes the classical wave of electrodynamics when a large number of photons have the same wave function. In this case, the wave function itself is an observable, i.e., the fields $\boldsymbol{E}$ and $\boldsymbol{B}$ (and not only the squared magnitude of the field). This applies, for example, to radio waves or laser light.

In Young's experiment, natural light was used. Therefore the experiment shows the interference of the quantum mechanical wave function of individual photons, but not the interference of a classical electromagnetic wave. This difference (quantum mechanical wave function–classical electromagnetic wave) concerns the interpretation of the function ψ.

Shadow Formation and Diffraction

The rectilinear propagation of light, in particular the formation of shadows behind an obstacle, can be understood with the help of Huygens' principle. We discuss this qualitatively for a slit with a width d that is large compared to the wavelength λ (Figure 36.2).

According to Huygens' principle, a spherical wave emanates from each point of the slit opening. The surfaces of equal phase are spheres (depicted as circles in Figure 36.2). As indicated in Figure 36.2 these circles have an enveloping which corresponds approximately to the gap shifted to the right. On this envelope the spherical waves of neighboring starting points interfere positively. As a result, the wave has an overall wavefront that corresponds roughly to the slit width. At other points, however, the spherical waves arrive with different phases and average out.

This is a qualitative explanation of the shadow formation behind the aperture. The explanation and the wave fronts sketched in Figure 36.2 suggest that the shadow boundary is not sharp. In fact, the light at the slit boundary is deflected by an angle of the size $\varDelta\theta$. This deflection can be understood as follows: The slit cuts our a wave packet of the size d (perpendicular to $\boldsymbol{k}$). According to (20.46), the wave vector then has an uncertainty $\varDelta k/k \sim \lambda/d$ in the direction of the slit. From this follows the angular uncertainty $\varDelta\theta \sim \varDelta k/k$ or

$$\varDelta\theta \sim \frac{\lambda}{d} \quad \text{(diffraction)} \tag{36.5}$$

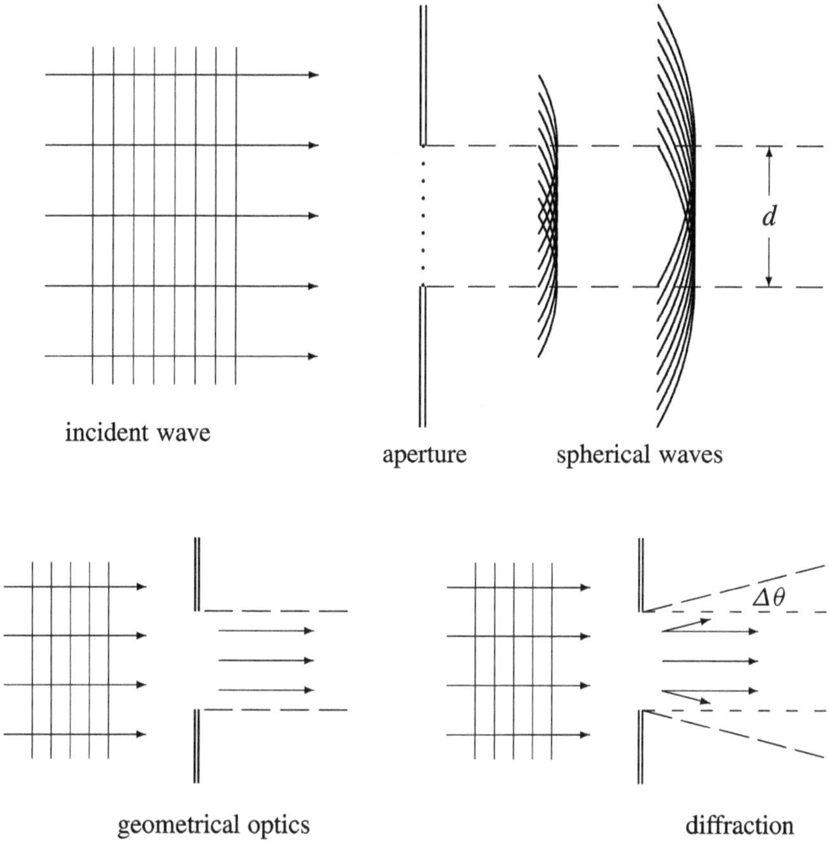

Figure 36.2 A wave falls on an aperture opening. The width d of the gap shall be much larger than the wavelength, $d \gg \lambda$. According to Huygens' principle, from each point of the opening a spherical wave propagates (top). For two different radii, the fronts of a number of these spherical waves are shown. They have an envelope, which can be recognized as an amplified blackening. At the position of the envelope the sphere waves interfere positively, at other positions destructive. The envelope roughly corresponds to the slit, when it is shifted to the right. This corresponds to the rectilinear propagation of light (bottom left), i.e., the limiting case of geometric optics. The picture illustrates that the boundary of the light beam is not sharp. Rather, the light beam is bent slightly (bottom right), namely b the angle range $\varDelta\theta \sim \lambda/d \ll 1$. For a slit with $d = 5\,\mathrm{mm}$ and light in the visible range, $\varDelta\theta$ is about two arc minutes. The picture illustrates the relationship between the interference (the wave character) and the other phenomena discussed in this chapter, i.e., the shadow, destructive and constructive interference, the diffraction and the rectilinear propagation of light.

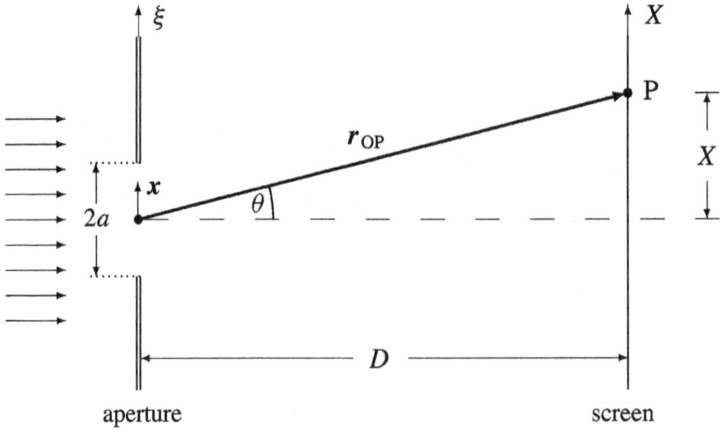

Figure 36.3 Light falls perpendicular to a slit of width $2a$. The resulting intensity I is observed on a screen at a distance $D \gg a$. The intensity (Figure 36.4) has a dominant maximum in the forward direction.

In the range of this angle, the boundary of the light behind the aperture is blurred (Figure 36.2, bottom right). This effect is called *diffraction*. Neglecting diffraction leads to the limiting case of geometric optics (Chapter 38).

For simple geometries the Huygens principle allows for a quantitative treatment of the discussed effects, the shadow formation and the diffraction. For a rectangular slit, this is done in the following section.

Historically, shadow formation has been used as an argument against the wave theory of light; it seemed much more understandable in a model based on particle beams (Newton). The clarification of this point with the help of Huygens' principle and the observation of interference effects led to the acknowledgement of the wave theory of light in the 19th century. Quantum effects (photo effect, Compton effect) made it clear at the beginning of the 20th century that light has a particle character, too. Light is subject to the wave–particle duality.

Diffraction at a Slit

In the Fraunhofer approximation we calculate the diffraction at a rectangular opening. In the opening area A_{opening} we use the coordinates $\boldsymbol{x} = (\xi, \eta)$:

$$A_{\text{opening}} = \big\{\xi, \eta \quad |\xi| \le a, \ |\eta| \le b\big\} \tag{36.6}$$

For the distance D of the screen, the condition for the Fraunhofer approximation (35.19) shall be fulfilled,

$$D \gg \frac{\pi a^2}{\lambda}, \quad D \gg \frac{\pi b^2}{\lambda} \tag{36.7}$$

We evaluate (35.19):

$$\psi(\boldsymbol{r}') = C \int_{\text{opening}} dA\ \exp\left(-\,\mathrm{i}\,\boldsymbol{k}_{\text{OP}} \cdot \boldsymbol{x}\right) \tag{36.8}$$

$$= C \int_{-a}^{a} d\xi \int_{-b}^{b} d\eta\ \exp\big(-\,\mathrm{i}\,(\boldsymbol{k}_{\text{OP}} \cdot \boldsymbol{e}_x)\,\xi\big)\ \exp\big(-\,\mathrm{i}\,(\boldsymbol{k}_{\text{OP}} \cdot \boldsymbol{e}_y)\,\eta\big)$$

The prefactor was abbreviated by $C = C' \exp(\mathrm{i}k r_{\text{OP}})/r_{\text{OP}} \approx$ const. The vector $\boldsymbol{k}_{\text{OP}}$ points from the center of the opening to the observation point. We use

$$\boldsymbol{k}_{\text{OP}} \cdot \boldsymbol{e}_x \overset{(35.17)}{=} k\,\frac{\boldsymbol{r}_{\text{OP}} \cdot \boldsymbol{e}_x}{r_{\text{OP}}} = k\,\frac{X}{r_{\text{OP}}} \approx \frac{k X}{D} \tag{36.9}$$

and correspondingly $\boldsymbol{k}_{\text{OP}} \cdot \boldsymbol{e}_y = kY/D$; here X and Y are the Cartesian coordinates of the observation point on the screen. The integral (36.8) results in

$$\psi(X, Y) = 4Cab\,\frac{\sin(\alpha X)}{\alpha X}\,\frac{\sin(\beta Y)}{\beta Y} \tag{36.10}$$

where

$$\alpha = \frac{ka}{D} \quad \text{and} \quad \beta = \frac{kb}{D} \tag{36.11}$$

The X-Y-dependence of the intensity on the screen is then

$$I(X, Y) = |\psi|^2 = I(0, 0)\,\frac{\sin^2(\alpha X)}{\alpha^2 X^2}\,\frac{\sin^2(\beta Y)}{\beta^2 Y^2} \tag{36.12}$$

In the limiting case $b \to \infty$ (slit instead of rectangular opening) the Y-dependence disappears. The intensity on the screen then becomes

$$\boxed{\frac{I(X)}{I(0)} = \frac{\sin^2(\alpha X)}{\alpha^2 X^2} \approx \frac{\sin^2(ka\theta)}{(ka\theta)^2} \quad \text{diffraction at a slit}} \tag{36.13}$$

The intensity $|I(X)|^2$ is depicted in Figure 36.4. The last expression can be used for small angles $\theta \approx X/D$. The width of the angular distribution can

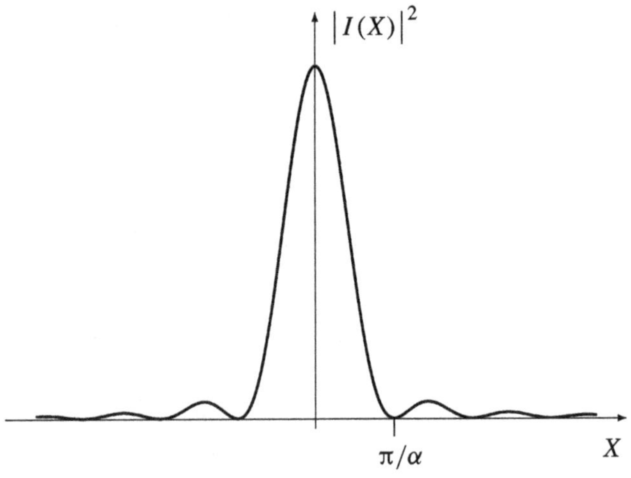

Figure 36.4 Interference pattern caused by the diffraction at a slit. The picture shows the intensity $I(X)$ calculated in (36.13). The interference pattern has a main maximum at $X = 0$, and secondary maxima at $\alpha X = \pm(2n + 1)\pi/2$.

be characterized by the angle $\Delta\theta$ of the first zero at $\alpha X = kaX/D = \pi$,

$$\Delta\theta \approx \frac{\pi}{ka} = \frac{\lambda}{2a} \tag{36.14}$$

The angular deviation from rectilinear propagation depends on λ/a; this has already been found in (36.5).

The width of the main maximum is inversely proportional to the size of the slit. This relationship can be found in many scattering experiments, where the slit size is replaced by the extension of a scattering object. For example, the form factor F in (25.27) is the Fourier transform of the density of the scattering centers; the width of the main peak of the Fourier is inversely proportional to the width of the function itself.

Neglecting diffraction, one obtains a rectilinear propagation of the light. The rectilinear propagation corresponds to the dominant maximum of (36.13) at $\theta = 0$. For the finite slit, the light rays are cut out to the width $2a$. For an obstacle in the light beam, it means the (geometric) formation of shadows.

Exercises

36.1 *Scattering at diffraction grating*

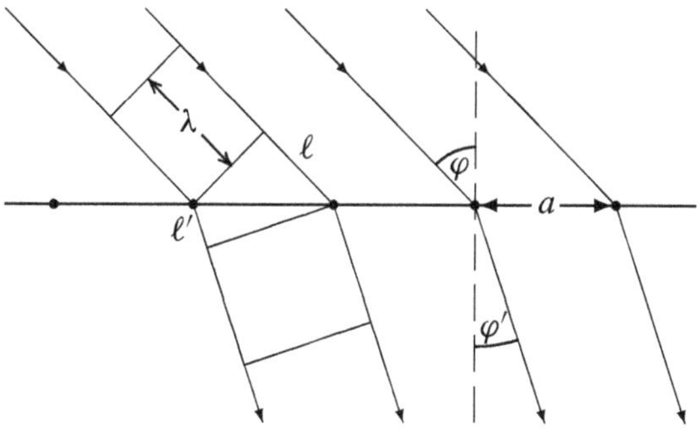

A plane wave (parallel line with arrows, incoming from the left upper part) is scattered on a *line grating* or diffraction grating. This could be visible light that is scattered onto a glass pane with incised lines (perpendicular to the image surface, marked by thick dots). According to Huygens' principle, a cylindrical wave emanates from each line.

Determine the phase difference $\varDelta$ of two neighboring cylindrical waves as a function of the incidence angle φ', the lattice spacing a and the wavelength λ. Show that for N lines, the superposition of the individual scalar waves leads to the intensity

$$I \propto \frac{\sin^2(N\varDelta/2)}{\sin^2(\varDelta/2)} \tag{36.15}$$

Sketch this as a function of $\varDelta$. Specify the conditions for φ, φ', a, and λ that lead to the primary scattering maxima.

Chapter 37

Reflection and Refraction

Two homogeneous media, such as air and water, shall have a plane interface, Figure 37.1. In one medium, a plane electromagnetic wave runs towards this border area. The wave is partially reflected at the interface and partially transmitted into the other medium. The transmitted ray is refracted. i.e., it propagates with a changed directions. The process is described by the wave solutions in matter (Chapter 33). As a result, the angles and intensities are obtained for the reflected and refracted wave.

The concepts *diffraction* (used in the last chapters) and *refraction* (considered in this chapter) both mean the bending of light rays. Refraction is specifically used for the abrupt bending at an interface between two media. In contrast to this, diffraction is the more general term; it includes, for example, the blurring of a shadow boundary (like in Figure 36.2), or a curved ray trajectory in an inhomogeneous medium (like in Figure 38.3).

For the optical constants of the two media, we assume

$$\begin{aligned} &\text{medium 1: } \mu = 1, \quad n = \sqrt{\varepsilon} = \text{real} \\ &\text{medium 2: } \mu' = 1, \quad n' = \sqrt{\varepsilon'} = n'_{\mathrm{r}} + \mathrm{i}\,\kappa' \end{aligned} \tag{37.1}$$

According to Chapter 32, most materials have a permeability $\mu \approx 1$. In contrast, $\varepsilon(\omega)$ often amounts to several units. The medium 1 in which the wave falls in shall be transparent.

The interface between the media is the x-y plane (Figure 37.1). In medium 1, the wave

$$\boldsymbol{E} = \boldsymbol{E}_0 \exp\big[\mathrm{i}(\boldsymbol{k}\cdot\boldsymbol{r} - \omega t)\big], \quad \boldsymbol{B} = \frac{\boldsymbol{k}\times\boldsymbol{E}}{k_0} \quad (z < 0) \tag{37.2}$$

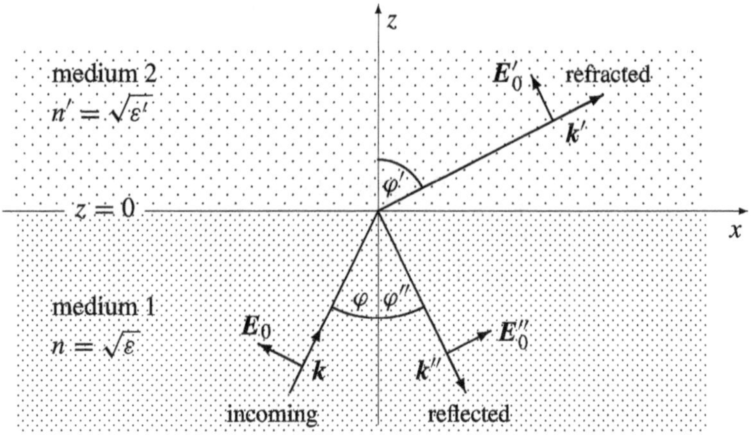

Figure 37.1 A light beam falls on the interface $z = 0$ between two homogeneous media. The beam is partially reflected (angle φ'') and partially transmitted (φ'). In the sketch, $n > n'$ is assumed; the medium 1 could be glass, medium 2 could be air. If the angle of incidence φ increases, then φ' of the refracted wave approaches $\pi/2$. If φ is increased further, the incident beam is totally reflected.

approaches towards the border area (the interface). This wave transports energy in the direction of the real wave vector $\boldsymbol{k}$. Figure 37.1 shows specifically the ray which leads to the origin of the coordinate system.

The macroscopic Maxwell equations imply boundary conditions that were established in Chapter 30. Without surfaces charges or currents, these conditions are:

$$\boldsymbol{e}_z \cdot \boldsymbol{D}, \quad \boldsymbol{e}_z \cdot \boldsymbol{B}, \quad \boldsymbol{e}_z \times \boldsymbol{E} \quad \text{and} \quad \boldsymbol{e}_z \times \boldsymbol{H} \quad \text{are continuous at } z = 0 \tag{37.3}$$

In order to fulfill these boundary conditions, there must be a further field in addition to (37.2) (with the same time dependence). We use the approach

$$\boldsymbol{E}'' = \boldsymbol{E}_0'' \exp\left[\mathrm{i}\,(\boldsymbol{k}'' \cdot \boldsymbol{r} - \omega t)\right], \qquad \boldsymbol{B}'' = \frac{\boldsymbol{k}'' \times \boldsymbol{E}''}{k_0} \qquad (z < 0) \tag{37.4}$$

for the reflected wave and

$$\boldsymbol{E}' = \boldsymbol{E}_0' \exp\left[\mathrm{i}\,(\boldsymbol{k}' \cdot \boldsymbol{r} - \omega t)\right], \qquad \boldsymbol{B}' = \frac{\boldsymbol{k}' \times \boldsymbol{E}'}{k_0} \qquad (z > 0) \tag{37.5}$$

for the transmitted or refracted wave. The real part formation is not explicitly written on. The approaches follow from (33.2), (33.6) and

(33.11). Equation (33.7) implies

$$\frac{\boldsymbol{k}^2}{n^2} = \frac{\boldsymbol{k}'^2}{n'^2} = \frac{\boldsymbol{k}''^2}{n^2} = k_0^2, \quad k_0 = \frac{\omega}{c} = \text{real} \tag{37.6}$$

From this follows

$$k = n\,k_0, \quad k' = \sqrt{\boldsymbol{k}' \cdot \boldsymbol{k}'} = n'\,k_0, \quad k'' = n\,k_0 \tag{37.7}$$

Due to (37.1), the quantities $\boldsymbol{k}$, $\boldsymbol{k}''$, k and k'' are real, whereas $\boldsymbol{k}'$ and k' are generally complex. Here, k' is the length of the vector $\boldsymbol{k}'$.

Each of the waves (37.2), (37.4) and (37.5) solves the macroscopic Maxwell equations in the respective range. Due to the linearity of the Maxwell equations, $\boldsymbol{E} + \boldsymbol{E}''$ is also solution for $z < 0$. At the interface the Maxwell equations imply the boundary conditions (37.3). If we satisfy the boundary conditions with the waves (37.2), (37.4) and (37.5), then we have found a solution to Maxwell's equations in the whole space.

The incident wave (with the wave vector $\boldsymbol{k}$ and the amplitude $\boldsymbol{E}_0$) is given by the experimental setup. With the help of the boundary conditions, we determine the properties of the reflected and the transmitted wave, i.e., we calculate quantities $\boldsymbol{k}'$, $\boldsymbol{k}''$, $\boldsymbol{E}_0'$ and $\boldsymbol{E}_0''$ as a function of $\boldsymbol{k}$ and $\boldsymbol{E}_0$.

Angle Relations

The conditions (37.3) must be fulfilled at all positions in the x-y-plane and at all times t. Then the waves (37.4) and (37.5) must have the same x-, y- and t-dependence as (37.2). Therefore, $\omega'' = \omega' = \omega$. For the same spatial dependence in the x-y plane, we must require

$$\left(\boldsymbol{k} \cdot \boldsymbol{r}\right)_{z=0} = \left(\boldsymbol{k}' \cdot \boldsymbol{r}\right)_{z=0} = \left(\boldsymbol{k}'' \cdot \boldsymbol{r}\right)_{z=0} \tag{37.8}$$

Written out, this means

$$k_x\,x + k_y\,y = k_x'\,x + k_y'\,y = k_x''\,x + k_y''\,y \tag{37.9}$$

The wave vector of the incident wave and the normal vector of the boundary surface define the *plane of incidence*. We choose the coordinate system such that the plane of incidence is the x-z plane (the image plane in Figure 37.1). A reflection or refraction out of this plane is not possible (because the boundary surface is symmetrical with respect to this plane).

This means that all wave vectors lie in this plane. Therefore, $k_y = k'_y = k''_y = 0$, and (37.9) becomes

$$k_x = k'_x = k''_x \tag{37.10}$$

We define the angles φ, φ' and φ'' by

$$k_x = k\,\sin\varphi, \quad k'_x = k'\,\sin\varphi', \quad k''_x = k''\,\sin\varphi'' \tag{37.11}$$

The real vectors $\boldsymbol{k}$ and $\boldsymbol{k}''$ enclose the angles φ and φ'' with the z-axis (Figure 37.1). The same interpretation applies to φ' if n' is real. For complex n', however, (37.11) results in a complex value for φ'; this case is discussed in the following.

Reflection law

With (37.11), $k_x = k''_x$ becomes

$$k\,\sin\varphi = k''\,\sin\varphi'' \tag{37.12}$$

With $k = k''$, (37.7), it follows that

$$\boxed{\varphi = \varphi'' \quad \text{reflection law}} \tag{37.13}$$

The incidence angle is equal to the reflection angle.

Diffraction law

With (37.11), $k_x = k'_x$ becomes

$$k\,\sin\varphi = k'\,\sin\varphi' \tag{37.14}$$

With $k/n = k'/n'$, (37.7), it follows that

$$\boxed{\frac{\sin\varphi'}{\sin\varphi} = \frac{n}{n'} \quad \text{refraction law}} \tag{37.15}$$

The law of refraction is also called Snellius' law. If the medium 2 is transparent (n' and $\boldsymbol{k}'$ are real), the law of refraction determines a real angle φ' for the direction of the refracted wave. The angle shown in Figure 37.1 refers to this case.

Refraction law for absorbing medium

The medium 1 is transparent ($\kappa = \mathrm{Im}\, n = 0$), the medium 2 absorbs light ($\kappa' = \mathrm{Im}\, n' \neq 0$). Then (37.15) results in a complex angle φ'. We discuss the geometric meaning of this angle.

With n', the wave vector $\boldsymbol{k}'$ is also complex. We write this vector as $\boldsymbol{k}' = \boldsymbol{k}'_{\mathrm{r}} + \mathrm{i}\,\boldsymbol{k}'_{\mathrm{i}}$ with real vectors $\boldsymbol{k}'_{\mathrm{r}}$ and $\boldsymbol{k}'_{\mathrm{i}}$. In the absorbing medium, the wave is proportional to

$$\exp(\mathrm{i}\boldsymbol{k}' \cdot \boldsymbol{r}) = \exp(\mathrm{i}\,\boldsymbol{k}'_{\mathrm{r}} \cdot \boldsymbol{r}) \exp(-\boldsymbol{k}'_{\mathrm{i}} \cdot \boldsymbol{r}) \tag{37.16}$$

From $k'_y = 0$ and $k'_x = k_x = \mathrm{rel}$ we get $\boldsymbol{k}'_{\mathrm{i}} \parallel \boldsymbol{e}_z$. This means that the damping factor only depends on the distance to the interface.

We evaluate $\boldsymbol{k}'^2 = k_0^2\, n'^2$ explicitly:

$$\boldsymbol{k}'^2 = \left(\boldsymbol{k}'_{\mathrm{r}} + \mathrm{i}\,\boldsymbol{k}'_{\mathrm{i}}\right)^2 = k'^2_{\mathrm{r}} - k'^2_{\mathrm{i}} + 2\,\mathrm{i}\,k'_{\mathrm{r}} k'_{\mathrm{i}} \cos\phi' = k_0^2 \left(n'^2_{\mathrm{r}} - \kappa'^2 + 2\,\mathrm{i}\,n'_{\mathrm{r}}\kappa'\right) \tag{37.17}$$

Since $\boldsymbol{k}'_{\mathrm{i}} \parallel \boldsymbol{e}_z$, ϕ' is the angle between $\boldsymbol{k}'_{\mathrm{r}}$ and the z-axis. For weak damping, $\kappa' \ll n'_{\mathrm{r}}$, we obtain from this

$$k'_{\mathrm{r}} \approx k_0\, n'_{\mathrm{r}}, \quad k'_{\mathrm{i}} \sim k_0\, \kappa' \quad (\kappa' \ll n'_{\mathrm{r}}) \tag{37.18}$$

Within a wavelength, the exponent of the attenuation factor in (37.16) changes by $k'_{\mathrm{i}}\, \lambda' \sim \kappa'/n'_{\mathrm{r}} \ll 1$; the wave (37.16) thus propagates across many wavelengths. Then, define $\boldsymbol{k}'_{\mathrm{r}} = \mathrm{Re}\, \boldsymbol{k}'$ as the observable direction of the refracted beam. From

$$\sin\phi' = \frac{\boldsymbol{k}'_{\mathrm{r}} \cdot \boldsymbol{e}_x}{k'_{\mathrm{r}}} = \frac{\mathrm{Re}\, k'_x}{k_0\, n'_{\mathrm{r}}} = \frac{k_x}{k_0\, n'_{\mathrm{r}}} = \frac{k_0\, n\, \sin\varphi}{k_0\, n'_{\mathrm{r}}} \tag{37.19}$$

we obtain the law of refraction

$$\boxed{\frac{\sin\phi'}{\sin\varphi} = \frac{n}{n'_{\mathrm{r}}} \quad \text{law of refraction} \ (\kappa' \ll n'_{\mathrm{r}})} \tag{37.20}$$

This determines the angle ϕ' that the refracted ray forms with the z-axis. In Figure 37.1, the angle φ' is to be replaced by ϕ'.

For a strongly absorbing medium, the conditions are theoretically and experimentally more complicated. The theoretical treatment is based on the exact relation (37.17).

General validity of the law of reflection and refraction

The derivation of the law of reflection and refraction only used the waveform $\exp\,\mathrm{i}(\boldsymbol{k}\cdot\boldsymbol{r}-\omega t)$ and the condition (37.8). No use was made of the special properties of the electromagnetic waves or of the specific form of the boundary conditions (37.3). Therefore, the refraction and reflection laws also apply to other waves.

Intensity Relations

In contrast to the law of refraction and reflection, the intensity and polarization of the reflected and refracted wave depend on the vector character of the wave. We first write on the boundary conditions (37.3) more explicitly, where we take into account $\mu = \mu' = 1$:

$$\left(\varepsilon\,(\boldsymbol{E}_0+\boldsymbol{E}_0'') - \varepsilon'\,\boldsymbol{E}_0'\right)\cdot\boldsymbol{e}_z = 0 \tag{37.21}$$

$$\left(\boldsymbol{k}\times\boldsymbol{E}_0+\boldsymbol{k}''\times\boldsymbol{E}_0''-\boldsymbol{k}'\times\boldsymbol{E}_0'\right)\cdot\boldsymbol{e}_z = 0 \tag{37.22}$$

$$\left(\boldsymbol{E}_0+\boldsymbol{E}_0''-\boldsymbol{E}_0'\right)\times\boldsymbol{e}_z = 0 \tag{37.23}$$

$$\left(\boldsymbol{k}\times\boldsymbol{E}_0+\boldsymbol{k}''\times\boldsymbol{E}_0''-\boldsymbol{k}'\times\boldsymbol{E}_0'\right)\times\boldsymbol{e}_z = 0 \tag{37.24}$$

The polarization vector $\boldsymbol{E}_0$ of the incident wave can be divided into the components parallel and perpendicular to the plane of incidence:

$$\boldsymbol{E}_0 = \boldsymbol{E}_{0\parallel} + \boldsymbol{E}_{0\perp} \quad \text{(relative to the plane of incidence)} \tag{37.25}$$

The interface is symmetrical under mirroring at the plane of incidence. By reflection or refraction, the field vector can thus not be rotated out of the plane of incidence, nor can its orthogonality to this plane can be changed. Therefore the two parts in (37.29) can be treated separately.

In the following, we restrict ourselves to the case $\boldsymbol{E}_0 = \boldsymbol{E}_{0\parallel}$; the other case $\boldsymbol{E}_0 = \boldsymbol{E}_{0\perp}$ is dealt with in Exercise 37.3. Then

$$\boldsymbol{E}_0\cdot\boldsymbol{e}_y = \boldsymbol{E}_0'\cdot\boldsymbol{e}_y = \boldsymbol{E}_0''\cdot\boldsymbol{e}_y = 0 \quad (\text{for } \boldsymbol{E}_0 = \boldsymbol{E}_{0\parallel}) \tag{37.26}$$

By this and by

$$\boldsymbol{E}_0\cdot\boldsymbol{k} = \boldsymbol{E}_0'\cdot\boldsymbol{k}' = \boldsymbol{E}_0''\cdot\boldsymbol{k}'' = 0 \tag{37.27}$$

the vectors $\boldsymbol{E}_0$, $\boldsymbol{E}_0'$ and $\boldsymbol{E}_0''$ are defined. These vectors have been plotted in Figure 37.1.

We evaluate the boundary conditions (37.21)–(37.24) using (37.26), (37.27). The condition (37.22) is trivially fulfilled because all $\boldsymbol{k}$ and $\boldsymbol{E}$ vectors lie in the plane of incidence; the cross products are then parallel to $\boldsymbol{e}_y$. In the remaining conditions (37.21), (37.23) and (37.24), we insert $\varepsilon = n^2$, $\varepsilon' = n'^2$, $k'' = k$, $k' = k\,n'/n$, the angles defined in (37.11) and $\varphi'' = \varphi$:

$$n^2\left(E_0 + E_0''\right)\sin\varphi - n'^2\,E_0'\,\sin\varphi' = 0 \tag{37.28}$$

$$\left(E_0 - E_0''\right)\cos\varphi - E_0'\,\cos\varphi' = 0 \tag{37.29}$$

$$n\left(E_0 + E_0''\right) - n'\,E_0' = 0 \tag{37.30}$$

The amplitudes $E_0 = \sqrt{\boldsymbol{E}_0^2}$, E_0' and E_0'' are generally complex. Due to (37.15), Equation (37.28) is equivalent to (37.30). It is therefore sufficient to consider the two equations (37.29) and (37.30).

The quantities E_0 and φ are determined by the incident wave; and φ' is determined by the law of refraction. Then (37.29) and (37.30) represent two equations for the determination of the two unknowns E_0' and E_0''. With these two equations all boundary conditions (37.3) are fulfilled, so that (37.2), (37.4) and (37.5) solve Maxwell's equations in the entire space. The solution of (37.29) and (37.30) for E_0' and E_0'' yields the *Fresnel formulas*:

$$\left(\frac{E_0'}{E_0}\right)_{\parallel} = \frac{2\,n\,n'\cos\varphi}{n'^2\cos\varphi + n\sqrt{n'^2 - n^2\sin^2\varphi}}$$
$$\left(\frac{E_0''}{E_0}\right)_{\parallel} = \frac{n'^2\cos\varphi - n\sqrt{n'^2 - n^2\sin^2\varphi}}{n'^2\cos\varphi + n\sqrt{n'^2 - n^2\sin^2\varphi}} \tag{37.31}$$

The index $\parallel$ denotes the polarization investigated here, i.e., $\boldsymbol{E}_0 = \boldsymbol{E}_{0\parallel}$.

The intensities of the incident and reflected waves are proportional to $|E_0|^2$ and to $|E_0''|^2$. Since they refer to the same medium, the proportionality coefficients are the same. Thus

$$R(\varphi) = \left|\frac{E_0''}{E_0}\right|^2 \qquad \text{(reflection coefficient)} \tag{37.32}$$

determines the fraction of the reflected intensity. Since no intensity is lost at the interface, the fraction of the refracted (or transmitted, letter T)

intensity becomes $T = 1 - R$. In general these coefficients are to be calculated as ratios of energy current densities (29.32). The time averaged components are perpendicular to the interface are the relevant ones, i.e., $R = \langle S_z''\rangle/\langle S_z\rangle$ and $T = \langle S_z'\rangle/\langle S_z\rangle$. When evaluating these expressions, the different optical constants in the two media must be taken into account.

Figures 37.2 and 37.3 display the angular dependencies of $R_{\|}(\varphi)$, (37.32) with (37.31), and $R_{\perp}(\varphi)$, (37.32) with (37.41), for real n and n'.

Specifically for perpendicular incidence ($\varphi = 0$), we get $R_{\|}(0) = R_{\perp}(0) = R(0)$. With (37.31) we obtain:

$$R(0) = \left|\frac{E_0''}{E_0}\right|^2 = \left|\frac{n' - n}{n' + n}\right|^2 = \frac{(n_r' - n)^2 + \kappa'^2}{(n_r' + n)^2 + \kappa'^2} \tag{37.33}$$

An interface between air ($n \approx 1$) and glass ($n_r' \approx 1.5$ and $\kappa' \approx 0$) yields in $R(0) = 4\%$ regardless of the medium in which the light falls in.

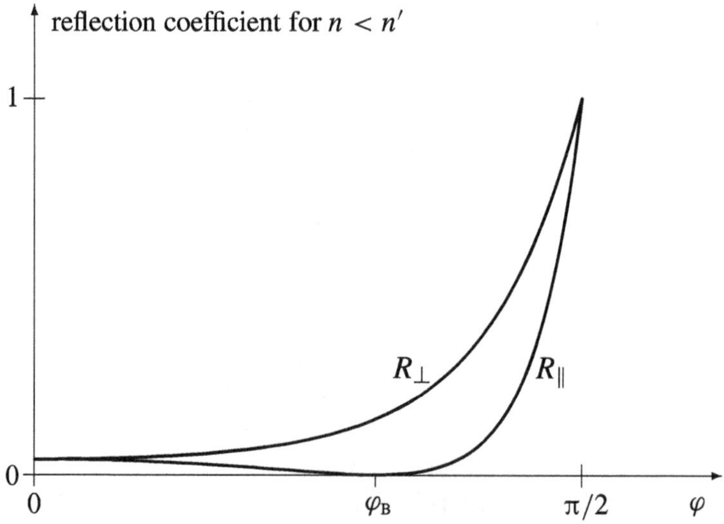

Figure 37.2 The reflection coefficients $R_{\|}$ and $R_{\perp}$ as a function of the angle of incidence φ for $n'/n = 1.5$. For a steep incidence ($\varphi \approx 0$) the reflection is small (you can see the bottom of a lake through the calm surface). For grazing incidence ($\varphi \to \pi/2$) the reflection is large (the opposite mountains are reflected in the lake). At the Brewster angle $\varphi_B = \arctan(n'/n) \approx 57°$ the coefficient $R_{\|}$ vanishes, and the reflected beam is linearly polarized.

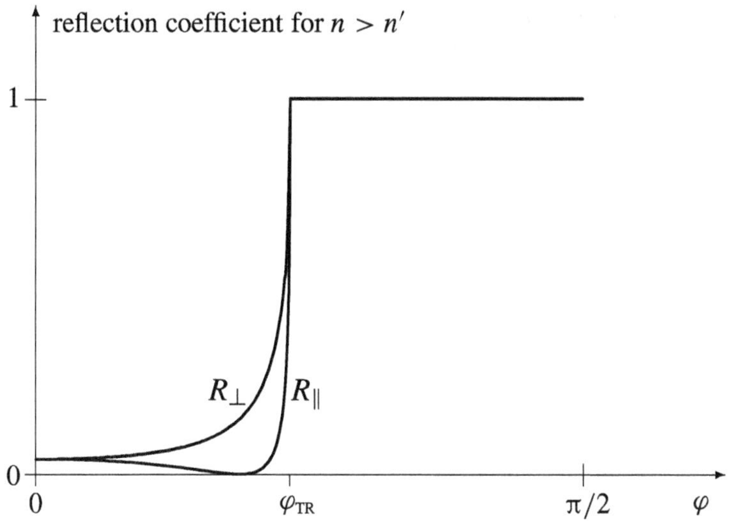

Figure 37.3 The reflection coefficients $R_\parallel$ and $R_\perp$ as a function of the incidence angle φ for $n/n' = 1.5$. The range $\varphi = 0 \ldots \varphi_{TR}$ results in $\varphi' = 0 \ldots \pi/2$ for the refracted beam. In the cone $\varphi < \varphi_{TR}$ a diver therefore may see the events outside the lake (frog's eye view). For $\varphi > \varphi_{TR}$ total reflection (TR) occurs; the diver sees the diver sees the mirrored lake bottom.

Visible light shall fall perpendicularly onto a flat metal or water surface. Let the medium 1 be air with $n \approx 1$. Then we obtain

$$\begin{aligned} &\text{water: } n'_r \approx 1.33, \ \kappa' \ll n'_{\mathrm{r}} \longrightarrow R(0) \approx 0.02 \\ &\text{metal: } n' \approx i\kappa' \qquad\qquad\quad \longrightarrow R(0) \approx 1 \end{aligned} \tag{37.34}$$

Due to $T = 1 - R \approx 1$ one can see vertically through a calm water surface. According to Figure 37.2, this also applies to angles that are not too flat; however, the different intensities of the reflected (sky) and refracted (seabed) light must be taken into account. Due to $R \approx 1$, a metal surface can serve as a mirror because $n' \approx i\kappa'$ the light does not propagate into the metal, and $R \approx 1$ is valid for any angle of incidence.

As an example of the frequency dependence of the reflection coefficient $R(\varphi, \omega)$ we consider the perpendicular incidence of light onto a metal surface. To do this, we insert the refractive indices n' from (34.22) and

$n \approx 1$ for air into (37.33):

$$R(0,\omega) = \left|\frac{n-n'}{n+n'}\right|^2 = \begin{cases} 1 & \left(\omega < \omega_{\rm P}/\sqrt{\varepsilon_{\rm o}}\right) \\ \dfrac{\left(1-\sqrt{\varepsilon_{\rm o}-\omega_{\rm P}^2/\omega^2}\right)^2}{\left(1+\sqrt{\varepsilon_{\rm o}-\omega_{\rm P}^2/\omega^2}\right)^2} & \left(\omega > \omega_{\rm P}/\sqrt{\varepsilon_{\rm o}}\right) \end{cases} \tag{37.35}$$

The resulting frequency dependence of the reflection coefficient is shown in Figure 34.6 on the right (for $\varepsilon_{\rm o}(\omega) = 1$). In the Lorentz model as well as in real materials, the dielectric function $\varepsilon_{\rm o}(\omega)$ has a small imaginary part around $\omega \sim \omega_{\rm P}/\sqrt{\varepsilon_{\rm o}}$. Therefore, the abrupt transition from reflective to transparent behavior (in Figure 34.6 at $\omega_{\rm P}$) is somewhat smeared out.

In our calculation, the interface between the two media was the plane $z = 0$. A physical boundary surface, on the other hand, has a finite thickness d; for very smooth surfaces (polished metal mirror) a thickness d of a few ångström is feasible (for comparison $\lambda_{0,\,\rm visible} \approx 4\ldots 7.5 \cdot 10^3$ Å). The neglect of the finite thickness of the boundary layer is justified for $\lambda = \lambda_0/n_{\rm r} \gg d$. For $\lambda \ll d$, the propagation of light in the boundary layer itself must be investigated; this can be done in the context of geometrical optics (Chapter 38). For $\lambda \sim d$, the wave and vector nature of the light is important; for example, interference can occur. For example, an additional layer with $d \sim \lambda$ can significantly influence the reflection (anti-reflective coating of spectacle lenses may be obtained with $d = \lambda/4$).

Brewster angle

As seen in Figure 37.2, there is an angle at which the reflection $R_\parallel$ disappears (n and n' are real). From (37.31) and $E_0''/E_0 = 0$ follows

$$\varphi_{\rm B} = \arctan\frac{n'}{n} \tag{37.36}$$

This *Brewster angle* exists both for $n > n'$ and for $n < n'$. The following applies to the Brewster angle (see also Exercise 37.4)

$$\varphi_{\rm B} + \varphi' = \frac{\pi}{2} \tag{37.37}$$

This means that the reflected beam (with $\varphi'' = \varphi_{\rm B}$) and the refracted ray (with φ') form a right angle. The induced electric dipoles in the boundary surface oscillate perpendicular to the direction of the refracted beam (and parallel to the image plane in Figure 37.1). For the reflected beam, the electric field vector would then be parallel to the wave vector. However, this is not possible for a transverse wave.

A non-polarized wave can be written as a linear combination (37.25) of the two polarization directions. If the incident wave has the angle φ_B, then the amplitude $\boldsymbol{E}''_{0\parallel}$ disappears; the reflected wave is therefore linearly polarized. However, the reflected (polarized) beam only has a low intensity, for example $R_\perp(\varphi_B) \approx 15\%$ in Figure 37.2. In order to generate polarized light, it is more effective to use the partial polarization of the refracted beam and several interfaces in succession.

It follows from (37.31) that the ratio $(E''_0/E_0)_\parallel$ changes its sign at φ_B. This means an abrupt phase jump of 180° for the reflected wave. A finite damping smoothes this jump.

Birefringence

The polarization of a solid body can depend on the direction of the electric field relative to the crystal axes. In most cases, the permittivity (described by the dielectric function) is the relevant effect.

Two polarization directions, $\boldsymbol{E}_{0\parallel}$ and $\boldsymbol{E}_{0\perp}$ then result in different values for the refractive index and therefore in different refraction angles. A non-polarized incident light beam is refracted into two beams which have a mutually perpendicular, linear polarization. This effect is called birefringence or double refraction. This is a simple and efficient process for the generation of polarized light.

Total reflection

We consider two transparent (n, n' real) media with $n > n'$. From the law of refraction follows

$$\sin\varphi = \frac{n'}{n}\,\sin\varphi' \le \frac{n'}{n} < 1 \tag{37.38}$$

This results in a restriction for the angle of incidence:

$$\varphi \le \varphi_{\text{TR}} = \arcsin\frac{n'}{n} \quad (n > n') \tag{37.39}$$

If the angle of incidence φ is equal to the angle $\varphi_{\rm TR}$, the angle φ' of the refracted approaches the maximum possible value of 90°, see also Figure 37.1. For larger angles of incidence, the incident beam becomes is totally reflected, Figure 37.3. In the expression of E_0''/E_0 in (37.31), the root is then purely imaginary. This means that $|E_0''/E_0| = 1$ and

$$R = 1 \quad \text{for } \varphi > \varphi_{\rm TR} \tag{37.40}$$

Applications of this total reflection are as follows:

- If light is sent into a thin, long glass cylinder (glass fiber) at a flat angle to the cylinder axis, it will be totally reflected again and again at the boundary surface (the interface to air). This turns the glass fiber into a light guide (or optical fiber).
- There are no suitable lenses for X-ray light. In the satellite Rosat, which takes images in the X-ray range, the imaging system is based on total reflection. For X-ray light, the refractive index of metal is slightly lower than 1 (for high frequencies (34.21) applies with $\varepsilon_{\rm o} \approx 1$, i.e., $n^2 = \varepsilon = 1 - \omega_{\rm P}^2/\omega^2 < 1$). Therefore, X-rays with grazing incidence are totally reflected at a metal surface. By a suitable arrangement of parabolically shaped metal surfaces a focussing of a Röntgen beam can be achieved.

Exercises

37.1 *Complex refractive index*

Two media are separated by a plane interface (see figure in Exercise 37.3). One refractive index n is real, the other one complex, $n' = n_{\rm r}' + {\rm i}\kappa'$ with $\kappa' \ll n_{\rm r}'$. Determine the complex angle φ' resulting from the law of refraction. Set $\varphi' = \phi' + {\rm i}\,\phi''$ and calculate the real angles ϕ' and ϕ''.

37.2 *Total reflection*

For two transparent media (n, n' real, see figure in Exercise 37.3) with $n > n'$ the law of refraction yields

$$\varphi \le \varphi_{\rm TR} = \arcsin\left(n'/n\right) \quad (n > n')$$

Show that for $\varphi > \varphi_{\rm TR}$ is the z-component of the transmitted wave vector $\boldsymbol{k}'$ is purely imaginary. What does this mean for the wave in medium 2?

37.3 *Fresnel's formulas for polarized light*

In order to determine the intensity relations (Fresnel's formulas) for refraction and reflection, the electric field is decomposed into the parts parallel and perpendicular to the interface:

$$\boldsymbol{E}_0 = \boldsymbol{E}_{0\parallel} + \boldsymbol{E}_{0\perp}$$

The perpendicular case shall be considered. The corresponding electric field vectors (perpendicular to the image plane) are indicated by small circles with a central point.

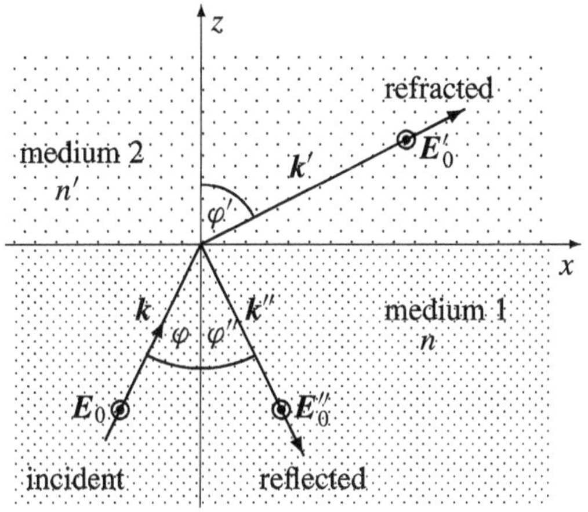

Derive the Fresnel formulas

$$\left(\frac{E_0'}{E_0}\right)_\perp = \frac{2n\cos\varphi}{n\cos\varphi + n'\cos\varphi'}, \quad \left(\frac{E_0''}{E_0}\right)_\perp = \frac{n\cos\varphi - n'\cos\varphi'}{n\cos\varphi + n'\cos\varphi'} \tag{37.41}$$

Using $n'\cos\varphi' = \sqrt{n'^2 - n^2\sin^2\varphi}$ these formulas can be expressed solely by the angle of incidence.

37.4 *Alternative form of Fresnel's formulae*

We continue to consider the refraction and reflection sketched in the figure of Exercise 37.3. Show that the Fresnel's formulas can also be written in the following form:

$$\left(\frac{E_0'}{E_0}\right)_{\parallel} = \frac{2\cos\varphi\sin\varphi'}{\sin(\varphi+\varphi')\cos(\varphi-\varphi')}\,, \qquad \left(\frac{E_0''}{E_0}\right)_{\parallel} = \frac{\tan(\varphi-\varphi')}{\tan(\varphi+\varphi')}$$

$$\left(\frac{E_0'}{E_0}\right)_{\perp} = -\frac{2\cos\varphi\sin\varphi'}{\sin(\varphi+\varphi')}\,, \qquad \left(\frac{E_0''}{E_0}\right)_{\perp} = -\frac{\sin(\varphi-\varphi')}{\sin(\varphi+\varphi')}$$

What results for the Brewster angle $\tan\varphi_B = n'/n$ in the case of parallel polarization?

37.5 *Rainbow*

A rainbow is created by the refraction and reflection of sun rays on water drops. Optical paths are depicted without reflection (left), or with one or two reflections (below). The single reflection is responsible for the primary arc of the rainbow, the first secondary arc is due to a twofold reflection.

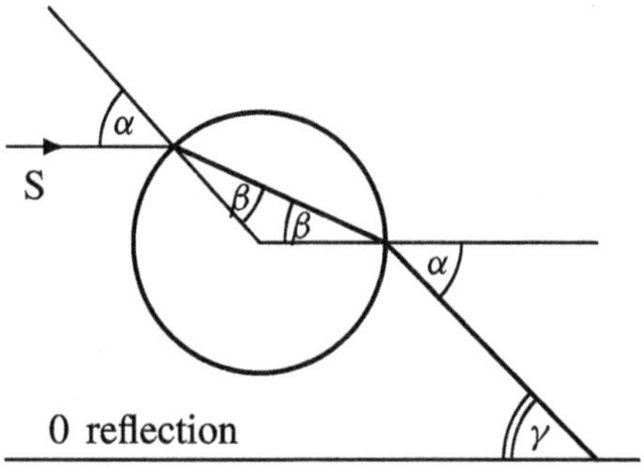

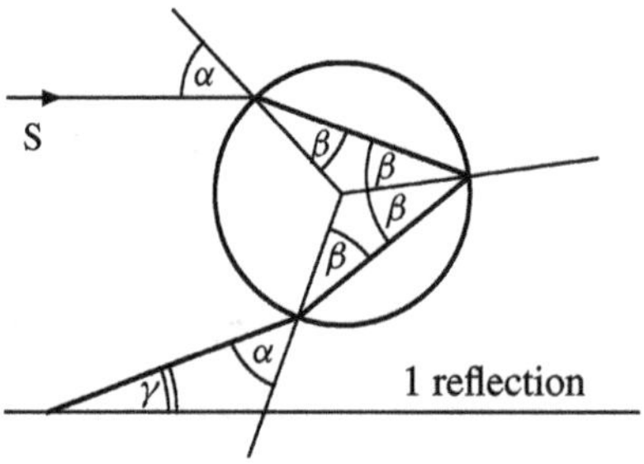

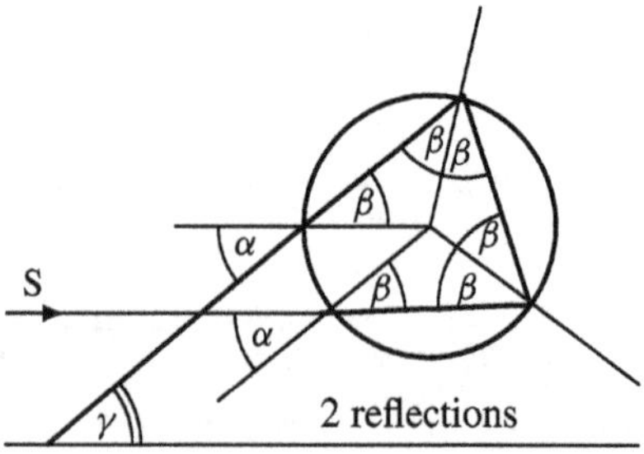

The sunlight (S) comes from the left and hits with various incidence angles of α onto the droplets surface. Relative to the direction of incidence (horizontal), the beam is finally observed at the angle γ. A clear intensification occurs if a (small) α interval contributes to the same observation angle $\gamma = \gamma(\alpha)$; this condition $d\gamma/d\alpha = 0$ leads to the angle γ_{extr}. The rainbow then appears as a circle, which is formed at the angle γ_{extr} relative to the sun-observer axis.

The refractive index of air is $n' \approx 1$, that of water is $n \approx 1.33$. Due to the slightly different frequency dependence $n = n(\omega)$, the angles γ_{extr} vary with the frequency (or colors).

Determine the observation angles γ_{extr} for the primary (1 reflection) and the first secondary rainbow (2 reflections). What is the color sequence in the two rainbows if the refractive indices for red and violet light are $n_{\text{red}} = 1.331$ and $n_{\text{violet}} = 1.334$? Which of the two rainbows appears wider?

Investigate the intensity ratio of the primary and secondary rainbows with the help of Fresnel's formulae. It is sufficient to consider light which is polarized perpendicular to the image plane.

37.6 *Alternative derivation of refraction law*

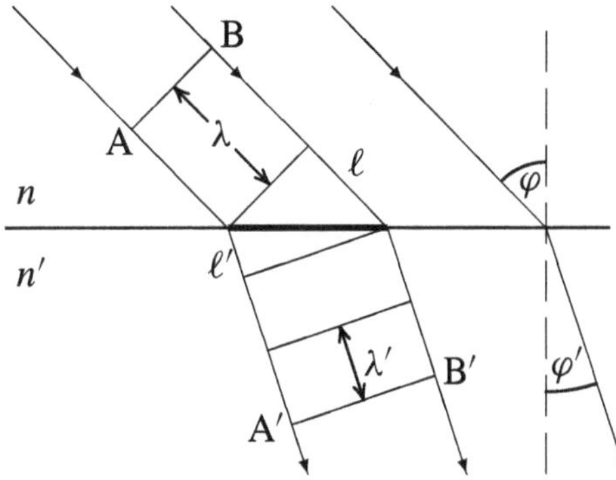

The refraction at the interface of two transparent media shall be calculated (refractive indices n and n'). An incident plane wave (frequency ω) has the same phase at points A and B. To ensure that this also applies to A′ and B′, the phase difference must be the same for the paths ℓ and ℓ'. What follows from this for the relation between the angles φ and φ'?

Chapter 38

Geometric Optics

Geometrical optics or ray optics refers to the limiting case in which the wave nature of the light plays no role. We summarize the basic features of geometrical optics. The generalization of the law of refraction to inhomogeneous media leads to the eikonal equation.

Basics

In the limiting case of short wavelengths

$$\lambda \to 0 \quad \text{(geometrical optics)} \tag{38.1}$$

diffraction and interference effects can be neglected. Here, $\lambda \to 0$ means that the wavelengths are small compared to the relevant dimensions of the experimental setup. According to Figures 36.2 and 36.4, a light beam is slightly diffracted by a slit of width d; hereby the angle deflection is of the size $\Delta\theta \sim \lambda/d$. A pronounced interference pattern (Figure 36.1) requires that if the slit distance is comparable to the wavelength. These diffraction and interference effects disappear in the limit (38.1).

Just as in Kirchhoff's theory (Chapter 35) we neglect polarization effects, i.e., effects that result from the vector property of the electromagnetic field. Therefore we consider a scalar wave field $\psi(\boldsymbol{r}, t) = \text{Re}\,\psi(\boldsymbol{r})\exp(-\mathrm{i}\omega t)$ only. This field represents the components of $\boldsymbol{E}$ and $\boldsymbol{B}$, and the wave equations (33.5) reduce to

$$\left(\Delta + k^2\right)\psi(\boldsymbol{r}) = 0 \tag{38.2}$$

where

$$k = \frac{\omega}{c}\, n = k_0\, n \tag{38.3}$$

In this chapter, we restrict ourselves to a real refractive index n.

The simplifications (38.1) and (38.2) are the basis of the geometrical optics. The essential properties of the beam propagation in geometric optics are as follows:

1. Rectilinear propagation in a homogeneous medium
2. Law of reflection and refraction
3. Superposition of rays
4. Reversibility of the ray path

(38.4)

Electromagnetic waves have these properties, provided that diffraction, interference and polarization effects can be neglected. In particular, these properties are compatible with the wave nature of light; this becomes clear in the following.

The term *ray* is understood to mean a continuous, *directional* matter or energy flow. We initially consider straight rays. About the intermediate step of a multiple refracted beam (Figure 38.3) we also allow for a curved ray trajectory. In contrast to this, in mathematics a ray is defined a one-directional straight line.

We first explain the compatibility of the properties (38.4) with the wave description of light. To do this, we assign rays to the waves. A possible solution of (38.2) is the plane wave

$$\psi(\boldsymbol{r}) = C\, \exp(\mathrm{i}\boldsymbol{k}\cdot\boldsymbol{r}) \tag{38.5}$$

The wave (38.5) transports energy at every point in the direction of $\boldsymbol{k}$. This means that the light propagates along straight, parallel rays.

A spherical wave emanates from a point source:

$$\psi(\boldsymbol{r}) = C'\, \frac{\exp(\mathrm{i}kr)}{r} \tag{38.6}$$

This could, for example, be a component of the radiation field (24.21) of an oscillating dipole. A radiation field is always proportional to $1/r$ because only then the radiation power ($\propto 4\pi r^2\, |\psi|^2$) through the spherical surface

$4\pi r^2$ is independent of r. The spherical wave can be approximated locally by a plane wave. To do this, we set $\boldsymbol{r} = \boldsymbol{r}_0 + \boldsymbol{r}'$ and expand for small r':

$$\frac{\exp(\mathrm{i}kr)}{r} = \frac{\exp(\mathrm{i}k|\boldsymbol{r}_0 + \boldsymbol{r}'|)}{|\boldsymbol{r}_0 + \boldsymbol{r}'|} \approx \frac{\exp(\mathrm{i}kr_0)}{r_0}\exp(\mathrm{i}\boldsymbol{k}\cdot\boldsymbol{r}') \quad (r' \ll r_0) \tag{38.7}$$

Here, $\boldsymbol{k} = k\,\boldsymbol{r}_0/r_0$. In the vicinity of $\boldsymbol{r}_0$ this is a plane wave $\exp(\mathrm{i}\boldsymbol{k}\cdot\boldsymbol{r}')$ with the amplitude $\exp(\mathrm{i}kr_0)/r_0$. The energy transport at the point $\boldsymbol{r}_0$ is directed towards $\boldsymbol{k} = k\,\boldsymbol{r}_0/r_0$, i.e., along straight rays outwards (Figure 38.1).

If one combines the star-shaped propagation of the wave from a point source (Figure 38.1, right) with the law of refraction, the imaging property by a lens can be understood (Figure 38.2). The light emitted by Q solid is bent or refracted by the lens' interfaces. As a result, Q is imaged to B. A photographic plate may captures this image at the point B. The intensity at B is proportional to the radiation emitted by Q. This construction and its generalizations lead to the camera, the telescope or the microscope. The basis of the description in all these cases is the rectilinear propagation in a homogeneous medium and the law of refraction at interfaces. The genuine wave effects (like diffraction, interference, polarization) are not taken into account; mostly they play no role for these everyday optical devices. Absorption and dispersion (Chapter 34) can, however, not be

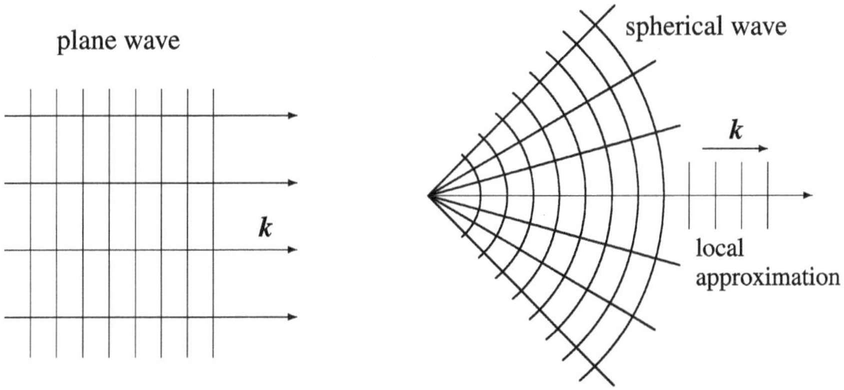

Figure 38.1 Rays can be assigned to a plane wave, or also to a spherical wave which may locally be approximated by a plane wave (right). The beam direction is perpendicular to surfaces of equal phase. These surfaces are planes on the left (shown as vertical lines) and spheres (shown as circles) on the right.

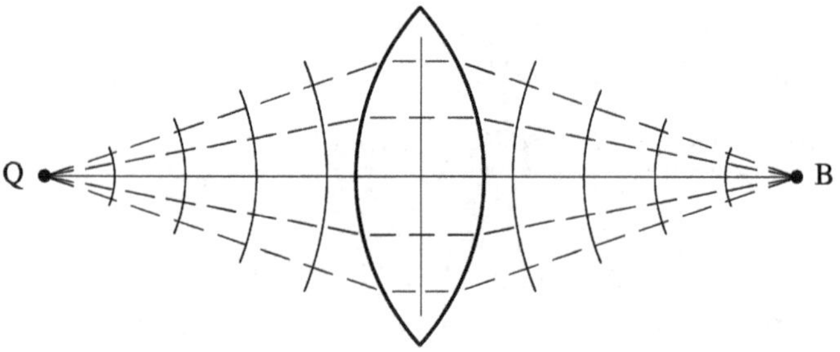

Figure 38.2 A lens (thick lines) collects the light emitted by the point source Q into an image point B. Shown are the rectilinear light rays and the circles with equal phase. At the lens surface, the law of refraction determines the ray trajectories. Geometrical optics is used for the construction of optical instruments, such as cameras, microscopes or telescopes.

neglected without further ado. For example, the frequency dependence of the refractive index implies that different color rays emanating from Q lead to different image points. This leads to imaging errors in a camera.

We summarize the considerations that explain the compatibility of the electromagnetic waves with the four properties (38.4) geometrical optics:

1. The waves can be assigned to rectilinear rays, (38.7) and Figure 38.1. Diffraction effects disappear in the limiting case (38.1).
2. The reflection and refraction laws were described in the previous chapter on the basis of waves. These laws determine the continuation of light rays at interfaces.
3. The superposition of waves follows from the linearity of the Maxwell's equations. In the limiting case (38.1), interferences are insignificant. Then the superposition of the waves leads to a superposition of the intensities. Various rays propagate independent from each other.
4. Replacing $\boldsymbol{k}$ with $-\boldsymbol{k}$ in (38.5) lead again to a solution of the wave equation (38.2). This can be generalized to an arbitrary radiation field (because it can be written as a superposition of plane waves). In the new solution, the direction of the energy current density is reversed at every point. So the ray path is reversible.

Eikonal Equation

We generalize geometrical optics (38.4) to inhomogeneous media. This generalization applies in particular to the law of refraction. For this

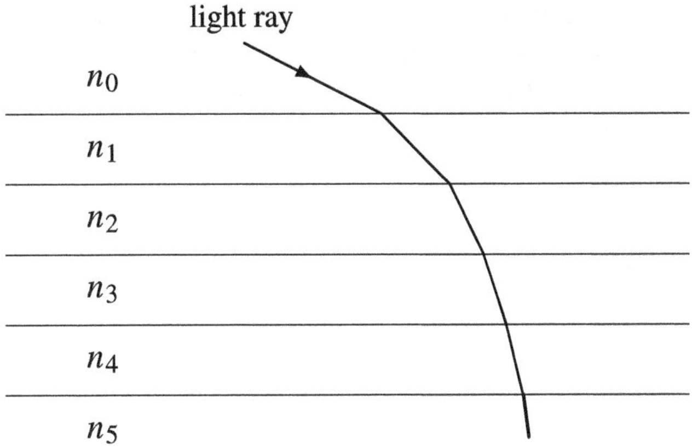

Figure 38.3 For homogeneous, parallel layers, the ray trajectory results from the multiple application of the law of refraction; for the sketch we assumed $n_0 < n_1 < n_2 \ldots$. If the layer thickness goes to zero, the ray trajectory is determined by a continuous refractive index $n(\boldsymbol{r})$. A demonstration experiment for this is shown in Figure 38.5. The beam trajectory obeys the eikonal equation.

purpose, we first consider in Figure 38.3 an arrangement of parallel layers with the refractive indices $n_0, n_1, n_2, \ldots$. Using the law of refraction, we can easily construct the multiply refracted beam. If the layer thickness approaches zero, we obtain a position dependent refractive index

$$n = n(\boldsymbol{r}) = \text{real} \tag{38.8}$$

At the same time, the multiply refracted beam becomes a curved path. For these trajectories, i.e., for the ray paths in inhomogeneous media, we derive a differential equation. Hereby we restrict ourselves to transparent media, i.e., to real functions $n(\boldsymbol{r})$.

The prerequisite (38.1) implies that the refractive index varies only slightly in the range of a wavelength:

$$\lambda \left|\operatorname{grad} n(\boldsymbol{r})\right| \ll n \tag{38.9}$$

In addition to slow changes of $n(\boldsymbol{r})$, discontinuities at interfaces are permitted. At these points, (38.9) is not fulfilled; but the kink in the rays follows from the refraction law.

In the derivation of the wave equations (33.5) from Maxwell's equations, the material variables ε and μ were drawn in front of the spatial derivatives, for example $\operatorname{div}(\varepsilon \boldsymbol{E}) = \varepsilon \operatorname{div} \boldsymbol{E}$. For this purpose, it was assumed that these material parameters do not depend on the position ($n = \text{const.}$). For position-dependent material parameters, we may use $\text{div}\,(\varepsilon \boldsymbol{E}) \approx \varepsilon\, \text{div}\, \boldsymbol{E}$, if the variability of the material parameters is much less than that of the fields. Since the fields vary on the scale of a wave length, (38.9) is the prerequisite for the validity of this approximation. With this approximation we obtain the wave equation (38.5) or (38.2) with a position dependent refractive index $n(\boldsymbol{r})$:

$$\left[\, \Delta + k(\boldsymbol{r})^2 \,\right] \psi(\boldsymbol{r}) = 0 \quad \text{with} \quad k(\boldsymbol{r}) = k_0\, n(\boldsymbol{r}) \tag{38.10}$$

This equation is the starting point for the following analysis. A general solution approach is

$$\psi(\boldsymbol{r}) = A(\boldsymbol{r}) \exp\big(\mathrm{i} k_0\, S(\boldsymbol{r})\big) \tag{38.11}$$

The amplitude $A(\boldsymbol{r})$ and the phase $S(\boldsymbol{r})$ are real functions; the phase $S(\boldsymbol{r})$ is also called *eikonal* (a Greek word for image). For $n = \text{const.}$, the plane wave (38.5) is a solution, with $A = \text{const.}$ and $S = \boldsymbol{k} \cdot \boldsymbol{r}/k_0$ or $\text{grad}\, S = \boldsymbol{k}/k_0 = \text{const.}$ If $n(\boldsymbol{r})$ varies slowly, then

$$A(\boldsymbol{r}) \;\text{ and }\; \text{grad}\, S(\boldsymbol{r}) \;\text{ are slowly varying} \tag{38.12}$$

This is equivalent to the condition (38.9). For evaluate the wave equation, we calculate

$$\frac{\partial^2 \psi}{\partial x^2} = -k_0^2\, \psi \left(\frac{\partial S}{\partial x}\right)^2 + \mathrm{i} k_0\, \psi \left(\frac{\partial^2 S}{\partial x^2} + \frac{2}{A}\frac{\partial A}{\partial x}\frac{\partial S}{\partial x}\right) + \cdots \tag{38.13}$$

The first term on the right-hand side is the leading one; it fulfills the wave equation for $n = \text{const.}$ with $\text{grad}\, S = \text{const.}$ The next term is small because it contains first derivatives of the weakly varying quantities (38.12). The third term, which is no longer shown, contains the second derivatives of the slowly changing quantities; compared to the second term, it will be neglected.

We substitute (38.13) and the corresponding derivatives with respect to y and z into (38.10):

$$0 = (\Delta + k^2)\, \psi \approx k_0^2\, \psi \left(-(\text{grad}\, S)^2 + n^2\right) + \mathrm{i} k_0\, \psi \left(\Delta S + \frac{2}{A}\, \text{grad}\, A \cdot \text{grad}\, S\right) \tag{38.14}$$

Two equations follow from this:

$$\boxed{\big(\text{grad}\, S(\boldsymbol{r})\big)^2 = n(\boldsymbol{r})^2 \quad \text{eikonal equation}} \tag{38.15}$$

$$2\ \text{grad}\,\big(\ln A(\boldsymbol{r})\big) \cdot \text{grad}\, S(\boldsymbol{r}) = -\Delta S(\boldsymbol{r}) \tag{38.16}$$

The eikonal equation determines the phase $S(\boldsymbol{r})$ of the solution (38.11). Once the phase is known, Equation (38.16) determines the amplitude $A(\boldsymbol{r})$.

For relating of the beam path to $S(\boldsymbol{r})$, we expand (38.11) around the point $\boldsymbol{r}_0$:

$$\psi(\boldsymbol{r}) = \psi(\boldsymbol{r}_0 + \boldsymbol{r}') \approx A_0 \exp\big[\mathrm{i}k_0\,(\text{grad}\, S(\boldsymbol{r}_0)) \cdot \boldsymbol{r}'\big] \tag{38.17}$$

Here, $A_0 = A(\boldsymbol{r}_0) \exp \mathrm{i}k_0 S(\boldsymbol{r}_0)$. The function (38.17) has the form of a plane wave with the wave vector $\boldsymbol{k} = k_0\,\text{grad}\, S$. The phase field $S(\boldsymbol{r})$ determines the beam direction, the surfaces of equal phase and the optical paths:

$$\begin{array}{ll} \text{beam direction at } \boldsymbol{r}_0\text{:} & \text{grad}\, S(\boldsymbol{r}) \\ \text{equal phase surface:} & S(\boldsymbol{r}) = \text{const.} \\ \text{optical path:} & \delta S = S(\boldsymbol{r}_{\text{P}'}) - S(\boldsymbol{r}_{\text{P}}) \end{array} \tag{38.18}$$

The rays follow curves for which grad S is a tangent at every point. They are perpendicular to the *equal phase surface* $S =$ const., i.e., to the areas of the same phase. The *optical path* between two points P and P′ plays an important role in Fermat's principle (Exercises 38.1 and 38.2).

We present two simple solutions of the eikonal equation for $n =$ const. For

$$S = n\,(\boldsymbol{e} \cdot \boldsymbol{r}) \quad \text{with} \quad |\boldsymbol{e}| = 1 \tag{38.19}$$

we get grad $S = n\,\boldsymbol{e}$ and $(\text{grad}\, S)^2 = n^2$. The beam direction grad $S = n\,\boldsymbol{e} =$ const. is everywhere the same; the rays are therefore straight lines. The surfaces $S =$ const. are planes that are perpendicular to $\boldsymbol{e}$. From (38.16) and $\Delta S = 0$ we obtain $A =$ const. This makes (38.11) to a plane wave.

Another solution for $n =$ const. is given by

$$S = n\,\sqrt{x^2 + y^2 + z^2} = n\,|\boldsymbol{r}| \tag{38.20}$$

From this follows

$$\operatorname{grad} S = n \, \frac{\boldsymbol{r}}{r} \tag{38.21}$$

and $(\operatorname{grad} S)^2 = n^2$. The rays start from $r = 0$ radially in all directions. This corresponds to a point source at $r = 0$. From (38.16) follows $A = \text{const.}/r$. This turns (38.11) into an outgoing spherical wave.

Refraction of the sunlight

One application of the eikonal equation is the calculation of the rays of the setting sun. The density of the atmosphere decreases with altitude, this also applies to the refractive index. Inclined incident solar rays are therefore refracted towards the surface of the earth. A ray pattern similar to that shown in Figure 38.3, but with a much smaller curvature of the ray. For given $n(\boldsymbol{r})$, the ray path can be calculated from the eikonal equation.

The deviation from the rectilinear trajectory leads to the following effect: You can still see the sun about five minutes after it has actually set. "Set" here means that a straight line from the observer to the sun passes through the earth. For a straight line propagation, the sun would already be invisible.

In the laboratory, this effect can be demonstrated in a sugar solution (Figure 38.4). A further example for this is investigated in Exercise 38.2.

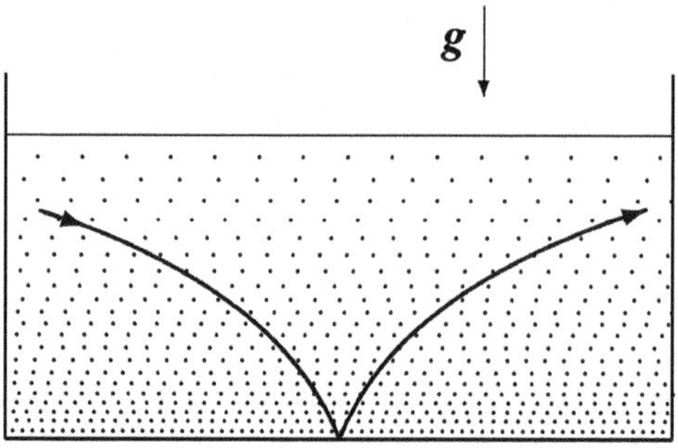

Figure 38.4 In a sugar solution, the concentration and therefore the refractive index decrease with the height. This may lead to the outlined beam trajectory. At the bottom of the vessel, a mirror reflects the rays.

Shadow

Equation (38.16) contains the scalar product grad $A \cdot$ grad S. This determines the change of A in the beam direction (grad S); about the behavior of A perpendicular to it (38.16) makes no statement. Perpendicular to the beam direction, A can also be discontinuous. This is important for the description of the *shadow* in geometric optics. In Figure 38.5, the rays hit an aperture, i.e., an opaque surface with openings. A linear continuation of the rays creates areas with and without light behind the aperture. The shadow area is limited by the rectilinear rays that just pass at an end of an opening. In the vicinity of these rays, the amplitude drops abruptly from a finite value (in the light) to zero (in the shadow); this is consistent with the equations (38.15, 38.16) of geometrical optics. The blurring of the shadow border by diffraction ($\Delta\theta \sim \lambda/d$, (36.5) and (36.14)) disappears in the limit (38.1) considered here.

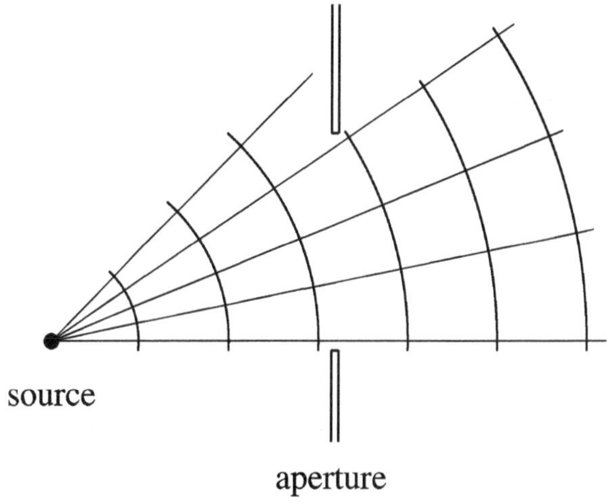

Figure 38.5 The shadow formation behind an aperture can be understood in geometrical optics. On one hand, the equation for a constant refractive index results in a rectilinear trajectory of the rays. On the other hand, (38.16) admits that the amplitude A is discontinuous perpendicular to the beam direction. Such a discontinuity occurs for the beam that just grazes the aperture opening.

Exercises

38.1 *Law of refraction from Fermat's principle*

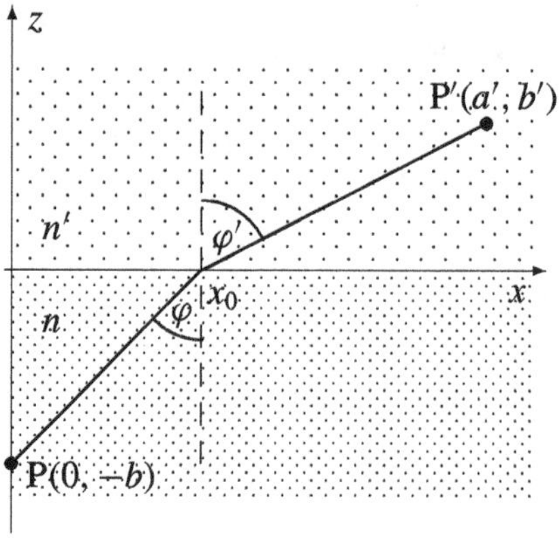

The *Fermat principle* states that light rays between two points P and P′ propagate such that the optical path ΔS is minimal:

$$\Delta S = \int_{\mathrm{P}}^{\mathrm{P}'} d\boldsymbol{r} \cdot \operatorname{grad} S(\boldsymbol{r}) = \text{minimal}$$

The phase $S(\boldsymbol{r})$ is determined by the eikonal equation $(\operatorname{grad} S)^2 = n^2$ with the real refractive index $n(\boldsymbol{r})$. Derive the law of refraction from this.

38.2 *Spatially dependent refractive index*

In a medium, the refractive index shall increase linearly in the y-direction:

$$n(\boldsymbol{r}) = n(y) = n_0 + n_1\, y \quad (n_0 > 0,\ n_1 > 0)$$

What is the trajectory of a light ray of light in the x-y-plane, which runs from the origin to the point $(x_0, 0)$? Use the Fermat principle (38.22). Discuss in particular the case $n_1 x_0 \ll n_0$.

Appendix A

SI Unit System

In principle, the Gaussian system of units is used in this book. As explained in detail in Chapter 5, this has the advantage that relativistic effect show up clearly. For numerical estimations, on the other hand, the SI (Système International d'Unités) is more practical. A somewhat older designation is "MKSA system", where the capital letters stand for meter, kilogram, second and ampere; these are also the units of SI. This appendix presents the most relevant formulas of electrodynamics in the SI unit system.

The SI uses the basic units meter (m), kilogram (kg), second (s) and ampere (A). The basic units of the Gaussian system are centimeters (cm), grams (g) and seconds (s); it is therefore also called cgs system. In mechanics, the differences between the SI (or MKSA system) and cgs system are trivial powers of ten. In particular $\text{dyn} \equiv \text{g}\,\text{cm}/\text{s}^2 = 10^{-5}\,\text{kg}\,\text{m}/\text{s}^2 = 10^{-5}\,\text{N}$ and

$$\text{erg} \equiv \frac{\text{g}\,\text{cm}^2}{\text{s}^2} = 10^{-7}\,\frac{\text{kg}\,\text{m}^2}{\text{s}^2} = 10^{-7}\,\text{J} \tag{A.1}$$

Due to the trivial conversion factors, one may use m, kg, N (newton) and J (joule) parallel to cm, g, dyn and erg.

Unit of Charge and Current

The SI or MKSA system defines the unit ampere for the current as the fourth basic unit. One considers two parallel wires at a distance of 1 m, where the current I flows through each of them. This current I is equal to 1 A if the force $2 \cdot 10^{-7}$ N acts per 1 m wire length. According to (5.6), the constant k in the Coulomb law follows from this:

$$k = \frac{1}{4\pi\varepsilon_0} \overset{\text{def}}{=} 10^{-7}\,\frac{\mathrm{N}\,c^2}{\mathrm{A}^2} \quad \text{(definition of the ampere)} \tag{A.2}$$

In the SI, the constant k is denoted by $1/4\pi\varepsilon_0$. The Coulomb law (5.1) with this constant and with $k = 1$ (Gaussian system) reads:

$$\boldsymbol{F}_1 = q_1\,q_2\,\frac{\boldsymbol{r}_1 - \boldsymbol{r}_2}{|\boldsymbol{r}_1 - \boldsymbol{r}_2|^3}\cdot\begin{cases}\dfrac{1}{4\pi\varepsilon_0} & \text{SI or MKSA system}\\[1ex] 1 & \text{Gaussian system}\end{cases} \tag{A.3}$$

In the SI or MKSA system, the charges are measured in the unit coulomb ($\mathrm{C} = \mathrm{A\,s}$). Table A.1 summarizes the most important units of electrodynamics in the SI.

In the Gaussian system, the unit of charge was defined via (A.3): Two charges of size esu (electrostatic unit) at a distance of 1 cm exert the force $1\,\mathrm{dyn} = \mathrm{g\,cm/s^2}$ on each other. The unit of charge esu is therefore a unit derived from g, cm and s.

Table A.1 Derived units of electrodynamics in the SI. The unit A (ampere) of the current I is a basic unit. The electric and magnetic field are defined by their force effect, (A.7) and (A.10). The following definitions apply in both the Gaussian unit system and the SI: Voltage $U = \int d\boldsymbol{r}\cdot\boldsymbol{E}$, work or energy $A = W = UQ$, magnetic flux $\Phi_\mathrm{m} = \int d\boldsymbol{a}\cdot\boldsymbol{B}$, resistance $R = U/I$, capacity $C = Q/U$ and inductance $L = -U/(dI/dt)$.

Quantity	Symbol	Unit	Designation
Charge	Q	$\mathrm{C} = \mathrm{A\,s}$	coulomb
Voltage	U	$\mathrm{V} = \mathrm{J/C}$	volt
Work, energy	A, W	$\mathrm{J} = \mathrm{Ws} = \mathrm{CV}$	joule
Electric field strength	$\boldsymbol{E}$	$\mathrm{V/m}$	volt/meter
Magnetic induction	$\boldsymbol{B}$	$\mathrm{T} = \mathrm{Vs/m^2}$	tesla
Magnetic flux	Φ_m	$\mathrm{Wb} = \mathrm{Vs}$	weber
Resistance	R	$\Omega = \mathrm{V/A}$	ohm
Capacity	C	$\mathrm{F} = \mathrm{C/V}$	farad
Inductance	L	$\mathrm{H} = \mathrm{Vs/A}$	henry

We denote the charge unit esu in SI by Q_{esu}. For this, we equate the forces for 2 such charges at a distance of 1 cm that follow from (A.3):

$$1\,\text{dyn} = \frac{1}{4\pi\,\varepsilon_0}\,\frac{Q_{\text{esu}}^{\,2}}{\text{cm}^2} \tag{A.4}$$

With 1 dyn $= 10^{-5}$ N and (A.2), this results in

$$Q_{\text{esu}} = 10\,\frac{\text{cm}}{c}\,\text{A} \approx 3.3 \cdot 10^{-10}\,\text{C} \tag{A.5}$$

A charge of size 1 C in the SI corresponds to then about $3 \cdot 10^9$ esu in the Gaussian system. The force between two charges of this size are the same in the SI and in the Gaussian system. From this follows

$$\frac{\text{C}}{\sqrt{4\pi\,\varepsilon_0}} \approx 3 \cdot 10^9\,\text{esu} \tag{A.6}$$

The electrostatic unit (esu) is also called statcoulomb (statC).

Electrostatic

The definition of the electric field strength by

$$\boldsymbol{F} = q\boldsymbol{E} \tag{A.7}$$

also applies in the SI. This determines the unit of the electric field strength,

$$E = \frac{\text{N}}{\text{C}} = \frac{\text{J}}{\text{C}\,\text{m}} = \frac{\text{V}}{\text{m}} \tag{A.8}$$

All formulas of electrostatics were derived from the Coulomb law (A.3) and the field definition (A.7). From (A.3) follow the modifications in the SI as compared to the Gaussian system. The field equations of electrostatics in the SI are

$$\operatorname{div}\boldsymbol{E}(\boldsymbol{r}) = \frac{\varrho(\boldsymbol{r})}{\varepsilon_0}, \qquad \operatorname{curl}\boldsymbol{E}(\boldsymbol{r}) = 0 \tag{A.9}$$

Magnetostatics

The magnetic induction $\boldsymbol{B}$ is determined in the SI by

$$d\boldsymbol{F}(\boldsymbol{r}) = I\,d\boldsymbol{\ell} \times \boldsymbol{B}(\boldsymbol{r}) \tag{A.10}$$

This means that $\boldsymbol{B}$ has the dimension

$$\mathrm{B} = \frac{\mathrm{N}}{\mathrm{A\,m}} = \frac{\mathrm{J}}{\mathrm{A\,m^2}} = \frac{\mathrm{V\,s}}{\mathrm{m^2}} \tag{A.11}$$

In contrast to (13.15), a factor $1/c$ is missing in (A.10). Accordingly, the Lorentz force is $\boldsymbol{F}_{\mathrm{L}} = q\,(\boldsymbol{E} + \boldsymbol{v} \times \boldsymbol{B})$. Instead of (13.20), we then have

$$d\boldsymbol{B}(\boldsymbol{r}_1) = \frac{\mu_0}{4\pi}\, I\, d\boldsymbol{\ell} \times \frac{\boldsymbol{r}_1 - \boldsymbol{r}_2}{|\boldsymbol{r}_1 - \boldsymbol{r}_2|^3} \tag{A.12}$$

After $\boldsymbol{B}$ has been determined as an observable by (A.10), the constant $\mu_0/4\pi$ follows from the experiment. For the so-determined permeability constant (of the vacuum) μ_0 and the dielectric constant (of the vacuum) ε_0 in (A.3) applies

$$\mu_0 = \frac{1}{\varepsilon_0\, c^2} \tag{A.13}$$

The field equations of magnetostatics are as follows:

$$\operatorname{div} \boldsymbol{B}(\boldsymbol{r}) = 0, \qquad \operatorname{curl} \boldsymbol{B}(\boldsymbol{r}) = \mu_0\, \boldsymbol{j}(\boldsymbol{r}) \tag{A.14}$$

Maxwell Equations

The generalization of (A.9) and (A.14) to time-dependent phenomena leads to the Maxwell equations:

$$\begin{aligned} &\operatorname{div} \boldsymbol{E}(\boldsymbol{r}, t) = \frac{\varrho(\boldsymbol{r}, t)}{\varepsilon_0}, && \operatorname{curl} \boldsymbol{E}(\boldsymbol{r}, t) + \frac{\partial \boldsymbol{B}(\boldsymbol{r}, t)}{\partial t} = 0 \\ &\operatorname{curl} \boldsymbol{B}(\boldsymbol{r}, t) - \frac{1}{c^2}\frac{\partial \boldsymbol{E}(\boldsymbol{r}, t)}{\partial t} = \mu_0\, \boldsymbol{j}(\boldsymbol{r}, t), && \operatorname{div} \boldsymbol{B}(\boldsymbol{r}, t) = 0 \end{aligned} \tag{A.15}$$

The macroscopic Maxwell equations (29.14) in matter become

$$\begin{aligned} &\operatorname{div} \boldsymbol{D} = \varrho_{\mathrm{ext}}, && \operatorname{curl} \boldsymbol{E} + \frac{\partial \boldsymbol{B}}{\partial t} = 0 \\ &\operatorname{curl} \boldsymbol{H} - \frac{\partial \boldsymbol{D}}{\partial t} = \boldsymbol{j}_{\mathrm{ext}}, && \operatorname{div} \boldsymbol{B} = 0 \end{aligned} \tag{A.16}$$

The response functions ε and μ occur as factors to the vacuum constants ε_0 and μ_0,

$$\boldsymbol{D} = \varepsilon\, \varepsilon_0\, \boldsymbol{E} \quad \text{and} \quad \boldsymbol{H} = \frac{\boldsymbol{B}}{\mu\, \mu_0} \tag{A.17}$$

Appendix B

Physical Constants

We compile some physical constants in the SI (where we feel free to use also cm instead of m, or erg instead of joule). In this book, these constants have often been used in estimations.

Table B.1 begins with the most important physical constants for electrodynamics. These are the speed of light c, the elementary charge e and (for applications in the atom or solid body) Planck's quantum of action $\hbar$. After that we compile frequently used lengths, frequencies and energies.

In Chapter 5, the advantages of the Gaussian system for the theoretical electrodynamics has been explained. For concrete estimations, however, we prefer the SI or MKSA system (Appendix A). This means that actual values of the electrical field strength are given in volt/centimeter = V/cm, and energy quantities often in the unit *electron volt* (eV) instead of erg. In eV, e is always the elementary charge e_{SI} in the SI:

$$\mathrm{eV} \equiv e_{\mathrm{SI}}\,\mathrm{volt} \approx 1.6 \cdot 10^{-19}\,\mathrm{C\,V} = 1.6 \cdot 10^{-12}\,\mathrm{erg} \tag{B.1}$$

The conversion from $\mathrm{J} = \mathrm{C\,V} = \mathrm{N\,m}$ to erg is given in (A.1). The lower part of Table B.1 gives some energy quantities in electron volts, which are useful for many applications.

The temperature T occurred in electrodynamics in matter. The temperature T is the measure for the energy per degree of freedom in the thermal equilibrium. For historical reasons, however, T is not expressed in energy units but in Kelvin (K). The proportionality constant between the Kelvin and energy scales is the Boltzmann constant k_{B}:

$$k_{\mathrm{B}}\,\mathrm{K} = 1.38 \cdot 10^{-23}\,\mathrm{J} \tag{B.2}$$

Table B.1 Physical constants, lengths, frequencies (see also Table 20.2) and energy quantities used in this book.

Name	Symbol	Value
Speed of light	c	$2.998 \cdot 10^{10}\ \dfrac{\text{cm}}{\text{s}}$
Elementary charge (Gaussian system)	e	$4.803 \cdot 10^{-10}\ \dfrac{\text{g}^{1/2}\,\text{cm}^{3/2}}{\text{s}}$
Elementary charge (SI unit system)	e_{SI}	$1.602 \cdot 10^{-19}\,\text{C}$
Planck's quantum of action	$\hbar$	$1.055 \cdot 10^{-27}$ erg s
Fine structure constant	$\alpha = \dfrac{e^2}{\hbar c}$	$\dfrac{1}{137.0}$
Ångström	Å	$1\,\text{Å} \equiv 10^{-8}\,\text{cm}$
Fermi	fm	$1\,\text{fm} \equiv 10^{-13}\,\text{cm}$
Bohr radius	$a_{\text{B}} = \dfrac{\hbar^2}{m_{\text{e}} e^2}$	$5.3 \cdot 10^{-9}\,\text{cm} = 0.53\,\text{Å}$
Wavelength of visible light ($\hbar\omega = 2...3\,\text{eV}$)	λ_{vis}	$4 \cdot 10^{-5} \ldots 8 \cdot 10^{-5}\,\text{cm}$
Atomic frequency	$\omega_{\text{at}} = \dfrac{v_{\text{at}}}{a_{\text{B}}} = \dfrac{\alpha c}{a_{\text{B}}}$	$4.1 \cdot 10^{16}\,\text{s}^{-1}$
Frequency of visible light	$\omega_{\text{vis}} = 2\pi\,\dfrac{c}{\lambda_{\text{vis}}}$	$2 \ldots 5 \cdot 10^{15}\,\text{s}^{-1}$
Electron mass	$m_{\text{e}} c^2$	0.51 MeV
Proton mass	$m_{\text{p}} c^2$	0.94 GeV
Atomic energy unit	$E_{\text{at}} = \dfrac{e^2}{a_{\text{B}}} = \dfrac{\hbar^2}{m_{\text{e}} a_{\text{B}}^2}$	27.2 eV
Scale of the Coulomb energy in the atom	$\dfrac{e^2}{\text{Å}}$	14.4 eV
Scale of the Coulomb energy in the nucleus	$\dfrac{e^2}{\text{fm}}$	1.44 MeV
Room temperature, $T \approx 290\,\text{K}$	$k_{\text{B}} T$	$290\,k_{\text{B}}\,\text{K} \approx \dfrac{\text{eV}}{40}$

Appendix C

Vector Operations

The explicit forms of the common vector operations are compiled for Cartesian coordinates x, y, z, cylindrical coordinates ρ, φ, z and spherical coordinates r, θ, ϕ. The effect of the nabla operator on some products of fields is worked out.

From (1.19)–(1.21) and (1.17) follows:

$$\operatorname{grad}\Phi = \begin{cases} \dfrac{\partial \Phi}{\partial x}\,\boldsymbol{e}_x + \dfrac{\partial \Phi}{\partial y}\,\boldsymbol{e}_y + \dfrac{\partial \Phi}{\partial z}\,\boldsymbol{e}_z & \text{(Cartesian coordinates)} \\[2ex] \dfrac{\partial \Phi}{\partial \rho}\,\boldsymbol{e}_\rho + \dfrac{1}{\rho}\,\dfrac{\partial \Phi}{\partial \varphi}\,\boldsymbol{e}_\varphi + \dfrac{\partial \Phi}{\partial z}\,\boldsymbol{e}_z & \text{(cylindrical coordinates)} \\[2ex] \dfrac{\partial \Phi}{\partial r}\,\boldsymbol{e}_r + \dfrac{1}{r}\,\dfrac{\partial \Phi}{\partial \theta}\,\boldsymbol{e}_\theta + \dfrac{1}{r\,\sin\theta}\,\dfrac{\partial \Phi}{\partial \phi}\,\boldsymbol{e}_\phi & \text{(spherical coordinates)} \end{cases} \tag{C.1}$$

$$\operatorname{div}\boldsymbol{V} = \begin{cases} \dfrac{\partial V_x}{\partial x} + \dfrac{\partial V_y}{\partial y} + \dfrac{\partial V_z}{\partial z} \\[2ex] \dfrac{1}{\rho}\,\dfrac{\partial(\rho\, V_\rho)}{\partial \rho} + \dfrac{1}{\rho}\,\dfrac{\partial V_\varphi}{\partial \varphi} + \dfrac{\partial V_z}{\partial z} \\[2ex] \dfrac{1}{r^2}\,\dfrac{\partial(r^2\, V_r)}{\partial r} + \dfrac{1}{r\,\sin\theta}\,\dfrac{\partial(\sin\theta\, V_\theta)}{\partial \theta} + \dfrac{1}{r\,\sin\theta}\,\dfrac{\partial V_\phi}{\partial \phi} \end{cases} \tag{C.2}$$

$$\operatorname{curl} \boldsymbol{V} = \begin{cases} \left(\frac{\partial V_z}{\partial y} - \frac{\partial V_y}{\partial z}\right) \boldsymbol{e}_x + \left(\frac{\partial V_x}{\partial z} - \frac{\partial V_z}{\partial x}\right) \boldsymbol{e}_y + \left(\frac{\partial V_y}{\partial x} - \frac{\partial V_x}{\partial y}\right) \boldsymbol{e}_z \\ \left(\frac{1}{\rho}\frac{\partial V_z}{\partial \varphi} - \frac{\partial V_\varphi}{\partial z}\right) \boldsymbol{e}_\rho + \left(\frac{\partial V_\rho}{\partial z} - \frac{\partial V_z}{\partial \rho}\right) \boldsymbol{e}_\varphi \\ \qquad + \frac{1}{\rho}\left(\frac{\partial(\rho V_\varphi)}{\partial \rho} - \frac{\partial V_\rho}{\partial \varphi}\right) \boldsymbol{e}_z \\ \frac{1}{r \sin\theta}\left(\frac{\partial(\sin\theta\, V_\phi)}{\partial \theta} - \frac{\partial V_\theta}{\partial \phi}\right) \boldsymbol{e}_r + \frac{1}{r}\left(\frac{1}{\sin\theta}\frac{\partial V_r}{\partial \phi} - \frac{\partial (r V_\phi)}{\partial r}\right) \boldsymbol{e}_\theta \\ \qquad + \frac{1}{r}\left(\frac{\partial(r V_\theta)}{\partial r} - \frac{\partial V_r}{\partial \theta}\right) \boldsymbol{e}_\phi \end{cases} \tag{C.3}$$

From (1.24) with (1.17) follows

$$\Delta \Phi = \begin{cases} \frac{\partial^2 \Phi}{\partial x^2} + \frac{\partial^2 \Phi}{\partial y^2} + \frac{\partial^2 \Phi}{\partial z^2} \\ \frac{1}{\rho}\frac{\partial}{\partial \rho}\left(\rho \frac{\partial \Phi}{\partial \rho}\right) + \frac{1}{\rho^2}\frac{\partial^2 \Phi}{\partial \varphi^2} + \frac{\partial^2 \Phi}{\partial z^2} \\ \frac{1}{r}\frac{\partial^2 (r\Phi)}{\partial r^2} + \frac{1}{r^2 \sin\theta}\frac{\partial}{\partial \theta}\left(\sin\theta \frac{\partial \Phi}{\partial \theta}\right) + \frac{1}{r^2\sin^2\theta}\frac{\partial^2 \Phi}{\partial \phi^2} \end{cases} \tag{C.4}$$

For spherical coordinates, the radial derivative can also be written differently:

$$\frac{1}{r}\frac{\partial^2 (r\Phi)}{\partial r^2} = \frac{\partial^2 \Phi}{\partial r^2} + \frac{2}{r}\frac{\partial \Phi}{\partial r} = \frac{1}{r^2}\frac{\partial}{\partial r}\left(r^2 \frac{\partial \Phi}{\partial r}\right) \tag{C.5}$$

For curvilinear coordinates, the evaluation $\Delta \boldsymbol{V}$ is the coordinate independency of the basis vectors must be taken into account; in particular, $\Delta \boldsymbol{V} \neq \sum_i (\Delta V_i)\, \boldsymbol{e}_i$. For calculating $\Delta \boldsymbol{V}$, one usually starts form (2.27):

$$\Delta \boldsymbol{V} = \operatorname{grad}(\operatorname{div} \boldsymbol{V}) - \operatorname{curl}(\operatorname{curl} \boldsymbol{V}) = \boldsymbol{\nabla}(\boldsymbol{\nabla} \cdot \boldsymbol{V}) - \boldsymbol{\nabla} \times (\boldsymbol{\nabla} \times \boldsymbol{V}) \tag{C.6}$$

The following relations also contain two nabla operators:

$$\operatorname{curl}\operatorname{grad} \Phi = \boldsymbol{\nabla} \times (\boldsymbol{\nabla} \Phi) = 0 \tag{C.7}$$

$$\operatorname{div}\operatorname{curl} \boldsymbol{V} = \boldsymbol{\nabla} \cdot (\boldsymbol{\nabla} \times \boldsymbol{V}) = 0 \tag{C.8}$$

Finally, we summarize the effect of the nabla operator on some products of fields:

$$\begin{aligned}
\boldsymbol{\nabla}(\Phi\,\Psi) &= (\boldsymbol{\nabla}\,\Phi)\,\Psi + \Phi\,(\boldsymbol{\nabla}\,\Psi) \\
\boldsymbol{\nabla}\cdot(\Phi\,\boldsymbol{V}) &= \boldsymbol{V}\cdot\boldsymbol{\nabla}\Phi + \Phi\,\boldsymbol{\nabla}\cdot\boldsymbol{V} \\
\boldsymbol{\nabla}\times(\Phi\,\boldsymbol{V}) &= (\boldsymbol{\nabla}\,\Phi)\times\boldsymbol{V} + \Phi\,\boldsymbol{\nabla}\times\boldsymbol{V} \\
\boldsymbol{\nabla}(\boldsymbol{V}\cdot\boldsymbol{W}) &= (\boldsymbol{V}\cdot\boldsymbol{\nabla})\boldsymbol{W} + (\boldsymbol{W}\cdot\boldsymbol{\nabla})\boldsymbol{V} + \boldsymbol{V}\times(\boldsymbol{\nabla}\times\boldsymbol{W}) + \boldsymbol{W}\times(\boldsymbol{\nabla}\times\boldsymbol{V}) \\
\boldsymbol{\nabla}\cdot(\boldsymbol{V}\times\boldsymbol{W}) &= \boldsymbol{W}\cdot(\boldsymbol{\nabla}\times\boldsymbol{V}) - \boldsymbol{V}\cdot(\boldsymbol{\nabla}\times\boldsymbol{W}) \\
\boldsymbol{\nabla}\times(\boldsymbol{V}\times\boldsymbol{W}) &= \boldsymbol{V}(\boldsymbol{\nabla}\cdot\boldsymbol{W}) - \boldsymbol{W}(\boldsymbol{\nabla}\cdot\boldsymbol{V}) + (\boldsymbol{W}\cdot\boldsymbol{\nabla})\boldsymbol{V} - (\boldsymbol{V}\cdot\boldsymbol{\nabla})\boldsymbol{W}
\end{aligned} \tag{C.9}$$

The relationships result from the chain rule, the cyclic interchangeability of the factors in the scalar triple product $\boldsymbol{a}\cdot(\boldsymbol{b}\times\boldsymbol{c}) = \boldsymbol{b}\cdot(\boldsymbol{c}\times\boldsymbol{a}) = \boldsymbol{c}\cdot(\boldsymbol{a}\times\boldsymbol{b})$ and from

$$\boldsymbol{a}\times(\boldsymbol{b}\times\boldsymbol{c}) = (\boldsymbol{a}\cdot\boldsymbol{c})\,\boldsymbol{b} - (\boldsymbol{a}\cdot\boldsymbol{b})\,\boldsymbol{c} \tag{C.10}$$

Another useful relation is

$$(\boldsymbol{a}\times\boldsymbol{b})\cdot(\boldsymbol{c}\times\boldsymbol{d}) = (\boldsymbol{a}\cdot\boldsymbol{c})(\boldsymbol{b}\cdot\boldsymbol{d}) - (\boldsymbol{a}\cdot\boldsymbol{d})(\boldsymbol{b}\cdot\boldsymbol{c}) \tag{C.11}$$

Index

Abbreviations

CONS	complete ONS
IS	inertial system
LT	Lorentz transformation
ONS	orthonormalized set of states (or functions)
p.i.	partial integration
S	coordinate system
SI	Système International d'Unités) (also MKSA system)

Units

Å	ångström, 1 Å $= 10^{-10}$ m
A	ampere
C	coulomb
eV	electron volt
fm	fermi, 1 fm $= 10^{-15}$ m
GeV	giga electron volt 1 GeV $= 10^9$ eV
J	joule, 1 J = 1 N m = 1 C V
kg	kilogram, mass unit
N	newton, 1 N = 1 kg m/s^2
m	meter, length unit
s	second, time unit
V	volt
W	watt, 1 W = J/s

Symbols

$[X]$	dimension of X e.g. [q] = esu or C
$=$ const.	equal to a constant or defined by
$=$	represented by, for e.g. $\boldsymbol{r} = (x, y, z)$
$\overset{(5.20)}{=}$	yields with the help of equation (5.20)
$\widehat{=}$	corresponds to
$\propto$	proportional to
$\sim$	of similar size
$= \mathcal{O}(\ldots)$	of the order of magnitude

A

B

T

U

V

W

X

Y